International
REVIEW OF
Neurobiology
Volume 101

International

REVIEW OF

Neurobiology

Volume 101

SERIES EDITORS

Biomarkers of Neurological and Psychiatric Disease

EDITED BY

PAUL C. GUEST

Department of Chemical Engineering and Biotechnology
University of Cambridge
Cambridge, United Kingdom

SABINE BAHN

Department of Neuroscience Erasmus Medical Centre
Rotterdam, The Netherlands
Department of Chemical Engineering and Biotechnology
University of Cambridge
Cambridge, United Kingdom

ELSEVIER

AMSTERDAM • BOSTON • HEIDELBERG • LONDON
NEW YORK • OXFORD • PARIS • SAN DIEGO
SAN FRANCISCO • SINGAPORE • SYDNEY • TOKYO
Academic Press is an imprint of Elsevier

Academic Press is an imprint of Elsevier
32 Jamestown Road, London NW1 7BY, UK
Radarweg 29, PO Box 211, 1000 AE Amsterdam, The Netherlands
Linacre House, Jordan Hill, Oxford OX2 8DP, UK
225 Wyman Street, Waltham, MA 02451, USA
525 B Street, Suite 1900, San Diego, CA 92101-4495, USA

First edition 2011

ISBN: 978-0-12-387718-5
ISSN: 0074-7742

For information on all Academic Press publications
visit our website at elsevierdirect.com

Printed and bound by CPI Group (UK) Ltd, Croydon, CR0 4YY

Transferred to Digital Print 2012

CONTENTS

The Utility of Gene Expression in Blood Cells for Diagnosing Neuropsychiatric Disorders

Christopher H. Woelk, Akul Singhania, Josué Pérez-Santiago, Stephen J. Glatt and Ming T. Tsuang

Proteomic Technologies for Biomarker Studies in Psychiatry: Advances and Needs

Daniel Martins-de-Souza, Paul C. Guest, Natacha Vanattou-Saifoudine, Laura W. Harris and Sabine Bahn

Converging Evidence of Blood-Based Biomarkers for Schizophrenia: An update

Man K. Chan, Paul C. Guest, Yishai Levin, Yagnesh Umrania, Emanuel Schwarz, Sabine Bahn and Hassan Rahmoune

Abnormalities in Metabolism and Hypothalamic–Pituitary–Adrenal Axis Function in Schizophrenia

Paul C. Guest, Daniel Martins-de-Souza, Natacha Vanattou-Saifoudine, Laura W. Harris and Sabine Bahn

Immune and Neuroimmune Alterations in Mood Disorders and Schizophrenia

Roosmarijn C. Drexhage, Karin Weigelt, Nico van Beveren, Dan Cohen, Marjan A. Versnel, Willem A. Nolen and Hemmo A. Drexhage

Behavioral and Molecular Biomarkers in Translational Animal Models for Neuropsychiatric Disorders

Zoltán Sarnyai, Murtada Alsaif, Sabine Bahn, Agnes Ernst, Paul C. Guest, Eva Hradetzky, Wolfgang Kluge, Viktoria Stelzhammer and Hendrik Wesseling

Stem Cell Models for Biomarker Discovery in Brain Disease

Alan Mackay-Sim, George Mellick and Stephen Wood

The Application of Multiplexed Assay Systems for Molecular Diagnostics

Emanuel Schwarz, Nico J.M. VanBeveren, Paul C. Guest, Rauf Izmailov and Sabine Bahn

Algorithm Development for Diagnostic Biomarker Assays

Rauf Izmailov, Paul C. Guest, Sabine Bahn and Emanuel Schwarz

Challenges of Introducing New Biomarker Products for Neuropsychiatric Disorders into the Market

Sabine Bahn, Richard Noll, Anthony Barnes, Emanuel Schwarz and Paul C. Guest

Toward Personalized Medicine in the Neuropsychiatric Field

Erik H.F. Wong, Jayne C. Fox, Mandy Y.M. Ng and Chi-Ming Lee

Clinical Utility of Serum Biomarkers for Major Psychiatric Disorders

Nico J.M. van Beveren and Witte J.G. Hoogendijk

The Future: Biomarkers, Biosensors, Neuroinformatics, and E-Neuropsychiatry

CHRISTOPHER R. LOWE

CONTRIBUTORS

Numbers in parentheses indicate the pages on which the authors' contributions begin.

Murtada Alsaif (203), Department of Chemical Engineering and Biotechnology, University of Cambridge, Cambridge, United Kingdom

Sabine Bahn (65, 95, 145, 203, 259, 279, 299), Department of Chemical Engineering and Biotechnology, University of Cambridge, Cambridge, United Kingdom; Department of Neuroscience, Erasmus Medical Centre, Rotterdam, The Netherlands

Anthony Barnes (299), Rules Based Medicine, Austin, Texas, USA

Man K. Chan (95), Department of Chemical Engineering and Biotechnology, University of Cambridge, Cambridge, United Kingdom

Dan Cohen (169), Department of Epidemiology, University Medical Center, Groningen, The Netherlands; Department of Severe Mental Illness, Mental Health Care Organization, North-Holland North, Heerhugowaard, The Netherlands

Hemmo A. Drexhage (169), Department of Immunology, Erasmus MC, Rotterdam, The Netherlands

Roosmarijn C. Drexhage (169), Department of Immunology, Erasmus MC, Rotterdam, The Netherlands

Agnes Ernst (203), Department of Chemical Engineering and Biotechnology, University of Cambridge, Cambridge, United Kingdom

Michaela D. Filiou (1), Proteomics and Biomarkers, Max Planck Institute of Psychiatry, Munich, Germany

Jayne C. Fox (329), AstraZeneca Pharmaceuticals, Personalized Health Care and Biomarkers, Alderley Park, Macclesfield, Cheshire, United Kingdom

Stephen J. Glatt (41), Departments of Psychiatry and Behavioral Sciences & Neuroscience and Physiology, Psychiatric Genetic Epidemiology & Neurobiology Laboratory (PsychGENe Lab), Medical Genetics Research Center, SUNY Upstate Medical University, Syracuse, New York, USA

Paul C. Guest (65, 95, 145, 203, 259, 279, 299), Department of Chemical Engineering and Biotechnology, University of Cambridge, Cambridge, United Kingdom

Laura W. Harris (65, 145), Department of Chemical Engineering and Biotechnology, University of Cambridge, Cambridge, United Kingdom

Witte J.G. Hoogendijk (351), Department of Psychiatry, Erasmus University Medical Center, Rotterdam, The Netherlands

Eva Hradetzky (203), Department of Chemical Engineering and Biotechnology, University of Cambridge, Cambridge, United Kingdom

Rauf Izmailov (259, 279), Rules-Based Medicine, Inc., Austin, Texas, USA

Wolfgang Kluge (203), Department of Chemical Engineering and Biotechnology, University of Cambridge, Cambridge, United Kingdom

Chi-Ming Lee (329), AstraZeneca Pharmaceuticals, Discovery Enabling Capabilities & Sciences, Alderley Park, Macclesfield, Cheshire, United Kingdom

Yishai Levin (95), Department of Chemical Engineering and Biotechnology, University of Cambridge, Cambridge, United Kingdom

Christopher R. Lowe (375), Department of Chemical Engineering and Biotechnology, Institute of Biotechnology, University of Cambridge, Cambridge, United Kingdom

Alan Mackay-Sim (239), National Centre for Adult Stem Cell Research, Eskitis Institute for Cell and Molecular Therapies, Griffith University, Brisbane, Queensland, Australia

Daniel Martins-de-Souza (65, 145), Department of Chemical Engineering and Biotechnology, University of Cambridge, Cambridge, United Kingdom; Lab. de Neurociências (LIM-27), Inst. Psiquiatria, Fac. de Medicina da Universidade de Sao Paulo, Sao Paulo, Brazil

Paul M. Matthews (19), GSK Clinical Imaging Centre, Hammersmith Hospital, London, United Kingdom; Centre for Neuroscience, Imperial College, London, United Kingdom

George Mellick (239), National Centre for Adult Stem Cell Research, Eskitis Institute for Cell and Molecular Therapies, Griffith University, Brisbane, Queensland, Australia

Mandy Y.M. Ng (329), AstraZeneca Pharmaceuticals, Personalized Health Care and Biomarkers, Alderley Park, Macclesfield, Cheshire, United Kingdom

Willem A. Nolen (169), Department of Psychiatry, University Medical Center, University of Groningen, Groningen, The Netherlands

Richard Noll (299), DeSales University, Center Valley, Pennsylvania, USA

David R.J. Owen (19), Division of Experimental Medicine, Imperial College, Hammersmith Hospital, London, United Kingdom; GSK Clinical Imaging Centre, Hammersmith Hospital, London, United Kingdom

Josué Pérez-Santiago (41), Department of Medicine, University of California San Diego, La Jolla, California, USA

Hassan Rahmoune (95), Department of Chemical Engineering and Biotechnology, University of Cambridge, Cambridge, United Kingdom

Zoltán Sarnyai (203), Department of Pharmacology, University of Cambridge, Cambridge, United Kingdom

Emanuel Schwarz (95, 259, 279, 299), Department of Chemical Engineering and Biotechnology, University of Cambridge, Cambridge, United Kingdom

Akul Singhania (41), Veterans Affairs San Diego Healthcare System, San Diego, California, USA

Viktoria Stelzhammer (203), Department of Chemical Engineering and Biotechnology, University of Cambridge, Cambridge, United Kingdom

Ming T. Tsuang (41), Department of Psychiatry and Institute of Genomic Medicine, Center for Behavioral Genomics, University of California San Diego, La Jolla, California, USA

Christoph W. Turck (1), Proteomics and Biomarkers, Max Planck Institute of Psychiatry, Munich, Germany

Yagnesh Umrania (95), Department of Chemical Engineering and Biotechnology, University of Cambridge, Cambridge, United Kingdom

Nico J.M. van Beveren (351), Department of Psychiatry, Erasmus University Medical Center, Rotterdam, The Netherlands

Nico van Beveren (169), Department of Psychiatry, Erasmus MC, Rotterdam, The Netherlands

Natacha Vanattou-Saifoudine (65, 145), Department of Chemical Engineering and Biotechnology, University of Cambridge, Cambridge, United Kingdom

Nico J.M. VanBeveren (259), Department of Psychiatry, Erasmus University, Medical Centre, Rotterdam, The Netherlands

Marjan A. Versnel (169), Department of Immunology, Erasmus MC, Rotterdam, The Netherlands

Karin Weigelt (169), Department of Immunology, Erasmus MC, Rotterdam, The Netherlands

Hendrik Wesseling (203), Department of Chemical Engineering and Biotechnology, University of Cambridge, Cambridge, United Kingdom

Christopher H. Woelk (41), Department of Medicine, University of California San Diego, La Jolla, California, USA; Veterans Affairs San Diego Healthcare System, San Diego, California, USA

Erik H.F. Wong (329), AstraZeneca Pharmaceuticals, External Science, CNS-Pain Innovative Medicine Unit, Wilmington, Delaware, USA

Stephen Wood (239), National Centre for Adult Stem Cell Research, Eskitis Institute for Cell and Molecular Therapies, Griffith University, Brisbane, Queensland, Australia

PREFACE

This volume of the *International Review of Neurobiology* describes the state of the art and the future of biomarkers in neurological and psychiatric diseases. Currently, the diagnosis for all neuropsychiatric disorders is carried out by psychiatrists via interview, observation, and classification of patients who typically have heterogeneous symptoms and medical histories. The recent emergence of molecular and image-based biomarkers for these conditions would therefore greatly facilitate disease diagnosis and stratification. This may require deconstruction of the existing long-standing procedures aimed at classification of broad patient categories in favor of identifying biomarker-defined disease subtypes. Ultimately, this will assist in personalized medicine approaches and may be facilitated by developments in the areas of biosensors, neuroinformatics, and e-neuropsychiatry.

The first chapter by Filiou and Turck introduces the content of the volume by describing how biomarkers are now in demand in neuropsychiatric research for diagnosis, treatment response monitoring, and development of novel therapeutics. However, biomarker discovery in this field is challenging due to the fact that these are complex disorders, information on the affected molecular pathways is scarce, and there is considerable interpatient heterogeneity within a given disorder and overlap of symptoms across different conditions. Because of this disease complexity, a panel of biomarkers derived from multiple platforms will be needed to define these conditions at the molecular level. Ultimately, the coordinated effort of researchers, physicians, funding organizations, and standardization initiatives will be needed to overcome these challenges.

The second chapter by Owen and Matthews describes recent advances in imaging technologies to study neuroinflammatory, neurodegenerative, and neuropsychiatric conditions such as multiple sclerosis, Alzheimer's disease, Parkinson's disease, stroke, and schizophrenia. The main technique described, positron emission tomography (PET), is used to study the proliferation of microglia in these conditions as this is a stereotyped response after a variety of pathological insults. There has been significant interest in quantifying microglial density *in vivo* in research and clinical decision making. However, this has been hindered by the lack of appropriate radioligands. With recent development of several new generation ligands with improved specific binding, this now seems possible and should enable PET to become a more valuable tool for use in the clinical studies of such neuropsychiatric disorders.

The third chapter by Woelk *et al.* covers the use of gene expression analysis of blood cells for diagnosis of neuropsychiatric disorders. The authors carry out a review of studies which have analyzed gene expression in blood cells from patients with neuropsychiatric disorders with an emphasis on developing diagnostics for schizophrenia. The authors also discuss the future directions of the field including using microRNA expression for developing diagnostic classifiers and the potential use of blood cell gene expression patterns to tailor antipsychotic medications to individual patients. They also describe the likely future impact of next-generation sequencing technologies. As the costs diminish and software tools for analysis of this high-content data becomes more available, the possibility of developing more accurate classification tools for neuropsychiatric disorders will increase.

The fourth chapter by Martins-de-Souza *et al.* describes the state of the art and possible future developments in the use of proteomic technologies for the study of neuropsychiatric conditions and for the development of novel molecular diagnostic/prognostic tests. Such advances have already been partly achieved for illnesses such as cancer although they have had a less profound impact in the case of neuropsychiatric disorders such as schizophrenia. The authors discuss the pressing need for more sensitive and accurate technologies with overall importance on technologies which can be used for validation and implementation of the resulting biomarkers as simple and effective tests for use in the clinical environment. Such advances will put proteomics closer to clinical applications in the neuropsychiatry field.

The fifth chapter by Chan *et al.* gives an update on emerging evidence for identification of blood-based molecular biomarkers in schizophrenia. The authors have combined a review of the literature with the results of a comprehensive in-house study showing the identification of candidate blood-based biomarkers for schizophrenia and for antipsychotic drug response. Taken together, the findings suggest that there are effects on the immune system and inflammation response in schizophrenia. The findings also suggested that there is an activation of the stress response, as shown by increased levels of cortisol and activation of the hypothalamic–pituitary–adrenal (HPA) axis in patients. It is expected that such biomarkers will prove useful as an additional means of characterizing specific immune, metabolic, or hormonal pathways in schizophrenia, which should pave the way for development of future patient stratification and personalized medicine strategies.

The sixth chapter by Guest *et al.* describes the finding of abnormalities in metabolism and hormonal function in patients with schizophrenia. The authors describe decades of research converging on the fact that the pathogenesis of schizophrenia can involve perturbations in metabolic and HPA axis pathways in some patients. The observed differences in manifestation of these effects could be related to differences in symptoms between patients and in responses to antipsychotic treatments. The authors describe the identification of circulating molecular biomarkers in schizophrenia patients including changes in vital hormones and related molecules such as insulin, proinsulin, proinsulin conversion intermediates, C-peptide, chromogranin A, pancreatic polypeptide, cortisol, adrenocorticotrophic hormone, growth hormone, prolactin, and progesterone. Stratification of subjects according to molecular phenotype reflecting the disease state or trait could help to improve existing treatments through application of novel personalized medicine strategies which target patients with metabolic or hormonal abnormalities.

The seventh chapter by Drexhage *et al.* describes the occurrence of immune and neuroimmune alterations in neuropsychiatric conditions such as mood disorders and schizophrenia. The authors review the literature over the past 20 years which has implicated alterations in immune system function in such patients. These findings indicate that there is a proinflammatory state of the cytokine network which can induce the psychopathological symptoms and may also be involved in the pathophysiology of these neuropsychiatric illnesses. The authors also present the results of their recent studies, which relate immune activation to present theories on the influence of activated immune cells on brain function. Increased understanding in this area could help in the development of novel treatment strategies and improved clinical management of mental disorders. This is important as a proinflammatory state may not affect all patients but may be a feature of a subset of patients.

The eighth chapter by Sarnyai *et al.* carries out a review and presents novel findings on behavioral and molecular biomarkers in translational animal models for neuropsychiatric disorders. Modeling neuropsychiatric disorders in animals presents significant challenges due to the subjective nature of classifying symptoms and the lack of empirical biomarkers and understanding of the pathophysiology. Successful translation of preclinical models to clinical research requires carefully characterized animal models which are informative about disease mechanisms and therapeutic targets. In this light, the authors review behavioral, neurobiological, and molecular findings from selected animal models for schizophrenia, bipolar disorder, and major depressive disorder. Importantly, they focus this assessment on the use of appropriate statistical tools and newly developed Research Domain Criteria (RDoC) to link biomarkers from animal models with the human disease. They argue that this approach will lead to development of "validated" animal models for specific neuropsychiatric disorders and may ultimately lead to better understanding of the pathophysiology and to identification of novel biomarkers and therapeutic targets.

The ninth chapter by Mackay-Sim *et al.* presents the use of stem cell models for biomarker discovery in neuropsychiatric disorders. Patient-derived stem cells have significant potential as disease models. Stem cells can be proliferated, stored, and then thawed for use in genomic, proteomic, and functional studies. Patient-derived induced pluripotent stem cells and adult stem cells from the olfactory tissue in the nose have already been used to provide novel insights into a number of brain diseases. This work was originally inspired by the observation that the sense of smell is impaired in many brain diseases, including neurodegenerative diseases, such as Alzheimer's disease and Parkinson's disease, and neuropsychiatric disorders, such as schizophrenia. These findings suggest that biomarker discovery may be possible from investigating such disease-associated cells in the form of patient-derived stem cells. Such cellular models carry the disease phenotypes and span the variability encountered across patient populations. They also provide potential experimental tools to identify novel molecular biomarkers that distinguish patients from controls and may therefore lead to development of empirical tests for differential diagnosis or for monitoring disease progression.

The 10th chapter by Schwarz *et al.* describes the recent advances in applying multiplexed immunoassay systems to identify molecular diagnostics for psychiatric disorders. The authors describe approaches to identify disease-related molecular abnormalities when there is uncertainty regarding the validity of the clinical diagnosis. They also present an introduction to the multiplex immunoassay approach that facilitates identification and quantitation of molecular biomarkers and also for extending these molecular findings into the realms of identifying the associated functional consequences. As chronic sufferers of neuropsychiatic diseases are likely to have a poor prognosis, an accurate molecular test may lead to early intervention and thereby improve patient outcomes. A molecular test would also open up the possibility of stratifying patients more accurately which is crucial for personalized medicine approaches.

The 11th chapter by Izmailov *et al.* gives a description of algorithm development for diagnostic biomarker assays. As a test case, they present the groundbreaking development of a serum-based test to help confirm the diagnosis of schizophrenia. They identified a multiplex panel of 51 immunoassays which allowed reproducible identification of schizophrenia patients compared to controls with high performance. Validation of this test involved development of a linear support vector machine decision rule and they tested the performance of this using cross-validation. This resulted in readjustment of the panel and algorithm to a smaller set of assays, and they developed a simple procedure for maintenance and recalibration of the assays across time. The resulting decision rule delivered a sensitive and specific test for presence of schizophrenia compared to controls. The next stage will be to carry out large-scale clinical validation studies using samples from more diverse psychiatric patient populations and settings in a series of prospective studies for translation to the clinical setting.

The 12th chapter by Bahn *et al.* describes the challenges of introducing new biomarker products for neuropsychiatric disorders into the market place. The general opinion is that improvements over the current subjective tests are essential. Despite this there is a reluctance to accept the possibility that identification of peripheral biomarkers can be of any benefit. In addition, psychiatrists find it difficult to accept that peripheral molecules such as blood-based proteins and small molecules can reflect what is happening in the brain. However, the health and regulatory authorities now consider that biomarkers are important for the future of drug development and have called for efforts to modernize methods, tools and techniques for this purpose. The authors describe the development of the first ever molecular blood test for schizophrenia and the reactions of research scientists and psychiatrists to this development, as a case in point. There is now reason for optimism that further technological advancements and interdisciplinary approaches in biomarker research will overcome current limitations and help to advance our ability to treat patients with neuropsychiatric disorders.

The 13th chapter by Wong *et al.* addresses the need and movement toward personalized medicine in the neuropsychiatric field. Advances in human genetics and molecular innovations in neuroscience have prompted the pharmaceutical industry to move beyond the treatment of broad spectrum diseases to more targeted (personalized) treatment approaches. Recurring failure in converting scientific discoveries in neuroscience to novel efficacious drugs has precipitated a crisis in the industry. A targeted and consistent investment is needed to restore confidence in translating science into clinical success. There are now movements for cross-pharmaceutical company and globally coordinated efforts for discovery of better, therapy-linked patient stratification, as exemplified by the European Union Innovative Medicine Initiatives project entitled: "New Medications in Depression and Schizophrenia—NEWMEDS." That fact that such efforts are now being made by individual pharmaceutical companies, suggests that the time and opportunity for a fresh approach in this area are now welcome.

The 14th chapter by van Beveren and Hoogendijk covers the clinical aspects of major neuropsychiatric disorders and the need for a paradigm shift to biomarker-assisted diagnostic tests. Thus far, the identification and application of such biomarker tests have been sparse. This is likely to be due to the fact that the existing diagnostic methods are based on long-standing heterogeneous concepts in psychiatry. In addition, a shift to using biomarkers for conditions which have been categorized for decades based on clinical phenomenology would not be clinically useful. However, there is a pressing need for biomarkers which can be used as an aid to the normal procedure to classify at-risk patients, such as young people with prodromal symptoms for psychosis and existing patients who are likely to progress to more severe states. The authors also stress that there is a need for better classification of patient subtypes and to deconstruct the traditional diagnoses in favor of using biomarker-assisted strategies to accomplish this.

The 15th chapter by Lowe describes the potential future of biomarkers in neuropsychiatric diseases which may include developments in the areas of biosensors, neuroinformatics, and e-neuropsychiatry. The emergence of molecular and image-based biomarkers for neuropsychiatric conditions is on the verge of causing a paradigm shift in the current diagnostic procedures. The development of biomarkers has created the prospect of producing more sensitive and specific tests to replace the traditional psychiatric interview-based approach. In the future, the emergence of biosensor technologies, point-of-care testing and fusion of biomarker, electroencephalogram and magnetic resonance imaging data with patient medical histories and biopatterns could lead to the development of personalized bioprofiles or fingerprints for neuropsychiatric patients. Also, the application of mobile communication technologies may help to facilitate disease prediction, diagnosis, prognosis, and compliance monitoring. It is anticipated that ultimately such mobile devices will usher in the next generation of personalized medicine strategies.

In summary, a number of various biomarkers and biomarker algorithms have emerged which may allow better classification and monitoring of patients with neuropsychiatric disorders. However, it will take considerable time before these advances can be translated and implemented in the clinic. There is now optimism that further technological advancements, interdisciplinary, and collaborative approaches will help to overcome the current limitations in the field and enable the concept of personalized medicine for subjects with these debilitating conditions.

GENERAL OVERVIEW: BIOMARKERS IN NEUROSCIENCE RESEARCH

Michaela D. Filiou and Christoph W. Turck

Proteomics and Biomarkers, Max Planck Institute of Psychiatry, Munich, Germany

Abstract

Biomarkers are in demand for disease diagnosis, treatment response monitoring, and development of novel therapeutics. Biomarker discovery in neuroscience is challenging due to absence of robust molecular correlates and the interpatient heterogeneity that characterizes neuropsychiatric disorders. Because of the complexity of these disorders, a panel of biomarkers derived from different platforms will be required to precisely reflect disease-related alterations. Animal models of psychiatric phenotypes as well as -omics and imaging methodologies are important tools for biomarker discovery. However, the limitations of current research concerning sample handling and collection, candidate biomarker validation, and a lack of interdisciplinary approaches need to be addressed. Ultimately, the coordinated effort of relevant stakeholders including researchers, physicians,

INTERNATIONAL REVIEW OF
NEUROBIOLOGY, VOL. 101
DOI: 10.1016/B978-0-12-387718-5.00001-8

and funding organizations together with standardization initiatives will be vital to overcome the present challenges and to advance personalized health care using sensitive and specific biomarkers.

I. The Quest for Biomarkers in Neuroscience

The field of biomarker research has received increasing attention from both the scientific community and funding organizations. According to the official definition by the National Institutes of Health (NIH), "a biomarker is a characteristic that is objectively measured and evaluated as an indicator of normal biological processes, pathogenic processes, or pharmacological responses to a therapeutic intervention" (Biomarkers Definitions Working Group *et al.*, 2001).

A. Biomarkers in Clinical Practice

The utilization of biomarkers for brain disorders is not a recent concept. In the nineteenth century, Kraepelin established a writing scale to stratify patients suffering from psychiatric disorders by measuring their writing pressure curves (Kraepelin, 1899). Due to the phenotypic heterogeneity and the lack of quantitative measures for disease symptoms, biomarker discovery in the field of neuroscience has been confronted with considerable challenges. This holds true especially for neuropsychiatric disorders where, despite tremendous progress in understanding brain function, the exact molecular underpinnings of mental dysfunction remain elusive. Because biomarkers can differentiate between distinct biological states, their availability is critical in clinical settings for premorbid diagnosis, patient stratification, and monitoring of disease progression and treatment. In this regard, established biomarkers in other areas of medicine including human chorionic gonadotropin to determine pregnancy (Spadoni *et al.*, 1964), serum ferritin to measure anemia (Pasricha *et al.*, 2010), and cholesterol to predict cardiovascular disease risk (Kannel *et al.*, 1979) have significantly simplified clinical practice.

Currently, the diagnosis for all psychiatric disorders is symptomatic and relies on interview-based communication between the patient and the physician. The only means for disease categorization is the *Diagnostic and Statistical Manual of Mental Disorders* (American Psychiatric Association, 2000). Although this manual may thoroughly describe the symptomatology of different mental disorders, it does not provide molecular correlates nor does it address the underlying disease etiology. In addition to the lack of any measurable molecular entities, disease classification is often confounded by symptomatic expressions because multiple psychiatric disorders that exhibit similar indications can coexist (Turck *et al.*, 2008).

The availability of molecular biomarkers for psychiatric disorders would, therefore, greatly facilitate disease diagnosis and stratification.

B. Biomarkers for the Development of Novel Therapeutics and in Basic Research

Current medications for neuropsychiatric disorders suffer from a plethora of side effects often so serious that they can perturb the basic aspects of everyday life including employment and family relationships. Further, drug efficacy can be delayed by several weeks, and the resistance to treatment is common in a substantial number of patients (Bystritsky, 2006). Thus, the introduction of novel therapeutics is required for optimal patient treatment. In this regard, biomarker availability would contribute not only to the identification of novel drug targets but also to the drug development pipeline by providing surrogate markers. A surrogate marker is an outcome that can be observed at an earlier time point, at a lower cost, and preferably in a less invasive manner than the true outcome, and enables valid inferences about the intervention effect of the true outcome (Staner, 2006). Surrogate biomarkers can accelerate drug discovery by assessing drug efficacy, thereby reducing clinical trial costs and development time (Schwarz and Bahn, 2008; see Chapter "Challenges of introducing new biomarker products for neuropsychiatric disorders into the market" by Bahn *et al.*). Although in the field of psychiatric disorders, surrogate measures are still at an early stage and each candidate surrogate marker should be thoroughly tested prior to large-scale use (Katz, 2004; Staner, 2006), new therapies for psychiatric disorders could substantially benefit from the use of surrogate markers.

In basic research, the search for biomarkers can contribute to the elucidation of pathogenetic molecular mechanisms by revealing affected pathways and processes. Currently, biomarker discovery studies are pursued at an increasing pace, and such studies attract generous financial support from funding organizations. The field of neuropsychiatric disorders is eagerly awaiting the availability of accurate and quantifiable biomarkers, whose implementation could eventually revolutionize neuroscience research by providing opportunities for diagnosis and treatment based on objective and measurable molecular characteristics rather than on solely variable and subjective clinical criteria.

II. Tools for Biomarker Discovery in Neuroscience

Relevant biomarkers for neuropsychiatric disorders can be derived from a number of discovery platforms and can include "wet" biomarkers, such as proteins and metabolites (see Chapter "Proteomic technologies for biomarker studies in psychiatry: Advances and needs" by Martins-de-Souza *et al.*), as well as "dry" biomarkers, such as brain images (see Chapter "Imaging brain microglial activation using positron emission tomography and translocator protein-specific radioligands" by Owen and Matthews). Regardless of the biomarker discovery platform type, appropriate quantitative assays need to be developed to translate the experimental workflows into clinical practice.

The starting point of biomarker discovery for neuropsychiatric disorders is either patient material or animal models that represent psychiatric endophenotypes. Although human material is the most relevant specimen for analysis, the interindividual variability together with the low sample amounts available can pose serious challenges for analytical efforts. Not surprisingly, during the exploratory phase of the biomarker discovery pipeline, animal model-based studies tend to have higher success rates compared to human-based studies due to the controlled genetic background, the limited heterogeneity, and the large sample cohorts that can be achieved in laboratory-bred animal populations (Turck *et al.*, 2005). However, because of the lack of defined lesions for most neuropsychiatric conditions and the fact that the whole spectrum of neuropsychiatric disorders in humans cannot be fully recapitulated in lower organisms, most of the existing animal models aim to capture only specific disease characteristics or endophenotypes (see Chapter "Behavioral and molecular biomarkers in translational animal models for neuropsychiatric disorders" by Sarnyai *et al.*). The study of endophenotypes has provided useful insights into the psychopathology of psychiatric disorders (Amann *et al.*, 2010; Kendler and Neale, 2010; Puls and Gallinat, 2008) and is a promising approach to identify biomarkers indicative of disease progression or a given disease symptom. Nevertheless, care should be taken when extrapolating conclusions drawn from animal models to humans.

The availability of -omics methods (genomics, transcriptomics, proteomics, and metabolomics) and new powerful *in vivo* imaging technologies have improved the understanding of psychiatric disorder pathophysiology by comprehensively interrogating disease states at the molecular level. As a result of the development of these holistic approaches, a shift from hypothesis-driven to hypothesis-free studies has occurred, raising the possibility of identifying novel molecular entities and affected brain circuits that constitute candidate biomarkers.

Genomic analyses have provided useful insights into genes conferring susceptibility to complex neuropsychiatric diseases (Gill *et al.*, 2010) as well as genes associated with treatment resistance and efficacy (Binder *et al.*, 2004; Foster *et al.*, 2010; Möller and Rujescu, 2010). Given that multiple genetic lesions, which may additionally vary among individuals, can cause psychiatric disorders, many disease-related genes have a low penetrance and do not exhibit an effect on the phenotype in a predictable and quantifiable manner (Schwarz and Bahn, 2008). This gap between a genetic lesion and an effect on the behavioral phenotype can be bridged by proteomics and metabolomics. Proteomic signatures are dynamic and have the potential to reflect different disease states and reveal mechanisms of drug action (Turck *et al.*, 2008; see Chapter "Proteomic technologies for biomarker studies in psychiatry: Advances and needs" by Martins-de-Souza *et al.*). Importantly, the majority of existing antipsychotic medications seem to primarily target protein entities rather than genes (Holsboer, 2008). Moreover, metabolomic signatures reflect the status of diverse biochemical pathways in health and disease, and in combination with other -omics information, they can lead to a systemic approach to shed light on disease pathogenesis and discover novel biomarkers (Kaddurah-Daouk *et al.*, 2008).

While microarrays have enabled the comparison of two different disease states at the whole genome level, quantitative proteomics has not yet been able to interrogate proteomes in a comprehensive manner and frequently only the most highly abundant proteins in plasma and tissue can be identified and quantified. Yet, recent methodological advances have allowed for increased sensitivity, specificity, and proteome coverage, and have already been applied to the study of neuropsychiatric disorders (Filiou *et al.*, 2011a). Proteomic tools including label-free (Huang *et al.*, 2007b; Levin *et al.*, 2007) as well as stable isotope labeling-based technologies [isotope-coded protein label (ICPL), isobaric tag for relative and absolute quantification (iTRAQ)] (Maccarrone *et al.*, 2010; Martins-de-Souza *et al.*, 2009a,b, 2010) have been used to analyze the brain tissue of schizophrenia patients. *In vivo* metabolic stable isotope labeling with amino acids in cell culture (SILAC) and global labeling have been applied to animal models for psychopathologies (Frank *et al.*, 2009; Filiou *et al.*, 2011b; Liao *et al.*, 2008; Zhang *et al.*, in press). By focusing on subproteomes of interest (i.e., synaptosomes and postsynaptic density; Filiou *et al.*, 2010; van de Bayés *et al.*, 2011), sample complexity can be reduced to result in a thorough quantitative analysis.

The large data sets that are routinely generated by high-throughput -omics approaches have highlighted the need for computational data analysis workflows and integrated network-based approaches, thus adding bioinformatics, biostatistics, and systems biology to the researchers' toolbox for biomarker discovery. An increasing number of bioinformatic solutions for quantitative proteomic and metabolomic analyses (Cox and Mann, 2008; Haegler *et al.*, 2009; Hiller *et al.*, 2009; Pan *et al.*, 2006; Zhang *et al.*, 2009) have greatly contributed to the relative protein quantification speed and accuracy. At the same time, *in silico* approaches have been used to model and to *de novo* identify affected networks in major neuropsychiatric disorders (Gormanns *et al.*, 2011; Sun *et al.*, 2010).

III. Advancements in Biomarker Discovery in Neuroscience

A. Mouse Models

Animal model studies, which have been mainly based on characterizing behavioral changes in rodents (Landgraf *et al.*, 2007; Otte *et al.*, 2009), have provided useful insights into our understanding of the molecular networks involved in psychiatric disorders (see Chapter "Behavioral and molecular biomarkers in translational animal models for neuropsychiatric disorders" by Sarnyai *et al.*). Among the most prominent examples is the implication of oxidative stress in anxiety disorders (Bouayed *et al.*, 2007; Rammal *et al.*, 2008). Glyoxylase 1, a protein exerting a neuroprotective role against oxidative damage, has been found to be consistently dysregulated in the amygdala, cortex, and hypothalamus as well as in red blood cells in a mouse model of anxiety-related behavior (Ditzen *et al.*, 2006; Frank *et al.*, 2009; Krömer *et al.*, 2005). Notably, the glyoxylase 1 gene copy number has been associated with anxiety-related behaviors (Williams *et al.*, 2009). Together with glyoxylase 1, glutathione reductase 1, which is also involved in the antioxidant defense pathway, has been reported to regulate anxiety in mice (Gingrich, 2005; Hovatta *et al.*, 2005). Another candidate biomarker is the enzyme enolase phosphatase, which is present in two different isoforms in high and low anxiety-related behavior mouse lines. The line-specific isoforms are the result of two nonsynonymous single nucleotide polymorphisms (SNPs) that affect the enzymatic activity of enolase phosphatase as well as downstream enzymatic reactions (Ditzen *et al.*, 2006, 2010).

Several studies in rodents have implicated energy metabolism alterations in pathways such as glycolysis, oxidative phosphorylation, and the Krebs cycle in anxiety or depression-like behaviors (Kedracka-Krok *et al.*, 2010; Marais *et al.*, 2009; Szego *et al.*, 2010). Most cellular systems are flexible in shifting their metabolism between different energy pathways according to energetic demands or nutrient availability. Interestingly, the therapeutic potential of modulating this reallocation has been demonstrated for several diseases (Chen *et al.*, 2007; Huber *et al.*, 2004; Riepe *et al.*, 1997). Because nutrient-sensitized screening has shown that several FDA-approved drugs are able to redirect energy metabolism pathways (Gohil *et al.*, 2010), such pathway shifts may provide sensitive biomarker information.

B. Human Data

1. *Postmortem* Brain Tissue Studies

Postmortem brain tissue is commonly used for genetic, transcriptomic, and proteomic biomarker discovery studies. Similar to the aforementioned studies in animal models, data generated from a number of proteomic studies on human brain tissue have implicated energy metabolism alterations and synaptic pathology in schizophrenia, bipolar, and anxiety disorders (Beasley *et al.*, 2006; Clark *et al.*, 2006; Filiou *et al.*, 2011b; Johnston-Wilson *et al.*, 2000; Prabakaran *et al.*, 2004; Pennington *et al.*, 2008; Martins-de-Souza *et al.*, 2010). Although the outcomes of these studies have not been a clearly defined set of candidate biomarkers, there is a remarkable convergence between the results across different proteomic analyses that are also in considerable agreement with data derived from genetic linkage and association studies (English *et al.*, 2011).

2. *Cerebrospinal Fluid Studies*

As *postmortem* brain tissue is rare and access to brain biobanks is subject to a legislation framework that may not allow unlimited use of material for a given research project, body fluids have been used as an alternative to brain tissue for biomarker discovery efforts. Cerebrospinal fluid (CSF) is the most relevant biofluid for research in brain disorders due to its close proximity to the "site of action." As CSF surrounds the brain tissue, it contains a great number of molecular entities that either mediate or are products of brain function. CSF reflects the brain metabolic state and thus has the ability to reveal pathophysiological alterations which occur in the brain (Turck *et al.*, 2005). Moreover, due to existing standard protocols for CSF acquisition from patients (i.e., lumbar puncture), sample collection can be implemented in a reproducible manner across different clinical settings. However, the low amount of starting material that is typically acquired limits extensive CSF analyses, while the presence of high-abundance proteins (i.e., albumin) hinders the detection and quantification of low-abundance brain-derived proteins of interest. Apart from these technical considerations, another concern is the possibility of an infiltration of serum proteins into CSF due to a leakage in the blood brain barrier. Blood brain barrier alterations are common in patients suffering from brain disorders (Stolp and Dziegielewska, 2009), and the blood–CSF interchange may result in the identification of proteins in CSF that are of unknown origin and thus provide inconclusive data for brain pathology evaluations.

To explore the complexity of the CSF proteome, optimized methodologies have been established to address technical limitations by extensive fractionation and the depletion of abundant proteins (Maccarrone *et al.*, 2004; Pan *et al.*, 2007; Schutzer *et al.*, 2010). A CSF proteome profile of patients suffering from schizophrenia was published (Jiang *et al.*, 2003) and several quantitative studies on CSF have reported neuropeptide and neurotransmitter alterations in major psychiatric disorders (Asberg, 1997; Heilig *et al.*, 2004; Nikisch *et al.*, 2005; Sah *et al.*, 2009). A comparison of CSF proteome profiles has also been performed to identify differences associated with suicidal behavior (Brunner *et al.*, 2005). In addition, the CSF metabolic profiles of drug-naive (or minimally treated) patients versus patients with first-onset paranoid schizophrenia were compared and the identified alterations in drug-naive patients were validated in a test sample set (Holmes *et al.*, 2006). Interestingly, combined proteomic and metabolomic approaches have revealed characteristic signatures of the initial psychosis stage and have shown that schizophrenia-related biochemical disease processes can be traced in the CSF of prodromal psychosis patients (Huang *et al.*, 2007a).

3. *Plasma Studies*

Plasma constitutes the specimen of choice for the implementation of a biomarker assay in clinical settings. Despite its easy and noninvasive acquisition at relatively high amounts from patients, plasma analysis presents researchers with serious technical challenges, largely due to the great dynamic range of its protein constituents (Jacobs *et al.*, 2005). Nevertheless, a multidisciplinary study in a large cohort of well-characterized schizophrenic and major depressive patients has resulted in identification of molecular candidate plasma biomarker signatures (Domenici *et al.*, 2010). Toward the clinical implementation of plasma biomarkers, there have been attempts to provide molecular "kits" to aid in the diagnosis or risk assessment for major psychiatric disorders. These mainly involve tests based on genetic susceptibility to predict risk for developing a psychiatric disorder. Although not yet commercially available, efforts to develop such tests are ongoing (Couzin, 2008). Recently, a multiplex protein immunoassay-based plasma/serum diagnostic test for schizophrenia was launched, assessing the levels of 51 molecules for identification of patients with schizophrenia compared to healthy control subjects (Schwarz *et al.*, 2010, 2011; see Chapters "The application of multiplexed assay systems for molecular diagnostics" by Schwarz *et al.* and "Algorithm development for diagnostic biomarker assays" by Izmailov *et al.*). These kits provide promising tools for the prognosis and classification of neuropsychiatric disorders and constitute a first step toward biomarker-based molecular diagnostics, indicating the potential of nonhypothesis-driven -omics approaches for biomarker discovery.

C. Future Directions

Biomarker research laboratories now have access to advanced analytical instrumentation and computational power. Hence, the search for novel biomarkers will intensify and yield new information on subtle molecular changes associated with disease etiology and progression. Twin studies can address the pivotal environmental contribution in neuropsychiatric disorders. Genome-wide microarray (Matigian *et al.*, 2007) and epigenetic (Kuratomi *et al.*, 2008) analyses of twins discordant for major psychiatric disorders have provided useful insights into disease-specific alterations in a shared genetic background and may also be a valuable starting point for studies using sophisticated proteomic and metabolomic platforms.

Another research area that has not yet been extensively studied in the context of psychiatric disorders is posttranslational modifications. Protein modifications, such as phosphorylation, are involved in neurotransmission, regulate the active and inactive forms of proteins, and serve as signaling regulators in pathways involved in disease pathogenesis (Smart, 1997; Takahashi *et al.*, 2003). For example, changes in the phosphorylation levels of the cAMP response element-binding (CREB) signaling protein have been observed in patients who respond to psychotherapeutic or psychopharmacological treatments compared to those who do not respond to treatment (Koch *et al.*, 2002, 2009). Along with the technological advances in mass spectrometry-based methods, the high-throughput analysis and quantification of posttranslational modifications is now feasible and may provide new directions for applications in biomarker discovery.

At a systemic level, structural brain abnormalities usually in the form of cell loss (e.g., loss of dopaminergic neurons in the *substantia nigra* in Parkinson's disease) or volume alterations (e.g., atrophy) are common in brain disorders. Notably, one of the most consistent alterations in psychiatric disorders is the reduction of hippocampal volume in patients and this has been recapitulated in mouse models of posttraumatic stress disorder (Golub *et al.*, 2011; Karl *et al.*, 2006). These types of morphological/morphometrical alterations may provide useful information for monitoring disease progression and treatment efficacy. In this regard, data from imaging techniques may complement molecular markers for an accurate assessment of disease (see Chapter "Imaging brain microglial activation using positron emission tomography and translocator protein-specific radioligands" by Owen and Matthews).

IV. Considerations for Biomarker Discovery and Translation in Neuroscience

Despite the overwhelming advancements in method development and instrumentation, no biomarkers for neuropsychiatric disorders have been successfully translated to clinical practice. This poor success rate can be attributed to disease complexity, sample quality and collection, limitations of validation procedures, and the lack of interdisciplinary approaches from the involved stakeholders.

A. Disease Complexity

Psychiatric disorders are characterized by a great complexity at the genome level in the form of DNA copy variations, multiple gene interactions, and epigenetic reprogramming. Adding to this complexity, genes, proteins, and metabolites are also influenced by a multidimensional interplay with environmental factors with unpredictable and uncontrollable effects on the behavioral phenotype. It is therefore highly unlikely that single biomarkers with high specificity and sensitivity will be the answer of biomarker discovery research for neuropsychiatric disorders.

B. Sample Quality and Collection

One of the primary considerations in study design for biomarker discovery is the material used for analysis. In animal research, it is critical that face, construct, and predictive validity are clearly defined. In human research, clinically and pharmacologically well-characterized patient populations are required to achieve meaningful exploratory analyses. Without adequate information on confounding pathologies, epidemiological parameters as well as family and medication history, the data from these studies will be difficult, if not impossible, to interpret. Interindividual variability between patient and control groups should be assessed *a priori* taking into account age, ethnicity, family history, body mass index, confounding pathologies, and lifestyle (Rifai *et al.*, 2006). Moreover, the lack of standard operating procedures for tissue and biofluid acquisition may affect the study outcome, as samples collected from different clinical settings are frequently not processed in the same manner, resulting in artifactual data. Such problems could be addressed through the establishment of standard operating procedures for sample collection and storage, along with similar procedures for sample analysis and data processing.

C. Candidate Biomarker Validation

Without a doubt, the bottleneck of biomarker discovery has been their validation. Typically, the output of -omics studies consists of a long list of genes, transcripts, proteins, or metabolites whose levels differ between disease and control states. Yet, only a few of these candidates can realistically be validated in a reasonable time frame and at a justifiable cost. In the majority of cases, a candidate biomarker qualifies for validation based on a subjective evaluation by the research team. Consequently, the "usual suspects" tend to be chosen for validation, and candidates that do not seem to be relevant to the disease phenotype are excluded. For biomarker validation efforts, it is therefore critical that objective criteria are applied to select the most robust candidates based on technical parameters.

Other important considerations are sensitivity and specificity of validation methodologies. Methods need to be evaluated with regard to their technical limitations, analytical reproducibility, and detection ranges to distinguish true biological differences from analytical variability. For routine clinical use, high-throughput capacity and cost-effectivity should also be taken into account. Clinical translation does not only involve the transition from animal models to human specimens, but it also frequently entails testing a candidate biomarker initially identified in the brain tissue, in accessible peripheral specimens such as CSF or plasma. Unfortunately, differences that are found in the brain are often not mirrored or are too low in abundance in plasma, which makes the validation of biomarkers rather challenging. During the biomarker discovery phase, a well-characterized, homogeneous patient cohort of limited size with adequately matching controls is often used. In the validation phase though, larger cohorts are studied where heterogeneity and intragroup-related variation are markedly higher. This reduces the probability that a biomarker candidate will be verified, minimizing its chances of qualifying for use in a clinical trial. Hence, large-scale studies are required to ensure the specificity of a candidate biomarker and its relevance to the disease under examination.

D. Systemic Approaches and Biomarker Initiatives

Due to the complex interplay of genetic and environmental factors in neuropsychiatric disorders, efforts toward applying network-oriented approaches rather than focusing on single molecules are essential for biomarker research. Moving from single entities to pathways and networks necessitates the integration of multiomics approaches as well as a systemic framework of data analysis and interpretation. This underlines the need for the interdisciplinary cooperation between clinical psychiatry, neuroscience, analytical biochemistry, bioinformatics, and biostatistics to investigate alterations that are relevant to brain disease in a global manner. Several biomarker initiatives have been established to combine information from a number of fields and provide a coordinated approach for biomarker discovery. Examples include *The Foundation for the National Institutes of Health (FNIH) Biomarkers Consortium* (http://www.biomarkersconsortium.org), consisting of disease-specific committees that include psychiatric disorders, the *Critical Path Initiative* (http://www.fda.gov/ScienceResearch/SpecialTopics/CriticalPathInitiative/default.htm), a Food and Drug Administration (FDA) consortium to promote the discovery and evaluation of novel medical products, the *Innovative Medicines Initiative* (IMI) of the European Union (http://www.imi.europa.eu), which seeks to establish partnerships between industry and academia for novel drug discovery, and the *Alzheimer's Disease Neuroimaging Initiative* (http://www.adni-info.org), which aims to stage disease progression and create a data repository platform. In 2010, a report from the US *Institute of Medicine* was issued that recommended a framework for biomarker evaluation of chronic diseases (http://www.iom.edu/Reports/2010/Evaluation-of-Biomarkers-and-Surrogate-Endpoints-in-Chronic-Disease.aspx). The ultimate goal of all these initiatives is to engage clinical personnel, medical doctors, and researchers from different fields in an orchestrated effort to discover and translate biomarkers into the clinic. Establishment of standard operating procedures to ensure interlab reproducibility (as mentioned above), the training of scientific staff, and the integration of well-defined validation steps will greatly contribute to the advancement of the field.

V. Outlook—The Perspective of Personalized Medicine

It is now widely accepted that a single biomarker will not be able to unequivocally distinguish between clinical neuropsychiatric phenotypes. A panel of biomarkers that are able to depict a disease state more accurately is required for complex diseases such as anxiety disorders, depression, and schizophrenia (Turck *et al.*, 2008). In this context, integrating -omics and imaging data will provide a more comprehensive characterization of different disease states through a combination of biomarkers (e.g., altered protein levels and functional magnetic resonance imaging (fMRI) profiles combined with genetic information), which in turn will enable more accurate disease diagnoses and patient-specific treatment options. Currently, no molecular biomarker is available for neuropsychiatric disorders, and there is still a long path ahead toward clinical implementation. However, there are reasons for optimism that further technological advancements and interdisciplinary approaches will overcome the current limitations in the field and will finally enable the concept of personalized medicine (see Chapter "Toward personalized medicine in the neuropsychiatric field" by Wong *et al.*).

Acknowledgments

The authors would like to thank all members of the 'Proteomics and Biomarkers Research Group' at the *Max Planck Institute of Psychiatry* for insightful discussions.

References

Amann, L.C., Gandal, M.J., Halene, T.B., Ehrlichman, R.S., White, S.L., McCarren, H.S., and Siegel, S.J. (2010). Mouse behavioral endophenotypes for schizophrenia. *Brain Res. Bull.* **83**(3–4), 147–161.

American Psychiatric Association (2000). Diagnostic and statistical manual of mental disorders. 4th edn., text revision. American Psychiatric Association, Washington, DC, USA.

Asberg, M. (1997). Neurotransmitters and suicidal behavior. The evidence from cerebrospinal fluid studies. *Ann. N. Y. Acad. Sci.* **836**, 158–181.

Beasley, C.L., Pennington, K., Behan, A., Wait, R., Dunn, M.J., and Cotter, D. (2006). Proteomic analysis of the anterior cingulate cortex in the major psychiatric disorders: evidence for disease-associated changes. *Proteomics* **6**(11), 3414–3425.

Binder, E.B., Salyakina, D., Lichtner, P., Wochnik, G.M., Ising, M., Pütz, B., Papiol, S., Seaman, S., Lucae, S., Lucae, S., Kohli, M.A., Nickel, T., *et al.* (2004). Polymorphisms in FKBP5 are associated with increased recurrence of depressive episodes and rapid response to antidepressant treatment. *Nat. Genet.* **36**(12), 1319–1325.

Biomarkers Definitions Working Group, Atkinson, A.J., Colburn, W.A., DeGruttola, V.G., DeMets, D.L., Downing, G.J., Hoth, D.F., Oates, J.A., Peck, C.C., Schooley, R.T., Spilker, B.A., Woodcock, J., *et al.* (2001). Biomarkers and surrogate endpoints: preferred definitions and conceptual framework. *Clin. Pharmacol. Ther.* **69**(3), 89–95.

Bouayed, J., Rammal, H., Younos, C., and Soulimani, R. (2007). Positive correlation between peripheral blood granulocyte oxidative status and level of anxiety in mice. *Eur. J. Pharmacol.* **564** (1–3), 146–149.

Brunner, J., Bronisch, T., Uhr, M., Ising, M., Binder, E., Holsboer, F., and Turck, C.W. (2005). Proteomic analysis of the CSF in unmedicated patients with major depressive disorder reveals alterations in suicide attempters. *Eur. Arch. Psychiatry Clin. Neurosci.* **255**(6), 438–440.

Bystritsky, A. (2006). Treatment-resistant anxiety disorders. *Mol. Psychiatry* **11**(9), 805–814.

Chen, Q., Camara, A.K.S., Stowe, D.F., Hoppel, C.L., and Lesnefsky, E.J. (2007). Modulation of electron transport protects cardiac mitochondria and decreases myocardial injury during ischemia and reperfusion. *Am. J. Physiol. Cell Physiol.* **292**(1), C137–C147.

Clark, D., Dedova, I., Cordwell, S., and Matsumoto, I. (2006). A proteome analysis of the anterior cingulate cortex gray matter in schizophrenia. *Mol. Psychiatry* **11**(5), 459–470.

Couzin, J. (2008). Science and commerce. Gene tests for psychiatric risk polarize researchers. *Science* **319**(5861), 274–277.

Cox, J., and Mann, M. (2008). MaxQuant enables high peptide identification rates, individualized p.p.b.-range mass accuracies and proteome-wide protein quantification. *Nat. Biotechnol.* **26**(12), 1367–1372.

Ditzen, C., Jastorff, A.M., Kessler, M.S., Bunck, M., Teplytska, L., Erhardt, A., Krömer, S.A., Varadarajulu, J., Targosz, B.S., Sayan-Ayata, E.F., Holsboer, F., Landgraf, R., *et al.* (2006). Protein biomarkers in a mouse model of extremes in trait anxiety. *Mol. Cell. Proteomics* **5**(10), 1914–1920.

Ditzen, C., Varadarajulu, J., Czibere, L., Gonik, M., Targosz, B.S., Hambsch, B., Bettecken, T., Kessler, M.S., Frank, E., Bunck, M., Teplytska, L., Erhardt, A., *et al.* (2010). Proteomic-based genotyping in a mouse model of trait anxiety exposes disease-relevant pathways. *Mol. Psychiatry* **15** (7), 702–711.

Domenici, E., Willé, D.R., Tozzi, F., Prokopenko, I., Miller, S., McKeown, A., Brittain, C., Rujescu, D., Giegling, I., Turck, C.W., Holsboer, F., Bullmore, E.T., *et al.* (2010). Plasma protein biomarkers for depression and schizophrenia by multi analyte profiling of case-control collections. *PLoS One* **5**(2), e9166.

English, J.A., Pennington, K., Dunn, M.J., and Cotter, D.R. (2011). The neuroproteomics of schizophrenia. *Biol. Psychiatry* **69**(2), 163–172.

Filiou, M.D., Bisle, B., Reckow, S., Teplytska, L., Maccarrone, G., and Turck, C.W. (2010). Profiling of mouse synaptosome proteome and phosphoproteome by IEF. *Electrophoresis* **31**(8), 1294–1301.

Filiou, M.D., Turck, C.W., and Martins-de-Souza, D. (2011a). Quantitative proteomics for investigating psychiatric disorders. *Proteomics Clin. Appl.* **5**(1–2), 38–49.

Filiou, M.D., Zhang, Y., Teplytska, L., Reckow, S., Gormanns, P., Maccarrone, G., Frank, E., Kessler, M.S., Hambsch, B., Nussbaumer, M., Bunck, M., Ludwig, T., *et al.* (2011b). Proteomics and metabolomics analysis of a trait anxiety mouse model reveals divergent mitochondrial pathways. *Biol. Psychiatry* Jul 24, Epub ahead of print. doi: 10.1016/j.biopsych.2011.06.009.

Foster, A., Miller, del D., and Buckley, P. (2010). Pharmacogenetics and schizophrenia. *Clin. Lab. Med.* **30**(4), 975–993.

Frank, E., Kessler, M.S., Filiou, M.D., Zhang, Y., Maccarrone, G., Reckow, S., Bunck, M., Heumann, H., Turck, C.W., Landgraf, R., and Hambsch, B. (2009). Stable isotope metabolic labeling with a novel ^{15}N-enriched bacteria diet for improved proteomic analyses of mouse models for psychopathologies. *PLoS One* **4**(11), e7821.

Gill, M., Donohoe, G., and Corvin, A. (2010). What have the genomics ever done for the psychoses? *Psychol. Med.* **40**(4), 529–540.

Gingrich, J.A. (2005). Oxidative stress is the new stress. *Nat. Med.* **11**(12), 1281–1282.

Gohil, V.M., Sheth, S.A., Nilsson, R., Wojtovich, A.P., Lee, J.H., Perocchi, F., Chen, W., Clish, C.B., Ayata, C., Brookes, P.S., and Mootha, V.K. (2010). Nutrient-sensitized screening for drugs that shift energy metabolism from mitochondrial respiration to glycolysis. *Nat. Biotechnol.* **28**(3), 249–255.

Golub, Y., Kaltwasser, S.F., Mauch, C.P., Herrmann, L., Schmidt, U., Holsboer, F., Czisch, M., and Wotjak, C.T. (2011). Reduced hippocampus volume in the mouse model of Posttraumatic Stress Disorder. *J. Psychiatr. Res.* **45**(5), 650–659.

Gormanns, P., Mueller, N.S., Ditzen, C., Wolf, S., Holsboer, F., and Turck, C.W. (2011). Phenome-transcriptome correlation unravels anxiety and depression related pathways. *J. Psychiatr. Res.* **45**(7), 973–979.

Haegler, K., Mueller, N.S., Maccarrone, G., Hunyadi-Gulyas, E., Webhofer, C., Filiou, M.D., Zhang, Y., and Turck, C.W. (2009). QuantiSpec-Quantitative mass spectrometry data analysis of ^{15}N-metabolically labeled proteins. *J. Proteomics* **71**(6), 601–608.

Heilig, M., Zachrisson, O., Thorsell, A., Ehnvall, A., Mottagui-Tabar, S., Sjögren, M., Asberg, M., Ekman, R., Wahlestedt, C., and Agren, H. (2004). Decreased cerebrospinal fluid neuropeptide Y (NPY) in patients with treatment refractory unipolar major depression: preliminary evidence for association with preproNPY gene polymorphism. *J. Psychiatr. Res.* **38**(2), 113–121.

Hiller, K., Hangebrauk, J., Jäger, C., Spura, J., Schreiber, K., and Schomburg, D. (2009). MetaboliteDetector: comprehensive analysis tool for targeted and nontargeted GC/MS based metabolome analysis. *Anal. Chem.* **81**(9), 3429–3439.

Holmes, E., Tsang, T.M., Huang, J.T., Leweke, F.M., Koethe, D., Gerth, C.W., Nolden, B.M., Gross, S., Schreiber, D., Nicholson, J.K., and Bahn, S. (2006). Metabolic profiling of CSF: evidence that early intervention may impact on disease progression and outcome in schizophrenia. *PLoS Med.* **3**(8), e327.

Holsboer, F. (2008). How we can realize the promise of personalized antidepressant medicines? *Nat. Rev. Neurosci.* **9**(8), 638–646.

Hovatta, I., Tennant, R.S., Helton, R., Marr, R.A., Singer, O., Redwine, J.M., Ellison, J.A., Schadt, E. E., Verma, I.M., Lockhart, D.J., and Barlow, C. (2005). Glyoxalase 1 and glutathione reductase 1 regulate anxiety in mice. *Nature* **438**(7068), 662–666.

Huang, J.T., Leweke, F.M., Tsang, T.M., Koethe, D., Kranaster, L., Gerth, C.W., Gross, S., Schreiber, D., Ruhrmann, S., Schultze-Lutter, F., Klosterkötter, J., Holmes, E., *et al.* (2007a). CSF metabolic and proteomic profiles in patients prodromal for psychosis. *PLoS One* **2**(1), e756.

Huang, J.T., McKenna, T., Hughes, C., Leweke, F.M., Schwarz, E., and Bahn, S. (2007b). CSF biomarker discovery using label-free nano-LC-MS based proteomic profiling: technical aspects. *J. Sep. Sci.* **30**(2), 214–225.

Huber, R., Spiegel, T., Buchner, M., and Riepe, M.W. (2004). Graded reoxygenation with chemical inhibition of oxidative phosphorylation improves posthypoxic recovery in murine hippocampal slices. *J. Neurosci. Res.* **75**(3), 441–449.

Jacobs, J.M., Adkins, J.N., Qian, W.J., Liu, T., Shen, Y., Camp, D.G. 2nd, and Smith, R.D. (2005). Utilizing human blood plasma for proteomic biomarker discovery. *J. Proteome Res.* **4**(4), 1073–1085.

Jiang, L., Lindpaintner, K., Li, H.F., Gu, N.F., Langen, H., He, L., and Fountoulakis, M. (2003). Proteomic analysis of the cerebrospinal fluid of patients with schizophrenia. *Amino Acids* **25**(1), 49–57.

Johnston-Wilson, N.L., Sims, C.D., Hofmann, J.P., Anderson, L., Shore, A.D., Torrey, E.F., and Yolken, R.H. (2000). Disease-specific alterations in frontal cortex brain proteins in schizophrenia, bipolar disorder, and major depressive disorder. *Mol. Psychiatry* **5**(2), 142–149.

Kaddurah-Daouk, R., Kristal, B.S., and Weinshilboum, R.M. (2008). Metabolomics: a global biochemical approach to drug response and disease. *Annu. Rev. Pharmacol. Toxicol.* **48**, 653–683.

Kannel, W.B., Castelli, W.P., and Gordon, T. (1979). Cholesterol in the prediction of atherosclerotic disease. New perspectives based on the Framingham study. *Ann. Intern. Med.* **90**(1), 85–91.

Karl, A., Schaefer, M., Malta, L.S., Dörfel, D., Rohleder, N., and Werner, A. (2006). A meta-analysis of structural brain abnormalities in PTSD. *Neurosci. Biobehav. Rev.* **30**(7), 1004–1031.

Katz, R. (2004). Biomarkers and surrogate markers: an FDA perspective. *NeuroRx.* **1**(2), 189–195.

Kedracka-Krok, S., Fic, E., Jankowska, U., Jaciuk, M., Gruca, P., Papp, M., Kusmider, M., Solich, J., Debski, J., Dadlez, M., and Dziedzicka-Wasylewska, M. (2010). Effect of chronic mild stress and imipramine on the proteome of the rat dentate gyrus. *J. Neurochem.* **113**(4), 848–859.

Kendler, K.S., and Neale, M.C. (2010). Endophenotype: a conceptual analysis. *Mol. Psychiatry* **15**(8), 789–797.

Koch, J.M., Kell, S., Hinze-Selch, D., and Aldenhoff, J.B. (2002). Changes in CREB-phosphorylation during recovery from major depression. *J. Psychiatr. Res.* **36**(6), 369–375.

Koch, J.M., Hinze-Selch, D., Stingele, K., Huchzermeier, C., Goder, R., Seeck-Hirschner, M., and Aldenhoff, J.B. (2009). Changes in CREB phosphorylation and BDNF plasma levels during psychotherapy of depression. *Psychother. Psychosom.* **78**(3), 187–192.

Kraepelin, E. (1899). Psychiatry Textbook, Vol. II, 6th edn. Verlag von Johann Ambrosius Barth, Leipzig, Germany.

Krömer, S.A., Kessler, M.S., Milfay, D., Birg, I.N., Bunck, M., Czibere, L., Panhuysen, M., Putz, B., Deussing, J.M., Holsboer, F., Landgraf, R., and Turck, C.W. (2005). Identification of glyoxalase-I as a protein marker in a mouse model of extremes in trait anxiety. *J. Neurosci.* **25**(17), 4375–4384.

Kuratomi, G., Iwamoto, K., Bundo, M., Kusumi, I., Kato, N., Iwata, N., Ozaki, N., and Kato, T. (2008). Aberrant DNA methylation associated with bipolar disorder identified from discordant monozygotic twins. *Mol. Psychiatry* **13**(4), 429–441.

Landgraf, R., Kessler, M.S., Bunck, M., Murgatroyd, C., Spengler, D., Zimbelmann, M., Nussbaumer, M., Czibere, L., Turck, C.W., Singewald, N., Rujescu, D., and Frank, E. (2007). Candidate genes of anxiety-related behavior in HAB/LAB rats and mice: focus on vasopressin and glyoxalase-I. *Neurosci. Biobehav. Rev.* **31**(1), 89–102.

Levin, Y., Schwarz, E., Wang, L., Leweke, F.M., and Bahn, S. (2007). Label-free LC-MS/MS quantitative proteomics for large-scale biomarker discovery in complex samples. *J. Sep. Sci.* **30** (14), 2198–2203.

Liao, L.J., Park, S.K., Xu, T., Vanderklish, P., and Yates, J.R. (2008). Quantitative proteomic analysis of primary neurons reveals diverse changes in synaptic protein content in fmr1 knockout mice. *Proc. Natl. Acad. Sci. USA* **105**(40), 15281–15286.

Maccarrone, G., Milfay, D., Birg, I., Rosenhagen, M., Grimm, R., Bailey, J., Jolotarjova, N., Holsboer, F., and Turck, C.W. (2004). Mining the human CSF proteome by immunodepletion and shotgun mass spectrometry. *Electrophoresis* **25**(14), 2402–2412.

Maccarrone, G., Turck, C.W., and Martins-de-Souza, D. (2010). Shotgun mass spectrometry workflow combining IEF and LC-MALDI-TOF/TOF. *Protein J.* **29**(2), 99–102.

Marais, L., Hattingh, S.M., Stein, D.J., and Daniels, W.M. (2009). A proteomic analysis of the ventral hippocampus of rats subjected to maternal separation and escitalopram treatment. *Metab. Brain Dis.* **24**(4), 569–586.

Martins-de-Souza, D., Gattaz, W.F., Schmitt, A., Rewerts, C., Maccarrone, G., Dias-Neto, E., and Turck, C.W. (2009a). Prefrontal cortex shotgun proteome analysis reveals altered calcium homeostasis and immune system imbalance in schizophrenia. *Eur. Arch. Psychiatry Clin. Neurosci.* **259**(3), 151–163.

Martins-de-Souza, D., Gattaz, W.F., Schmitt, A., Rewerts, C., Marangoni, S., Novello, J.C., Maccarrone, G., Turck, C.W., and Dias-Neto, E. (2009b). Alterations in oligodendrocyte proteins, calcium homeostasis and new potential markers in schizophrenia anterior temporal lobe are revealed by shotgun proteome analysis. *J. Neural Transm.* **116**(3), 275–289.

Martins-de-Souza, D., Maccarrone, G., Wobrock, T., Zerr, I., Gormanns, P., Reckow, S., Falkai, P., Schmitt, A., and Turck, C.W. (2010). Proteome analysis of the thalamus and cerebrospinal fluid reveals glycolysis dysfunction and potential biomarkers candidates for schizophrenia. *J. Psychiatr. Res.* **44**(16), 1176–1189.

Matigian, N., Windus, L., Smith, H., Filippich, C., Pantelis, C., McGrath, J., Mowry, B., and Hayward, N. (2007). Expression profiling in monozygotic twins discordant for bipolar disorder reveals dysregulation of the WNT signalling pathway. *Mol. Psychiatry* **12**(9), 815–825.

Möller, H.J., and Rujescu, D. (2010). Pharmacogenetics-genomics and personalized psychiatry. *Eur. Psychiatry* **25**(5), 291–293.

Nikisch, G., Agren, H., Eap, C.B., Czernik, A., Baumann, P., and Mathé, A.A. (2005). Neuropeptide Y and corticotropin-releasing hormone in CSF mark response to antidepressive treatment with citalopram. *Int. J. Neuropsychopharmacol.* **8**(3), 403–410.

Otte, D.M., Bilkei-Gorzó, A., Filiou, M.D., Turck, C.W., Yilmaz, Ö., Holst, M.I., Schilling, K., Abou-Jamra, R., Schumacher, J., Benzel, I., Kunz, W.S., Beck, H., *et al.* (2009). Behavioral changes in G72/G30 transgenic mice. *Eur. Neuropsychopharmacol.* **19**(5), 339–348.

Pan, C., Kora, G., McDonald, W.H., Tabb, D.L., VerBerkmoes, N.C., Hurst, G.B., Pelletier, D.A., Samatova, N.F., and Hettich, R.L. (2006). ProRata: a quantitative proteomics program for accurate protein abundance ratio estimation with confidence interval evaluation. *Anal. Chem.* **78** (20), 7121–7131.

Pan, S., Zhu, D., Quinn, J.F., Peskind, E.R., Montine, T.J., Lin, B., Goodlett, D.R., Taylor, G., Eng, J., and Zhang, J. (2007). A combined dataset of human cerebrospinal fluid proteins identified by multi-dimensional chromatography and tandem mass spectrometry. *Proteomics* **7**(3), 469–473.

Pasricha, S.R., Flecknoe-Brown, S.C., Allen, K.J., Gibson, P.R., McMahon, L.P., Olynyk, J.K., Roger, S.D., Savoia, H.F., Tampi, R., Thomson, A.R., Wood, E.M., and Robinson, K.L. (2010). Diagnosis and management of iron deficiency anaemia: a clinical update. *Med. J. Aust.* **193**(9), 525–532.

Pennington, K., Beasley, C.L., Dicker, P., Fagan, A., English, J., Pariante, C.M., Wait, R., Dunn, M.J., and Cotter, D.R. (2008). Prominent synaptic and metabolic abnormalities revealed by proteomic analysis of the dorsolateral prefrontal cortex in schizophrenia and bipolar disorder. *Mol. Psychiatry* **13**(12), 1102–1117.

Prabakaran, S., Swatton, J.E., Ryan, M.M., Huffaker, S.J., Huang, J.T., Griffin, J.L., Wayland, M., Freeman, T., Dudbridge, F., Lilley, K.S., Karp, N.A., Hester, S., *et al.* (2004). Mitochondrial dysfunction in schizophrenia: evidence for compromised brain metabolism and oxidative stress. *Mol. Psychiatry* **9**(2), 684–697.

Puls, I., and Gallinat, J. (2008). The concept of endophenotypes in psychiatric diseases meeting the expectations? *Pharmacopsychiatry* **41**(Suppl. 1), S37–S43.

Rammal, H., Bouayed, J., Younos, C., and Soulimani, R. (2008). Evidence that oxidative stress is linked to anxiety-related behaviour in mice. *Brain Behav. Immun.* **22**(8), 1156–1159.

Riepe, M.W., Esclaire, F., Kasischke, K., Schreiber, S., Nakase, H., Kempski, O., Ludolph, A.C., Dirnagl, U., and Hugon, J. (1997). Increased hypoxic tolerance by chemical inhibition of oxidative phosphorylation: "chemical preconditioning". *J. Cereb. Blood Flow Metab.* **17**(3), 257–264.

Rifai, N., Gillette, M.A., and Carr, S.A. (2006). Protein biomarker discovery and validation: the long and uncertain path to clinical utility. *Nat. Biotechnol.* **24**(8), 971–983.

Sah, R., Ekhator, N.N., Strawn, J.R., Sallee, F.R., Baker, D.G., Horn, P.S., and Geracioti, T.D. Jr. (2009). Low cerebrospinal fluid neuropeptide Y concentrations in posttraumatic stress disorder. *Biol. Psychiatry* **66**(7), 705–707.

Schutzer, S.E., Liu, T., Natelson, B.H., Angel, T.E., Schepmoes, A.A., Purvine, S.O., Hixson, K.K., Lipton, M.S., Camp, D.G., Coyle, P.K., Smith, R.D., and Bergquist, J. (2010). Establishing the proteome of normal human cerebrospinal fluid. *PLoS One* **5**(6), e10980.

Schwarz, E., and Bahn, S. (2008). Biomarker discovery in psychiatric disorders. *Electrophoresis* **29**(13), 2884–2890.

Schwarz, E., Izmailov, R., Spain, M., Barnes, A., Mapes, J.P., Guest, P.C., Rahmoune, H., Pietsch, S., Leweke, F.M., Rothermundt, M., Steiner, J., Koethe, D., *et al.* (2010). Validation of a blood-based laboratory test to aid in the confirmation of a diagnosis of schizophrenia. *Biomark Insights* **12**(5), 39–47.

Schwarz, E., Guest, P.C., Rahmoune, H., Harris, L.W., Wang, L., Leweke, F.M., Rothermundt, M., Bogerts, B., Koethe, D., Kranaster, L., Ohrmann, P., Suslow, T., *et al.* (2011). Identification of a biological signature for schizophrenia in serum. *Mol. Psychiatry* Apr 12. (Epub ahead of print).

Smart, T.G. (1997). Regulation of excitatory and inhibitory neurotransmitter-gated ion channels by protein phosphorylation. *Curr. Opin. Neurobiol.* **7**(3), 358–367.

Spadoni, L.R., Mclean, R.B., and Herrmann, W.L. (1964). A rapid immunological test for the detection of early pregnancy. *West. J. Surg. Obstet. Gynecol.* **72**, 92–97.

Staner, L. (2006). Surrogate outcomes in neurology, psychiatry, and psychopharmacology. *Dialogues Clin. Neurosci.* **8**(3), 345–352.

Stolp, H.B., and Dziegielewska, K.M. (2009). Review: role of developmental inflammation and blood-brain barrier dysfunction in neurodevelopmental and neurodegenerative diseases. *Neuropathol. Appl. Neurobiol.* **35**(2), 132–146.

Sun, J., Jia, P., Fanous, A.H., van den Oord, E., Chen, X., Riley, B.P., Amdur, R.L., Kendler, K.S., and Zhao, Z. (2010). Schizophrenia gene networks and pathways and their applications for novel candidate gene selection. *PLoS One* **5**(6), e11351.

Szego, E.M., Janáky, T., Szabó, Z., Csorba, A., Kompagne, H., Müller, G., Lévay, G., Simor, A., Juhász, G., and Kékesi, K.A. (2010). A mouse model of anxiety molecularly characterized by altered protein networks in the brain proteome. *Eur. Neuropsychopharmacol.* **20**(2), 96–111.

Takahashi, M., Itakura, M., and Kataoka, M. (2003). New aspects of neurotransmitter release and exocytosis: regulation of neurotransmitter release by phosphorylation. *J. Pharmacol. Sci.* **93**(1), 41–45.

Turck, C.W., Maccarrone, G., Sayan-Ayata, E., Jacob, A.M., Ditzen, C., Kronsbein, H., Birg, I., Doertbudak, C.C., Haegler, K., Lebar, M., Teplytska, L., Kolb, N., *et al.* (2005). The quest for brain disorder biomarkers. *J. Med. Invest.* **52**(Suppl.), 231–235.

Turck, C.W., Ditzen, C., and Sayan-Ayata, E. (2008). Proteomic strategies for biomarker discovery: from differential expression to isoforms to pathways. In: C.W. Turck (Ed.), *Biomarkers for Psychiatric Disorders.* Springer, New Yorkpp. 57–74.

van de Bayés, A., Lagemaat, L.N., Collins, M.O., Croning, M.D., Whittle, I.R., Choudhary, J.S., and Grant, S.G. (2011). Characterization of the proteome, diseases and evolution of the human postsynaptic density. *Nat. Neurosci.* **14**(1), 19–21.

Williams, R. 4th, Lim, J.E., Harr, B., Wing, C., Walters, R., Distler, M.G., Teschke, M., Wu, C., Wiltshire, T., Su, A.I., Sokoloff, G., Tarantino, L.M., *et al.* (2009). A common and unstable copy number variant is associated with differences in Glo1 expression and anxiety-like behavior. *PLoS One* **4**, e4649.

Zhang, Y., Webhofer, C., Reckow, S., Filiou, M.D., Maccarrone, G., and Turck, C.W. (2009). A MS data search method for improved ^{15}N-labeled protein identification. *Proteomics* **9**(17), 4265–4270.

Zhang, Y., Filiou, M.D., Reckow, S., Gormanns, P., Maccarrone, G., Kessler, M.S., Frank, E., Hambsch, B., Holsboer, F., Landgraf, R., and Turck, CW. Proteomic and metabolomic profiling of a trait anxiety mouse model implicate affected pathways. *Mol. Cell. Proteomics*, in press.

IMAGING BRAIN MICROGLIAL ACTIVATION USING POSITRON EMISSION TOMOGRAPHY AND TRANSLOCATOR PROTEIN-SPECIFIC RADIOLIGANDS

David R.J. Owen[1,2] and Paul M. Matthews[2,3]

[1]Division of Experimental Medicine, Imperial College, Hammersmith Hospital, London, United Kingdom
[2]GSK Clinical Imaging Centre, Hammersmith Hospital, London, United Kingdom
[3]Centre for Neuroscience, Imperial College, London, United Kingdom

Abstract

Microglia are rapidly activated by a wide range of neuropathological insults. Quantifying microglial density *in vivo* would allow a new, potentially important range of clinic-pathological correlations. Microglia express the 18 kDa translocator protein (TSPO) which can be quantified by the positron emission tomography (PET) ligand $[^{11}C]$PK11195, although signal quantification is limited by nonspecific binding. New generation TSPO radioligands with an improved signal-to-noise ratio are now available, but variation in their binding affinity for the TSPO between subjects complicates their use. This review describes the principles of PET imaging, the rationale and challenges in targeting the TSPO as

INTERNATIONAL REVIEW OF NEUROBIOLOGY, VOL. 101
DOI: 10.1016/B978-0-12-387718-5.00002-X

means of quantifying microglial activation *in vivo*, and disease applications that have been studied with TSPO-PET hitherto.

I. Introduction

Microglia are brain resident macrophages which continuously sample their local environment by sending out and retracting extensions (Gehrmann *et al.*, 1995; Kreutzberg, 1996). Microglia are believed to be central effectors of neuroinflammation, neurodegeneration, and brain repair, and they are rapidly activated by a wide range of insults (including trauma, ischemia, inflammation, neurodegeneration, and infection). When activated, microglia adopt an amoeboid shape and express cytokines regulating an inflammatory response (Gehrmann *et al.*, 1995; Kreutzberg, 1996; Venneti *et al.*, 2006). There is therefore considerable interest in developing imaging techniques to quantify microglial activation *in vivo*, because such a technique would allow new and potentially important ranges of clinic-pathological correlations.

Microglia express the 18 kDa translocator protein (TSPO), which is found in many cell types throughout the body but with relatively low background expression in the healthy brain (Doble *et al.*, 1987) which can be quantified by positron emission tomography (PET) imaging (Cagnin *et al.*, 2007). However, signal quantification is limited by the poor specific signal-to-background noise ratio (SBR) of the TSPO targeting radioligand, [^{11}C]PK11195. New generation TSPO ligands with an improved SBR relative to [^{11}C]PK11195 are now available (Chauveau *et al.*, 2008), but variation in their binding affinity for the TSPO between subjects may complicate their use (Owen *et al.*, 2011). This review describes the principles of PET imaging analysis, along with the rationale and challenges in targeting the TSPO as a means of quantifying microglial activation *in vivo*. In addition, we describe several applications for the study of neuropsychiatric and neurodegenerative diseases which have been studied with TSPO-PET imaging.

II. Principles of PET Imaging

PET imaging studies require the design of a ligand which binds with high specificity to a desired target, but with minimal nonspecific binding to other structures. The ligand is labeled with a positron emitting radioisotope with a short half-life ($t_{1\backslash 2}$), commonly ^{11}C ($t_{1\backslash 2}$ ~20 min) or ^{18}F ($t_{1\backslash 2}$ ~110 min). Following intravenous administration of the radiolabeled ligand (radioligand), the emitted positrons will collide with nearby electrons resulting in the production of pairs of photons that travel at 180° to each other. The photon pairs are detected by γ-detectors surrounding the subject, allowing the spatial distribution of the radioligand to be reconstructed, once corrections have been made for scatter, attenuation, random detection events, dead time, and detector efficiencies. To relate spatial distribution to anatomy, PET images are coregistered with structural images from computed tomography (CT) or magnetic resonance imaging (MRI) images. The effective spatial resolution of a PET scan is approximately 4 mm.

Development of useful, new PET radioligands, however, is nontrivial. A sufficiently specific molecule target must be chosen. The affinity at which the radioligand binds to the target must be sufficiently high to produce a detectable signal, but the radioligand must dissociate from the target quickly enough to allow the binding equilibrium to be approached within the timeframe of the scan (1–2 h). Lipid solubility is required to allow the radioligand to cross the blood brain barrier (BBB), but radioligands which are too lipid soluble will bind nonspecifically to cell membranes and produce a high background signal (Guo *et al.*, 2009). Radioligands can be rapidly metabolized following administration. If the metabolites also are radiolabeled and can cross the BBB, the resultant combined signal can be too complex to interpret quantitatively. Finally, synthesis of the radioligand must be simple enough to be performed rapidly (and to pharmaceutical standards), given the short half-lives of the isotopes.

Various methods can be used to derive quantitative data for PET. Clinical PET scans, such as ^{18}F-fluorodeoxyglucose (FDG) scans to detect tumors, are usually reported based on the signal to background contrast obtained from a single three-dimensional image. The standardized uptake value (SUV) is a crude means of signal quantification, calculated by normalizing the signal measured by the PET camera within the region of interest to the injected dose and body weight (Thie, 2004). However, SUVs do not distinguish between molecules of the radioligand which are specifically bound to the target (which is the measurement of interest) compared to the radioligand molecules which are bound nonspecifically to other structures or unbound in the blood or tissue.

More complex analyses, based on kinetic modeling of the dynamically acquired signal over time, allow for the component of the signal which represents specific binding of the radioligand to the target to be extracted. The modeling is simplified when there is a "reference region" within the brain which is devoid of the target. Such reference regions are useful for the determination of nonspecific binding. In cases where targets are expressed throughout the brain and therefore do not have a reference region (such as with the TSPO), more complex approaches are needed to estimate the nonspecific binding (Turkheimer *et al.*, 2007).

III. TSPO for Assessment of Microglial Expression

Microglia express TSPO. Quantifying TSPO expression therefore provides a means of estimating microglial density, particularly since baseline expression of TSPO in most other cells in the healthy human brain is low (Doble *et al.*, 1987). Indeed, a number of *in vitro* studies using *postmortem* human brain tissue have shown increased TSPO density (measured as [^{3}H]PK11195 binding) in diseases characterized by microglial proliferation. These include studies of multiple sclerosis (MS; Banati *et al.*, 2000; Vowinckel *et al.*, 1997), Alzheimer's disease (AD; Diorio *et al.*, 1991; Gulyas *et al.*, 2009; Venneti *et al.*, 2008), Huntington's disease (HD; Messmer and Reynolds, 1998), stroke (Venneti *et al.*, 2008), frontotemporal dementia (Venneti *et al.*, 2008), and amyotrophic lateral sclerosis (Sitte *et al.*, 2001; Venneti *et al.*, 2008). However, the mechanisms controlling TSPO expression in microglia are not clear. It is possible that the increase in TSPO signal in these studies merely reflect an increase in microglial number. It may also be that TSPO expression increases within each microglial cell as it becomes activated. This has been demonstrated in Leydig cells (Rey *et al.*, 2000) and in pancreatic islet cells (Trincavelli *et al.*, 2002), both of which show increases in TSPO and TSPO mRNA in response to activation by cytokines. However, no such data is available for microglia.

It should be noted that not all microglial cells express TSPO (Takaya *et al.*, 2007) and significant TSPO expression is found in other cell types. For example, astrocytes express TSPO under certain conditions in culture (Itzhak *et al.*, 1993) and in animal models (Ji *et al.*, 2008; Maeda *et al.*, 2007). Nevertheless, *in vitro* studies with *postmortem* tissue demonstrate that TSPO expression follows the distribution of microglial and not astrocyte activation (Conway *et al.*, 1998; Dubois *et al.*, 1988; Myers *et al.*, 1991; Raghavendra *et al.*, 2000; Stephenson *et al.*, 1995). In high-resolution microautoradiography studies, TSPO expression has been found in infiltrating blood borne cells and microglia, but not in astrocytes (Banati *et al.*, 2000; Shah *et al.*, 1994). Further, hippocampal sclerosis, a disease characterized by astrocytosis, shows no signal increase in a clinical PET study measuring the TSPO (Banati *et al.*, 1999). Hence, it is widely accepted that monocyte-derived microglia and macrophages are the dominant cell types responsible for TSPO expression in the brain in the majority of circumstances.

IV. Challenges Facing PET Imaging of the TSPO

PK11195 is a TSPO antagonist with nanomolar affinity (Shah *et al.*, 1994), which was first labeled with ^{11}C for use as a PET radioligand in humans in 1986 (Charbonneau *et al.*, 1986). Since then [^{11}C]PK11195 has been used in PET studies to investigate various brain diseases (Matthews and Comley, 2009), as well as in studies of systemic innate immune responses, owing to its expression in peripheral macrophages as well as microglia.

However, it is well recognized that *in vivo* PET applications of the [^{11}C]PK11195 ligand are limited by difficulty in modeling the signal, chiefly due to its poor SBR (Banati *et al.*, 2000), which severely limits accurate quantification. This difficulty in modeling the [^{11}C]PK11195 signal is multifactorial. Some factors relate to targeting the TSPO *per se*. First, unlike many central nervous system (CNS) PET targets, TSPO is expressed widely throughout the body (Anholt *et al.*, 1985; Gavish *et al.*, 1999) and therefore only a small fraction of administered radioligand is available for binding in the CNS (Petit-Taboue *et al.*, 1991). Second, there is no area within the brain without any TSPO expression which can act as a reference region to help characterize the contribution of the signal that represents nonspecific binding (Kropholler *et al.*, 2006). Third, modeling the input function is complicated by the high expression of TSPO on blood cells (Canat *et al.*, 1993). Other factors, however, are specific to modeling with the [^{11}C]PK11195 ligand. In particular, modeling is hampered by high nonspecific binding, which is likely due to the high lipophilicity of this molecule. This represents the greatest challenge to accurate quantification of the signal (Petit-Taboue *et al.*, 1991; Shah *et al.*, 1994), although this property has less effect on use of the ligand *in vitro*, as nonspecific interactions are typically weak and can be dislodged by washing tissue samples. Additionally, PK11195 adheres to plastic and glass making accurate plasma measurements difficult.

More complex modeling techniques have been applied to [^{11}C]PK11195 PET data to overcome some of these challenges. Specifically, in the absence of a true reference region, a data-driven signal clustering technique has been used to derive estimates of nonspecific binding and input function (Turkheimer *et al.*, 2007). However, this method of modeling may not be ideal for longitudinal studies as reference regions may change over time (Matthews and Comley, 2009). A modified reference tissue model has also been applied to account for the expression of TSPO within brain vasculature (both endothelial and smooth muscle cells; Tomasi *et al.*, 2008).

Despite these advances in modeling, extracting accurate physiological parameters from [^{11}C]PK11195 PET data remains challenging. There has therefore been considerable interest in developing a new generation of high-affinity TSPO radioligands with low *in vivo* nonspecific binding compared to that seen with [^{11}C]PK11195. Because of the fundamental physiochemical and pharmacological characteristics required to make a successful PET radioligand, as previously discussed, hundreds of new generation tracers have been synthesized and approximately 40 of these have been evaluated in preclinical studies (Chauveau *et al.*, 2008). Several promising candidate molecules are being evaluated in man, including [^{18}F]PBR06 (Fujimura *et al.*, 2009), [^{18}F]FEPPA (Wilson *et al.*, 2008), [^{11}C]DAA1106 (Ikoma *et al.*, 2007), [^{11}C]DPA-713 (Endres *et al.*, 2009), [^{18}F]PBR111 (Fookes *et al.*, 2008; Van *et al.*, 2010), and [^{11}C]PBR28 (Imaizumi *et al.*, 2008).

Recent clinical PET studies using these radioligands have confirmed that they have substantially improved SBR compared to that with [^{11}C]PK11195, but revealed an unexpected complication to their use. With the [^{11}C]PBR28 ligand, approximately 10% of human subjects do not show specific TSPO binding (Venneti *et al.*, 2008). We have recently clarified that such subjects express the TSPO, but that it shows an approximate 50-fold reduction in affinity for PBR28. We have called such subjects low-affinity binders (LABs; Owen *et al.*, 2010). Among those remaining subjects with a measurable specific PET signal using [^{11}C]PBR28, two further groups can be identified: high-affinity binders (HABs), who express a single class of high-affinity binding sites, and mixed-affinity binders (MABs), who express approximately equal proportions of high- and low-affinity binding sites (Fig. 1). This phenomenon is not limited to [^{11}C]PBR28 as we have also shown that all new generation TSPO tracers recognize these three binding classes (Owen *et al.*, 2011). The reason this may have gone undetected with these

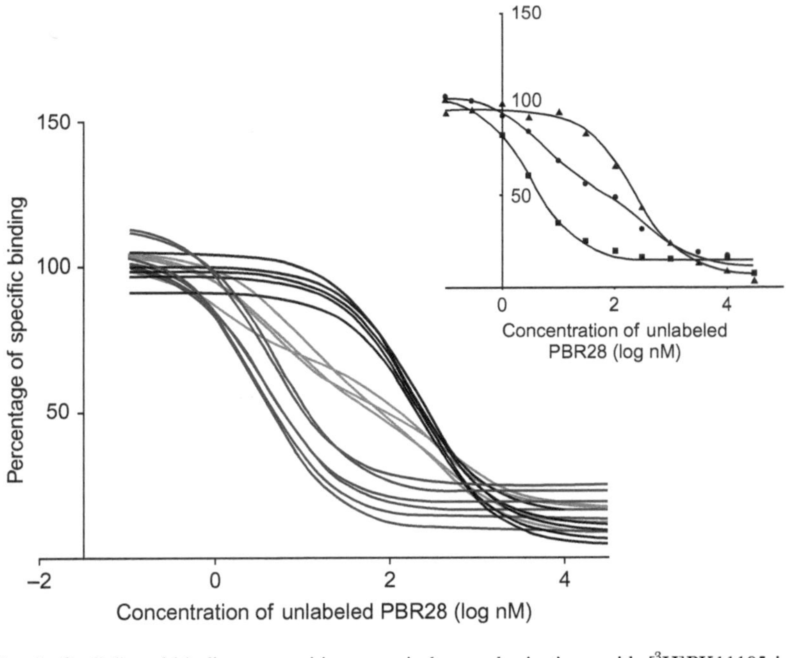

FIG. 1. Radioligand binding competition assay in human brain tissue with [^{3}H]PK11195 in the presence of increasing concentrations of unlabeled PBR28. Curve fit is shown for each sample ($n = 15$). Inset—one representative curve with data points for each binding category. Black, low-affinity binder; Red, high-affinity binder; Green, mixed-affinity binder. (For interpretation of the references to color in this figure legend, the reader is referred to the Web version of this chapter.)

tracers hitherto is that, with the exception of PBR28, the reduction in affinity with LABs compared to HABs is relatively small (four- to sixfold). Crucially, however, in such studies the PET signal will substantially underestimate TSPO expression in subjects who express the low-affinity receptor (namely LABs and MABs).

Complicating matters further is the fact that binding affinity in the brain cannot be ascertained from a single PET scan. We have shown, however, that measuring binding affinity in platelets isolated from whole blood is feasible and that the HAB, MAB, and LAB binding patterns seen in the brain are also present in platelets (Owens, R.D.J., Matthews, P.M., Rabiner, E.A., Parker, C.A., and Gunn, R.N., unpublished findings). Platelet binding data could therefore potentially be used to determine a subject's TSPO-binding affinity. With this information, the binding potential measurements obtained with PET could be corrected to allow valid comparisons between subjects of different binding classes. However, further work is first required to confirm the correspondence of platelet and brain binding affinities.

V. Disease Applications

A. Neuroinflammatory Diseases

Numerous [^{3}H]PK11195 *in vitro* radioligand binding studies in human *post-mortem* tissues have documented an increase in TSPO expression associated with lesions in MS patients. Tissue samples containing white matter lesions express TSPO at levels three to four times greater than the levels in normal white matter (Banati *et al.*, 2000; Venneti *et al.*, 2008), with the majority of binding seen at the periphery of the plaque (Owen *et al.*, 2010; Vowinckel *et al.*, 1997; Fig. 2). However, even in histologically normal-appearing white and gray matter (GM) from subjects with MS, an increase in [^{3}H]PK11195 signal colocalizing with microglia markers can be detected (Banati *et al.*, 2000). In acute active lesions, it appears that monocyte-derived microglia or macrophages are the only cell types contributing to TSPO expression (Banati *et al.*, 2000; Cosenza-Nashat *et al.*, 2009; Venneti *et al.*, 2008), although in chronic silent MS lesions, reactive astrocytes can also express TSPO (Cosenza-Nashat *et al.*, 2009).

In PET studies in MS patients using the [^{11}C]PK11195 ligand, increased signal colocalizes with areas of focal pathology identified by MRI as gadolinium-enhancing lesions (Banati *et al.*, 2000). However, as anticipated from *in vitro* data, uptake of the ligand is also detected in areas of the brain defined by MRI as normal-appearing white matter (NAWM), with higher uptake by NAWM reported in patients with greater disability (Debruyne *et al.*, 2003). [^{11}C]PK11195 uptake in NAWM also appears to correlate with brain atrophy (Versijpt *et al.*, 2005).

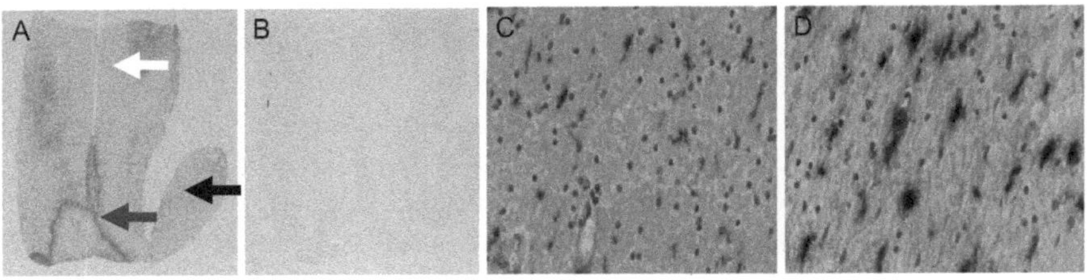

FIG. 2. TSPO distribution in an acute demyelinating lesion in postmortem brain tissue from a subject with MS. (A) Binding with TSPO ligand [^{3}H]PBR28 showing high signal (red arrow) around the periphery of the lesion and lower signal in normal white (white arrow) and gray (black arrow) matter. (B) Binding of neighboring section with TSPO ligand [^{3}H]PBR28 in the presence of an excess of unlabeled PK11195. Because almost all TSPO is bound to unlabeled PK11195, no signal from [^{3}H]PBR28 is evident demonstrating the specificity of [^{3}H]PBR28. (C) Low MHC II expression from normal-appearing white matter (400×) colocalizing with low [^{3}H]PBR28 binding (e.g., white arrow). (D) High MHC II expression from white matter lesion (400×) corresponding with high [^{3}H]PBR28 binding (e.g., red arrow). (For interpretation of the references to color in this figure legend, the reader is referred to the Web version of this chapter.)

Although MS was traditionally thought of as a disease associated with white matter, studies over the past decade have revealed substantial GM involvement, both at the cortical (Kutzelnigg *et al.*, 2005) and subcortical (Cifelli *et al.*, 2002) levels. Indeed, cognitive features in MS such as memory impairment and attention deficits are common (Rao *et al.*, 1991) and not explained by isolated white matter focal demyelination, which itself is poorly correlated with disease progression (Barkhof, 2002). Imaging GM lesions with MRI is possible (Mainero *et al.*, 2009), but it is challenging because there is poor MRI contrast between lesions and normal GM using conventional sequences. However, GM lesions are characterized by pronounced activation of macrophages and microglia (Vercellino *et al.*, 2007) making these amenable to identification using TSPO-PET analyses. Indeed, in a recent PET study, large increases in [^{11}C]PK11195-specific binding were detected in cortical GM of patients with secondary progressive MS relative to healthy controls (Politis *et al.*, 2010). The signal correlated with both the MS impact scale (MSIS-29) and the expanded disability status scale (EDSS), although these studies were performed using small numbers of subjects and, as such, correlations can provide only limited evidence. TSPO-PET analyses may provide an important new index of GM which is pathologically relevant to the progression of clinical disability.

A recent study using the new generation TSPO tracer [^{11}C]PBR28 showed similar findings to [^{11}C]PK11195 studies in that areas of high [^{11}C]PBR28 signal corresponded well to gadolinium-enhancing lesions (Oh *et al.*, 2010). Of note, in some cases, serial scanning showed that the [^{11}C]PBR28 signal preceded the appearance of MRI contrast enhancement, suggesting that glial activation may be an early event in MS lesion formation (Oh *et al.*, 2010; Fig. 3). Surprisingly,

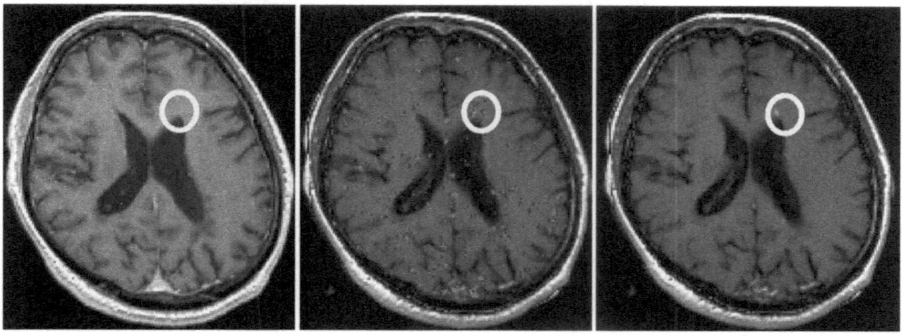

FIG. 3. Serial scanning in multiple sclerosis patients showed that the [^{11}C]PBR28 signal can precede the appearance of MRI contrast enhancement, suggesting that glial activation may be an early event in MS lesion formation. Postcontrast T1-weighted MRI (left) and coregistered [^{11}C]PBR28 VT difference map (center). Gadolinium enhancement of the same region is detected a month later (right; Oh *et al.*, 2010).

however, although a significant but weak correlation was found between global [^{11}C]PBR28 binding levels and disease duration, global binding did not differ significantly between groups of healthy volunteers and subjects with MS (Oh *et al.*, 2010). This deserves further study.

B. Neurodegenerative Diseases

It is well recognized that the amyloid deposition characteristic of AD is associated with activation of microglia (Cameron and Landreth, 2010). The beta-amyloid precursor protein increases markers of activation in microglia and enhances microglial production of neurotoxins (Barger and Harmon, 1997). Microglia are found in large numbers surrounding AD plaques (Joshi and Crutcher, 1998) and it is believed that the microglial inflammatory response contributes to damage in AD (Akiyama *et al.*, 2000).

There has therefore been interest in determining whether imaging the TSPO in AD may be a useful strategy. Initial *in vitro* autoradiographic studies with TSPO ligands, comparing *postmortem* AD brain samples to those of controls, have produced encouraging results. [^{3}H]PK11195 binding has been shown to be increased in the temporal lobe of AD brains compared to that seen in controls, and the increase colocalizes with decreased choline acetyltransferase activity (Diorio *et al.*, 1991). TSPO density has been shown to be approximately fourfold greater in the frontal lobe of AD brains compared to that in controls using both the [^{3}H]DAA1106 and [^{3}H]PK11195 tracers (Venneti *et al.*, 2008). Further, in a recent paper which showed a similar increase in the [^{3}H]DAA1106 signal in temporal and parietal lobe, TSPO binding was found to colocalize with immunohistochemical microglial markers (Gulyas *et al.*, 2009).

PET data also has been encouraging. The first report of *in vivo* microglial activation in AD was in 2001, in which [^{11}C]PK11195 binding was significantly increased relative to that in control subjects, and the spatial distribution of increased binding was consistent with the characteristic pattern of AD. Interestingly, binding was also increased in a subject with mild cognitive impairment (MCI), suggesting that microglial activation may be detectable prior to the development of overt AD. However, there was substantial overlap in signal between the AD group and controls which would severely limit the utility of the scan in clinical practice (Cagnin *et al.*, 2001).

Recent studies using both [^{11}C]PK11195 to image microglia and [^{11}C]PIB to assess amyloid load have confirmed an increase in the [^{11}C]PK11195 signal in several cortical regions in patients with AD (Edison *et al.*, 2008; Yokokura *et al.*, 2011). In these studies, there was no correlation between microglial activation and amyloid load (measured as the [^{11}C]PIB signal), suggesting that these processes are not linked in a simple way. In both studies, the dementia score was inversely correlated with [^{11}C]PK11195, but not with the [^{11}C]PIB signal, suggesting that microglial activation becomes more pronounced as the dementia progresses. This is consistent with data from studies of subjects with MCI (Okello *et al.*, 2009). It may also explain why one recent study found no difference in the [^{11}C]PK11195 signal between AD, MCI, and control subjects, as the AD subjects in this study had mild disease only (Wiley *et al.*, 2009).

The relatively small changes in [^{11}C]PK11195 signal between control, MCI, and mild AD subjects may be difficult to discriminate, given the poor SBR of the [^{11}C]PK11195 tracer. Studies using new generation TSPO ligands, with significantly enhanced SNR profiles, are likely to help clarify the situation and may show greater separation between the two groups. Although such a study has been performed in AD and controls subjects with [^{11}C]DAA1106, this study did not take into account the effect of variable binding affinity with new generation ligands (Owen *et al.*, 2010, 2011). Therefore, although a significant increase in signal in the AD group was detected, there was substantial overlap in the signal between the two groups (Yasuno *et al.*, 2008).

The lack of correlation between amyloid deposition and microglial activation in AD suggests that other forms of dementia might also be characterized by increased TSPO binding. Indeed, as with AD, autoradiography studies of *postmortem* brains have shown that in frontotemporal demential (FTD), TSPO density is elevated in affected regions compared to that seen in controls (Venneti *et al.*, 2008). Further, increased binding was also seen in a PET study with [^{11}C]DAA1106 in three subjects with no neurologic abnormalities, but who had been diagnosed with preclinical FTD based on their genotype (Miyoshi *et al.*, 2010).

C. Movement Disorders

1. *Parkinson's Disease*

The rationale for using TSPO-PET analyses to quantify neuroinflammation in Parkinson's disease (PD) is based on immunohistochemical data showing large numbers of microglia within the substantia nigra region of *postmortem* brain samples from subjects with PD. By contrast, control subjects have few microglia in the substantia nigra (Banati *et al.*, 1998; McGeer *et al.*, 1988). Interestingly, the microglial activation in these studies appears to be independent of disease duration and severity (Banati *et al.*, 1998). Studies in patients with PD plus syndromes have demonstrated the feasibility of imaging the basal ganglia with the [^{11}C] PK11195 tracer in multisystem atrophy (Gerhard *et al.*, 2003), progressive supranuclear palsy (Gerhard *et al.*, 2006a,b), and corticobasal degeneration (Gerhard *et al.*, 2004; Henkel *et al.*, 2004). [^{11}C]PK11195 PET analyses revealed patterns of increased microglial activation that corresponds well with the known distribution of microglia in these diseases.

Two PET studies with [^{11}C]PK11195 have together confirmed that quantifying microglial activation *in vivo* in patients with PD is possible (Gerhard *et al.*, 2006a,b; Ouchi *et al.*, 2005). Ouchi *et al.* studied drug-naive patients with mild disease and reported an increased [^{11}C]PK11195 signal in the midbrain in patients compared to the signal in controls. Although the [^{11}C]PK11195 signal did not correlate with disease duration, there were positive correlations found with measures of severity of motor deficits and loss of dopaminergic projections in the putamen region of the dorsal striatum (Ouchi *et al.*, 2005). Gerhard *et al.* studied patients with more severe disease and found more global changes including an increased [^{11}C]PK11195 signal in the striatum, pallidum, and pons, as well as in cortical areas (Gerhard *et al.*, 2006a,b). However, a correlation neither between signal and duration of disease nor between [^{11}C]PK11195 signal and disease severity or loss of dopaminergic projections was found. The independence of the microglial TSPO signal and clinical symptoms was further confirmed by follow-up scans approximately 2 years later in a subset of the study population. No change in [^{11}C]PK11195 signal was observed, despite clinical deterioration and a further degeneration of dopaminergic neurons. A recent pilot study assessing the pharmacodynamic effects of anti-inflammatory treatment in PD also showed no significant differences at baseline between groups of PD patients (both drug naive and advanced disease) and controls (Bartels *et al.*, 2010).

The reasons underlying the discrepancies between these three studies are not clear. It is significant that the results within each study varied greatly with different methods of image analysis, highlighting the difficulties in accurately quantifying [^{11}C]PK11195 binding *in vivo*. Theoretically, based on the fact that substantial microglial activation is known to occur in PD, TSPO-PET analysis should be a useful tool with which to explore dynamic clinico-pathological correlations. We believe that use of newer radioligands with improved SBR should enable more informative experiments. Autoradiography studies with *postmortem* PD brain samples and PET studies with new generation ligands may help to clarify whether TSPO-PET will have a role in investigating PD in the future.

2. *Huntington's Disease*

In HD, reactive microglia are present at all stages and accumulate with severity of disease in proportion to the degree of neuronal loss (Sapp *et al.*, 2001). TSPO-binding sites are increased in the HD brain, with a 70% increase in binding in the putamen, a 25% increase in the frontal lobe, and a 10% increase in the temporal lobe. A first TSPO-PET study in HD found increased [^{11}C]PK11195 binding in the striatum and cortex of patients relative to that in healthy controls. Further, the binding correlated with clinical severity as assessed by motor scores and correlated inversely with striated [^{11}C]raclopride binding, a measure dopamine receptor density that provides an index of striated neurodegeneration. A voxel-based analysis suggested colocalization of the increased [^{11}C]PK11195 binding with the decreased [^{11}C]raclopride binding (Pavese *et al.*, 2006). A study assessing presymptomatic HD gene carriers found a similar increase in [^{11}C]PK11195 binding and colocalization with decreased [^{11}C]raclopride binding, suggesting that microglial activation is an early event in the evolution of disease pathology (Tai *et al.*, 2007).

D. Stroke

Microglia are activated by neuronal injury and death and a robust microglial response follows cerebrovascular infarction (Wang *et al.*, 2007; Wood, 1995). Large increases in monocyte-derived cells are found in the ischemic core within 48 h of infarction (Krupinski *et al.*, 1996; Tomimoto *et al.*, 1996) and there is an approximate threefold increase in TSPO binding in *postmortem* human brain samples in the core of infarcts (Venneti *et al.*, 2008).

Most evidence suggests that activated microglial cells may contribute to injury following infarction by release of proinflammatory cytokines and directly by cytotoxic molecules. In a rodent model of ischemia, inhibition of microglial activation with minocycline has a neuroprotective effect (Yrjanheikki *et al.*, 1998, 1999). Likewise, in neuron and oligodendrocyte cultures, microglia/macrophages are associated with greater injury during ischemia (Giulian *et al.*, 1993; Lehnardt *et al.*, 2003; Zhang *et al.*, 1997). However, other studies provide unequivocal evidence that microglia/macrophages, or their secreted factors, can also protect cells against neuronal damage (Watanabe *et al.*, 2000). The extent to which the microglial response following brain ischemia is deleterious, or only in part beneficial, is currently not clear (Wang *et al.*, 2007). Microglial phenotypes may change with time or may be spatially heterogeneous within lesions. Further, like macrophages, microglia differ phentotypically and it is likely that different phenotypes may have different effects on the resolution of lesions.

Two patterns of microglial activation have been observed following photochemically induced focal ischemia of the rat cortex (Schroeter *et al.*, 1999). Initially, phagocytic microglia expressing major histocompatibility (MHC) class I receptors are present in the core and periphery of the ischemic lesion. Subsequently, after a delay of days, microglia with low phagocytic activity but with high MHC Class II expression appear along degenerating fiber tracts with connections to the infarct. This remote activation of microglia is thought to represent Wallerian degeneration as a consequence of focal damage.

Evidence consistent with both patterns of microglial activation was reported with the first use of the [^{11}C]PK11195 ligand to measure TSPO expression following cerebral ischemia in humans (Ramsay *et al.*, 1992). In this case, a single patient was studied following ischemic stroke. An increased signal was identified adjacent to the lesion 6 days following stroke. When the patient was imaged again 1 week later, there was also evidence of tracer binding in areas remote from the lesion. Similar findings were reported in a study of a series of seven patients with middle cerebral artery infarcts. All patients showed increased [^{11}C]PK11195 binding in the thalamus ipsilateral to the infarct, that persisted at 24 months of follow-up (Pappata *et al.*, 2000).

The kinetics of the signal change was investigated in a small series of patients scanned between 3 and 150 days following cerebral ischemia. Increased [^{11}C]PK11195 binding around the lesion was observed as early as 3 days after ischemia. Subsequently, both the primary lesion and areas distant from the primary lesion site began to show increases in signal (Gerhard *et al.*, 2005). In another series, signal changes were examined over three time points ($<$ 72 h, week 2, week 4). In this study, minimal binding was found within 72 h in the core of the lesion, but binding rose significantly by week 2 before reducing slightly by week 4. As with previous studies, binding was present from early on in the core and periphery of the lesion and extended to remote zones after 7–10 days (Price *et al.*, 2006).

A recent study tested the hypothesis that signal changes in areas remote to the lesion are a direct response to the ischemic insult. The [11C]PK11195 signal was measured in the pyramidal tract (PT) of patients with acute subcortical ischemia, and an increased signal was only found in patients in whom the PT was affected by stroke. In stroke patients in which the lesioned brain area did not project to the PT, no signal change in the PT was detected (Radlinska *et al.*, 2009). In a similar experiment in which patients were investigated serially, the same authors found that while microglial activity in the infarct returns to baseline levels with time, microglia in the remote regions persisted throughout the 6-month follow-up period. Although greater microglial activation in the lesion was correlated with poor clinical outcome, remote microglial activation was correlated with *improved* clinical outcome (Thiel *et al.*, 2010).

E. Neuropsychiatric Diseases

Schizophrenia also has been investigated using TSPO-PET analyses (Hulshoff Pol *et al.*, 2001; Rapoport *et al.*, 1999). However, the association of microglial activation with schizophrenia is controversial (Arnold *et al.*, 1998; Kurumaji *et al.*, 1997; Radewicz *et al.*, 2000; Steiner *et al.*, 2006).

Two recent PET studies with [^{11}C]PK11195 have been performed. The first study included patients with a disease duration of less than 5 years and found a small but significant increase in binding for total GM in patients relative to that in controls (van Berckel *et al.*, 2008). The second study included seven patients who were recovering from an episode of psychosis and reported significantly higher binding in the hippocampus of patients compared to that in healthy volunteers. No other regions showed significant changes, although the mean binding was 30% higher in the whole-brain GM of schizophrenic patients (Doorduin *et al.*, 2009). These studies tentatively support the claim that TSPO may be upregulated in the schizophrenic brain, although much further work is required to substantiate the observations and to understand their significance.

VI. Conclusion

Because microglial proliferation is a stereotyped response following a wide variety of pathological insults, there has been great interest in quantifying microglial density *in vivo* both as a research tool and as an aid to clinical decision making. TSPO imaging with PET analysis is potentially helping us to take the first steps toward this goal. However, *in vivo* studies to date have been hindered by the lack of an appropriate radioligand for robust quantification of specific binding. With the recent development of several new generation TSPO ligands with improved SBR relative to [^{11}C]PK11195, accurate quantification of TSPO expression now seems possible. Challenges are posed because of the variation in binding affinity across human subjects, but we anticipate that practical approaches to addressing these will be met, enabling TSPO-PET to become an even more valuable tool for neuroscience research, even in the near term.

Acknowledgments

DRJO has been funded by the Wellcome Trust-GSK Translational Medicine Training Programme in Imperial College London. P. M. M. is a full time employee of GlaxoSmithKline.

References

Akiyama, H., Barger, S., Barnum, S., Bradt, B., Bauer, J., Cole, G.M., Cooper, N.R., Eikelenboom, P., Emmerling, M., Fiebich, B.L., Finch, C.E., Frautschy, S., *et al.* (2000). Inflammation and Alzheimer's disease. *Neurobiol. Aging* **21**, 383–421.

Anholt, R.R., De Souza, E.B., Oster-Granite, M.L., and Snyder, S.H. (1985). Peripheral-type benzodiazepine receptors: autoradiographic localization in whole-body sections of neonatal rats. *J. Pharmacol. Exp. Ther.* **233**, 517–526.

Arnold, S.E., Trojanowski, J.Q., Gur, R.E., Blackwell, P., Han, L.Y., and Choi, C. (1998). Absence of neurodegeneration and neural injury in the cerebral cortex in a sample of elderly patients with schizophrenia. *Arch. Gen. Psychiatry* **55**, 225–232.

Banati, R.B., Daniel, S.E., and Blunt, S.B. (1998). Glial pathology but absence of apoptotic nigral neurons in long-standing Parkinson's disease. *Mov. Disord.* **13**, 221–227.

Banati, R.B., Goerres, G.W., Myers, R., Gunn, R.N., Turkheimer, F.E., Kreutzberg, G.W., Brooks, D.J., Jones, T., and Duncan, J.S. (1999). [11C](R)-PK11195 positron emission tomography imaging of activated microglia in vivo in Rasmussen's encephalitis. *Neurology* **53**, 2199–2203.

Banati, R.B., Newcombe, J., Gunn, R.N., Cagnin, A., Turkheimer, F., Heppner, F., Price, G., Wegner, F., Giovannoni, G., Miller, D.H., Perkin, G.D., Smith, T., *et al.* (2000). The peripheral benzodiazepine binding site in the brain in multiple sclerosis: quantitative in vivo imaging of microglia as a measure of disease activity. *Brain* **123**(Pt 11), 2321–2337.

Barger, S.W., and Harmon, A.D. (1997). Microglial activation by Alzheimer amyloid precursor protein and modulation by apolipoprotein E. *Nature* **388**, 878–881.

Barkhof, F. (2002). The clinico-radiological paradox in multiple sclerosis revisited. *Curr. Opin. Neurol.* **15**, 239–245.

Bartels, A.L., Willemsen, A.T., Doorduin, J., de Vries, E.F., Dierckx, R.A., and Leenders, K.L. (2010). [11C]-PK11195 PET: quantification of neuroinflammation and a monitor of anti-inflammatory treatment in Parkinson's disease? *Parkinsonism Relat. Disord.* **16**, 57–59.

Cagnin, A., Brooks, D.J., Kennedy, A.M., Gunn, R.N., Myers, R., Turkheimer, F.E., Jones, T., and Banati, R.B. (2001). In-vivo measurement of activated microglia in dementia. *Lancet* **358**, 461–467.

Cagnin, A., Kassiou, M., Meikle, S.R., and Banati, R.B. (2007). Positron emission tomography imaging of neuroinflammation. *Neurotherapeutics* **4**, 443–452.

Cameron, B., and Landreth, G.E. (2010). Inflammation, microglia, and Alzheimer's disease. *Neurobiol. Dis.* **37**, 503–509.

Canat, X., Carayon, P., Bouaboula, M., Cahard, D., Shire, D., Roque, C., Le, F.G., and Casellas, P. (1993). Distribution profile and properties of peripheral-type benzodiazepine receptors on human hemopoietic cells. *Life Sci.* **52**, 107–118.

Charbonneau, P., Syrota, A., Crouzel, C., Valois, J.M., Prenant, C., and Crouzel, M. (1986). Peripheral-type benzodiazepine receptors in the living heart characterized by positron emission tomography. *Circulation* **73**, 476–483.

Chauveau, F., Boutin, H., Van, C.N., Dolle, F., and Tavitian, B. (2008). Nuclear imaging of neuroinflammation: a comprehensive review of [11C]PK11195 challengers. *Eur. J. Nucl. Med. Mol. Imaging* **35**, 2304–2319.

Cifelli, A., Arridge, M., Jezzard, P., Esiri, M.M., Palace, J., and Matthews, P.M. (2002). Thalamic neurodegeneration in multiple sclerosis. *Ann. Neurol.* **52**, 650–653.

Conway, E.L., Gundlach, A.L., and Craven, J.A. (1998). Temporal changes in glial fibrillary acidic protein messenger RNA and [3H]PK11195 binding in relation to imidazoline-I2-receptor and alpha 2-adrenoceptor binding in the hippocampus following transient global forebrain ischaemia in the rat. *Neuroscience* **82**, 805–817.

Cosenza-Nashat, M., Zhao, M.L., Suh, H.S., Morgan, J., Natividad, R., Morgello, S., and Lee, S.C. (2009). Expression of the translocator protein of 18 kDa by microglia, macrophages and astrocytes based on immunohistochemical localization in abnormal human brain. *Neuropathol. Appl. Neurobiol.* **35**, 306–328.

Debruyne, J.C., Versijpt, J., Van Laere, K.J., De, V.F., Keppens, J., Strijckmans, K., Achten, E., Slegers, G., Dierckx, R.A., Korf, J., and De Reuck, J.L. (2003). PET visualization of microglia in multiple sclerosis patients using [11C]PK11195. *Eur. J. Neurol.* **10**, 257–264.

Diorio, D., Welner, S.A., Butterworth, R.F., Meaney, M.J., and Suranyi-Cadotte, B.E. (1991). Peripheral benzodiazepine binding sites in Alzheimer's disease frontal and temporal cortex. *Neurobiol. Aging* **12**, 255–258.

Doble, A., Malgouris, C., Daniel, M., Daniel, N., Imbault, F., Basbaum, A., Uzan, A., Gueremy, C., and Le, F.G. (1987). Labelling of peripheral-type benzodiazepine binding sites in human brain with [3H]PK 11195: anatomical and subcellular distribution. *Brain Res. Bull.* **18**, 49–61.

Doorduin, J., de Vries, E.F., Willemsen, A.T., de Groot, J.C., Dierckx, R.A., and Klein, H.C. (2009). Neuroinflammation in schizophrenia-related psychosis: a PET study. *J. Nucl. Med.* **50**, 1801–1807.

Dubois, A., Benavides, J., Peny, B., Duverger, D., Fage, D., Gotti, B., MacKenzie, E.T., and Scatton, B. (1988). Imaging of primary and remote ischaemic and excitotoxic brain lesions: An autoradiographic study of peripheral type benzodiazepine binding sites in the rat and cat. *Brain Res.* **445**, 77–90.

Edison, P., Archer, H.A., Gerhard, A., Hinz, R., Pavese, N., Turkheimer, F.E., Hammers, A., Tai, Y.F., Fox, N., Kennedy, A., Rossor, M., and Brooks, D.J. (2008). Microglia, amyloid, and cognition in Alzheimer's disease: an [11C](R)PK11195-PET and [11C]PIB-PET study. *Neurobiol. Dis.* **32**, 412–419.

Endres, C.J., Pomper, M.G., James, M., Uzuner, O., Hammoud, D.A., Watkins, C.C., Reynolds, A., Hilton, J., Dannals, R.F., and Kassiou, M. (2009). Initial evaluation of 11C-DPA-713, a novel TSPO PET ligand, in humans. *J. Nucl. Med.* **50**, 1276–1282.

Fookes, C.J., Pham, T.Q., Mattner, F., Greguric, I., Loc'h, C., Liu, X., Berghofer, P., Shepherd, R., Gregoire, M.C., and Katsifis, A. (2008). Synthesis and biological evaluation of substituted [18F] imidazo[1,2-a]pyridines and [18F]pyrazolo[1,5-a]pyrimidines for the study of the peripheral benzodiazepine receptor using positron emission tomography. *J. Med. Chem.* **51**, 3700–3712.

Fujimura, Y., Zoghbi, S.S., Simeon, F.G., Taku, A., Pike, V.W., Innis, R.B., and Fujita, M. (2009). Quantification of translocator protein (18 kDa) in the human brain with PET and a novel radioligand, (18)F-PBR06. *J. Nucl. Med.* **50**, 1047–1053.

Gavish, M., Bachman, I., Shoukrun, R., Katz, Y., Veenman, L., Weisinger, G., and Weizman, A. (1999). Enigma of the peripheral benzodiazepine receptor. *Pharmacol. Rev.* **51**, 629–650.

Gehrmann, J., Matsumoto, Y., and Kreutzberg, G.W. (1995). Microglia: intrinsic immuneffector cell of the brain. *Brain Res. Brain Res. Rev.* **20**, 269–287.

Gerhard, A., Banati, R.B., Goerres, G.B., Cagnin, A., Myers, R., Gunn, R.N., Turkheimer, F., Good, C.D., Mathias, C.J., Quinn, N., Schwarz, J., and Brooks, D.J. (2003). [11C](R)-PK11195 PET imaging of microglial activation in multiple system atrophy. *Neurology* **61**, 686–689.

Gerhard, A., Watts, J., Trender-Gerhard, I., Turkheimer, F., Banati, R.B., Bhatia, K., and Brooks, D.J. (2004). In vivo imaging of microglial activation with [11C](R)-PK11195 PET in corticobasal degeneration. *Mov. Disord.* **19**, 1221–1226.

Gerhard, A., Schwarz, J., Myers, R., Wise, R., and Banati, R.B. (2005). Evolution of microglial activation in patients after ischemic stroke: a [11C](R)-PK11195 PET study. *Neuroimage* **24**, 591–595.

Gerhard, A., Pavese, N., Hotton, G., Turkheimer, F., Es, M., Hammers, A., Eggert, K., Oertel, W., Banati, R.B., and Brooks, D.J. (2006a). In vivo imaging of microglial activation with [11C](R)-PK11195 PET in idiopathic Parkinson's disease. *Neurobiol. Dis.* **21**, 404–412.

Gerhard, A., Trender-Gerhard, I., Turkheimer, F., Quinn, N.P., Bhatia, K.P., and Brooks, D.J. (2006b). In vivo imaging of microglial activation with [11C](R)-PK11195 PET in progressive supranuclear palsy. *Mov. Disord.* **21**, 89–93.

Giulian, D., Corpuz, M., Chapman, S., Mansouri, M., and Robertson, C. (1993). Reactive mononuclear phagocytes release neurotoxins after ischemic and traumatic injury to the central nervous system. *J. Neurosci. Res.* **36**, 681–693.

Gulyas, B., Makkai, B., Kasa, P., Gulya, K., Bakota, L., Varszegi, S., Beliczai, Z., Andersson, J., Csiba, L., Thiele, A., Dyrks, T., Suhara, T., *et al.* (2009). A comparative autoradiography study in post mortem whole hemisphere human brain slices taken from Alzheimer patients and age-matched controls using two radiolabelled DAA1106 analogues with high affinity to the peripheral benzodiazepine receptor (PBR) system. *Neurochem. Int.* **54**, 28–36.

Guo, Q., Brady, M., and Gunn, R.N. (2009). A biomathematical modeling approach to central nervous system radioligand discovery and development. *J. Nucl. Med.* **50**, 1715–1723.

Henkel, K., Karitzky, J., Schmid, M., Mader, I., Glatting, G., Unger, J.W., Neumaier, B., Ludolph, A. C., Reske, S.N., and Landwehrmeyer, G.B. (2004). Imaging of activated microglia with PET and [11C]PK 11195 in corticobasal degeneration. *Mov. Disord.* **19**, 817–821.

Hulshoff Pol, H.E., Schnack, H.G., Mandl, R.C., van Haren, N.E., Koning, H., Collins, D.L., Evans, A.C., and Kahn, R.S. (2001). Focal gray matter density changes in schizophrenia. *Arch. Gen. Psychiatry* **58**, 1118–1125.

Ikoma, Y., Yasuno, F., Ito, H., Suhara, T., Ota, M., Toyama, H., Fujimura, Y., Takano, A., Maeda, J., Zhang, M.R., Nakao, R., and Suzuki, K. (2007). Quantitative analysis for estimating binding potential of the peripheral benzodiazepine receptor with [(11)C]DAA1106. *J. Cereb. Blood Flow Metab.* **27**, 173–184.

Imaizumi, M., Briard, E., Zoghbi, S.S., Gourley, J.P., Hong, J., Fujimura, Y., Pike, V.W., Innis, R.B., and Fujita, M. (2008). Brain and whole-body imaging in nonhuman primates of [11C]PBR28, a promising PET radioligand for peripheral benzodiazepine receptors. *Neuroimage* **39**, 1289–1298.

Itzhak, Y., Baker, L., and Norenberg, M.D. (1993). Characterization of the peripheral-type benzodiazepine receptors in cultured astrocytes: evidence for multiplicity. *Glia* **9**, 211–218.

Ji, B., Maeda, J., Sawada, M., Ono, M., Okauchi, T., Inaji, M., Zhang, M.R., Suzuki, K., Ando, K., Staufenbiel, M., Trojanowski, J.Q., Lee, V.M., *et al.* (2008). Imaging of peripheral benzodiazepine receptor expression as biomarkers of detrimental versus beneficial glial responses in mouse models of Alzheimer's and other CNS pathologies. *J. Neurosci.* **28**, 12255–12267.

Joshi, S.N., and Crutcher, K.A. (1998). Rat microglia exhibit increased density on Alzheimer's plaques in vitro. *Exp. Neurol.* **149**, 42–50.

Kreutzberg, G.W. (1996). Microglia: a sensor for pathological events in the CNS. *Trends Neurosci.* **19**, 312–318.

Kropholler, M.A., Boellaard, R., Schuitemaker, A., Folkersma, H., van Berckel, B.N., and Lammertsma, A.A. (2006). Evaluation of reference tissue models for the analysis of [11C](R)-PK11195 studies. *J. Cereb. Blood Flow Metab.* **26**, 1431–1441.

Krupinski, J., Kaluza, J., Kumar, P., and Kumar, S. (1996). Immunocytochemical studies of cellular reaction in human ischemic brain stroke. MAB anti-CD68 stains macrophages, astrocytes and microglial cells in infarcted area. *Folia Neuropathol.* **34**, 17–24.

Kurumaji, A., Wakai, T., and Toru, M. (1997). Decreases in peripheral-type benzodiazepine receptors in postmortem brains of chronic schizophrenics. *J. Neural Transm.* **104**, 1361–1370.

Kutzelnigg, A., Lucchinetti, C.F., Stadelmann, C., Bruck, W., Rauschka, H., Bergmann, M., Schmidbauer, M., Parisi, J.E., and Lassmann, H. (2005). Cortical demyelination and diffuse white matter injury in multiple sclerosis. *Brain* **128**, 2705–2712.

Lehnardt, S., Massillon, L., Follett, P., Jensen, F.E., Ratan, R., Rosenberg, P.A., Volpe, J.J., and Vartanian, T. (2003). Activation of innate immunity in the CNS triggers neurodegeneration through a Toll-like receptor 4-dependent pathway. *Proc. Natl. Acad. Sci. USA* **100**, 8514–8519.

Maeda, J., Higuchi, M., Inaji, M., Ji, B., Haneda, E., Okauchi, T., Zhang, M.R., Suzuki, K., and Suhara, T. (2007). Phase-dependent roles of reactive microglia and astrocytes in nervous system injury as delineated by imaging of peripheral benzodiazepine receptor. *Brain Res.* **1157**, 100–111.

Mainero, C., Benner, T., Radding, A., van der, K.A., Jensen, R., Rosen, B.R., and Kinkel, R.P. (2009). In vivo imaging of cortical pathology in multiple sclerosis using ultra-high field MRI. *Neurology* **73**, 941–948.

Matthews, P.M., and Comley, R. (2009). Advances in the molecular imaging of multiple sclerosis. *Expert Rev. Clin. Immunol.* **5**, 765–777.

McGeer, P.L., Itagaki, S., Boyes, B.E., and McGeer, E.G. (1988). Reactive microglia are positive for HLA-DR in the substantia nigra of Parkinson's and Alzheimer's disease brains. *Neurology* **38**, 1285–1291.

Messmer, K., and Reynolds, G.P. (1998). Increased peripheral benzodiazepine binding sites in the brain of patients with Huntington's disease. *Neurosci. Lett.* **241**, 53–56.

Miyoshi, M., Shinotoh, H., Wszolek, Z.K., Strongosky, A.J., Shimada, H., Arakawa, R., Higuchi, M., Ikoma, Y., Yasuno, F., Fukushi, K., Irie, T., Ito, H., *et al.* (2010). In vivo detection of neuropathologic changes in presymptomatic MAPT mutation carriers: a PET and MRI study. *Parkinsonism Relat. Disord.* **16**, 404–408.

Myers, R., Manjil, L.G., Frackowiak, R.S., and Cremer, J.E. (1991). [3H]PK 11195 and the localisation of secondary thalamic lesions following focal ischaemia in rat motor cortex. *Neurosci. Lett.* **133**, 20–24.

Oh, U., Fujita, M., Ikonomidou, V.N., Evangelou, I.E., Matsuura, E., Harberts, E., Ohayon, J., Pike, V. W., Zhang, Y., Zoghbi, S.S., Innis, R.B., and Jacobson, S. (2010). Translocator protein PET imaging for glial activation in multiple sclerosis. *J. Neuroimmune Pharmacol.* **6**, 354–361.

Okello, A., Edison, P., Archer, H.A., Turkheimer, F.E., Kennedy, J., Bullock, R., Walker, Z., Kennedy, A., Fox, N., Rossor, M., and Brooks, D.J. (2009). Microglial activation and amyloid deposition in mild cognitive impairment: a PET study. *Neurology* **72**, 56–62.

Ouchi, Y., Yoshikawa, E., Sekine, Y., Futatsubashi, M., Kanno, T., Ogusu, T., and Torizuka, T. (2005). Microglial activation and dopamine terminal loss in early Parkinson's disease. *Ann. Neurol.* **57**, 168–175.

Owen, D.R., Howell, O.W., Tang, S.P., Wells, L.A., Bennacef, I., Bergstrom, M., Gunn, R.N., Rabiner, E.A., Wilkins, M.R., Reynolds, R., Matthews, P.M., and Parker, C.A. (2010). Two binding sites for [(3)H]PBR28 in human brain: implications for TSPO PET imaging of neuroinflammation. *J. Cereb. Blood Flow Metab.*

Owen, D.R., Gunn, R.N., Rabiner, E.A., Bennacef, I., Fujita, M., Kreisl, W.C., Innis, R.B., Pike, V.W., Reynolds, R., Matthews, P.M., and Parker, C.A. (2011). Mixed-affinity binding in humans with 18-kDa translocator protein ligands. *J. Nucl. Med.* **52**, 24–32.

Pappata, S., Levasseur, M., Gunn, R.N., Myers, R., Crouzel, C., Syrota, A., Jones, T., Kreutzberg, G. W., and Banati, R.B. (2000). Thalamic microglial activation in ischemic stroke detected in vivo by PET and [11C]PK1195. *Neurology* **55**, 1052–1054.

Pavese, N., Gerhard, A., Tai, Y.F., Ho, A.K., Turkheimer, F., Barker, R.A., Brooks, D.J., and Piccini, P. (2006). Microglial activation correlates with severity in Huntington disease: a clinical and PET study. *Neurology* **66**, 1638–1643.

Petit-Taboue, M.C., Baron, J.C., Barre, L., Travere, J.M., Speckel, D., Camsonne, R., and MacKenzie, E.T. (1991). Brain kinetics and specific binding of [11C]PK 11195 to omega 3 sites in baboons: positron emission tomography study. *Eur. J. Pharmacol.* **200**, 347–351.

Politis, M., Giannetti, P., Su, P., Turkheimer, F., Keihaninejad, S., Wu, K., Waldman, A., Reynolds, R., Nicholas, R., and Piccini, P. (2010). Cortical microglial activation is associated with disability in Secondary progressive Multiple Sclerosis: an in vivo imaging study. *Neurology* **74**(Suppl. 2), A290.

Price, C.J., Wang, D., Menon, D.K., Guadagno, J.V., Cleij, M., Fryer, T., Aigbirhio, F., Baron, J.C., and Warburton, E.A. (2006). Intrinsic activated microglia map to the peri-infarct zone in the subacute phase of ischemic stroke. *Stroke* **37**, 1749–1753.

Radewicz, K., Garey, L.J., Gentleman, S.M., and Reynolds, R. (2000). Increase in HLA-DR immunoreactive microglia in frontal and temporal cortex of chronic schizophrenics. *J. Neuropathol. Exp. Neurol.* **59**, 137–150.

Radlinska, B.A., Ghinani, S.A., Lyon, P., Jolly, D., Soucy, J.P., Minuk, J., Schirrmacher, R., and Thiel, A. (2009). Multimodal microglia imaging of fiber tracts in acute subcortical stroke. *Ann. Neurol.* **66**, 825–832.

Raghavendra, R.V., Dogan, A., Bowen, K.K., and Dempsey, R.J. (2000). Traumatic brain injury leads to increased expression of peripheral-type benzodiazepine receptors, neuronal death, and activation of astrocytes and microglia in rat thalamus. *Exp. Neurol.* **161**, 102–114.

Ramsay, S.C., Weiller, C., Myers, R., Cremer, J.E., Luthra, S.K., Lammertsma, A.A., and Frackowiak, R.S. (1992). Monitoring by PET of macrophage accumulation in brain after ischaemic stroke. *Lancet* **339**, 1054–1055.

Rao, S.M., Leo, G.J., Bernardin, L., and Unverzagt, F. (1991). Cognitive dysfunction in multiple sclerosis. I. Frequency, patterns, and prediction. *Neurology* **41**, 685–691.

Rapoport, J.L., Giedd, J.N., Blumenthal, J., Hamburger, S., Jeffries, N., Fernandez, T., Nicolson, R., Bedwell, J., Lenane, M., Zijdenbos, A., Paus, T., and Evans, A. (1999). Progressive cortical change during adolescence in childhood-onset schizophrenia. A longitudinal magnetic resonance imaging study. *Arch. Gen. Psychiatry* **56**, 649–654.

Rey, C., Mauduit, C., Naureils, O., Benahmed, M., Louisot, P., and Gasnier, F. (2000). Up-regulation of mitochondrial peripheral benzodiazepine receptor expression by tumor necrosis factor alpha in testicular leydig cells. Possible involvement in cell survival. *Biochem. Pharmacol.* **60**, 1639–1646.

Sapp, E., Kegel, K.B., Aronin, N., Hashikawa, T., Uchiyama, Y., Tohyama, K., Bhide, P.G., Vonsattel, J.P., and DiFiglia, M. (2001). Early and progressive accumulation of reactive microglia in the Huntington disease brain. *J. Neuropathol. Exp. Neurol.* **60**, 161–172.

Schroeter, M., Jander, S., Witte, O.W., and Stoll, G. (1999). Heterogeneity of the microglial response in photochemically induced focal ischemia of the rat cerebral cortex. *Neuroscience* **89**, 1367–1377.

Shah, F., Hume, S.P., Pike, V.W., Ashworth, S., and McDermott, J. (1994). Synthesis of the enantiomers of [N-methyl-11C]PK 11195 and comparison of their behaviours as radioligands for PK binding sites in rats. *Nucl. Med. Biol.* **21**, 573–581.

Sitte, H.H., Wanschitz, J., Budka, H., and Berger, M.L. (2001). Autoradiography with [3H]PK11195 of spinal tract degeneration in amyotrophic lateral sclerosis. *Acta Neuropathol.* **101**, 75–78.

Steiner, J., Mawrin, C., Ziegeler, A., Bielau, H., Ullrich, O., Bernstein, H.G., and Bogerts, B. (2006). Distribution of HLA-DR-positive microglia in schizophrenia reflects impaired cerebral lateralization. *Acta Neuropathol.* **112**, 305–316.

Stephenson, D.T., Schober, D.A., Smalstig, E.B., Mincy, R.E., Gehlert, D.R., and Clemens, J.A. (1995). Peripheral benzodiazepine receptors are colocalized with activated microglia following transient global forebrain ischemia in the rat. *J. Neurosci.* **15**, 5263–5274.

Tai, Y.F., Pavese, N., Gerhard, A., Tabrizi, S.J., Barker, R.A., Brooks, D.J., and Piccini, P. (2007). Microglial activation in presymptomatic Huntington's disease gene carriers. *Brain* **130**, 1759–1766.

Takaya, S., Hashikawa, K., Turkheimer, F.E., Mottram, N., Deprez, M., Ishizu, K., Kawashima, H., Akiyama, H., Fukuyama, H., Banati, R.B., and Roncaroli, F. (2007). The lack of expression of the peripheral benzodiazepine receptor characterises microglial response in anaplastic astrocytomas. *J. Neurooncol* **85**, 95–103.

Thie, J.A. (2004). Understanding the standardized uptake value, its methods, and implications for usage. *J. Nucl. Med.* **45**, 1431–1434.

Thiel, A., Radlinska, B.A., Paquette, C., Sidel, M., Soucy, J.P., Schirrmacher, R., and Minuk, J. (2010). The temporal dynamics of poststroke neuroinflammation: a longitudinal diffusion tensor imaging-guided PET study with 11C-PK11195 in acute subcortical stroke. *J. Nucl. Med.* **51**, 1404–1412.

Tomasi, G., Edison, P., Bertoldo, A., Roncaroli, F., Singh, P., Gerhard, A., Cobelli, C., Brooks, D.J., and Turkheimer, F.E. (2008). Novel reference region model reveals increased microglial and reduced vascular binding of 11C-(R)-PK11195 in patients with Alzheimer's disease. *J. Nucl. Med.* **49**, 1249–1256.

Tomimoto, H., Akiguchi, I., Wakita, H., Kinoshita, A., Ikemoto, A., Nakamura, S., and Kimura, J. (1996). Glial expression of cytokines in the brains of cerebrovascular disease patients. *Acta Neuropathol.* **92**, 281–287.

Trincavelli, M.L., Marselli, L., Falleni, A., Gremigni, V., Ragge, E., Dotta, F., Santangelo, C., Marchetti, P., Lucacchini, A., and Martini, C. (2002). Upregulation of mitochondrial peripheral benzodiazepine receptor expression by cytokine-induced damage of human pancreatic islets. *J. Cell. Biochem.* **84**, 636–644.

Turkheimer, F.E., Edison, P., Pavese, N., Roncaroli, F., Anderson, A.N., Hammers, A., Gerhard, A., Hinz, R., Tai, Y.F., and Brooks, D.J. (2007). Reference and target region modeling of [11C]-(R)-PK11195 brain studies. *J. Nucl. Med.* **48**, 158–167.

van Berckel, B.N., Bossong, M.G., Boellaard, R., Kloet, R., Schuitemaker, A., Caspers, E., Luurtsema, G., Windhorst, A.D., Cahn, W., Lammertsma, A.A., and Kahn, R.S. (2008). Microglia activation in recent-onset schizophrenia: a quantitative (R)-[11C]PK11195 positron emission tomography study. *Biol. Psychiatry* **64**, 820–822.

Van, C.N., Boisgard, R., Kuhnast, B., Theze, B., Viel, T., Gregoire, M.C., Chauveau, F., Boutin, H., Katsifis, A., Dolle, F., and Tavitian, B. (2010). In vivo imaging of neuroinflammation: a comparative study between [(18)F]PBR111, [(11)C]CLINME and [(11)C]PK11195 in an acute rodent model. *Eur. J. Nucl. Med. Mol. Imaging.*

Venneti, S., Lopresti, B.J., and Wiley, C.A. (2006). The peripheral benzodiazepine receptor (Translocator protein 18kDa) in microglia: from pathology to imaging. *Prog. Neurobiol.* **80**, 308–322.

Venneti, S., Wang, G., Nguyen, J., and Wiley, C.A. (2008). The positron emission tomography ligand DAA1106 binds with high affinity to activated microglia in human neurological disorders. *J. Neuropathol. Exp. Neurol.* **67**, 1001–1010.

Vercellino, M., Merola, A., Piacentino, C., Votta, B., Capello, E., Mancardi, G.L., Mutani, R., Giordana, M.T., and Cavalla, P. (2007). Altered glutamate reuptake in relapsing-remitting and secondary progressive multiple sclerosis cortex: correlation with microglia infiltration, demyelination, and neuronal and synaptic damage. *J. Neuropathol. Exp. Neurol.* **66**, 732–739.

Versijpt, J., Debruyne, J.C., Van Laere, K.J., De, V.F., Keppens, J., Strijckmans, K., Achten, E., Slegers, G., Dierckx, R.A., Korf, J., and De Reuck, J.L. (2005). Microglial imaging with positron emission tomography and atrophy measurements with magnetic resonance imaging in multiple sclerosis: a correlative study. *Mult. Scler.* **11**, 127–134.

Vowinckel, E., Reutens, D., Becher, B., Verge, G., Evans, A., Owens, T., and Antel, J.P. (1997). PK11195 binding to the peripheral benzodiazepine receptor as a marker of microglia activation in multiple sclerosis and experimental autoimmune encephalomyelitis. *J. Neurosci. Res.* **50**, 345–353.

Wang, Q., Tang, X.N., and Yenari, M.A. (2007). The inflammatory response in stroke. *J. Neuroimmunol.* **184**, 53–68.

Watanabe, H., Abe, H., Takeuchi, S., and Tanaka, R. (2000). Protective effect of microglial conditioning medium on neuronal damage induced by glutamate. *Neurosci. Lett.* **289**, 53–56.

Wiley, C.A., Lopresti, B.J., Venneti, S., Price, J., Klunk, W.E., DeKosky, S.T., and Mathis, C.A. (2009). Carbon 11-labeled Pittsburgh Compound B and carbon 11-labeled (R)-PK11195 positron emission tomographic imaging in Alzheimer disease. *Arch. Neurol.* **66**, 60–67.

Wilson, A.A., Garcia, A., Parkes, J., McCormick, P., Stephenson, K.A., Houle, S., and Vasdev, N. (2008). Radiosynthesis and initial evaluation of [18F]-FEPPA for PET imaging of peripheral benzodiazepine receptors. *Nucl. Med. Biol.* **35**, 305–314.

Wood, P.L. (1995). Microglia as a unique cellular target in the treatment of stroke: potential neurotoxic mediators produced by activated microglia. *Neurol. Res.* **17**, 242–248.

Yasuno, F., Ota, M., Kosaka, J., Ito, H., Higuchi, M., Doronbekov, T.K., Nozaki, S., Fujimura, Y., Koeda, M., Asada, T., and Suhara, T. (2008). Increased binding of peripheral benzodiazepine receptor in Alzheimer's disease measured by positron emission tomography with [11C]DAA1106. *Biol. Psychiatry* **64**, 835–841.

Yokokura, M., Mori, N., Yagi, S., Yoshikawa, E., Kikuchi, M., Yoshihara, Y., Wakuda, T., Sugihara, G., Takebayashi, K., Suda, S., Iwata, Y., Ueki, T., *et al.* (2011). In vivo changes in microglial activation and amyloid deposits in brain regions with hypometabolism in Alzheimer's disease. *Eur. J. Nucl. Med. Mol. Imaging* **38**, 343–351.

Yrjanheikki, J., Keinanen, R., Pellikka, M., Hokfelt, T., and Koistinaho, J. (1998). Tetracyclines inhibit microglial activation and are neuroprotective in global brain ischemia. *Proc. Natl. Acad. Sci. USA* **95**, 15769–15774.

Yrjanheikki, J., Tikka, T., Keinanen, R., Goldsteins, G., Chan, P.H., and Koistinaho, J. (1999). A tetracycline derivative, minocycline, reduces inflammation and protects against focal cerebral ischemia with a wide therapeutic window. *Proc. Natl. Acad. Sci. USA* **96**, 13496–13500.

Zhang, Z., Chopp, M., and Powers, C. (1997). Temporal profile of microglial response following transient (2 h) middle cerebral artery occlusion. *Brain Res.* **744**, 189–198.

THE UTILITY OF GENE EXPRESSION IN BLOOD CELLS FOR DIAGNOSING NEUROPSYCHIATRIC DISORDERS

Christopher H. Woelk[1,2], Akul Singhania[2], Josué Pérez-Santiago[1], Stephen J. Glatt[3] and Ming T. Tsuang[4]

[1]Department of Medicine, University of California San Diego, La Jolla, California, USA
[2]Veterans Affairs San Diego Healthcare System, San Diego, California, USA
[3]Departments of Psychiatry and Behavioral Sciences & Neuroscience and Physiology, Psychiatric Genetic Epidemiology & Neurobiology Laboratory (PsychGENe Lab), Medical Genetics Research Center, SUNY Upstate Medical University, Syracuse, New York, USA
[4]Department of Psychiatry and Institute of Genomic Medicine, Center for Behavioral Genomics, University of California San Diego, La Jolla, California, USA

Abstract

Objective diagnostic tools are required for neuropsychiatric disorders. Gene expression in blood cells may provide such a tool and has already been used to construct classifiers capable of diagnosing many human diseases. This chapter discusses the use of microarray gene expression data to construct diagnostic classifiers for neuropsychiatric disorders. The potential pitfalls of microarray gene expression analysis and the experimental design and methods suitable for classifier construction are described in detail. A review of studies that have analyzed gene expression in blood cells from patients with neuropsychiatric disorders is presented with an emphasis on the feasibility of generating a diagnostic classifier for schizophrenia. Finally, the future directions of the field are

INTERNATIONAL REVIEW OF NEUROBIOLOGY, VOL. 101
DOI: 10.1016/B978-0-12-387718-5.00003-1

discussed with respect to using blood gene expression to tailor antipsychotic medications to individual patients, applying microRNA expression for diagnostic purposes, as well as the implications of next-generation sequencing technologies for gene expression analysis.

I. Introduction

There are currently no objective genetic tests in widespread use for the diagnosis of any one of a large number of neuropsychiatric disorders. The exact pathogenesis of most neuropsychiatric disorders remains unclear with a prominent role for biological, genetic, and environmental factors (Glatt *et al.*, 2010). Subjective clinical evaluations of disorders such as schizophrenia, bipolar disorder, Parkinson's disease, posttraumatic stress disorder (PTSD), and major depressive disorder (MDD) may sometimes lead to inconsistencies in diagnosis between different psychiatrists especially when patients are followed longitudinally. Gene expression analysis of the entire transcriptome using microarray technology is now a common laboratory practice with demonstrated reproducibility (Shi *et al.*, 2006). This chapter investigates the feasibility of using gene expression to construct objective classifiers for the diagnosis of neuropsychiatric disorders.

The use of gene expression classifiers for disease prognosis has already met with approval from the United States Food and Drug Administration (FDA). In 2007, the FDA granted approval to MammaPrint®, a classifier constructed from the expression of 70 genes in tumors biopsied from women with breast cancer that is capable of predicting recurrence within 5 years (vant Veer *et al.*, 2002). MammaPrint® is currently being evaluated in the MINDACT (Microarray in Node-Negative Disease May Avoid Chemotherapy) trial to assess its utility for identifying patients with a low risk of recurrence, and who can therefore be spared the inconvenience and morbidity associated with adjuvant chemotherapy (Cardoso *et al.*, 2008). MammaPrint® and future gene expression classifiers are defined by the FDA as *in vitro* diagnostic multivariate index assays (IVDMIAs) which require a complex algorithm to calculate the diagnostic score. This is in contrast to traditional biomarkers for which a simple concentration (i.e., blood cholesterol) is used as a score for diagnosis (Kato, 2009; see Chapter "Algorithm development for diagnostic biomarker assays" by Izmailov *et al.*).

Due to its accessibility, peripheral blood has historically been a valuable tissue source for the derivation of biomarkers (Woelk and Burczynski, 2008). A plethora of studies have been published recently demonstrating the utility of constructing gene expression classifiers from blood cells in order to diagnose disease states, predict disease outcomes, and determine an individual's response to treatment

(Burczynski and Dorner, 2006; Woelk and Burczynski, 2008). With respect to neuropsychiatric disorders, it is important to define the difference between a mechanistic- and a classifier-based gene expression study. Mechanistic studies have focused on analyzing gene expression in brain tissue, which, in the case of schizophrenia, is normally derived from the prefrontal cortex (Garbett *et al.*, 2008; Glatt *et al.*, 2005; Maycox *et al.*, 2009; Narayan *et al.*, 2008; Prabakaran *et al.*, 2004), although other brain regions have also been analyzed, that is, the superior temporal gyrus (Bowden *et al.*, 2008). Methods of class comparison are then used in order to identify differentially expressed genes that are mapped on to biological pathways, gene ontologies, and protein networks in order to reveal the higher-level processes contributing to disease pathogenesis. In a classifier study, methods of class prediction are used to identify those genes whose expression can be used to predict whether a blinded sample came from an individual with a neuropsychiatric disorder or a healthy control. Obviously, sampling brain tissue for such a diagnostic purpose is not feasible (Struyf *et al.*, 2008) and many classifier studies have focused on analyzing gene expression in blood cells for this purpose (Table I).

Our own studies, among others, have focused on the overlap in gene expression between the blood and brain compartments in order to evaluate the utility of gene expression in the blood for diagnosing neuropsychiatric disorders (Glatt *et al.*, 2005; Shehadeh *et al.*, 2010; Sullivan *et al.*, 2006). For example, we previously identified 177 genes differentially expressed in the dorsolateral prefrontal cortex (DLPFC) between 19 schizophrenia and 27 control cases (Glatt *et al.*, 2005). When gene expression was analyzed in white blood cells (WBCs) derived from 30 schizophrenia and 24 control cases in an unrelated cohort, 123 genes were identified as being differentially expressed. An overlap of only two genes modulated in the same manner was found between compartments. The level of selenium binding protein 1 (*SELENBP1*) mRNA was increased, and that of major histocompatibility complex, class II, DR beta 1 (*HLA-DRB1*) was decreased, in both compartments. The differential expression of *SELENBP1* was confirmed at the mRNA transcript level by real-time quantitative PCR (RT-qPCR) and at the protein level by antibody staining. In addition, this finding was subsequently replicated in brain tissue samples from an independent cohort (Kanazawa *et al.*, 2008). Although only a small overlap in differential gene expression between schizophrenia and control cases across compartments was revealed in our study, it should not be taken as evidence that the blood compartment is of limited use for the derivation of diagnostic gene expression signatures. The prevailing hypothesis behind hemogenomic studies that attempt to construct gene expression classifiers is that a transcriptional signature exists in common to all neuropsychiatric cases that is distinct from all control samples (Woelk and Burczynski, 2008). If this is the case, then it is of no consequence whether the genes whose expression comprises the signature are expressed in brain tissue; they are simply being used as a diagnostic read out and not intended for mechanistic interpretation (Middleton *et al.*, 2005).

Table I

BLOOD GENE EXPRESSION STUDIES OF NEUROPSYCHIATRIC DISORDERS THAT DEVELOPED A DIAGNOSTIC CLASSIFIER.[a]

Disorder	Tissue source	Number of microarrays				Microarray platform	Classifier method	Validation	Classifier performance[b]	GEO	Reference
		Total	Disorder	Drug-naive	Control						
Bipolar disorder and psychotic disorders	Whole blood	78	37 (HM), 13 (IM), and 28 (LM)	0	NA	Affymetrix HG U133 Plus 2.0 GeneChip	Mood prediction score	Hold-out	PD test data set: HM—71.4% sensitivity, 62.5% specificity; LM—66.7% sensitivity, 61.9% specificity BPD test data set: HM—70% sensitivity, 66.7% specificity; LM—66.7% sensitivity, 61.5% specificity	No	Le-Niculescu *et al.* (2009)
Major depressive disorder	Whole blood (LPS-stimulated)	67	33	33	34	Agilent 44K Human whole genome array	Nearest shrunken centroid and MDD score	Hold-out	74.1%	GSE19738	Spijker *et al.* (2010)
Posttraumatic stress disorder	PBMCs	33	17	NR	16	Affymetrix HU95A GeneChip	Naive Bayesian	LOOCV	89% for M4 samples[c]	GDS1020	Segman *et al.* (2005)

Schizophrenia	Whole blood	101	52	44	49	CodeLink Human Whole Genome Bioarray	Artificial neural network	Threefold CV and hold-out	91.2%—training data set 87.9%—test data set	No	Takahashi *et al.* (2010)
Schizophrenia and bipolar disorder	WBCs	76	33 (SZ) and 5 (BPD)	NR	38	Affymetrix HG U133A GeneChip	Support vector machine	Hold-out	100%	No	Middleton *et al.* (2005)
	WBCs	54	30 (SZ) and 7 (BPD)	0	17	Affymetrix HG U133A and U133Plus 2.0 GeneChip	Logistic regression	NA[d]	0.960 (AUC for SZ) 0.948 (AUC for BPD)	No	Tsuang *et al.* (2005)

[a] *Drug-naive* is a subset of the subjects in the Disorder column and refers to the number of samples used for microarray analysis that were taken from patients with the disorder that were also drug-naive or drug-free. Shading is for display purposes only. *Abbreviations*: GEO, refers to the Gene Expression Omnibus (http://www.ncbi.nlm.nih.gov/geo/) and whether microarray data has been deposited in this public repository; Ref., refers to the literature reference for the microarray study; HM, high mood; IM, intermediate mood; LM, low mood; NR, not reported; NA, not applicable; PD, psychotic disorder; BPD, bipolar disorder; LPS, lipopolysaccharide; MDD, major depressive disorder; PBMCs, peripheral blood mononuclear cells; LOOCV, leave-one-out cross-validation; WBCs, white blood cells; SZ, schizophrenia; AUC, area under the curve from receiver operator curve analysis.

[b] Classification accuracy is presented unless otherwise stated.

[c] M4 samples were those taken 4 months after a traumatic event ($N = 9$).

[d] Analysis of microarray gene expression data was used to construct an eight gene classifier whose expression was then evaluated by RT-qPCR in some of the same samples used for microarray analysis. AUC values are given for the performance of this RT-qPCR classifier, which was not validated in a *k*-fold cross-validation or hold-out validation procedure.

II. Microarray Gene Expression Analysis

The typical workflow for a microarray gene expression study that aims to construct a diagnostic gene expression classifier is outlined in Fig. 1. Previous gene expression studies have analyzed many different subsets of cells from the blood

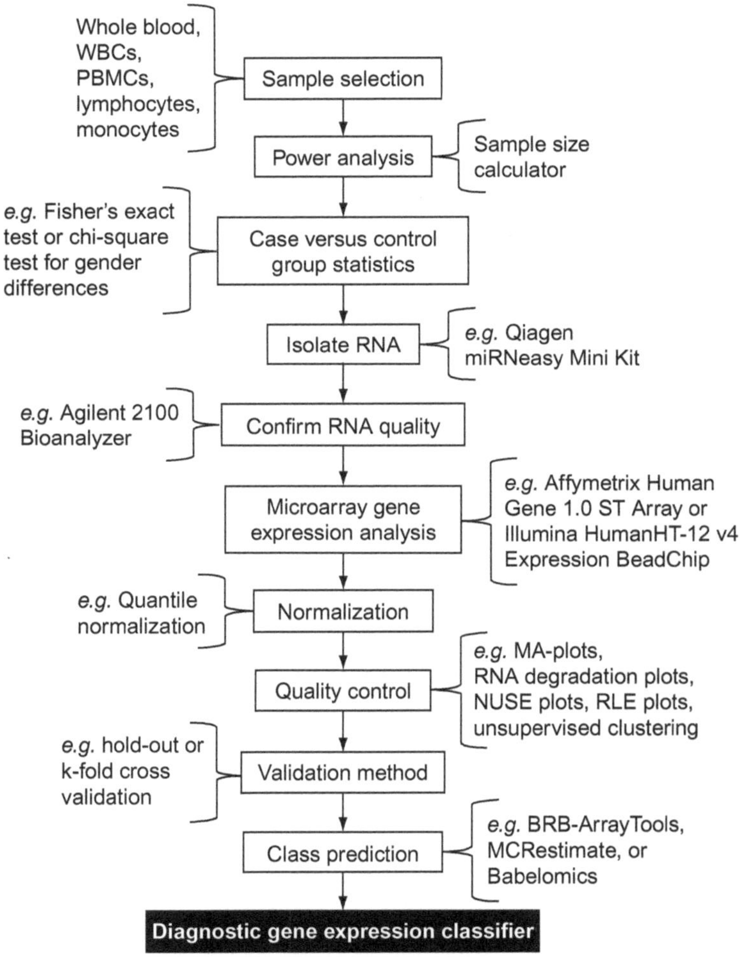

FIG. 1. A typical workflow for a microarray gene expression study aiming to construct a diagnostic gene expression classifier. Sample size calculator refers to the method developed by Dobbin *et al.* (2008) available online (http://linus.nci.nih.gov/brb/samplesize/samplesize4GE.html). Abbreviations: MA, log-intensity *M*-values versus log-intensity averages or *A*-values; WBCs, white blood cells; PBMCs, peripheral blood mononuclear cells; NUSE, normalized unscaled standard error; RLE, relative log expression.

including monocytes, lymphocytes, peripheral blood mononuclear cells (PBMCs), and WBCs, but the use of whole blood is by far the most common (Tables I and II). This is probably due to the utility of PAXgene tubes that can stabilize RNA in whole blood for up to 50 months when stored below −20 °C. However, the vast majority of whole blood studies did not perform a globin RNA reduction step prior to assessing gene expression with microarrays. Globin RNA constitutes 70% of all expressed transcripts in the blood and can confound the accurate measurement of gene expression by microarrays by decreasing the specificity of transcript detection and increasing signal variation (Field *et al.*, 2007; Liu *et al.*, 2006; Winn *et al.*, 2010). Commercial kits such as GLOBINclear™ (Applied Biosystems, Foster City, CA, USA; Ambion, Austin, TX, USA) are available and remove up to 95% of α- and β-globin transcripts.

Once the samples have been identified for microarray analysis, it is important to design the study correctly in order to construct an accurate diagnostic gene expression classifier. It is desirable to know how many samples are required in order to derive a gene expression classifier of optimal accuracy. In this regard, Dobbin *et al.* (2008) developed a power calculator for microarray classifier studies that uses the number of probes on the microarray platform, the likely proportion of samples in each class (i.e., schizophrenia vs. controls), and an estimate of the standardized fold change between classes based on the gene exhibiting the greatest difference. In addition to ensuring that enough samples are analyzed, it is best to have a balanced study design for which similar numbers of neuropsychiatric and control cases are compared (Dupuy and Simon, 2007). It is also essential that all variables are balanced between the two groups being compared. For example, if the neuropsychiatric group has a significantly greater number of males than occurs in the control group, it will be unclear as to whether the resulting classifier is diagnosing the patient's disease status or their gender (Takahashi *et al.*, 2010). Finally, most classifier studies in Table I compare cases of a single neuropsychiatric disorder to healthy control subjects. Ideally, to better evaluate the specificity of a diagnostic gene expression classifier, the control group should also contain neuropsychiatric cases related to the disorder under study but that can be distinguished by their own distinct clinical diagnosis. For example, while evaluating the utility of gene expression in whole blood for diagnosing Parkinson's disease, Scherzer *et al.* (2007) combined cases of Alzheimer's disease, progressive supranuclear palsy, and multiple system atrophy, among others, with healthy subjects in the control group.

After the experimental design has been finalized and samples identified for analysis, RNA must be extracted from the samples. It is now common practice to assess the quality of the RNA using an Agilent 2100 Bioanalyzer (Santa Clara, CA, USA) which produces an RNA integrity number (RIN) ranging from 1 (totally degraded) to 10 (good-quality RNA). An acceptable RIN threshold for the tissue type and the microarray platform under study can be calculated

Table II

BLOOD GENE EXPRESSION STUDIES OF NEUROPSYCHIATRIC DISORDERS THAT DID NOT DEVELOP A DIAGNOSTIC CLASSIFIER.[a]

		Number of microarrays						
Disorder	Tissue source	Total	Disorder	Drug-naive	Control	Microarray platform	GEO	Ref.
Parkinson's disease	Whole blood	105	50	NR[b]	55	Affymetrix HG U133A GeneChip	GSE6613	Scherzer *et al.* (2007)
Posttraumatic stress disorder	Monocytes	67	34	34	33	CodeLink Human Whole Genome Bioarray	No	Neylan *et al.* (2010)
	Whole blood	35	15	15	20	Affymetrix HG U133 Plus 2.0 GeneChip	No	Yehuda *et al.* (2009)
	Whole blood	16	8	8	8	Stress/immune chips (custom cDNA array)	No	Zieker *et al.* (2007)
Schizophrenia	WBCs	54	30	0	24	Affymetrix HG U133A GeneChip	No	Glatt *et al.* (2005)
	Whole blood	64	32	32	32	Affymetrix HG U133 Plus 2.0 GeneChip	No	Kuzman *et al.* (2009)
	LCL	15	8	7	7	Human Genome v.2.1 oligo arrays (custom cDNA array)	No	Matigian *et al.* (2008)
	LCL	14	5	NR	9	Research Genetics NIA-Neuroarray (custom cDNA array)	No	Vawter *et al.* (2004)
	Lymphocytes	23	13	13	10	Full Moon BioSystems cDNA slides (custom cDNA array)	No	Zvara *et al.* (2005)
Schizophrenia and bipolar disorder	PBMCs	71	47	1	24	Affymetrix Human Exon 1.0 ST GeneChip and Affymetrix HG U133A or U133Plus 2.0 GeneChip	No	Bousman *et al.* (2010)

[a] Column headings, table descriptions, and abbreviations are defined as for Table I except for LCL, which refers to lymphoblastoid cell line. Shading is for display purposes only.

[b] Scherzer *et al.* (2007) did not indicate if the same Parkinson's disease patient was treated with more than one therapy so the exact number of drug-naive or drug-free individuals could not be determined. For the 50 Parkinson's disease patients analyzed in the study, 31 were treated with L-dopa, 24 with dopamine agonist, and seven with selegiline.

empirically in a pilot analysis by correlating RIN with the quality of gene expression data obtained. However, a general rule of thumb is to only proceed with microarray gene expression analysis with samples that have a RIN of greater than seven (Schroeder *et al.*, 2006). Good-quality RNA is converted by reverse transcription to cDNA which itself is converted into labeled cRNA for hybridization to microarray platforms capable of detecting gene expression of the entire transcriptome. With respect to choice of microarray platform, Affymetrix (Santa Clara) and Illumina (San Diego, CA, USA) both offer comparably high-quality products (Barnes *et al.*, 2005), and the majority (9/16) of blood gene expression studies of neuropsychiatric disorders summarized in Tables I and II utilized microarray platforms from Affymetrix.

When raw microarray gene expression data have been generated, they must be subjected to normalization and quality control procedures. Quantile normalization is a popular method that can be implemented for Affymetrix data using the GeneChip Robust Multichip Average (GCRMA) algorithm (Wu *et al.*, 2004) in *gcrma*, a Bioconductor package in the R statistical programming language, and for Illumina data using the Bioconductor package *lumi* (Du *et al.*, 2008). A complete description of the quality control procedures for microarray data is beyond the scope of this chapter but has been reviewed extensively by Bolstad *et al.* (2005) and includes the following types of plots that can be used to identify outlier arrays that should be removed from analysis: box, histogram, *MA* (log-intensity *M*-values vs. log-intensity averages or *A*-values), RNA degradation, normalized unscaled standard error (NUSE), and relative log expression (RLE). Attention should be drawn to using methods of class discovery (i.e., unsupervised clustering) for quality control, a step that is often missing from gene expression studies. Gene expression data should be filtered to remove those genes not expressed in any of the samples analyzed, and then samples clustered in an unsupervised manner (i.e., with no regard to sample label) using tools such as CLUSTER (Eisen *et al.*, 1998). The resulting clustergram can be visualized with TREEVIEW (Eisen *et al.*, 1998) and used to evaluate whether the samples cluster as expected (i.e., do drug treated patients cluster distinctly from drug-naive patients) or based on technical batch effects. Technical variation can be introduced through the use of different reagent lots, RNA isolation protocols, instrument settings, and even technician handling (Johnson *et al.*, 2007). In our studies, we have noted that technical batch effects resulted in samples clustering based on the date that the microarray hybridization was performed and not based on biological signal (Woelk *et al.*, 2010). Therefore, it is highly recommended that all samples in a microarray study be run within a narrow time window and unsupervised clustering used to identify technical batch effects by labeling samples based on the technical variables listed above to determine if samples cluster based on these variables. If technical batch effects are identified in microarray data, they may be removed using ComBat, which utilizes an empirical Bayes method (Johnson *et al.*, 2007). Finally, once a microarray gene expression study has been published, the data should be deposited in a public repository such as ArrayExpress (http://www.ebi.ac.uk/arrayexpress/) at the European Bioinformatics Institute or the Gene Expression Omnibus (http://www.ncbi.nlm.nih.gov/geo/) at the National Center for Biotechnology Information. This facilitates reanalysis of the microarray data with novel methods as they emerge and also allows the merging of studies to facilitate meta-analyses with increased power. Discouragingly, the vast majority (13/16) of neuropsychiatric studies assessing gene expression in blood cells have not deposited their microarray data in public repositories (Tables I and II).

III. Diagnostic Gene Expression Classifiers

Many useful tools have been developed for the construction of gene expression classifiers and those that are freely available include BRB-Array Tools (Simon *et al.*, 2007), MCRestimate (Ruschhaupt *et al.*, 2004), and Babelomics (Al-Shahrour *et al.*, 2008). Each of these tools use methods of class prediction: a supervised approach that incorporates the sample label (e.g., schizophrenia or control) to identify the genes whose expression can be used to predict which group a blinded sample belongs to. Multiple methods have been used to classify samples and include support vector machines (SVMs), artificial neural networks (ANNs), logistic regression, and Bayesian methods (Middleton *et al.*, 2005; Segman *et al.*, 2005; Takahashi *et al.*, 2010; Tsuang *et al.*, 2005).

In order to evaluate classifier performance, a 2×2 contingency table needs to be populated so that the numbers of true positives (TP), true negatives (TN), false positives (FP), and false negatives (FN) may be estimated in order to calculate classifier accuracy (TP + TN/TP + TN + FP + FN), sensitivity (TP/TP + FN), and specificity (TN/TN + FP). A number of strategies exist for populating such a 2×2 contingency table but the most common are hold-out and *k*-fold cross-validation (Dupuy and Simon, 2007). In hold-out validation, a number of samples are removed from analysis as an independent test data set and the remaining samples used to construct and train the gene expression classifier. Then the 2×2 contingency table is populated by predicting the class of the samples in the test data set using the classifier constructed using the training data set. Hold-out validation is normally used when a large number of samples are available (Le-Niculescu *et al.*, 2009; Middleton *et al.*, 2005; Takahashi *et al.*, 1996), whereas most smaller pilot studies employ *k*-fold (commonly 1 or 10) cross-validation to assess classifier accuracy. In the case of leave-one-out cross-validation (LOOCV), a sample is removed from the training data, the remaining samples used to construct the classifier, and then this classifier is used to predict the class of the sample that was left out. This process is repeated until every sample has been left out at least once and the predictions of each sample are used to populate the 2×2 contingency table. In both these validation procedures, a set of genes must be selected whose expression is used for classification purposes. This set of genes is sometimes selected as all those with a *p*-value above a certain threshold but more commonly identified using a method of feature selection. For example, in BRB-ArrayTools it is possible to perform recursive feature elimination where the number of genes desired in the classifier is selected *a priori* and then a machine learning method (i.e., SVM) is used to remove genes from the classifier based on their lack of contribution to prediction performance (Simon *et al.*, 2009).

There are several essential steps that are required in order to construct a diagnostic gene expression classifier with clinical utility for neuropsychiatric studies. These steps include benefit, feasibility, internal validation, external validation, and a prospective trial. *Benefit* is an initial assessment of whether a diagnostic gene expression classifier is likely to improve on the accuracy or reduce the cost of any diagnostic tool that already exists for the disease. In the case of neuropsychiatric disorders, there are currently no diagnostic genetic tests in widespread use and clinical diagnosis requires a battery of tests incurring significant costs and substantial amounts of clinical time. Therefore, there is clearly a great need for diagnostic gene expression classifiers for neuropsychiatric disorders. *Feasibility* refers to a pilot study whereby the utility of measuring gene expression in the blood for diagnosing the neuropsychiatric disorder under study is established. A larger number of patients from the same cohort that was used in the pilot study should then be analyzed in the *internal validation* step. Once a diagnostic gene expression classifier has been developed, it needs to be subjected to *external validation* in an unrelated cohort. For example, if a diagnostic gene expression classifier was constructed using samples taken from schizophrenia and control cases in Japan (Takahashi *et al.*, 2010), the accuracy of the classifier would then need to be assessed for a similar cohort in the United States or Europe. Finally, a *prospective trial* should be initiated to evaluate the clinical utility of a diagnostic gene expression classifier for the neuropsychiatric disorder. Patients could be diagnosed with both a gene expression classifier and by a large panel of trained psychiatrists. When discordant diagnoses arise, the patients could then be followed up longitudinally to determine whether the classifier or the traditional diagnostic approach was more accurate. The genes whose expression are being used for classification purposes can be transferred onto a different platform, for example, a custom miniature microarray or an RT-qPCR platform (i.e., TaqMan® low density arrays, Applied Biosystems, Carlsbad, CA, USA). However, this new platform would need to be internally and externally validated; therefore there is an argument for performing this step earlier in the process of developing a diagnostic gene expression classifier (Spijker *et al.*, 2010; Tsuang *et al.*, 2005). All of the studies using blood gene expression to diagnose neuropsychiatric disorders presented in Table I should be considered pilot studies that have demonstrated feasibility. It will be extremely exciting to see internal and external validation of these studies and the emergence of diagnostic gene expression classifiers for neuropsychiatric disorders in the clinical setting.

IV. Blood Gene Expression Studies of Neuropsychiatric Disorders

A large number of studies have been performed that analyzed gene expression in blood cells derived from patients with neuropsychiatric disorders (Tables I and II). A subset of these studies have demonstrated the feasibility of using blood gene expression for diagnosing neuropsychiatric disorders and constructed classifiers of moderate to high accuracy (Table I). A potential obstacle to deriving a diagnostic gene expression classifier for neuropsychiatric disorders is the availability of drug-naive and drug-free patients. This does not appear to be a problem when analyzing data from PTSD cohorts but is a major confounder for schizophrenia studies. Blood gene expression studies have primarily focused on schizophrenia (9/16) where the majority of patients are often undergoing drug treatment or treatment status is not reported (Tables I and II). Therefore, when classifiers are constructed from such cohorts, it is unclear as to whether they are diagnosing disease status or the receipt of pharmacotherapy. In this respect, the studies of Kuzman *et al.* (2009) and Takahashi *et al.* (2010) warrant particular discussion. Kuzman *et al.* (2009) analyzed gene expression in whole blood from 32 treatment-naive patients presenting with their first psychotic episode suggestive of schizophrenia compared to the same number of age- and gender-matched control cases. Although a total of 180 probes were found to be differentially expressed between the psychosis and nonpsychosis control group, these probes were not used in conjunction with methods of class prediction to construct and evaluate the predictive accuracy of a diagnostic gene expression classifier. Further, since Kuzman *et al.* (2009) have not deposited their microarray gene expression data in a public repository, it is not possible to assess the accuracy with which these genes can distinguish psychosis cases from controls.

Takahashi *et al.* (2010) also analyzed gene expression in whole blood from 52 patients with schizophrenia or schizophreniform disorder recruited across six outpatient clinics in Japan and 49 age-matched controls. Although 44 of these patients were drug-naive or drug-free, eight patients were identified as antipsychotics-naive, as they were taking antidepressants, benzodiazepines, or mood stabilizers for prodromal symptoms. The samples were divided into a training data set (35 schizophrenia and 33 controls) and an independent test data set (17 schizophrenia and 16 controls) for hold-out validation. ANNs and forward feature selection were used to identify 14 probes whose expression in whole blood could discriminate schizophrenia cases from controls in the training data with 91.2% accuracy as assessed by threefold cross-validation. The 14 probes in this diagnostic gene expression classifier consisted of eight genes (*CINP, DAOA, INSL3, LIPH, MAP1D, NAF1, PGRMC1*, and *TDRD9*) and six expressed sequence tags. This diagnostic gene expression classifier was capable of correctly predicting 14/17 patients and 15/16 controls in the hold-out test data set resulting in a classification accuracy of 87.9%. This pilot study clearly demonstrates the feasibility of using blood gene expression to diagnosis schizophrenia; however, it still suffers from a number of limitations. First, it does not appear that a globin reduction procedure was employed when analyzing gene expression in whole blood. Second, Takahashi *et al.* (2010) stated that gender was significantly different between their schizophrenia and control groups in both the training ($p = 0.014$) and the test data sets ($p = 0.049$) as measured by a chi-square test. Although attempts were made to filter out gender-related genes, the authors stated that this study should be repeated in a gender-matched cohort to ensure that the disease itself is being diagnosed by the classifier (Takahashi *et al.*, 2010). Third, this study could be improved by the inclusion of schizophrenia-related neuropsychiatric disorders (e.g., bipolar disorder) into the control group to confirm the specificity of the classifier for schizophrenia. Finally, the diagnostic accuracy of this 14-probe classifier should be validated in a cohort outside Japan. The study of Kuzman *et al.* (2009), which analyzed schizophrenia patients in Croatia, would be well-suited to the external validation of this Japanese study if the microarray data from both studies were publicly available.

The maximization of sample size increases discriminatory and inferential power, which is essential for any biomarker study of complex and heterogeneous psychiatric disorders. As such, some groups have attempted to exploit the vast stocks of blood-derived lymphoblastoid cell lines (LCLs) that have been established from psychiatric patients over the past several decades as a potential source of biomarkers. For several reasons, though, this approach is not entirely advisable. Principally, this recommendation is based on scientific grounds, or more precisely, the lack of scientific evidence supporting the suitability of LCLs for some types of biomarker analysis. Clearly, blood cells immortalized by transformation with Epstein-Barr virus produce faithful copies of DNA suitable for genomic analysis—this is the premise on which such LCL repositories are founded—however, there is less evidence that the epigenomic and transcriptomic patterns of LCLs faithfully recapitulate profiles seen in fresh blood cells. Further, as LCLs are necessarily far removed from the time of blood draw and the corresponding clinical condition of the patient, LCL gene expression may have biomarker potential only in reflecting biological "traits" and not "states." On average, 31% of the variance in a gene's expression level in LCLs is heritable (with some being almost completely heritable and others entirely nonheritable;

McRae *et al.*, 2007). Thus, gene expression in LCLs may reflect, to some extent, the effects of regulatory polymorphisms and can therefore be examined as a partially heritable trait (but not state) in biomarker studies. Yet, Choy *et al.* (2008) have shown that gene expression in LCLs exhibits significant variability from day to day, as well as considerable susceptibility to nongenetic confounders, such as baseline growth rates, the metabolic state in culture, transformation efficiency, and biological noise (Choy *et al.*, 2008). Studies of gene expression in freshly drawn blood samples are certainly not immune to potential confounds, but most of those can be controlled through careful study planning and ascertainment, or detected and accounted for in statistical models.

An excellent and illuminating study by Min *et al.* (2010) was devised to establish the degree of comparability of gene expression in freshly drawn and transformed blood cell samples using multiple preparation protocols. Unfortunately, the results indicated that gene expression signatures differed broadly between each preparation, suggesting that the transcriptomic signature of LCLs is not a suitable proxy for that derived from fresh blood cell samples. We characterize this result as unfortunate, since we understand the desirability of utilizing our precious existing resources to maximize the output of biomarker discovery effort. Yet, the transcriptomic and regulatory disparities between fresh and transformed blood samples may, in part, explain why Matigian *et al.* (2008) found less than promising results in their study of schizophrenia, with no genes showing even nominally significant dysregulation at a twofold change in the patient group relative to a matched-control group. With an eye toward the future and anticipating the desirability of immediate laboratory-based identification of biomarkers in freshly drawn blood samples, coupled with the limitations of LCLs identified above, we recommend that biomarker samples should be obtained anew from patients at the time they are studied, and the precise clinical state of the patient at that time should be noted. This practice will ensure that putative biomarker discoveries may generalize more quickly to eventual clinical practice.

V. MicroRNA Expression Analysis

The expression of microRNAs (miRNAs) in blood cells was first identified as a valuable diagnostic and prognostic tool in the field of oncology (Mitchell *et al.*, 2008). miRNAs are a class of small (22 nucleotides) noncoding RNAs that negatively regulate gene expression at the posttranscriptional level by degrading or repressing target mRNAs (Chen, 2005). The human genome appears to code for over 1000 miRNAs (Griffiths-Jones *et al.*, 2006) which have been detected by RT-qPCR (e.g., TaqMan® Array MicroRNA Cards, Applied Biosystems, or microRNA Ready-to-Use PCR Panels, Exiqon, Woburn, MA, USA), microarray (e.g., miRCURY LNA™ microRNA Arrays, Exiqon), or next-generation sequencing (miRNA-Seq) platforms. miRNAs are a valuable source of biomarkers since they regulate a large number of mRNA targets (~30% of transcriptome) and hence a large number of biological pathways (Lewis *et al.*, 2005; Mack, 2007). Further, miRNAs are relatively resistant to RNAse degradation and thus stable in body fluids such as whole blood (Hausler *et al.*, 2010), serum (Gilad *et al.*, 2008; Mitchell *et al.*, 2008), cerebrospinal fluid (Cogswell *et al.*, 2008), and urine (Hanke *et al.*, 2010).

Despite the many appeals of analyzing miRNA expression in blood cells, this data source has not been extensively mined with respect to constructing diagnostic classifiers for neuropsychiatric disorders. Lai *et al.* (2011) recently identified a diagnostic classifier for schizophrenia consisting of 7 miRNAs that was able to discriminate cases from controls in a hold-out validation test set with accuracies as high as 80%. The majority of miRNA expression studies have focused on brain tissue where miRNAs have been shown to play a role in the mechanism of neuropsychiatric and neurological disorders including schizophrenia (Beveridge *et al.*, 2008; Perkins *et al.*, 2007), MDD (Tatro *et al.*, 2010), Parkinson's disease (Kim *et al.*, 2007), and Alzheimer's disease (Wang *et al.*, 2008). Recently, Cox *et al.* (2010) used the Illumina Human v1 MicroRNA Expression BeadChip to identify miRNAs differentially expressed in the whole blood between 59 multiple sclerosis patients and 37 healthy controls. Although a classifier was not constructed, this study did demonstrate that 26 miRNAs were decreased and 1 increased in the multiple sclerosis group compared to the control group of which miR-17 and miR-20a were confirmed by RT-qPCR. Multiple sclerosis is an immune-mediated neurodegenerative disease and this study from Cox *et al.* (2010) further indicates that miRNA expression classifiers may have great utility for diagnosing neuropsychiatric disorders. Therefore, it is recommended that any future neuropsychiatric studies analyzing gene expression in the blood use RNA isolation protocols that preserve the miRNA fraction (e.g., miRNeasy Mini Kit, Qiagen, Valencia, CA, USA).

VI. Pharmacogenomics

Pharmacogenomics uses genome-wide approaches to determine the inherited basis of differences between individuals in their response to drugs (Evans and McLeod, 2003). Although a full description of pharmacogenomic approaches to neuropsychiatric disorders is described in the chapter indicated above, it should be noted that blood gene expression could also be used to develop prognostic classifiers capable of predicting a patient's response to a particular drug treatment (Burczynski and Dorner, 2006; Woelk *et al.*, 2010). For example, gene expression in blood cells has already been used to predict treatment outcomes for glucocorticoid treatment of asthma (Hakonarson *et al.*, 2005), interferon-α and ribivarin treatment of hepatitis C virus (Lempicki *et al.*, 2006; Tateno *et al.*, 2007), angiogenesis inhibitor SU5416 treatment of colorectal cancer (DePrimo *et al.*, 2003), interferon-β treatment of multiple sclerosis (Baranzini *et al.*, 2005; van Baarsen *et al.*, 2008), and Torisel (CCI-779) treatment of renal cell carcinoma (RCC; Burczynski *et al.*, 2005). Specifically, Burczynski *et al.* (2005) identified 30 genes in PBMCs whose expression could be used to predict good versus poor outcome (i.e., time to progression) for RCC patients when subjected to Torisel treatment with 74% accuracy as assessed by LOOCV in the training data, and 85% accuracy in a test data set used for hold-out validation.

Neuropsychiatric disorders represent important targets for pharmacogenomic studies. In schizophrenia, for example, 25–40% of patients fail to respond to their first treatment option and 10–20% of those that do respond react with adverse effects of clinical importance (Broich and Moller, 2008). Pharmacological studies of neuropsychiatric disorders have primarily focused on associating single nucleotide polymorphisms (SNPs) with treatment outcome. The majority of these studies have used a candidate gene approach and are thus pharmacogenetic in scope rather than whole genome-based pharmacogenomic analyses (Alenius *et al.*, 2008; Kato *et al.*, 2008, 2009; Kwon *et al.*, 2009; Need *et al.*, 2009; Xu *et al.*, 2010; Yasui-Furukori *et al.*, 2006). Recently, genome-wide association studies (GWAS) have been used in true pharmacogenomic approaches to identify SNPs that could predict responses in schizophrenia patients when treated with antipsychotics (McClay *et al.*, 2011) and in patients suffering from depression when treated with antidepressants (Ising *et al.*, 2009). In addition to these GWAS approaches, the utility of gene expression in the blood for predicting treatment outcomes and adverse effects in neuropsychiatric patients should now be determined.

VII. Concluding Remarks

Constructing classifiers from gene and miRNA expression in whole blood holds promise for developing a diagnostic tool with clinical utility for neuropsychiatric disorders. Every effort should now be made to optimize the experimental design when attempting to construct diagnostic gene expression classifiers from the blood compartment. This includes identifying drug-naive or drug-free patients for which all variables except disease status are balanced between case and control groups. In addition, the specificity of the resulting diagnostic classifier can be established if the control group contains other neuropsychiatric disorders distinct from the primary disorder under study (Scherzer *et al.*, 2007). Finally, when RNA is isolated from whole blood it is recommended that a globin reduction step be performed (Liu *et al.*, 2006; Winn *et al.*, 2010).

It is unlikely that the ultimate diagnostic classifier for a particular neuropsychiatric disorder will rely solely on gene or miRNA expression data. The most accurate classifier will probably incorporate a whole range of potentially diagnostic criteria that may include any of the following: demographic variables, copy number variation data, methylation patterns, SNPs, and protein expression. With respect to the latter criteria, Schwarz *et al.* (2010) recently developed a diagnostic protein expression classifier for schizophrenia with a classification accuracy of 77% for drug-naive patients. This study is fully reviewed in Chapters "The application of multiplexed assay systems for molecular diagnostics" by Schwarz *et al.* and "Algorithm development for diagnostic biomarker assays" by Izmailov *et al.*

Genomics technologies are rapidly evolving. Soon, gene and miRNA expression analysis using microarrays will be replaced by next-generation sequencing technologies. These technologies include Illumina's HiSeq and Genome Analyzer platforms, Applied Biosystems' SOLiD™ system, and 454 Life Sciences' Genome Sequencer FLX system, which are used to perform RNA-Seq and miRNA-Seq for the assessment of gene and miRNA expression, respectively. RNA-Seq analysis of gene expression has several advantages for constructing diagnostic gene expression classifiers compared to microarray analysis including a greater dynamic range, the sequencing of SNPs in coding regions, and the identification of novel coding regions and alternatively spliced transcripts (Wang *et al.*, 2009). As the cost of sequencing continues to diminish and the software tools for analyzing next-generation sequencing become more available (Oshlack *et al.*, 2010), the possibility of constructing more accurate diagnostic gene expression classifiers for neuropsychiatric disorders will increase.

Acknowledgments

This work was performed with the support from "NARSAD: The Brain and Behavior Research Fund," in the form of a Young Investigator Award, and the Sidney R. Baer, Jr. Prize (S.J.G.) and Lieber Prize Award (M.T.T.) for Schizophrenia Research. Additional support was derived from the Genomics Core at the UCSD CFAR (AI36214), the San Diego Veterans Medical Research Foundation, and other research grants from the National Institutes of Health (MH081755, MH085240, AI087164). This chapter is based upon work supported in part by the Department of Veterans Affairs (VA), Veterans Health Administration, Office of Research and Development. The views expressed in this chapter are those of the authors and do not necessarily reflect the position or policy of the Department of

Veterans Affairs or the United States Government. This chapter is dedicated to the memory of the first author's grandmother, Violet Rozel Woelk née Werner (1920–2005), who was a long-time sufferer of a neuropsychiatric disorder.

References

Alenius, M., Wadelius, M., Dahl, M.L., Hartvig, P., Lindstrom, L., and Hammarlund-Udenaes, M. (2008). Gene polymorphism influencing treatment response in psychotic patients in a naturalistic setting. *J. Psychiatr. Res.* **42**, 884–893.

Al-Shahrour, F., Carbonell, J., Minguez, P., Goetz, S., Conesa, A., Tarraga, J., Medina, I., Alloza, E., Montaner, D., and Dopazo, J. (2008). Babelomics: advanced functional profiling of transcriptomics, proteomics and genomics experiments. *Nucleic Acids Res.* **36**, W341–W346.

Baranzini, S.E., Mousavi, P., Rio, J., Caillier, S.J., Stillman, A., Villoslada, P., Wyatt, M.M., Comabella, M., Greller, L.D., Somogyi, R., Montalban, X., and Oksenberg, J.R. (2005). Transcription-based prediction of response to IFNbeta using supervised computational methods. *PLoS Biol.* **3**, e2.

Barnes, M., Freudenberg, J., Thompson, S., Aronow, B., and Pavlidis, P. (2005). Experimental comparison and cross-validation of the Affymetrix and Illumina gene expression analysis platforms. *Nucleic Acids Res.* **33**, 5914–5923.

Beveridge, N.J., Tooney, P.A., Carroll, A.P., Gardiner, E., Bowden, N., Scott, R.J., Tran, N., Dedova, I., and Cairns, M.J. (2008). Dysregulation of miRNA 181b in the temporal cortex in schizophrenia. *Hum. Mol. Genet.* **17**, 1156–1168.

Bolstad, B.M., Collin, F., Brettschneider, J., Simpson, K., Cope, L., Irizarry, R., and Speed, T.P. (2005). Quality assessment of Affymetrix GeneChip data. In: R. Gentleman, R., V. Carey, V., W. Huber, W., R. Irizarry, R., and S. Dutoit, S. (Eds.), *Bioinformatics and Computational Biology Solutions Using R and Bioconductor.* Springer, Heidelberg, pp. 33–47.

Bousman, C.A., Chana, G., Glatt, S.J., Chandler, S.D., Lucero, G.R., Tatro, E., May, T., Lohr, J.B., Kremen, W.S., Tsuang, M.T., and Everall, I.P. (2010). Preliminary evidence of ubiquitin proteasome system dysregulation in schizophrenia and bipolar disorder: convergent pathway analysis findings from two independent samples. *Am. J. Med. Genet. B Neuropsychiatr. Genet.* **153B**, 494–502.

Bowden, N.A., Scott, R.J., and Tooney, P.A. (2008). Altered gene expression in the superior temporal gyrus in schizophrenia. *BMC Genomics* **9**, 199.

Broich, K., and Moller, H.J. (2008). Pharmacogenetics, pharmacogenomics and personalized psychiatry: are we there yet? *Eur. Arch. Psychiatry Clin. Neurosci.* **258**(Suppl. 1), 1–2.

Burczynski, M.E., and Dorner, A.J. (2006). Transcriptional profiling of peripheral blood cells in clinical pharmacogenomic studies. *Pharmacogenomics* **7**, 187–202.

Burczynski, M.E., Twine, N.C., Dukart, G., Marshall, B., Hidalgo, M., Stadler, W.M., Logan, T., Dutcher, J., Hudes, G., Trepicchio, W.L., Strahs, A., Immermann, F., *et al.* (2005). Transcriptional profiles in peripheral blood mononuclear cells prognostic of clinical outcomes in patients with advanced renal cell carcinoma. *Clin. Cancer Res.* **11**, 1181–1189.

Cardoso, F., Van't Veer, L., Rutgers, E., Loi, S., Mook, S., and Piccart-Gebhart, M.J. (2008). Clinical application of the 70-gene profile: the MINDACT trial. *J. Clin. Oncol.* **26**, 729–735.

Chen, C.Z. (2005). MicroRNAs as oncogenes and tumor suppressors. *N. Engl. J. Med.* **353**, 1768–1771.

Choy, E., Yelensky, R., Bonakdar, S., Plenge, R.M., Saxena, R., De Jager, P.L., Shaw, S.Y., Wolfish, C.S., Slavik, J.M., Cotsapas, C., Rivas, M., Dermitzakis, E.T., *et al.* (2008). Genetic analysis of human traits in vitro: drug response and gene expression in lymphoblastoid cell lines. *PLoS Genet.* **4**, e1000287.

Cogswell, J.P., Ward, J., Taylor, I.A., Waters, M., Shi, Y., Cannon, B., Kelnar, K., Kemppainen, J., Brown, D., Chen, C., Prinjha, R.K., Richardson, J.C., *et al.* (2008). Identification of miRNA changes in Alzheimer's disease brain and CSF yields putative biomarkers and insights into disease pathways. *J. Alzheimers Dis.* **14**, 27–41.

Cox, M.B., Cairns, M.J., Gandhi, K.S., Carroll, A.P., Moscovis, S., Stewart, G.J., Broadley, S., Scott, R.J., Booth, D.R., and Lechner-Scott, J. (2010). MicroRNAs miR-17 and miR-20a inhibit T cell activation genes and are under-expressed in MS whole blood. *PLoS One* **5**, e12132.

DePrimo, S.E., Wong, L.M., Khatry, D.B., Nicholas, S.L., Manning, W.C., Smolich, B.D., O'Farrell, A.M., and Cherrington, J.M. (2003). Expression profiling of blood samples from an SU5416 Phase III metastatic colorectal cancer clinical trial: a novel strategy for biomarker identification. *BMC Cancer* **3**, 3.

Dobbin, K.K., Zhao, Y., and Simon, R.M. (2008). How large a training set is needed to develop a classifier for microarray data? *Clin. Cancer Res.* **14**, 108–114.

Du, P., Kibbe, W.A., and Lin, S.M. (2008). lumi: a pipeline for processing Illumina microarray. *Bioinformatics* **24**, 1547–1548.

Dupuy, A., and Simon, R.M. (2007). Critical review of published microarray studies for cancer outcome and guidelines on statistical analysis and reporting. *J. Natl. Cancer Inst.* **99**, 147–157.

Eisen, M.B., Spellman, P.T., Brown, P.O., and Botstein, D. (1998). Cluster analysis and display of genome-wide expression patterns. *Proc. Natl. Acad. Sci. USA* **95**, 14863–14868.

Evans, W.E., and McLeod, H.L. (2003). Pharmacogenomics—drug disposition, drug targets, and side effects. *N. Engl. J. Med.* **348**, 538–549.

Field, L.A., Jordan, R.M., Hadix, J.A., Dunn, M.A., Shriver, C.D., Ellsworth, R.E., and Ellsworth, D.L. (2007). Functional identity of genes detectable in expression profiling assays following globin mRNA reduction of peripheral blood samples. *Clin. Biochem.* **40**, 499–502.

Garbett, K., Gal-Chis, R., Gaszner, G., Lewis, D.A., and Mirnics, K. (2008). Transcriptome alterations in the prefrontal cortex of subjects with schizophrenia who committed suicide. *Neuropsychopharmacol. Hung.* **10**, 9–14.

Gilad, S., Meiri, E., Yogev, Y., Benjamin, S., Lebanony, D., Yerushalmi, N., Benjamin, H., Kushnir, M., Cholakh, H., Melamed, N., Bentwich, Z., Hod, M., *et al.* (2008). Serum microRNAs are promising novel biomarkers. *PLoS One* **3**, e3148.

Glatt, S.J., Everall, I.P., Kremen, W.S., Corbeil, J., Sasik, R., Khanlou, N., Han, M., Liew, C.C., and Tsuang, M.T. (2005). Comparative gene expression analysis of blood and brain provides concurrent validation of SELENBP1 up-regulation in schizophrenia. *Proc. Natl. Acad. Sci. USA* **102**, 15533–15538.

Glatt, S.J., Faraone, S.V., and Tsuang, M.T. (2010). Mental health etiology: biological and genetic determinants. In: V. Patel, V., A. Woodward, A., V.L. Feigin, V.L., H.K. Heggenhougen, H.K., and S. Quah, S. (Eds.), *Mental and Neurological Public Health: A Global Perspective.* Elsevier Publishing Services, Chennai, India, pp. 107–113.

Griffiths-Jones, S., Grocock, R.J., van Dongen, S., Bateman, A., and Enright, A.J. (2006). miRBase: microRNA sequences, targets and gene nomenclature. *Nucleic Acids Res.* **34**, D140–D144.

Hakonarson, H., Bjornsdottir, U.S., Halapi, E., Bradfield, J., Zink, F., Mouy, M., Helgadottir, H., Gudmundsdottir, A.S., Andrason, H., Adalsteinsdottir, A.E., Kristjansson, K., Birkisson, I., *et al.* (2005). Profiling of genes expressed in peripheral blood mononuclear cells predicts glucocorticoid sensitivity in asthma patients. *Proc. Natl. Acad. Sci. USA* **102**, 14789–14794.

Hanke, M., Hoefig, K., Merz, H., Feller, A.C., Kausch, I., Jocham, D., Warnecke, J.M., and Sczakiel, G. (2010). A robust methodology to study urine microRNA as tumor marker: microRNA-126 and microRNA-182 are related to urinary bladder cancer. *Urol. Oncol.* **28**, 655–661.

Hausler, S.F., Keller, A., Chandran, P.A., Ziegler, K., Zipp, K., Heuer, S., Krockenberger, M., Engel, J.B., Honig, A., Scheffler, M., Dietl, J., and Wischhusen, J. (2010). Whole blood-derived miRNA profiles as potential new tools for ovarian cancer screening. *Br. J. Cancer* **103**, 693–700.

Ising, M., Lucae, S., Binder, E.B., Bettecken, T., Uhr, M., Ripke, S., Kohli, M.A., Hennings, J.M., Horstmann, S., Kloiber, S., Menke, A., Bondy, B., *et al.* (2009). A genomewide association study points to multiple loci that predict antidepressant drug treatment outcome in depression. *Arch. Gen. Psychiatry* **66**, 966–975.

Johnson, W.E., Li, C., and Rabinovic, A. (2007). Adjusting batch effects in microarray expression data using empirical Bayes methods. *Biostatistics* **8**, 118–127.

Kanazawa, T., Chana, G., Glatt, S.J., Mizuno, H., Masliah, E., Yoneda, H., Tsuang, M.T., and Everall, I.P. (2008). The utility of SELENBP1 gene expression as a biomarker for major psychotic disorders: replication in schizophrenia and extension to bipolar disorder with psychosis. *Am. J. Med. Genet. B Neuropsychiatr. Genet.* **147B**, 686–689.

Kato, K. (2009). Algorithm for in vitro diagnostic multivariate index assay. *Breast Cancer* **16**, 248–251.

Kato, M., Fukuda, T., Serretti, A., Wakeno, M., Okugawa, G., Ikenaga, Y., Hosoi, Y., Takekita, Y., Mandelli, L., Azuma, J., and Kinoshita, T. (2008). ABCB1 (MDR1) gene polymorphisms are associated with the clinical response to paroxetine in patients with major depressive disorder. *Prog. Neuropsychopharmacol. Biol. Psychiatry* **32**, 398–404.

Kato, M., Fukuda, T., Wakeno, M., Okugawa, G., Takekita, Y., Watanabe, S., Yamashita, M., Hosoi, Y., Azuma, J., Kinoshita, T., and Serretti, A. (2009). Effect of 5-HT1A gene polymorphisms on antidepressant response in major depressive disorder. *Am. J. Med. Genet. B Neuropsychiatr. Genet.* **150B**, 115–123.

Kim, J., Inoue, K., Ishii, J., Vanti, W.B., Voronov, S.V., Murchison, E., Hannon, G., and Abeliovich, A. (2007). A MicroRNA feedback circuit in midbrain dopamine neurons. *Science* **317**, 1220–1224.

Kuzman, M.R., Medved, V., Terzic, J., and Krainc, D. (2009). Genome-wide expression analysis of peripheral blood identifies candidate biomarkers for schizophrenia. *J. Psychiatr. Res.* **43**, 1073–1077.

Kwon, J.S., Joo, Y.H., Nam, H.J., Lim, M., Cho, E.Y., Jung, M.H., Choi, J.S., Kim, B., Kang, D.H., Oh, S., Park, T., and Hong, K.S. (2009). Association of the glutamate transporter gene SLC1A1 with atypical antipsychotics-induced obsessive-compulsive symptoms. *Arch. Gen. Psychiatry* **66**, 1233–1241.

Lai, C.Y., Yu, S.L., Hsieh, M.H., Chen, C.H., Chen, H.Y., Wen, C.C., Huang, Y.H., Hsiao, P.C., Hsiao, C.K., Liu, C.M., Yang, P.C., Hwu, H.G., *et al.* (2011). MicroRNA expression aberration as potential peripheral blood biomarkers for schizophrenia. *PLoS One* **6**(6), e21635.

Lempicki, R.A., Polis, M.A., Yang, J., McLaughlin, M., Koratich, C., Huang, D.W., Fullmer, B., Wu, L., Rehm, C.A., Masur, H., Lane, H.C., Sherman, K.E., *et al.* (2006). Gene expression profiles in Hepatitis C Virus (HCV) and HIV coinfection: class prediction analyses before treatment predict the outcome of Anti-HCV therapy among HIV-coinfected persons. *J. Infect. Dis.* **193**, 1172–1177.

Le-Niculescu, H., Kurian, S.M., Yehyawi, N., Dike, C., Patel, S.D., Edenberg, H.J., Tsuang, M.T., Salomon, D.R., Nurnberger, J.I. Jr., and Niculescu, A.B. (2009). Identifying blood biomarkers for mood disorders using convergent functional genomics. *Mol. Psychiatry* **14**, 156–174.

Lewis, B.P., Burge, C.B., and Bartel, D.P. (2005). Conserved seed pairing, often flanked by adenosines, indicates that thousands of human genes are microRNA targets. *Cell* **120**, 15–20.

Liu, J., Walter, E., Stenger, D., and Thach, D. (2006). Effects of globin mRNA reduction methods on gene expression profiles from whole blood. *J. Mol. Diagn.* **8**, 551–558.

Mack, G.S. (2007). MicroRNA gets down to business. *Nat. Biotechnol.* **25**, 631–638.

Matigian, N.A., McCurdy, R.D., Feron, F., Perry, C., Smith, H., Filippich, C., McLean, D., McGrath, J., Mackay-Sim, A., Mowry, B., and Hayward, N.K. (2008). Fibroblast and lymphoblast gene expression profiles in schizophrenia: are non-neural cells informative? *PLoS One* **3**, e2412.

Maycox, P.R., Kelly, F., Taylor, A., Bates, S., Reid, J., Logendra, R., Barnes, M.R., Larminie, C., Jones, N., Lennon, M., Davies, C., Hagan, J.J., *et al.* (2009). Analysis of gene expression in two large schizophrenia cohorts identifies multiple changes associated with nerve terminal function. *Mol. Psychiatry* **14**, 1083–1094.

McClay, J.L., Adkins, D.E., Aberg, K., Bukszar, J., Khachane, A.N., Keefe, R.S., Perkins, D.O., McEvoy, J.P., Stroup, T.S., Vann, R.E., Beardsley, P.M., Lieberman, J.A., *et al.* (2011). Genome-wide pharmacogenomic study of neurocognition as an indicator of antipsychotic treatment response in schizophrenia. *Neuropsychopharmacology* **36**, 616–626.

McRae, A.F., Matigian, N.A., Vadlamudi, L., Mulley, J.C., Mowry, B., Martin, N.G., Berkovic, S.F., Hayward, N.K., and Visscher, P.M. (2007). Replicated effects of sex and genotype on gene expression in human lymphoblastoid cell lines. *Hum. Mol. Genet.* **16**, 364–373.

Middleton, F.A., Pato, C.N., Gentile, K.L., McGann, L., Brown, A.M., Trauzzi, M., Diab, H., Morley, C.P., Medeiros, H., Macedo, A., Azevedo, M.H., and Pato, M.T. (2005). Gene expression analysis of peripheral blood leukocytes from discordant sib-pairs with schizophrenia and bipolar disorder reveals points of convergence between genetic and functional genomic approaches. *Am. J. Med. Genet. B Neuropsychiatr. Genet.* **136B**, 12–25.

Min, J.L., Barrett, A., Watts, T., Pettersson, F.H., Lockstone, H.E., Lindgren, C.M., Taylor, J.M., Allen, M., Zondervan, K.T., and McCarthy, M.I. (2010). Variability of gene expression profiles in human blood and lymphoblastoid cell lines. *BMC Genomics* **11**, 96.

Mitchell, P.S., Parkin, R.K., Kroh, E.M., Fritz, B.R., Wyman, S.K., Pogosova-Agadjanyan, E.L., Peterson, A., Noteboom, J., O'Briant, K.C., Allen, A., Lin, D.W., Urban, N., *et al.* (2008). Circulating microRNAs as stable blood-based markers for cancer detection. *Proc. Natl. Acad. Sci. USA* **105**, 10513–10518.

Narayan, S., Tang, B., Head, S.R., Gilmartin, T.J., Sutcliffe, J.G., Dean, B., and Thomas, E.A. (2008). Molecular profiles of schizophrenia in the CNS at different stages of illness. *Brain Res.* **1239**, 235–248.

Need, A.C., Keefe, R.S., Ge, D., Grossman, I., Dickson, S., McEvoy, J.P., and Goldstein, D.B. (2009). Pharmacogenetics of antipsychotic response in the CATIE trial: a candidate gene analysis. *Eur. J. Hum. Genet.* **17**, 946–957.

Neylan, T.C., Sun, B., Rempel, H., Ross, J., Lenoci, M., O'Donovan, A., and Pulliam, L. (2010). Suppressed monocyte gene expression profile in men versus women with PTSD. *Brain Behav. Immun.* **25**, 524–531.

Oshlack, A., Robinson, M.D., and Young, M.D. (2010). From RNA-seq reads to differential expression results. *Genome Biol.* **11**, 220.

Perkins, D.O., Jeffries, C.D., Jarskog, L.F., Thomson, J.M., Woods, K., Newman, M.A., Parker, J.S., Jin, J., and Hammond, S.M. (2007). microRNA expression in the prefrontal cortex of individuals with schizophrenia and schizoaffective disorder. *Genome Biol.* **8**, R27.

Prabakaran, S., Swatton, J.E., Ryan, M.M., Huffaker, S.J., Huang, J.T., Griffin, J.L., Wayland, M., Freeman, T., Dudbridge, F., Lilley, K.S., Karp, N.A., Hester, S., *et al.* (2004). Mitochondrial dysfunction in schizophrenia: evidence for compromised brain metabolism and oxidative stress. *Mol. Psychiatry* **9**(684–697), 643.

Ruschhaupt, M., Huber, W., Poustka, A., and Mansmann, U. (2004). A compendium to ensure computational reproducibility in high-dimensional classification tasks. *Stat. Appl. Genet. Mol. Biol.* **3,** Article 37.

Scherzer, C.R., Eklund, A.C., Morse, L.J., Liao, Z., Locascio, J.J., Fefer, D., Schwarzschild, M.A., Schlossmacher, M.G., Hauser, M.A., Vance, J.M., Sudarsky, L.R., Standaert, D.G., *et al.* (2007). Molecular markers of early Parkinson's disease based on gene expression in blood. *Proc. Natl. Acad. Sci. USA* **104**, 955–960.

Schroeder, A., Mueller, O., Stocker, S., Salowsky, R., Leiber, M., Gassmann, M., Lightfoot, S., Menzel, W., Granzow, M., and Ragg, T. (2006). The RIN: an RNA integrity number for assigning integrity values to RNA measurements. *BMC Mol. Biol.* **7**, 3.

Schwarz, E., Izmailov, R., Spain, M., Barnes, A., Mapes, J.P., Guest, P.C., Rahmoune, H., Pietsch, S., Leweke, F.M., Rothermundt, M., Steiner, J., Koethe, D., *et al.* (2010). Validation of a blood-based laboratory test to aid in the confirmation of a diagnosis of schizophrenia. *Biomark. Insights* **5**, 39–47.

Segman, R.H., Shefi, N., Goltser-Dubner, T., Friedman, N., Kaminski, N., and Shalev, A.Y. (2005). Peripheral blood mononuclear cell gene expression profiles identify emergent post-traumatic stress disorder among trauma survivors. *Mol. Psychiatry* **10**, 500–513, p. 425.

Shehadeh, L.A., Yu, K., Wang, L., Guevara, A., Singer, C., Vance, J., and Papapetropoulos, S. (2010). SRRM2, a potential blood biomarker revealing high alternative splicing in Parkinson's disease. *PLoS One* **5**, e9104.

Shi, L., Reid, L.H., Jones, W.D., Shippy, R., Warrington, J.A., Baker, S.C., Collins, P.J., de Longueville, F., Kawasaki, E.S., Lee, K.Y., Luo, Y., Sun, Y.A., *et al.* (2006). The MicroArray Quality Control (MAQC) project shows inter- and intraplatform reproducibility of gene expression measurements. *Nat. Biotechnol.* **24**, 1151–1161.

Simon, R.The BRB-Array Tools Development TeamThe EMMES Corporation (2009). BRB-Array Tools Version 3.8 User's Manual. Rockville, MD, USA.

Simon, R., Lam, A., Li, M.C., Ngan, M., Menenzes, S., and Zhao, Y. (2007). Analysis of gene expression data using BRB-ArrayTools. *Cancer Inform.* **3**, 11–17.

Spijker, S., Van Zanten, J.S., De Jong, S., Penninx, B.W., van Dyck, R., Zitman, F.G., Smit, J.H., Ylstra, B., Smit, A.B., and Hoogendijk, W.J. (2010). Stimulated gene expression profiles as a blood marker of major depressive disorder. *Biol. Psychiatry* **68**, 179–186.

Struyf, J., Dobrin, S., and Page, D. (2008). Combining gene expression, demographic and clinical data in modeling disease: a case study of bipolar disorder and schizophrenia. *BMC Genomics* **9**, 531.

Sullivan, P.F., Fan, C., and Perou, C.M. (2006). Evaluating the comparability of gene expression in blood and brain. *Am. J. Med. Genet. B Neuropsychiatr. Genet.* **141B**, 261–268.

Takahashi, I., Miyaji, H., Yoshida, T., Sato, S., and Mizukami, T. (1996). Selective inhibition of IL-2 gene expression by trichostatin A, a potent inhibitor of mammalian histone deacetylase. *J. Antibiot. (Tokyo)* **49**, 453–457.

Takahashi, M., Hayashi, H., Watanabe, Y., Sawamura, K., Fukui, N., Watanabe, J., Kitajima, T., Yamanouchi, Y., Iwata, N., Mizukami, K., Hori, T., Shimoda, K., *et al.* (2010). Diagnostic classification of schizophrenia by neural network analysis of blood-based gene expression signatures. *Schizophr. Res.* **119**, 210–218.

Tateno, M., Honda, M., Kawamura, T., Honda, H., and Kaneko, S. (2007). Expression profiling of peripheral-blood mononuclear cells from patients with chronic hepatitis C undergoing interferon therapy. *J. Infect. Dis.* **195**, 255–267.

Tatro, E.T., Scott, E.R., Nguyen, T.B., Salaria, S., Banerjee, S., Moore, D.J., Masliah, E., Achim, C.L., and Everall, I.P. (2010). Evidence for Alteration of Gene Regulatory Networks through MicroRNAs of the HIV-infected brain: novel analysis of retrospective cases. *PLoS One* **5**, e10337.

Tsuang, M.T., Nossova, N., Yager, T., Tsuang, M.M., Guo, S.C., Shyu, K.G., Glatt, S.J., and Liew, C.C. (2005). Assessing the validity of blood-based gene expression profiles for the classification of schizophrenia and bipolar disorder: a preliminary report. *Am. J. Med. Genet. B Neuropsychiatr. Genet.* **133B**, 1–5.

van Baarsen, L.G., Vosslamber, S., Tijssen, M., Baggen, J.M., van der Voort, L.F., Killestein, J., van der Pouw Kraan, T.C., Polman, C.H., and Verweij, C.L. (2008). Pharmacogenomics of interferon-beta therapy in multiple sclerosis: baseline IFN signature determines pharmacological differences between patients. *PLoS One* **3**, e1927.

vant Veer, L.J., Dai, H., van de Vijver, M.J., He, Y.D., Hart, A.A., Mao, M., Peterse, H.L., van der Kooy, K., Marton, M.J., Witteveen, A.T., Schreiber, G.J., Kerkhoven, R.M., *et al.* (2002). Gene expression profiling predicts clinical outcome of breast cancer. *Nature* **415**, 530–536.

Vawter, M.P., Ferran, E., Galke, B., Cooper, K., Bunney, W.E., and Byerley, W. (2004). Microarray screening of lymphocyte gene expression differences in a multiplex schizophrenia pedigree. *Schizophr. Res.* **67**, 41–52.

Wang, W.X., Rajeev, B.W., Stromberg, A.J., Ren, N., Tang, G., Huang, Q., Rigoutsos, I., and Nelson, P.T. (2008). The expression of microRNA miR-107 decreases early in Alzheimer's disease and may accelerate disease progression through regulation of beta-site amyloid precursor protein-cleaving enzyme 1. *J. Neurosci.* **28**, 1213–1223.

Wang, Z., Gerstein, M., and Snyder, M. (2009). RNA-Seq: a revolutionary tool for transcriptomics. *Nat. Rev. Genet.* **10**, 57–63.

Winn, M.E., Zapala, M.A., Hovatta, I., Risbrough, V.B., Lillie, E., and Schork, N.J. (2010). The effects of globin on microarray-based gene expression analysis of mouse blood. *Mamm. Genome* **21**, 268–275.

Woelk, C.H., and Burczynski, M.E. (2008). The clinical relevance of gene expression profiles in peripheral blood mononuclear cells. In: F. Columbus, F. (Ed.), *Oligonucleotide Array Sequence Analysis.* NOVA Publishing, Hauppage, NY, pp. 38–50.

Woelk, C.H., Beliakova-Bethell, N., Goicoechea, M., Zhao, Y., Du, P., Rought, S.E., Lozach, J., Perez-Santiago, J., Richman, D.D., Smith, D.M., and Little, S.J. (2010). Gene expression before HAART initiation predicts HIV-infected individuals at risk of poor CD4+ T-cell recovery. *AIDS* **24**, 217–222.

Wu, Z., Irizarry, R.A., Gentleman, R., Martinez-Murillo, F., and Spencer, F. (2004). A model-based background adjustment for oligonucleotide expression arrays. *J. Am. Stat. Assoc.* **99**, 909–917.

Xu, M., Li, S., Xing, Q., Gao, R., Feng, G., Lin, Z., St Clair, D., and He, L. (2010). Genetic variants in the BDNF gene and therapeutic response to risperidone in schizophrenia patients: a pharmacogenetic study. *Eur. J. Hum. Genet.* **18**, 707–712.

Yasui-Furukori, N., Saito, M., Nakagami, T., Kaneda, A., Tateishi, T., and Kaneko, S. (2006). Association between multidrug resistance 1 (MDR1) gene polymorphisms and therapeutic response to bromperidol in schizophrenic patients: a preliminary study. *Prog. Neuropsychopharmacol. Biol. Psychiatry* **30**, 286–291.

Yehuda, R., Cai, G., Golier, J.A., Sarapas, C., Galea, S., Ising, M., Rein, T., Schmeidler, J., Muller-Myhsok, B., Holsboer, F., and Buxbaum, J.D. (2009). Gene expression patterns associated with posttraumatic stress disorder following exposure to the World Trade Center attacks. *Biol. Psychiatry* **66**, 708–711.

Zieker, J., Zieker, D., Jatzko, A., Dietzsch, J., Nieselt, K., Schmitt, A., Bertsch, T., Fassbender, K., Spanagel, R., Northoff, H., and Gebicke-Haerter, P.J. (2007). Differential gene expression in peripheral blood of patients suffering from post-traumatic stress disorder. *Mol. Psychiatry* **12**, 116–118.

Zvara, A., Szekeres, G., Janka, Z., Kelemen, J.Z., Cimmer, C., Santha, M., and Puskas, L.G. (2005). Over-expression of dopamine D2 receptor and inwardly rectifying potassium channel genes in drug-naive schizophrenic peripheral blood lymphocytes as potential diagnostic markers. *Dis. Markers* **21**, 61–69.

PROTEOMIC TECHNOLOGIES FOR BIOMARKER STUDIES IN PSYCHIATRY: ADVANCES AND NEEDS

Daniel Martins-de-Souza[1,2], Paul C. Guest[1], Natacha Vanattou-Saifoudine[1], Laura W. Harris[1] and Sabine Bahn[1,3]

[1]Department of Chemical Engineering and Biotechnology, University of Cambridge, Cambridge, United Kingdom
[2]Lab. de Neurociências (LIM-27), Inst. Psiquiatria, Fac. de Medicina da Universidade de Sao Paulo, Sao Paulo, Brazil
[3]Department of Neuroscience, Erasmus Medical Centre, Rotterdam, The Netherlands

Abstract

In the postgenome era, proteomics has arisen as a promising tool for more complete comprehension of diseases and for biomarker discovery. Some of these objectives have already been partly achieved for illnesses such as cancer.

INTERNATIONAL REVIEW OF NEUROBIOLOGY, VOL. 101
DOI: 10.1016/B978-0-12-387718-5.00004-3

In the case of psychiatric conditions, however, proteomic advances have had a less profound impact. Here, we outline the necessity of improving and applying proteomic methods for biomarker discovery and validation in the field of psychiatric disorders. While proteomic-based applications in neurosciences have increased in accuracy and sensitivity over the past 10 years, the development of orthogonal validation technologies has fallen behind. These issues are discussed along with the importance of integrating systems biology approaches and combining proteomics with other research approaches. The future development of such technologies may put proteomics closer to clinical applications in psychiatry.

ABBREVIATIONS

ELISA	enzyme-linked immunosorbent assay
ESI	electrospray ionization
MALDI	matrix-assisted laser desorption/ionization
SELDI	surface-enhanced laser desorption/ionization
TOF	time of flight

I. Introduction

Most likely when Patrick O'Farrell first presented two-dimensional gel electrophoresis (2DE) in 1975, he did not expect that this tool would revolutionize a field of study and even give rise to a new scientific approach (O'Farrell, 1975). It is also likely that Marc Wilkins did not expect the word "proteome," which he coined while working on the concept as a PhD student in 1994, to give a name to this approach which we know today as proteomics (Wilkins *et al.*, 1996). At the beginning of genome era, 2DE and its numerous optimizations has remained the most used method for comparative global proteome analyses especially when combined with mass spectrometry (MS) for protein identification. However, most of the data generated by 2DE in neuroscience research has been more useful for increasing our understanding of biological processes than for revealing clinically useful biomarkers. The necessity of improving proteomic methods for biomarker and target discovery and for the characterization and expression analysis of low-abundance proteins has led to the development of more sensitive and accurate MS-based approaches. The concept of shotgun proteomics was then launched in 1999 by Andrew Link, increasing the capacity for proteome characterization (Link *et al.*, 1999). Proteomics in psychiatric studies followed the tendency of delivering accurate and sensitive technologies for biomarker discovery but did not pay sufficient attention to validation technologies. Although the usefulness of Western blot (WB) and immunoassay methods are recognized, newer high-throughput and more sensitive tools are needed. Recently, a validation method termed selective reaction monitoring (SRM) has emerged which shows more promise in this area. Besides the requirement of new validation technologies, new systems biology approaches of merging proteomics with transcriptomic and metabolomic methods are also needed. Here, we present and discuss current proteomics technologies and the future steps that will allow this relatively new scientific methodology to advance the identification, validation, and clinical deployment of biomarkers in the challenging field of psychiatric disorders.

II. The Social Impact of Psychiatric Disorders

According to the World Health Organization (http://www.who.int), mental disorders account for 4 out of the 10 leading causes of disability in developed countries. Psychiatric disorders commonly appear during young adulthood, resulting in reduced workforce participation and lower economic living standards (Gibb *et al.*, 2010). This is aggravated by the fact that in most of the cases there are long delays between onset of symptoms and clinical intervention due to late or inaccurate diagnosis, which can lead to a more severe illness that is consequently more difficult to treat. This results most likely from the current practice that diagnosis and treatment of psychiatric conditions are dependent on the outcome of psychiatric interviews, which are essentially subjective question-and-answer procedures. In terms of economical impact, depression can cost as much as heart disease to the U.S. economy, if appropriate treatment measures are not taken. Moreover, it has been projected that depression will be the main cause of disability in the world for women and children by the year 2020 (Florida Council for Community Mental Health, USA).

Even considering all worldwide efforts invested so far, psychiatric disorders are still neglected compared to other major health problems. This is likely to be due to the nonfatal characteristics of this group of diseases, apart from their association with an increased rate of suicide. This can also be demonstrated by the fact that PubMed/MedLine database searching for the term "psychiatry" currently (March, 2011) identifies approximately 232,000 articles. In contrast, searching for "heart disease" identifies around 820,000 articles and a search for "cancer" finds approximately 2.5 million articles. As a consequence, greater advances have been achieved in therapeutics and mortality reduction in cardiopathologies and cancer, despite the similar or even greater rates of the social and economic burden of psychiatric disorders.

III. The Role of Proteomics in Psychiatry

The proteome is defined as "the total set of expressed proteins by a cell, tissue or organism at a given time under a determined condition" (Wilkins *et al.*, 1996). Proteomics is the study of proteomes and may also include the study of posttranslational modifications and protein–protein interactions. In 1999, the first proteomics study of cancer was published (Banks *et al.*, 1999). As of March 2011, a PubMed search for "cancer and proteomics/proteome" returned approximately 4201 articles, showing the large investment into proteomic analyses of these conditions. Proteomic studies of cancer have thus far led to breakthroughs in diagnosis and treatment (see Chapter "Challenges of introducing new biomarker products for neuropsychiatric disorders into the market" by Bahn *et al.*), supporting the importance of the technology as a tool for the comprehension of biochemical pathways, biomarker discovery, and identification of therapeutic alternatives for diseases. The first proteomics paper in psychiatric disease was published in 2000 (Johnston-Wilson *et al.*, 2000) and a current PubMed search for the appropriate terms returns only 144 relevant articles.

These comparisons have led us to the conclusion that new applications are necessary in psychiatric studies. Biomarker studies on psychiatric disorders present peculiar hurdles such as the inherent difficulties in accessing relevant biological materials, since the main manifestations appear to be in the brain. Moreover, the heterogeneity of brain tissue may be challenging considering, for example, that decreases in synaptic proteins could result from cell-specific mechanisms or a functional change such as lack of input to a particular brain region. As a consequence, the identification of correlating biomarkers in the periphery is required before these can be clinically useful.

A. What Has Been Done So Far?

Almost all large-scale proteome studies in psychiatry have aimed to profile protein expression differences compared to control samples in a hypothesis-free manner. Most of these studies have been performed in human tissues obtained from schizophrenia patients. This includes analysis of several brain regions (Martins-de-Souza *et al.*, 2010a), peripheral tissues such as serum (Levin *et al.*, 2010), cerebrospinal fluid (CSF; Huang *et al.*, 2008; Martins-de-Souza *et al.*, 2010b), liver (Prabakaran *et al.*, 2007), fibroblasts (Wang *et al.*, 2010), and pituitary (Krishnamurthy *et al.*, 2011). These combined studies have led to a better biochemical comprehension of the pathophysiology and have also identified putative biomarkers (Levin *et al.*, 2010) which have already led to a diagnostic test for schizophrenia (Schwarz *et al.*, 2010—http://www.veripsych.com; see Chapters "The application of multiplexed assay systems for molecular diagnostics" by Schwarz *et al.* and "Algorithm development for diagnostic biomarker assays" by Izmailov and Schwarz). Proteome profiling of human brain tissues has also been carried out for bipolar disorder and major depressive disorder (Beasley *et al.*, 2006; Pennington *et al.*, 2008), although mainly as controls for schizophrenia studies (Martins-de-Souza *et al.*, 2010c). Anxiety disorders have also been studied using proteomics, but so far only in animal models of these conditions (Kromer *et al.*, 2005; Ditzen *et al.*, 2006).

The majority of the cited studies were performed using 2DE, the most traditional proteomic method. Despite the limitations of 2DE, which is discussed below, it is a useful tool for the relative comparison of proteomes. Most of the remaining studies were carried out using shotgun-MS techniques, which have higher sensitivity and accuracy for detection and quantification of proteomic differences. In proteomic biomarker studies, large-scale 2DE and shotgun-MS approaches are part of the so-called discovery phase of biomarker identification (Fig. 1).

After this, validation of the biomarker candidates is essential, considering that many of these have proven to be false positives (Phan *et al.*, 2006). In the validation stage, the biomarker candidates are tested further using alternative methods for confirmation of initial findings. A further validation stage involves analysis of larger sample sets in which the experiments are focused on biological reproducibility and sensitivity of potential biomarkers is confirmed.

Most validation experiments have been carried out so far in conjunction with biomarker identification studies using orthogonal antibody-based techniques such as WB and enzyme-linked immunosorbent assay (ELISA). There are two main factors which should be considered before applying these methods: (1) WB and ELISA methods depend on antibody availability and (2) neither of these approaches is useful for high-throughput or large-scale validation in the formats which are available typically. The latter is not an important factor when the aim is comprehension of disease pathology. However, these are crucial considerations for the development of biomarkers as prognostic or diagnostic agents. These approaches, as well as advances and new methodologies, are discussed below.

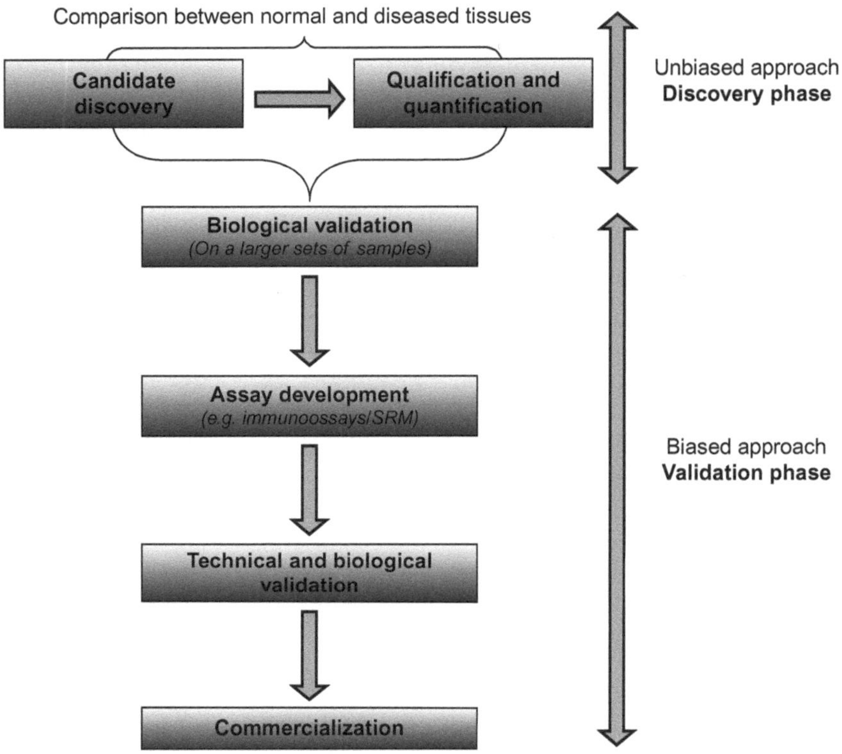

FIG. 1. Biomarkers discovery process from discovery to commercialization.

IV. Proteomic Studies in Psychiatry: What Methods Have Been Used to Date?

A. Sample Preparation

Sample preparation is the most important stage of all proteome studies. To obtain a satisfactory coverage of the proteome of interest, a protein solubilization method of sufficient quality is required in order to maintain proteins in the matrix. Proteins that are not extracted and solubilized from the tissue of interest will not be identified. Consequently, potential biomarkers can be lost at an early stage of the procedure. Proteomes from different tissues may be distinct and contain different classes of proteins. Hence, appropriate changes should be made in the extraction protocol, depending on the nature of the sample (Shaw and Riederer, 2003).

In studies of brain tissues, there may be a particular interest in identification of transmembrane receptors which normally contain multiple hydrophobic domains. Such proteins are often underrepresented in proteome analyses given their low abundance and technical difficulties involved in existing extraction procedures. Numerous protein extraction protocols for brain tissue have been published over the past 15 years, aiming for the comprehensive coverage of the proteome, including the wide variety of membrane proteins (Rabilloud, 1996; Shaw and Riederer, 2003; Lee, 2007; Roulhac *et al.*, 2011). Most of these methods have been developed to incorporate high concentrations of chaotropic agents such as urea, thiourea, or guanidine, and disulfide bond-reducing agents such as dithiothreitol (DTT), tributylphosphine (TBP) or beta-mercaptoethanol, and zwitterionic detergents, such as 3-[(3-cholamidopropyl)-dimethylammonio]-1-propanesulfonate (CHAPS) or aminosulfobetaine (ASB; Carboni *et al.*, 2002; Martins-de-Souza *et al.*, 2007; Everberg *et al.*, 2008). Depending on the membrane complexity of the sample, the use of sodium dodecyl sulfate (SDS) can be beneficial (Ericsson and Nistér, 2011), although the incompatibility with 2DE and MS methods diminishes its use. Commercially available protein extraction kits claim to solve the problem of membrane protein extraction. However, in many cases, the extraction buffers in these kits are not compatible with downstream methodologies of protein separation and identification. At this point, proteomic scientists must develop methods for circumventing these issues including optimization of buffers and running conditions, the decision of whether to prefractionate samples or the necessity of protein precipitation for sample concentration and buffer exchange purposes (Hirano *et al.*, 2006). Protein prefractionation protocols are also used in proteome studies. The main purpose is to reduce the complexity of the analyzed samples and increase proteome coverage by dissociating different classes of proteins such as cytosolic and membrane, high and low abundance, and proteins in different subcellular compartments (Görg *et al.*, 2002; Righetti *et al.*, 2005; Martosella *et al.*, 2006; Fang *et al.*, 2010).

In proteome analyses of body fluids such as serum, CSF, and urine, the solubilization step is typically not a factor, since the majority of the proteins are already in this state. However, considering that albumin and immunoglobulins may represent more than 75% of the total protein content (Urbas *et al.*, 2009), depletion of abundant proteins by immunoaffinity chromatography has become an indispensable stage in studies of these body fluids. The dynamic concentration range of serum proteins may vary by more than 10 orders of magnitude (Anderson and Anderson, 2002), obscuring the accurate identification and quantitation of low-abundance serum proteins. Given these difficulties, different chromatographic columns targeting various abundant proteins are now commercially available. Various methods for proteome prefractionation to suit most requirements are also available (Guerrier *et al.*, 2005).

B. Two-Dimensional Gel Electrophoresis and Mass Spectrometry

The resolution power of SDS-polyacrylamide gel electrophoresis (PAGE) developed in the 1950s/1960s (Smithies, 1955; Raymond, 1964; Shapiro *et al.*, 1967; Righetti, 2004) was satisfactory for separation of simple protein mixtures, such as those from bacteriophages. However, more complex systems such as those from eukaryotic cells required considerable improvement of existing methods. This necessity led to the development of 2DE which was finally achieved by O'Farrell (1975), based on previous ideas and experiments (Kaltschmidt and Wittmann, 1970; Orrick *et al.*, 1973). O'Farrell published an article showing an optimized tool for 2D separation of proteins according to their isoelectric point (p*I*) using isoelectrofocusing (IEF), followed by separation according to apparent molecular weight (MW) using SDS-PAGE (Fig. 2A). This method would eventually lead to the field currently known as "proteomics" and would become the most used method for proteome analyses so far.

Prior to the current format, the 2DE method underwent a remarkable number of optimization steps even prior to the proteomics era. The use of carrier ampholytes for IEF was replaced by immobilized pH gradients (IPG) in order to resolve the frequent reproducibility issues. This refinement alone led to the possibility of interlaboratory 2DE pattern matching (Bjellqvist *et al.*, 1982) and the implementation of other improvements such as protein disulfide bond reduction and alkylation after the IEF stage (Görg *et al.*, 1987). In addition, the development of IPG strips (Görg *et al.*, 1988) and narrow pH-range strips allowed the detailed analyses of subsets of proteins based on their p*I* (Görg *et al.*, 1985).

These steps led to increased resolution and reproducibility of 2DE, which is essential for comparative proteome analyses. After serial separation of proteins by IEF and SDS-PAGE, the gels are stained by Coomassie Blue or silver nitrate, for example, for visualization of the separated protein spots. After this, the gel images are scanned, protein spots are detected, quantities and corresponding MWs and p*I*s determined based on internal standards, using a 2DE analysis software. The resulting 2DE gel images are then divided into groups, and matched for comparison purposes. This is achieved by software which can be used for matching the same protein spots in different gel images, enabling statistical analyses to determine differences in protein expression levels based on spot volume intensities (Fig. 2B and C). Thanks to the development of an *in-gel* digestion protocol (Shevchenko *et al.*, 1996) and further optimizations, 2DE spots can be manually or automatically extracted from 2DE gels and digested *in situ* prior to MS analyses. The resolution power of 2DE can lead to the separation of more than 2000 protein spots on large format gels although the true power of the technique is not realized without its combined use with MS for protein identification (more on this below). 2DE followed by MS is the most used proteomic technique in studies of psychiatric disorders so far (Kromer *et al.*, 2005; Beasley *et al.*, 2006; Ditzen *et al.*, 2006; Pennington *et al.*, 2008; Martins-De-Souza *et al.*, 2010a,b).

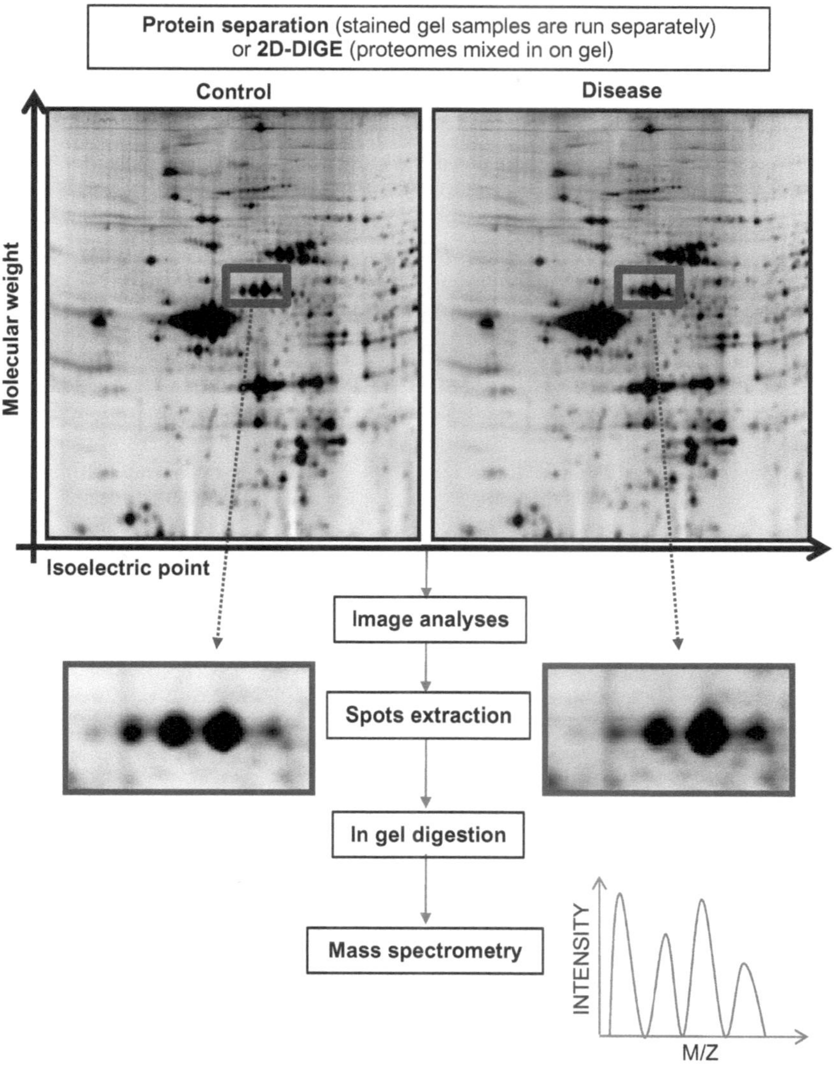

FIG. 2. Two-dimensional gel electrophoresis for proteomics.

The advent of fluorimetric methods in 2DE increased the sensitivity and reproducibility of the technique. Fluorescent 2DE difference gel electrophoresis (2D-DIGE; Unlü *et al.*, 1997) requires approximately 10-fold lower amounts of sample compared to the standard detection procedures and only the use of a single gel (as opposed to two) to detect proteome differences.

Prior to 2DE separation, protein samples to be compared are labeled using distinct cyanine-derived fluorophores (Cy2, Cy3, and Cy5), which are mass- and charge-matched and interact with the free amines on proteins. Next, labeled samples are run in the same gel, minimizing technical variation by allowing comparison of the two samples and an internal standard for normalization purposes and multigel comparisons. After 2DE, gels are digitalized in a fluorescence scanner, revealing the spot volumes of the individual samples, and the internal standard. This provides a volume ratio between sample and internal standard spots, generating more accurate quantification compared to the standard cross-gel comparison methods.

Briefly, a mass spectrometer is minimally composed of three parts: (1) an ionization source (e.g., matrix-assisted laser desorption/ionization (MALDI) and electrospray ionization (ESI)), (2) an analyzer (e.g., time of flight (TOF) and quadrupole), and (3) a detector. A mass spectrometer measures the mass-to-charge (m/z) ratio of ionized particles. In proteomics, the ionized particles are peptides, since intact proteins are too large to be measured reliably or accurately. The samples are digested with a protease such as trypsin and the m/z of the resulting peptides measured in a mass spectrometer. The unknown peptides can then be identified by comparison of the m/z values with those in a protein database which has been subjected to virtual proteolysis. This methodology was first described as "peptide mass fingerprinting" (Mann *et al.*, 1993; Henzel *et al.*, 1993) and is still the basis of protein identification by MS. Detailed information about the use of MS in proteomics has already been described in previous reviews (Ahmed, 2008, 2009).

One of the main advantages of the combined 2DE–MS approach is the fact that it can provide direct information on intact proteins and protein isoforms. However, there are limitations (Ong and Pandey, 2001). The 2DE stage of the analysis is laborious and difficult to automate. Also, there is a limited representation of proteins with extreme p*I*s (pH lower than 4 or higher than 9) or MWs (larger than 200 kDa and smaller than 10 kDa). Another concerning factor is the poor representation of low-abundance proteins such as transcription factors and hydrophobic receptor proteins (including neurotransmitter receptors) which are normally masked by the high abundance proteins. Many of these difficulties can be ameliorated by the use of narrow pH IEF gels, extensive sample prefractionation, and optimization of protein extraction protocols. However, some researchers have explored non-2DE methods for a better proteome representation.

C. Shotgun Proteomics

Shotgun proteomics can be represented by several terminologies including shotgun-MS, liquid chromatography–tandem MS (LC–MS/MS), and multidimensional protein identification technology (MudPIT). This is a high-throughput and automated MS-based approach which was found to be more sensitive and reproducible at the time of its introduction compared to the existing proteomic techniques (Aebersold and Mann, 2003). Shotgun-MS was originally designed as a nongel/MS-direct approach to avoid the 2DE proteomics drawbacks stated above (Link *et al.*, 1999; Peng and Gygi, 2001) although the use of gel-based techniques may be combined with shotgun-MS.

The approach of a shotgun-MS analysis is to digest the whole proteome of interest using specific enzymes and subsequently identify the resulting peptides by MS. Given that such a digestion step increases the complexity of a given proteome, one or more prefractionation steps are necessary either at the protein or peptide level prior to MS/MS analyses to improve proteome coverage. In the general shotgun-MS workflow, one of the fractionation steps consists of LC or other chromatography methods which are coupled online to the mass spectrometer for peptide separation (Link *et al.*, 1999). Offline LC can also be used although this may result in lower throughput (Maccarrone *et al.*, 2010). Other fractionation steps which are commonly used prior to LC–MS/MS include strong ion exchange chromatography, SDS-PAGE, and IEF. Two-dimensional online chromatography is also employed, although still not shown in psychiatric studies. For protein identification, there are several commonly used algorithms which allow automated assignment of MS/MS spectra by matching acquired data with spectra predicted on the basis of protein sequence databases (Fig. 3). However, proteome quantification and sample comparison are dependent on the type of methodology used as described below.

1. *Label-Free MS*

The theoretical assumption that the chromatographic peak area of a given peptide should correspond to its concentration (Chelius and Bondarenko, 2002) was experimentally proven through studies of peptide ion current intensities (Chelius and Bondarenko, 2002; Levin *et al.*, 2007). This is the main principle of label-free quantitative proteome analysis which seems to be the easiest way of comparing proteomes, since no labeling for proteome quantification is required and, in contrast to other proteomic profiling techniques, there are no limits regarding number of samples for analysis, which makes this approach well-suited for clinical studies. This is important as analysis of larger samples cohorts provides greater statistical power, which is essential in biomarker discovery projects to minimize false-positive results.

Label-free proteomics can be performed basically using four distinct approaches. Two of these, "spectral counting" and "data-dependent acquisition (DDA)-based ion counting," are data dependent, whereas the other two, "MS survey scan-based ion counting" and "data-independent analysis (MS^E)," are acquired in an MS survey scan (Levin and Bahn, 2010). Considering that in MS^E (Li *et al.*, 2009), intact peptides and fragmented peptides are intermittently measured in a single experiment at high sampling rate, this seems to be the most

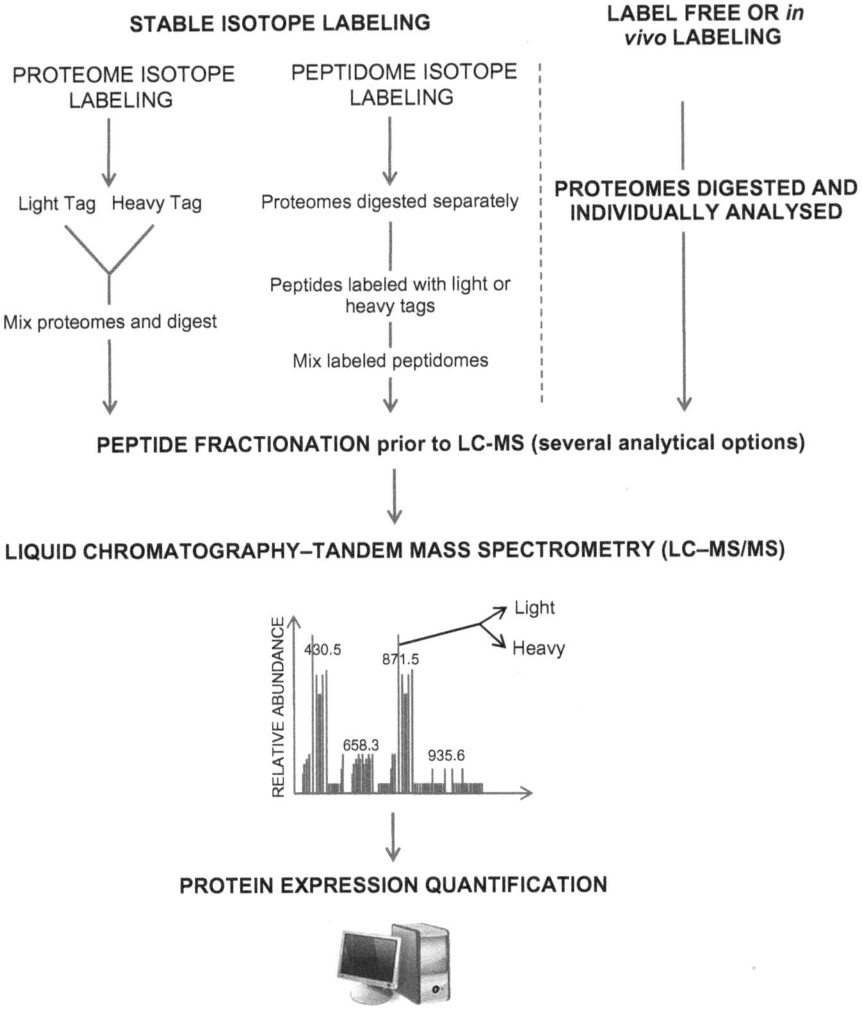

FIG. 3. Representation of shotgun proteomics methodology.

straightforward and precise label-free approach. It is true that label-free proteomics depends on precise algorithms and software so that scientists can interpret the obtained data, but these are largely available.

Label-free proteomics using the MS^E approach has already shown substantial impact in clinical studies of psychiatric disorders. We identified a serum biomarker panel capable of distinguishing first-onset drug-naive schizophrenia patients from control subjects. This final panel was validated by immunoassay and later adapted into the multiplex immunoassay platform MAP™ by Rules Based Medicine (Austin, TX, USA) leading to the first blood-based laboratory test for schizophrenia which has been marketed in the United States under the trade name VeriPsych™ (Schwarz *et al.*, 2010; see Chapters "Application of multiplexed assay systems for molecular diagnostics" by Schwarz *et al.* and "Algorithm development for diagnostic biomarker assays" by Izmailov and Schwarz). In addition, we have also successfully applied the label-free proteomic method in several other studies including identification of differences in (1) synaptic and neural development proteins in rats treated with antipsychotics (Ma *et al.*, 2009); (2) cell proliferation in dermal fibroblasts from schizophrenia patients (Wang *et al.*, 2010); (3) cell survival in peripheral blood mononuclear cells from bipolar disorder subjects (Herberth *et al.*, 2011); and (4) for increasing the known proteomic coverage of pituitary (Krishnamurthy *et al.*, 2011), dorsolateral prefrontal cortex (Martins-de-Souza *et al.*, 2011a,b,c), and hippocampus (Yang *et al.*, 2011), which are all anatomical regions known to be associated with schizophrenia.

2. *Stable Isotope Labeling*

Labeling chemical compounds with stable isotopes has been used for more than 60 years and currently is commonly employed in nuclear magnetic resonance (NMR) and MS analyses with the objective of tracking and quantifying specific molecules. Isotope labeling was first used for quantitative proteomics with the launching of isotope-coded affinity tags (ICAT; Gygi *et al.*, 1999a,b). The ICAT system uses heavy and light mass tags containing either eight or no deuterium atoms, respectively. These heavy and light tags can be bound via a thioester linkage to cysteine residues on proteins in the two samples under comparison. In this scenario, the first sample can be labeled with the light tag, for example, and the second can be labeled with the heavy tag prior to MS analysis. Heavy ICAT-labeled peptides will present a mass shift of 8 Da compared to light ICAT-labeled peptides. Using computational algorithms, the ratio of the heavy and light versions of the same peptides can be calculated, providing information on relative quantification of these peptides and, consequently, the corresponding proteins.

The basic principles are the same for other isotope labeling methods such as the Global Internal Standard Technology (GIST; Goodlett *et al.*, 2001) and isotope-coded protein labeling (ICPL; Schmidt *et al.*, 2005), as these only differ in the reactive sites on the targeted proteins and the mass differences used. Proteome quantification by isobaric tags for relative and absolute quantitation (iTRAQ; Ross *et al.*, 2004) represents a distinct approach. The eight available iTRAQ tags can be used to label up to eight different biological samples. In addition, each iTRAQ reagent contains a reporter group that is separable from the rest of the tag during MS fragmentation. Although all iTRAQ tags have the same mass, the reporter groups differ by 1 Da. Thus, the intensity of each of the reporter tags will correspond directly to the abundance of a given peptide in distinct samples.

The calculated ratios of different reporter tag intensities will provide relative quantitation for the peptide/protein in question. The larger number of mass tags also provides greater flexibility in experiments. For example, this will allow direct comparison of samples from the same subject in longitudinal studies of drug effects.

The ICPL and iTRAQ methodologies have been used successfully in recent proteome analyses of schizophrenia brain tissue (Martins-de-Souza *et al.*, 2009a,b, 2010d), providing evidence of proteins and biological pathways which are altered in the disease.

In general, stable isotope labeling methods can be used at the protein or peptide level and can be combined with most proteomic separation approaches, providing accurate quantification. The disadvantages of stable isotope labeling, however, is that the targeted proteins or peptides are labeled after sample preparation procedures such as protein extraction, subproteome isolation, or even generation of peptides, which can increase the probability of introduced experimental errors, resulting in decreased accuracy. Moreover, the maximum number of eight isotope tags is a limiting factor for most clinical applications, although different time points of the same patient can be measured in one experiment.

3. *In Vivo* Labeling

In vivo labeling methods avoid some of the drawbacks associated with the stable isotope labeling approach (Filiou *et al.*, 2011). Although these methods are not readily suitable for studies on living human tissues, they can be used easily in cell culture and animal model experiments. In this approach, stable isotopes such as ^{2}H, ^{13}C, ^{15}N, or ^{18}O are introduced *in vivo* such that they can be incorporated into newly synthesized proteins. Since these are natural isotopes, there should theoretically be no effects on the biological systems into which they are incorporated. In comparative proteomic studies using this method, labeled and unlabeled versions of the proteome extracts can be combined prior to any sample preparation avoiding the types of experimental errors that are seen with the postlabeling approaches (Gouw *et al.*, 2010).

In psychiatric studies, ^{15}N metabolic labeling has been applied to proteomic studies of a mouse model of trait anxiety (Kromer *et al.*, 2005). High, normal, or low anxiety-related behavior (denominated HAB, NAB, and LAB, respectively) mice were fed with ^{15}N-labeled diets from cultured algae or bacteria resulting in up to 92% of ^{15}N incorporation in plasma and brain tissue (Frank *et al.*, 2009). No effects on development, organ morphology, physiology, or reproductive ability due to the introduction of isotopes into the feed were detected (Wu *et al.*, 2004; Frank *et al.*, 2009). However, alterations in behavioral phenotype and in components of the proteome were observed (Frank *et al.*, 2009; Filiou *et al.*, 2011).

D. SELDI-TOF

The basic principle of ionization by MALDI is to mix the analyte of interest with a matrix that is capable of absorbing energy provided by a laser source through laser shots. Normally, MALDI MS is coupled to TOF analyzers, which consist of measuring masses of ionized molecules according to TOF in a flight tube of known length.

As a variation of MALDI-TOF MS, SELDI-TOF MS incorporates affinity capture chips that have been used in the proteome analyses of CSF from first-onset drug-naive schizophrenia patients and individuals with initial prodromal symptoms and healthy controls. This analysis led to identification of differences in the concentration of bioactive molecules including VGF and transthyretin (Huang *et al.*, 2007). Moreover, CSF, brain, and peripheral tissues from schizophrenia patients were studied using SELDI-TOF revealing alterations in apolipoproteins and proteins involved in immune response (Huang *et al.*, 2008).

E. Metabolomics

Proteomic studies, especially those on human brain tissues, have provided information about biochemical pathways involved in psychiatric disorders, leading to insights on the associated metabolites. Quantitative studies on metabolites can lead to a more complete overview of brain activities and function to allow better comprehension of the pathobiological processes involved in psychiatric conditions as well as providing an additional source of biomarkers.

Since key glycolytic enzymes have been found to be differentially expressed in schizophrenia brain tissue, this suggests that molecules such as NADPH and pyruvate may also be altered. Analysis of these molecules using an enzymatic assay confirmed alteration in the level of these metabolites, consistent with the proteomic data (Martins-de-Souza *et al.*, 2010d). Although neither of these metabolites was found to be altered in CSF from first-onset schizophrenia patients, which raises doubts about their use as diagnostic biomarkers, they have been informative for increasing our understanding of schizophrenia (Martins-de-Souza *et al.*, 2010d).

Based on previous proteomic profiling studies (Ditzen *et al.*, 2006), high-performance liquid chromatography (HPLC) has been used to quantify significant differences in the polyamines putrescine and spermidine in brain tissue extracts of HAB and LAB animal models described above (Ditzen *et al.*, 2010).

Metabolite profiling analyses have also been employed in psychiatric-related studies in a large-scale manner. High-resolution proton nuclear magnetic resonance spectroscopy (^{1}H NMR) has been used for multiplex analysis of metabolites in rat brain, related to the effects of psychotropic drug administration. Significant differences were found in *N*-acetylaspartate (NAA) levels (McLoughlin *et al.*, 2009). Changes associated with the drug treatment in *postmortem* brain tissue from bipolar disorder patients were also accessed using ^{1}H NMR, revealing increased glutamate levels (Lan *et al.*, 2009). In a comparative evolutionary study, changes in metabolite concentrations among human schizophrenia patients, healthy controls, chimpanzees, and rhesus macaques were found also using ^{1}H NMR, supporting the notion of altered metabolic processes in schizophrenia and raising the possibility that evolution of human cognitive abilities was accompanied by adaptive changes in brain metabolism (Khaitovich *et al.*, 2008).

F. Multiplex Analyte Profiling Approach

The recent introduction of fluorescent bead-based technologies in psychiatric studies allows the simultaneous measurement of multiple analytes in small-volume samples. This method is more targeted than those described thus far, since the identity and number of assays are known. However, this platform is potentially the most suitable for further development of accurate, sensitive, and specific diagnostic assays, given its high-throughput nature and relative ease of use in the clinical setting. Recently, we reported on the use of this platform for identification of candidate biomarkers for neuropsychiatric conditions such as Asperger syndrome (Schwarz *et al.*, 2011), bipolar disorder (Herberth *et al.*, 2011), and schizophrenia (Schwarz *et al.*, 2010). The latter study has now been developed further resulting in production of a diagnostic test for schizophrenia as described above and described in detail in Chapters "Application of multiplexed assay systems for molecular diagnostics" by Schwarz *et al.* and "Algorithm development for diagnostic biomarker assays" by Izmailov and Schwarz.

G. What Is the Best Method for Proteome Characterization?

All of the molecular profiling methods described here have their strengths and weaknesses, which have been described. Therefore, we recommend the combined use of different methodologies as this would not only maximize proteome coverage, but it would also result in more comprehensive information on the relevant molecular pathways involved in the disease process and give increased confidence in the biomarker candidates identified.

V. Underexplored Proteomic Methods in Psychiatry Studies

A. Phosphoproteomics

In contrast to studies of cancer and neurobiology (Semenov *et al.*, 2006; O'Hayre *et al.*, 2010), little work has been carried out in the pathways affected in psychiatric disorders using phosphoproteomic approaches. This involves the large-scale detection and quantification of phosphoproteins and phosphopeptides using proteomic techniques such as 2DE (Jacob and Turck, 2008) and MS (Zhou *et al.*, 2010).

Protein phosphorylation and dephosphorylation is a posttranslation modification which acts as a molecular switch in the regulation of a diverse range of cellular processes. This method is also known to be involved in the mechanism of action of some psychiatric medications (Martins-de-Souza *et al.*, 2011a,b,c). As the phosphorylation state provides information on cellular function, phosphoproteomic approaches can provide useful information in understanding the molecular mechanisms underlying the disease etiology and for identification of potential novel biomarker candidates. This method can also be used as a validation technique for confirming previous findings of proteomic profiling studies. Methods targeting other posttranslational modifications such as glycosylation, acetylation, and proteolysis may also be useful in studies of psychiatric disorders.

B. SILAC

Stable isotope labeling by/with amino acids in cell culture (SILAC) is an *in vivo* labeling method, which has been used in conjunction with proteomic studies. 7SILAC consists of growing the cells of interest in medium containing specific ^{13}C- or ^{15}N-labeled amino acids, such as lysine or arginine, for quantitative comparison of proteins in cells which have been cultured in unlabeled media (Ong *et al.*, 2002). Proteome quantification relies in the same principal as stable isotope labeling as described above.

SILAC could be used to provide precise quantitative proteomic information in cell culture models of psychiatric disorders as described previously (Steiner *et al.*, 2010; Martins-de-Souza *et al.*, 2011a,b). Primary cortical neurons from a mouse model of fragile X syndrome (FXS) and wild-type mice were cultured using the SILAC approach in order to identify changes in synaptic proteins (Liao *et al.*, 2008). Interestingly, the SILAC concept can also be applied to animal models by incorporation of ^{13}C-labeled lysine in the chow for subsequent quantitative comparisons (Krüger *et al.*, 2008).

C. MALDI Imaging

In MALDI imaging, the matrix is manually or automatically applied to slices of the tissue of interest. Then, ionization by laser provides the possibility of analyzing and quantifying proteins, peptides, lipids, nucleic acids, or drugs, depending on the type of matrix. Most importantly, the analyzed tissue can be visualized in real time, providing information on the localization and spatial distribution of the molecules under study (Caldwell and Caprioli, 2005). Three-dimensional molecular images are also possible (Andersson *et al.*, 2008) as well as the combination of this method with other imaging techniques such as magnetic resonance imaging *in vivo* (Sinha *et al.*, 2008). All the opportunities offered by MALDI imaging are suitable for psychiatric studies in a variety of human samples as well as tissue from animal models.

VI. The Importance of Validation Experiments in Proteomics for Biomarker Discovery in Psychiatry

The growing number of proteomic investigations has led recently to a stark increase in the number of identified differentially expressed proteins. These molecules are likely to be considered as potential biomarkers for several brain disorders and other diseases and constitute an important part of the global diagnostics market (Fig. 4). However, this can only be achieved if the identified molecules can be validated in repeat studies and in larger numbers of samples. All quantitative proteomic studies essentially require such validation steps to confirm or reject the importance of candidate biomarkers for use as clinical diagnostics. Such steps should be incorporated into all proteomic studies (Fig. 1) as a means of maximizing the impact of biomarker research in the field of psychiatric research and other disciplines.

The methods used in the discovery phase are well developed and widely applied in studies of psychiatric disorders, although the validation phase still requires improvement. Although validation technologies have been developed and applied during the past few years in psychiatric studies, antibody-based methods such as ELISA and WB have been the predominant methods of choice (Beasley *et al.*, 2006; Prabakaran *et al.*, 2007; Levin *et al.*, 2010; Martins-de-Souza *et al.*, 2010d; Wang *et al.*, 2010). However, the use of these techniques may be prohibitive as they require the availability of specific antibodies with high specificity and affinity. This is even more problematic for studies of proteins containing posttranslational modifications.

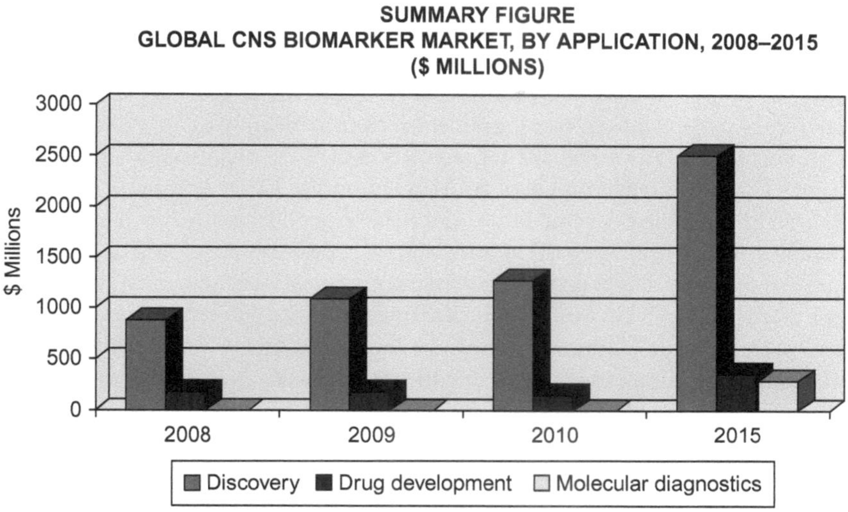

FIG. 4. Evolution of central nervous system biomarker marker (BBC Research, report: BIO074A, October 2010).

The analysis of differentially expressed proteins in different fields has led scientists to use more sensitive validation techniques such SRM. This method could be particularly useful if applied in psychiatric studies for uses such as validation of candidate biomarkers for diagnosis, prognosis, and treatment prediction/response. Other methods such as SILAC and MALDI imaging, as well as studies of posttranslation modifications such as phosphorylation, glycosylation, and proteolysis, could be helpful in the functional validation of proteomic findings. The use of RNA technologies, such as quantitative real-time PCR (qPCR), have already been employed in psychiatric research, although the findings may not be directly correlated with protein expression levels (Gygi *et al.*, 1999a,b).

A. Validation Technologies

1. *Selective Reaction Monitoring*

SRM, also known as multiple reaction monitoring (MRM), is an MS-based technique which can accurately measure the relative or absolute concentrations of particular molecules such as peptides/proteins, identified in proteomic profiling studies. SRM experiments are generally performed in triple-quadrupole (TQ) mass spectrometers, following few basic steps: (1) peptides coming from online nanoLC are ionized by nano-electrospray, (2) generated ions are aligned and peptides of interest are selected, (3) selected peptides are fragmented and the fragments (SRM transitions) are aligned, (4) SRM transitions are singly detected, and (5) data of individual SRM transitions are combined in order to provide a final intensity of the previous selected peptide, which consequently will provide quantitative information. The schematic representation of SRM analyses is represented in Fig. 5. The quantitative power of SRM relies on the fact that a single peptide is accurately quantified several times, since each peptide is broken into multiple fragments (Fig. 5, Q2). Each SRM transition (or peptide fragment) will be quantified and, together with data from the other fragments, provide accurate quantitative data for a given peptide. The combined quantitative data from different peptides from the same protein can then be used to derive the absolute and relative quantitation of that protein. It is important to highlight that the sensitivity of this method may reach the attomole range. Moreover, SRM allows precise characterization of several peptides and posttranslational modifications such as phosphorylation, acetylated lysine residues, ubiquitination, and glycosylation.

SRM experiments have been employed in proteomic studies (Armenta *et al.*, 2010; Zhang *et al.*, 2011) but have thus far not been applied in studies of psychiatric disorders. The high-throughput nature and multiplexing capability of SRM (it is possible to measure peptides from several proteins simultaneously) indicate that it could be useful as a potential technique in validation analyses of samples from large human cohorts. This technology is also of interest as it can be used in both preclinical and clinical trials, in which there is increased usage of biomarkers as surrogate endpoints. Moreover, the multiplexing capability makes SRM a useful methodology for use in clinical pipelines for detection of diagnostic, prognostic, and treatment-related biomarkers.

2. *ELISA*

ELISA is one of the main techniques used for detection and quantification of protein expression with high sensitivity and good reproducibility, which allows the analysis of protein samples in a microplate format. This technique requires the

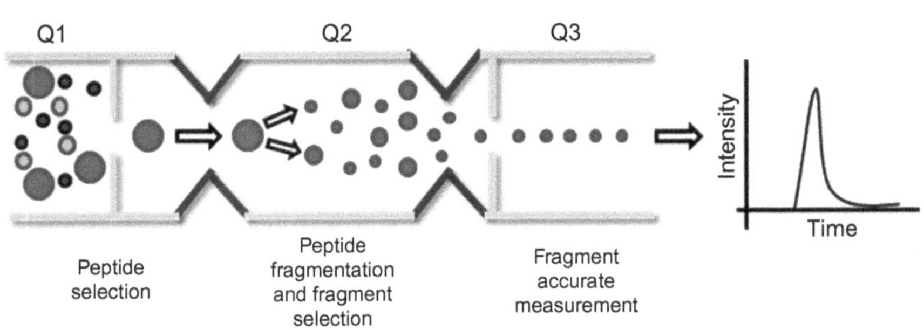

FIG. 5. Schematic representation of SRM for proteomics.

availability of two antibodies for each target protein. One antibody is used for the capture of the protein of interest and the other is used for detection. This technology can often be used easily off-the-shelf although it sometimes requires optimization steps when there are large numbers of variables including differences in antibody specificity, and reagent compatibility with conditions such as buffer components in the protein extracts or media of interest. This technique has already been used extensively in psychiatric and neurological disorders including the measurement of β-amyloid peptides in Alzheimer's disease (see review Humpel, 2010), profiling of insulin-related molecules in schizophrenia (Guest *et al.*, 2010), and for validation of biomarker candidates identified by LC–MS^E profiling of serum from schizophrenia patients (Craddock *et al.*, 2008; Levin *et al.*, 2010).

3. *Western Blot*

WB is another immunological method for protein detection and quantitation that it is often used in preference to ELISA in studies of tissue extracts. The method is only deemed semiquantitative although it has an advantage over ELISA in that WB affords visualization of protein bands, affording confidence in specificity of the antibodies used. This also has the added advantage that posttranslational changes such as proteolysis can be visualized by a change in apparent MW or p*I*, for example. Other advantages include the possibility of using one secondary antibody in the detection of several primary antibodies and the availability of secondary antibodies offering different types of detection. One main disadvantage of WB is that the secondary antibody may bind nonspecifically to other proteins, particularly immunoglobulin chains, in the extract. A potentially powerful application would involve combination of ELISA and WB methods as shown in a recent study (Carlino *et al.*, 2010). The ELISA method gave a result indicating a decrease in serum brain-derived neurotrophic factor (BDNF) in schizophrenia compared to control patients, and WB showed that this was associated with an increase in pro-BDNF and mature BDNF, with a concomitant decrease in a truncated form of the molecule. This combined analysis allowed the authors to correlate the expression of different forms of serum BDNF with cognitive performance.

4. *Functional Genomics*

Gene expression mediates cellular activity and first gives rise to synthesis of messenger RNA. In neuroscience and especially in neurological disorders, microarrays constitute the most common technology used, in association with qPCR validation. One example of multiplatform analysis is the characterization of *QKI* gene expression in suicide victims who suffered from major depressive disorder (Klempan *et al.*, 2009). The authors identified a reduction of *QKI* mRNA levels in cortical, hippocampic, and amygdala regions of suicide victims compared to control subjects. The microarray findings were confirmed by qPCR and also by reduced expression of the encoded proteins, as shown by WB analysis. There are several advantages of PCR arrays including high accuracy, flexibility of use, reproducibility, and specificity. However, combined analysis of messenger RNA and the encoded proteins might lead to a better understanding on the pathological mechanisms involved in neurological disorders. Also, messenger RNA levels do not necessarily correlate with the corresponding protein levels, as stated above.

5. *Tissue Microarray*

Despite the existence of many well defined molecular technologies such as ELISA, PCR, and WB, one main shared disadvantage of these methods is that they are subject to artifactual effects incurred by the tissue or body fluid before analysis. Moreover, proteins or genes might have differential expression regarding their subregional distribution and therefore identifying their specific localization is crucial. Tissue microarray (TMA) is a useful technique for providing information on intracellular distribution of molecules. This method consists of hundreds of tissue cores assembled as an array which allows the simultaneous analysis of multiple samples and biomarkers in a single study. Most of the assays used in TMA are immunohistochemical in nature but can also include *in situ* hybridization methods, mostly in cases involving tumor biopsy, and also in neurodegenerative and inflammatory diseases.

6. *Protein Arrays*

Protein arrays typically comprise multiple proteins or antibodies arrayed in separate locations on a microtiter plate or other surface, to allow simultaneous analysis of multiple protein targets. Such arrays provide a good platform for efficient profiling of protein expression (Table I) and also for identification of interactions between other proteins, antibodies, drugs, or ligands. Given that this is a multiplex format, other advantages include the need for lower samples volume and antibodies and, therefore, the high reproducibility and associated low costs.

Table I
Features of Technologies Currently Used for Biomarker Validation.

	WB	ELISA	Antibody array	MRM-MS
Target	Known	Known	Known	Known
Sensitivity		Nanomolar	Nanomolar	Attomolar
Number of proteins	Few	1	10,000	1000
Reproducibility	Medium	Good	Good	Very good
Measurement	Unique	Unique	Multiplex	Simultaneous
Volume	10–25 μl	≥100 μl	0.5–100 nl	Low

VII. Clinical Translation

Biomarkers validation is important in all fields, but the optimal analytic platform which allows their translation into the clinic is crucial. Findings at the proteomic level must be developed as user friendly and robust assays for ease of use in the clinic. Only in this way can biomarkers realize their full potential leading to improved diagnostic accuracy, better disease classification and, ultimately, to personalized medicine strategies for more effective treatment affected populations. To select the best candidate for clinical evaluation, optimization of the pharmacokinetic parameters is required (Sarker and Workman, 2007; Sarker *et al.*, 2007).

The use of validated biomarkers in clinical trials remains low. This could be important as only 5% of oncology drugs which have entered the clinical trial phase have been commercialized. Technologies and biomarkers are available for clinical trial use in areas such as cancer, including microRNAs (Bartels and Tsongalis, 2009; Scott *et al.*, 2007; Volinia *et al.*, 2006), SRM standards (Chen *et al.*, 2010), and gene transcripts (Rubinstein *et al.*, 2010). However, only few have been used in the area of psychiatric disorders. Recently, we launched a blood test for schizophrenia termed VeriPsych™, as described above. This test will help clinicians to confirm diagnosis of this pathology and give opportunities for the investigation and development of future treatment for this devastating disorder which affects more than 20 million people worldwide (World Health Organization). This represents a successful model of translational medicine applied in psychiatry, which bridged proteomic findings to result in a marketed product that will improve patients' lives.

VIII. Summary

Here we have covered most of proteomic technologies—from the most traditional to the most sophisticated—for biomarker discovery, highlighting their advantages and drawbacks. We also highlighted the importance of validating proteomic findings, which still need more attention in psychiatric studies in order to translate proteomic findings to bedside. Although some advances have been achieved in the diagnostic field, a lack still remains regarding prognostic markers and treatment efficacy. Optimized integrated machinery for biomarker discovery and validation will allow a rapid translation of research findings to clinical application. Consequently, not only overall healthcare services costs will be reduced but most importantly patients will experience a better quality of life.

Acknowledgments

This research was supported by the Stanley Medical Research Institute (SMRI) and the European Union FP7 SchizDX research programme (grant reference 223427).

References

Aebersold, R., and Mann, M. (2003). Mass spectrometry-based proteomics. *Nature* **422**(6928), 198–207.

Ahmed, F.E. (2008). Utility of mass spectrometry for proteome analysis: part I. Conceptual and experimental approaches. *Expert Rev. Proteomics* **5**(6), 841–864.

Ahmed, F.E. (2009). Utility of mass spectrometry for proteome analysis: part II. Ion-activation methods, statistics, bioinformatics and annotation. *Expert Rev. Proteomics* **6**(2), 171–197.

Anderson, N.L., and Anderson, N.G. (2002). *Mol. Cell. Proteomics* **1**, 845–867.

Andersson, M., Groseclose, M.R., Deutch, A.Y., and Caprioli, R.M. (2008). Imaging mass spectrometry of proteins and peptides: 3D volume reconstruction. *Nat. Methods* **5**(1), 101–108.

Armenta, J.M., Perez, M., Yang, X., Shapiro, D., Reed, D., Tuli, L., Finkielstein, C.V., and Lazar, I.M. (2010). Fast proteomic protocol for biomarker fingerprinting in cancerous cells. *J. Chromatogr. A* **1217**(17), 2862–2870.

Banks, R.E., Dunn, M.J., Forbes, M.A., Stanley, A., Pappin, D., Naven, T., Gough, M., Harnden, P., and Selby, P.J. (1999). The potential use of laser capture microdissection to selectively obtain distinct populations of cells for proteomic analysis—preliminary findings. *Electrophoresis* **20**(4–5), 689–700.

Bartels, C.L., and Tsongalis, G.J. (2009). MicroRNAs: novel biomarkers for human cancer. *Clin. Chem.* **55**, 623–631.

Beasley, C.L., Pennington, K., Behan, A., Wait, R., Dunn, M.J., and Cotter, D. (2006). Proteomic analysis of the anterior cingulate cortex in the major psychiatric disorders: evidence for disease-associated changes. *Proteomics* **6**(11), 3414–3425.

Bjellqvist, B., Ek, K., Righetti, P.G., Gianazza, E., Görg, A., Westermeier, R., and Postel, W. (1982). Isoelectric focusing in immobilized pH gradients: principle, methodology and some applications. *J. Biochem. Biophys. Methods* **6**(4), 317–339.

Caldwell, R.L., and Caprioli, R.M. (2005). Tissue profiling by mass spectrometry: a review of methodology and applications. *Mol. Cell. Proteomics* **4**(4), 394–401.

Carboni, L., Piubelli, C., Righetti, P.G., Jansson, B., and Domenici, E. (2002). Proteomic analysis of rat brain tissue: comparison of protocols for two-dimensional gel electrophoresis analysis based on different solubilizing agents. *Electrophoresis* **23**(24), 4132–4141.

Carlino, D., Leone, E., di Cola, F., Baj, G., Marin, R., Dinelli, G., Tongiorgi, E., and de Vanna, M. (2010). Low serum truncated-BDNF isoform correlates with higher cognitive impairment in schizophrenia. *J. Psychiatr. Res.* **45**, 273–279.

Chelius, D., and Bondarenko, P.V. (2002). Quantitative profiling of proteins in complex mixtures using liquid chromatography and mass spectrometry. *J. Proteome Res.* **1**, 317–323.

Chen, Y., Gruidl, M., Remily-Wood, E., Liu, R.Z., Eschrich, S., Lloyd, M., Nasir, A., Bui, M.M., Huang, E., Shibata, D., Yeatman, T., and koomen, J.M. (2010). Quantification of beta-catenin signaling components in colon cancer cell lines, tissue sections, and microdissected tumor cells using reaction monitoring mass spectrometry. *J. Proteome Res.* **9**, 4215–4227.

Craddock, R.M., Huang, J.T., Jackson, E., Harris, N., Torrey, E.F., Herberth, M., and Bahn, S. (2008). Increased alpha-defensins as a blood marker for schizophrenia susceptibility. *Mol. Cell. Proteomics* **7**, 1204–1213.

Ditzen, C., Jastorff, A.M., Kessler, M.S., Bunck, M., Teplytska, L., Erhardt, A., Krömer, S.A., Varadarajulu, J., Targosz, B.S., Sayan-Ayata, E.F., Holsboer, F., Landgraf, R., *et al.* (2006). Protein biomarkers in a mouse model of extremes in trait anxiety. *Mol. Cell. Proteomics* **5**(10), 1914–1920.

Ditzen, C., Varadarajulu, J., Czibere, L., Gonik, M., Targosz, B.S., Hambsch, B., Bettecken, T., Kessler, M.S., Frank, E., Bunck, M., Teplytska, L., Erhardt, A., *et al.* (2010). Proteomic-based genotyping in a mouse model of trait anxiety exposes disease-relevant pathways. *Mol. Psychiatry* **15** (7), 702–711.

Ericsson, C., and Nistér, M. (2011). Protein extraction from solid tissue. *Methods Mol. Biol.* **675**, 307–312.

Everberg, H., Gustavasson, N., and Tjerned, F. (2008). Enrichment of membrane proteins by partitioning in detergent/polymer aqueous two-phase systems. *Methods Mol. Biol.* **424**, 403–412.

Fang, Y., Robinson, D.P., and Foster, L.J. (2010). Quantitative analysis of proteome coverage and recovery rates for upstream fractionation methods in proteomics. *J. Proteome Res.* **9**(4), 1902–1912.

Filiou, M.D., Turck, C.W., and Martins-de-Souza, D. (2011). Quantitative proteomics for investigating psychiatric disorders. *Proteomics Clin. Appl.* **5**(1–2), 38–49.

Frank, E., Kessler, M.S., Filiou, M.D., Zhang, Y., *et al.* (2009). Stable isotope metabolic labeling with a novel 15N-enriched bacteria diet for improved proteomic analyses of mouse models for psychopathologies. *PLoS ONE* **4**, e7821.

Gibb, S.J., Fergusson, D.M., and Horwood, L.J. (2010). Burden of psychiatric disorder in young adulthood and life outcomes at age 30. *Br. J. Psychiatry* **197**, 122–127.

Goodlett, D.R., Keller, A., Watts, J.D., *et al.* (2001). Differential stable isotope labeling of peptides for quantitation and de novo sequence derivation. *Rapid Commun. Mass Spectrom.* **15**, 1214–1221.

Görg, A., Postel, W., Weser, J., Patutschnick, W., and Cleve, H. (1985). Improved resolution of PI (alpha 1-antitrypsin) phenotypes by a large-scale immobilized pH gradient. *Am. J. Hum. Genet.* **37**(5), 922–930.

Görg, A., Postel, W., Weser, J., Günther, S., Strahler, J.R., Hanash, S.M., and Somerlot, L. (1987). Elimination of point streaking on silver stained two-dimensional gels by addition of iodoacetamide to the equilibration buffer. *Electrophoresis* **8**, 122–124.

Görg, A., Postel, W., Domscheit, A., and Günther, S. (1988). Two-dimensional electrophoresis with immobilized pH gradients of leaf proteins from barley (*Hordeum vulgare*): method, reproducibility and genetic aspects. *Electrophoresis* **9**(11), 681–692.

Görg, A., Boguth, G., Kopf, A., *et al.* (2002). Sample prefractionation with Sephadex isoelectric focusing prior to narrow pH range two-dimensional gels. *Proteomics* **2**, 1652–1657.

Gouw, J.W., Krijgsveld, J., and Heck, A.J.R. (2010). Quantitative proteomics by metabolic labeling of model organisms. *Mol. Cell. Proteomics* **9**, 11–24.

Guerrier, L., Lomas, L., and Boschetti, E. (2005). A simplified monobuffer multidimensional chromatography for high-throughput proteome fractionation. *J. Chromatogr. A* **1073**(1–2), 25–33.

Guest, P.C., Wang, L., Harris, L.W., Burling, K., Levin, Y., Ernst, A., Wayland, M.T., Umrania, Y., Herberth, M., Koethe, D., van Beveren, J.M., Rothermundt, M., *et al.* (2010). Increased levels of circulating insulin-related peptides in first-onset, antipsychotic naïve schizophrenia patients. *Mol. Psychiatry* **15**(2), 118–119.

Gygi, S.P., Rist, B., Gerber, S.A., Turecek, F., Gelb, M.H., and Aebersold, R. (1999a). Quantitative analysis of complex protein mixtures using isotope-coded affinity tags. *Nat. Biotechnol.* **17**(10), 994–999.

Gygi, S.P., Rochon, Y., Franza, B.R., and Aebersold, R. (1999b). Correlation between protein and mRNA abundance in yeast. *Mol. Cell. Biol.* **19**(3), 1720–1730.

Henzel, W.J., Billeci, T.M., Stults, J.T., Wong, S.C., Grimley, C., and Watanabe, C. (1993). Identifying proteins from two-dimensional gels by molecular mass searching of peptide fragments in protein sequence databases. *Proc. Natl. Acad. Sci. USA* **90**(11), 5011–5015.

Herberth, M., Koethe, D., Levin, Y., Schwarz, E., Krzyszton, N.D., Schoeffmann, S., Ruh, H., Rahmoune, H., Kranaster, L., Schoenborn, T., Leweke, M.F., Guest, P.C., *et al.* (2011). Peripheral profiling analysis for bipolar disorder reveals markers associated with reduced cell survival. *Proteomics* **11**(1), 94–105.

Hirano, M., Rakwal, R., Shibato, J., Agrawal, G.K., Jwa, N.S., Iwahashi, H., and Masuo, Y. (2006). New protein extraction/solubilization protocol for gel-based proteomics of rat (female) whole brain and brain regions. *Mol. Cells* **22**(1), 119–125.

Huang, J.T., Leweke, F.M., Tsang, T.M., Koethe, D., Kranaster, L., Gerth, C.W., Gross, S., Schreiber, D., Ruhrmann, S., Schultze-Lutter, F., Klosterkötter, J., Holmes, E., *et al.* (2007). CSF metabolic and proteomic profiles in patients prodromal for psychosis. *PLoS ONE* **2**(1), e756.

Huang, J.T., Wang, L., Prabakaran, S., Wengenroth, M., Lockstone, H.E., Koethe, D., Gerth, C.W., Gross, S., Schreiber, D., Lilley, K., Wayland, M., Oxley, D., *et al.* (2008). Independent protein-profiling studies show a decrease in apolipoprotein A1 levels in schizophrenia CSF, brain and peripheral tissues. *Mol. Psychiatry* **13**(12), 1118–1128.

Humpel, C. (2010). Identifying and validating biomarkers for Alzheimer's disease. *Trends Biotechnol.* **29**, 26–32.

Jacob, A.M., and Turck, C.W. (2008). Detection of post-translational modifications by fluorescent staining of two-dimensional gels. *Methods Mol. Biol.* **446**, 21–32.

Johnston-Wilson, N.L., Sims, C.D., Hofmann, J.P., Anderson, L., Shore, A.D., Torrey, E.F., and Yolken, R.H. (2000). Disease-specific alterations in frontal cortex brain proteins in schizophrenia, bipolar disorder, and major depressive disorder. The Stanley Neuropathology Consortium. *Mol. Psychiatry* **5**(2), 142–149.

Kaltschmidt, E., and Wittmann, H.G. (1970). Ribosomal proteins. VII. Two-dimensional polyacrylamide gel electrophoresis for fingerprinting of ribosomal proteins. *Anal. Biochem.* **36**(2), 401–412.

Khaitovich, P., Lockstone, H.E., Wayland, M.T., Tsang, T.M., Jayatilaka, S.D., Guo, A.J., Zhou, J., Somel, M., Harris, L.W., Holmes, E., Pääbo, S., and Bahn, S. (2008). Metabolic changes in schizophrenia and human brain evolution. *Genome Biol.* **9**(8), R124.

Klempan, T.A., Ernst, C., Deleva, V., Labonte, B., and Turecki, G. (2009). Characterization of QKI gene expression, genetics, and epigenetics in suicide victims with major depressive disorder. *Biol. Psychiatry* **66**, 824–831.

Krishnamurthy, D., Levin, Y., Harris, L.W., Umrania, Y., Bahn, S., and Guest, P.C. (2011). Analysis of the human pituitary proteome by data independent label-free liquid chromatography tandem mass spectrometry. *Proteomics* **11**(3), 495–500.

Krishnamurthy, D., Harris, L.W., Levin, Y., Koutroukides, T.A., Rahmoune, H., Leweke, F.M., Guest, P.C., Bahn, S. (submitted). Metabolic, hormonal and stress-related molecular changes in post-mortem pituitary glands from schizophrenia subjects. *World J. Biol. Psychiatry.* (in press).

Kromer, S.A., Kessler, M.S., Milfay, D., Birg, I.N., Bunck, M., Czibere, L., *et al.* (2005). Identification of glyoxalase-I as a protein marker in a mouse model of extremes in trait anxiety. *J. Neurosci.* **25**, 4375–4384.

Krüger, M., Moser, M., Ussar, S., Thievessen, I., Luber, C.A., Forner, F., Schmidt, S., Zanivan, S., Fässler, R., and Mann, M. (2008). SILAC mouse for quantitative proteomics uncovers kindlin-3 as an essential factor for red blood cell function. *Cell* **134**(2), 353–364.

Lan, M.J., McLoughlin, G.A., Griffin, J.L., Tsang, T.M., Huang, J.T., Yuan, P., Manji, H., Holmes, E., and Bahn, S. (2009). Metabonomic analysis identifies molecular changes associated with the pathophysiology and drug treatment of bipolar disorder. *Mol. Psychiatry* **14**(3), 269–279.

Lee, C. (2007). Protein extraction from mammalian tissues. *Methods Mol. Biol.* **362**, 385–389.

Levin, Y., and Bahn, S. (2010). Quantification of proteins by label-free LC–MS/MS. *Methods Mol. Biol.* **658**, 217–231.

Levin, Y., Schwarz, E., Wang, L., Leweke, F.M., and Bahn, S. (2007). Label-free LC-MS/MS quantitative proteomics for large-scale biomarker discovery in complex samples. *J. Sep. Sci.* **30**, 2198–2203.

Levin, Y., Wang, L., Schwarz, E., Koethe, D., Leweke, F.M., and Bahn, S. (2010). Global proteomic profiling reveals altered proteomic signature in schizophrenia serum. *Mol. Psychiatry* **15**(11), 1088–1100.

Li, G.Z., Vissers, J.P., Silva, J.C., Golick, D., Gorenstein, M.V., and Geromanos, S.J. (2009). Database searching and accounting of multiplexed precursor and product ion spectra from the data independent analysis of simple and complex peptide mixtures. *Proteomics* **9**(6), 1696–1719.

Liao, L., Park, S.K., Xu, T., Vanderklish, P., and Yates, J.R. 3rd. (2008). Quantitative proteomic analysis of primary neurons reveals diverse changes in synaptic protein content in fmr1 knockout mice. *Proc. Natl. Acad. Sci. USA* **105**(40), 15281–15286.

Link, A.J., Eng, J., Schieltz, D.M., Carmack, E., Mize, G.J., Morris, D.R., Garvik, B.M., and Yates, J.R. 3rd (1999). Direct analysis of protein complexes using mass spectrometry. *Nat. Biotechnol.* **17**(7), 676–682.

Ma, D., Chan, M.K., Lockstone, H.E., Pietsch, S.R., *et al.* (2009). Antipsychotic treatment alters protein expression associated with presynaptic function and nervous system development in rat frontal cortex. *J. Proteome Res.* **8**, 3284–3297.

Maccarrone, G., Turck, C.W., and Martins-de-Souza, D. (2010). Shotgun mass spectrometry workflow combining IEF and LC-MALDI-TOF/TOF. *Protein J.* **29**(2), 99–102.

Mann, M., Højrup, P., and Roepstorff, P. (1993). Use of mass spectrometric molecular weight information to identify proteins in sequence databases. *Biol. Mass Spectrom.* **22**(6), 338–345.

Martins-de-Souza, D., Menezes de Oliveira, B., dos Santos Farias, A., Horiuchi, R.S., Crepaldi Domingues, C., de Paula, E., Marangoni, S., Gattaz, W.F., Dias-Neto, E., and Camillo Novello, J. (2007). The use of ASB-14 in combination with CHAPS is the best for solubilization of human brain proteins for two-dimensional gel electrophoresis. *Brief. Funct. Genomic. Proteomic.* **6**(1), 70–75.

Martins-de-Souza, D., Gattaz, W.F., Schmitt, A., Rewerts, C., Maccarrone, G., Dias-Neto, E., and Turck, C.W. (2009a). Prefrontal cortex shotgun proteome analysis reveals altered calcium homeostasis and immune system imbalance in schizophrenia. *Eur. Arch. Psychiatry Clin. Neurosci.* **259**(3), 151–163.

Martins-de-Souza, D., Gattaz, W.F., Schmitt, A., Rewerts, C., Marangoni, S., Novello, J.C., Maccarrone, G., Turck, C.W., and Dias-Neto, E. (2009b). Alterations in oligodendrocyte proteins, calcium homeostasis and new potential markers in schizophrenia anterior temporal lobe are revealed by shotgun proteome analysis. *J. Neural Transm.* **116**(3), 275–289.

Martins-De-Souza, D., Dias-Neto, E., Schmitt, A., Falkai, P., Gormanns, P., Maccarrone, G., Turck, C. W., and Gattaz, W.F. (2010a). Proteome analysis of schizophrenia brain tissue. *World J. Biol. Psychiatry* **11**(2), 110–120.

Martins-de-Souza, D., Harris, L.W., Guest, P.C., Turck, C.W., and Bahn, S. (2010b). The role of proteomics in depression research. *Eur. Arch. Psychiatry Clin. Neurosci.* **260**(6), 499–506.

Martins-de-Souza, D., Maccarrone, G., Wobrock, T., Zerr, I., Gormanns, P., Reckow, S., Falkai, P., Schmitt, A., and Turck, C.W. (2010c). Proteome analysis of the thalamus and cerebrospinal fluid reveals glycolysis dysfunction and potential biomarkers candidates for schizophrenia. *J. Psychiatr. Res.* **44**(16), 1176–1189.

Martins-De-Souza, D., Wobrock, T., Zerr, I., Schmitt, A., Gawinecka, J., Schneider-Axmann, T., Falkai, P., and Turck, C.W. (2010d). Different apolipoprotein E, apolipoprotein A1 and prostaglandin-H2 D-isomerase levels in cerebrospinal fluid of schizophrenia patients and healthy controls. *World J. Biol. Psychiatry* **11**(5), 719–728.

Martins-de-Souza, D., Guest, P.C., Steeb, H., Pietsch, S., Rahmoune, H., Harris, L.W., and Bahn, S. (2011a). Characterizing the proteome of the human dorsolateral prefrontal cortex by shotgun mass spectrometry. *Proteomics* **11**, 2347–2353.

Martins-de-Souza, D., Lebar, M., and Turck, C.W. (2011b). Proteome analyses of cultured astrocytes treated with MK-801 and clozapine: similarities with schizophrenia. *Eur. Arch. Psychiatry Clin. Neurosci* **261**, 217–228.

Martins-de-Souza, D., Guest, P.C., Vanattou-Saifoudine, N., Wesseling, H., Rahmoune, H., Bahn, S. (2011c). The need for phosphoproteomic approaches in psychiatric research. *J. Psychatr. Res.* May 24. [Epub ahead of print]

Martosella, J., Zolotarjova, N., Liu, H., *et al.* (2006). High recovery HPLC separation of lipid rafts for membrane proteome analysis. *J. Proteome Res.* **5**, 1301–1312.

McLoughlin, G.A., Ma, D., Tsang, T.M., Jones, D.N., Cilia, J., Hill, M.D., Robbins, M.J., Benzel, I.M., Maycox, P.R., Holmes, E., and Bahn, S. (2009). Analyzing the effects of psychotropic drugs on metabolite profiles in rat brain using ^{1}H NMR spectroscopy. *J. Proteome Res.* **8**(4), 1943–1952.

O'Farrell, P.H. (1975). High resolution two-dimensional electrophoresis of proteins. *J. Biol. Chem.* **250** (10), 4007–4021.

O'Hayre, M., Salanga, C.L., Kipps, T.J., Messmer, D., Dorrestein, P.C., and Handel, T.M. (2010). Elucidating the CXCL12/CXCR4 signaling network in chronic lymphocytic leukemia through phosphoproteomics analysis. *PLoS ONE* **5**(7), e11716.

Ong, S.E., Blagoev, B., Kratchmarova, I., Kristensen, D.B., Steen, H., Pandey, A., and Mann, M. (2002). Stable isotope labeling by amino acids in cell culture, SILAC, as a simple and accurate approach to expression proteomics. *Mol. Cell. Proteomics* **1**, 376–386.

Ong, S.E., and Pandey, A. (2001). An evaluation of the use of two-dimensional gel electrophoresis in proteomics. *Biomol. Eng.* **18**(5), 195–205.

Orrick, L.R., Olson, M.O., and Busch, H. (1973). Comparison of nucleolar proteins of normal rat liver and Novikoff hepatoma ascites cells by two-dimensional polyacrylamide gel electrophoresis. *Proc. Natl. Acad. Sci. USA* **70**(5), 1316–1320.

Peng, J., and Gygi, S.P. (2001). Proteomics: the move to mixtures. *J. Mass Spectrom.* **36**(10), 1083–1091.

Pennington, K., Beasley, C.L., Dicker, P., Fagan, A., English, J., Pariante, C.M., Wait, R., Dunn, M.J., and Cotter, D.R. (2008). Prominent synaptic and metabolic abnormalities revealed by proteomic analysis of the dorsolateral prefrontal cortex in schizophrenia and bipolar disorder. *Mol. Psychiatry* **13**(12), 1102–1117.

Phan, J.H., Quo, C.F., and Wang, M.D. (2006). Functional genomics and proteomics in the clinical neurosciences: data mining and bioinformatics. *Prog. Brain Res.* **158**, 83–108.

Prabakaran, S., Wengenroth, M., Lockstone, H.E., Lilley, K., Leweke, F.M., and Bahn, S. (2007). 2-D DIGE analysis of liver and red blood cells provides further evidence for oxidative stress in schizophrenia. *J. Proteome Res.* **6**(1), 141–149.

Rabilloud, T. (1996). Solubilization of proteins for electrophoretic analyses. *Electrophoresis* **17**, 813–829, Review.

Raymond, S. (1964). Acrylamide gel electrophoresis. *Ann. N. Y. Acad. Sci.* **121**, 350–365.

Righetti, P.G. (2004). Bioanalysis: its past, present, and some future. *Electrophoresis* **25**(14), 2111–2127.

Righetti, P.G., Castagna, A., Antonioli, P., and Boschetti, E. (2005). Prefractionation techniques in proteome analysis: the mining tools of the third millennium. *Electrophoresis* **26**(2), 297–319.

Ross, P.L., Huang, Y.N., Marchese, J.N., *et al.* (2004). Multiplexed protein quantitation in *Saccharomyces cerevisiae* using amine-reactive isobaric tagging reagents. *Mol. Cell. Proteomics* **3**, 1154–1169.

Roulhac, P.L., Ward, J.M., Thompson, J.W., Soderblom, E.J., Silva, M., Moseley, M.A. 3rd, and Jarvis, E.D. (2011). Microproteomics: quantitative proteomic profiling of small numbers of laser-captured cells. *Cold Spring Harb. Protoc* (1):pdb.prot5573.

Rubinstein, J.C., Tran, N., Ma, S., Halaban, R., and Krauthammer, M. (2010). Genome-wide methylation and expression profiling identifies promoter characteristics affecting demethylation-induced gene up-regulation in melanoma. *BMC Med. Genomics* **3**, 4.

Sarker, D., and Workman, P. (2007). Pharmacodynamic biomarkers for molecular cancer therapeutics. *Adv. Cancer Res.* **96**, 213–268.

Sarker, D., Pacey, S., and Workman, P. (2007). Use of pharmacokinetic/pharmacodynamic biomarkers to support rational cancer drug development. *Biomark. Med.* **1**, 399–417.

Schmidt, A., Kellermann, J., and Lottspeich, F. (2005). A novel strategy for quantitative proteomics using isotope-coded protein labels. *Proteomics* **5**, 4–15.

Schwarz, E., Izmailov, R., Spain, M., Barnes, A., Mapes, J.P., Guest, P.C., Rahmoune, H., Pietsch, S., Leweke, F.M., Rothermundt, M., Steiner, J., Koethe, D., *et al.* (2010). Validation of a blood-based laboratory test to aid in the confirmation of a diagnosis of schizophrenia. *Biomark. Insights* **5**, 39–47.

Schwarz, E., Guest, P.C., Rahmoune, H., Wang, L., Levin, Y., Ingudomnukul, E., Ruta, L., Kent, L., Spain, M., Baron-Cohen, S., and Bahn, S. (2011). Sex-specific serum biomarker patterns in adults with Asperger's syndrome. *Mol. Psychiatry*, Sep 28. [Epub ahead of print].

Scott, G.K., Goga, A., Bhaumik, D., Berger, C.E., Sullivan, C.S., and benz, C.C. (2007). Coordinate suppression of ERBB2 and ERBB3 by enforced expression of micro-RNA miR-125a or miR-125b. *J. Biol. Chem.* **282**, 1479–1486.

Semenov, A., Goldsteins, G., and Castrén, E. (2006). Phosphoproteomic analysis of neurotrophin receptor TrkB signaling pathways in mouse brain. *Cell. Mol. Neurobiol.* **26**(2), 163–175.

Shapiro, A.L., Viñuela, E., and Maizel, J.V. Jr. (1967). Molecular weight estimation of polypeptide chains by electrophoresis in SDS-polyacrylamide gels. *Biochem. Biophys. Res. Commun.* **28**(5), 815–820.

Shaw, M.M., and Riederer, B.M. (2003). Sample preparation for two-dimensional gel electrophoresis. *Proteomics* **3**, 1408–1417.

Shevchenko, A., Wilm, M., Vorm, O., and Mann, M. (1996). Mass spectrometric sequencing of proteins silver-stained polyacrylamide gels. *Anal. Chem.* **68**, 850–858.

Sinha, T.K., Khatib-Shahidi, S., Yankeelov, T.E., Mapara, K., Ehtesham, M., Cornett, D.S., Dawant, B.M., Caprioli, R.M., and Gore, J.C. (2008). Integrating spatially resolved three-dimensional MALDI IMS with in vivo magnetic resonance imaging. *Nat. Methods* **5**(1), 57–59.

Smithies, O. (1955). Zone electrophoresis in starch gels: group variations in the serum proteins of normal human adults. *Biochem. J.* **61**(4), 629–641.

Steiner, J., Schroeter, M.L., Schiltz, K., Bernstein, H.G., Müller, U.J., Richter-Landsberg, C., Müller, W.E., Walter, M., Gos, T., Bogerts, B., and Keilhoff, G. (2010). Haloperidol and clozapine decrease S100B release from glial cells. *Neuroscience* **167**(4), 1025–1031.

Unlü, M., Morgan, M.E., and Minden, J.S. (1997). Difference gel electrophoresis: a single gel method for detecting changes in protein extracts. *Electrophoresis* **18**(11), 2071–2077.

Urbas, L., Brne, P., Gabor, B., Barut, M., *et al.* (2009).*J. Chromatogr. A* **1216**, 2689–2694.

Volinia, S., Calin, G.A., Liu, C.G., Ambs, S., Cimmino, A., Petrocca, F., Visone, R., Iorio, M., Roldo, C., Ferracin, M., Prueitt, R.L., Yanaihara, N., *et al.* (2006). A microRNA expression signature of human solid tumors defines cancer gene targets. *Proc. Natl. Acad. Sci. USA* **103**, 2257–2261.

Wang, L., Lockstone, H.E., Guest, P.C., Levin, Y., Palotás, A., Pietsch, S., Schwarz, E., Rahmoune, H., Harris, L.W., Ma, D., and Bahn, S. (2010). Expression profiling of fibroblasts identifies cell cycle abnormalities in schizophrenia. *J. Proteome Res.* **9**(1), 521–527.

Wilkins, M.R., Sanchez, J.C., Gooley, A.A., Appel, R.D., Humphery-Smith, I., *et al.* (1996). Progress with proteome projects: why all proteins expressed by a genome should be identified and how to do it. *Biotechnol. Genet. Eng. Rev.* **13**, 19–50.

Wu, C.C., MacCoss, M.J., Howell, K.E., Matthews, D.E., and Yates, J.R. 3rd (2004). Metabolic labeling of mammalian organisms with stable isotopes for quantitative proteomic analysis. *Anal. Chem.* **76**, 4951–4959.

Yang, X., Levin, Y., Rahmoune, H., Ma, D., Schöffmann, S., Umrania, Y., Guest, P.C., and Bahn, S. (2011). Comprehensive two-dimensional liquid chromatography mass spectrometric profiling of the rat hippocampal proteome. *Proteomics* **11**(3), 501–505.

Zhang, G., Fang, B., Liu, R.Z., Lin, H., Kinose, F., Bai, Y., Oguz, U., Remily-Wood, E.R., Li, J., Altiok, S., Eschrich, S., Koomen, J., *et al.* (2011). Mass spectrometry mapping of epidermal growth factor receptor phosphorylation related to oncogenic mutations and tyrosine kinase inhibitor sensitivity. *J. Proteome Res.* **10**(1), 305–319.

Zhou, H., Albuquerque, C.P., Liang, J., Suhandynata, R.T., and Weng, S. (2010). Quantitative phosphoproteomics: new technologies and applications in the DNA damage response. *Cell Cycle* **9**(17), 3479–3484.

CONVERGING EVIDENCE OF BLOOD-BASED BIOMARKERS FOR SCHIZOPHRENIA: AN UPDATE

Man K. Chan[1], Paul C. Guest[1], Yishai Levin[1], Yagnesh Umrania[1], Emanuel Schwarz[1], Sabine Bahn[1,2] and Hassan Rahmoune[1]

[1]Department of Chemical Engineering and Biotechnology, University of Cambridge, Cambridge, United Kingdom
[2]Department of Neuroscience, Erasmus Medical Centre, Rotterdam, The Netherlands

Abstract

This chapter has carried out a review of the literature and combined this with the results of in-house studies to identify candidate blood-based biomarkers for schizophrenia and antipsychotic drug response. Literature searches retrieved 185 publications describing a total of 273 schizophrenia biomarkers identified in

INTERNATIONAL REVIEW OF
NEUROBIOLOGY, VOL. 101
DOI: 10.1016/B978-0-12-387718-5.00005-5

serum and/or plasma. Examination of seven in-house multicenter studies resulted in the identification of 137 serum/plasma biomarkers. Taken together, the findings suggested an ongoing immunological and inflammatory process in schizophrenia. This was accompanied by altered cortisol levels which suggested activated stress response and altered hypothalamic–pituitary–adrenal axis function in these patients. The authors conclude that such biomarkers may prove useful as additional parameters for characterizing specific immune and/or metabolic or hormonal subsystems in schizophrenia and might, therefore, facilitate the development of future patient stratification and personalized medicine strategies.

I. Introduction

Schizophrenia is a severe, chronic, and debilitating mental illness that affects advanced functions of the human brain (Wong and Van Tol, 2003). It represents a major source of suffering affecting over 1% of the world's population, and death by suicide occurs in 10% of cases (Thaker and Carpenter, 2001). The early onset of the disease and associated long-lasting social dysfunction has major implications for health and productivity throughout the world. The high rates of unemployment and hospitalization, high treatment costs, and the need to provide lifelong care impose a considerable financial burden on society that surpasses most other illnesses (Rice, 1999; Rice and Miller, 1995). The incidence of schizophrenia ranges from 8 to 43/100,000/year (McGrath *et al.*, 2004; Tandon *et al.*, 2008) with males more frequently affected than females (McGrath, 2006). The median lifetime prevalence (number of cases at any given time or time period) is estimated to be 4 per 1000 (Saha *et al.*, 2005).

Schizophrenia manifests over time and the development of frank psychotic symptoms marks the first onset of the disease. This usually begins in late adolescence or early adulthood, typically peaking at 20–25 years of age and followed by a decline with age reaching a minimum by 45–50 years (Jablensky, 2000). The spectrum of symptoms includes positive symptoms (e.g., hallucinations, delusions and thought disorder; Morris *et al.*, 2005), negative symptoms [e.g., depression, disturbances in social interaction, lack of motivation, and inability to experience pleasure (anhedonia) (Lewis, 2000)] and cognitive impairments (impaired executive function, selective attention, working memory, and mental flexibility; Weickert *et al.*, 2000). The course of schizophrenia can be episodic with one or more episodes (22% of patients) or without interepisode residual symptoms (35% of patients), although the majority of patients (43%) exhibit symptoms throughout all or most of the course of their illness with no return to normality (Wildenauer *et al.*, 2009).

Despite almost a century of research, knowledge regarding the pathogenesis and etiology of schizophrenia remains incomplete and diagnosis is still problematic.

Diagnosis currently relies on subjective assessments obtained from the patient's self-reported symptoms, mental status examinations, and the clinician's observations in line with the classifications listed in the Diagnostic and Statistical Manual of Mental Disorders 4th Edition (DSM-IV) or the International Statistical Classification of Diseases and Related Health Problems 10th Revision (ICD-10). Correct diagnosis may take months to years to complete and there is considerable room for error as symptoms may overlap with those in a number of psychiatric conditions (e.g., drug-induced psychoses, delirium, and dementia), psychotic mood disorders, and personality disorders (Lakhan and Kramer, 2009; Lakhan *et al.*, 2010). As a result, there has been a recent and rapid shift toward the study of specific and sensitive molecular biomarkers in psychiatry. The hope is that such biomarkers will complement the purely syndromal diagnostic procedures. Biomarker research has already had a positive impact in other branches of medicine including oncology, rheumatology, cardiology, and obstetrics (Cook, 2008). Recently, Psynova Neurotech successfully launched the first biomarker-based blood test designed to aid psychiatrists in the diagnosis of recent-onset schizophrenia (http://www.psynova-neurotech.com/default.htm; Schwarz *et al.*, 2010; see Chapter "The application of multiplexed assay systems for molecular diagnostics" by Schwarz *et al.*).

Relative to organ tissue (e.g., brain), body fluids such as blood, urine, and cerebral spinal fluid (CSF) represent more easily accessible sources for detection of systemic biomarkers. The blood, for example, can be sampled using standardized and routine clinical procedures without significant patient discomfort. The main advantages associated with using blood as a source of biomarkers include the fact that it is possible to design standardized sample collection procedures, it is available in sufficient quantities, and it can be sampled on multiple occasions with relative ease (Lakhan and Kramer, 2009). Biomarker studies in psychiatry have undergone a fundamental methodological shift from searching for cause to estimating the probability that a condition is present or may develop (Singh and Rose, 2009). Diagnostic biomarkers may be used for identification of the disease early in its course. This may be critical for remission since evidence suggests that delays in diagnosis and intervention lead to poorer prognoses (Lakhan and Kramer, 2009). Biomarkers with prognostic value may facilitate prediction of disease development, the likely course and outcome of illness, and certain disease-associated behaviors and personality traits. Further, biomarkers may also enable prediction of the type, timing, course, and response to treatment, and may ultimately enable disease subtyping and patient stratification. Integration of drug-response (DR) biomarkers into drug discovery programs may promote development of novel therapeutics along with a personalized medicine approach. Biomarkers may inform whether such therapeutics are worth pursuing and facilitate subsequent go/no go decisions regarding safety and efficacy. This information could be applied to drug design to facilitate selective targeting of relevant patient populations and aid dose selection/adjustment which will ultimately enable selection of potential drug responders (Singh and Rose, 2009; Tesch *et al.*, 2010).

The aim of this chapter was to compile a robust signature of blood-based biomarkers of schizophrenia with the overall aim of identifying correlated groups/ converging pathways associated with these molecules. For this purpose, we examined literature findings on peripheral biomarkers of schizophrenia using Pubmed (http://www.ncbi.nlm.nih.gov/pubmed/) between 1995 and April 2010, and analyzed our in-house and multicenter studies investigating serum/plasma protein profiles from first-onset drug-naive/free schizophrenia patients. It was anticipated that analysis of the latter would enable the identification of early stage diagnostic disease biomarkers. The in-house studies consisted of analysis of serum from five clinical centers using the Rules-Based Medicine (RBM) Discovery Multi-Analyte Profiling (MAP) platform (Schwarz *et al.*, 2010) and two independent cohorts analyzed by liquid chromatography–mass spectrometry (LC–MS) profiling (Levin and Bahn, 2010). One advantage of incorporating our in-house data into this study was the fact that stringent exclusion criteria were applied, and that patient and control subjects were matched for potential confounding factors. This was critical to ensure comparability, reliability, and reproducibility of the findings considering interindividual heterogeneities. The most influential confounding factors in biomarker studies include age, gender, illness duration, medication, disease stage and subtype (if known), medication, smoking, substance abuse, and body mass index (BMI). Washout periods should also be considered to minimize effects of prior medication (Drzyzga *et al.*, 2006). We focused solely on analysis of serum and plasma biomarkers identified from the literature and in-house studies.

II. Methodology

A. Compilation of Literature Serum/Plasma Biomarkers

The keywords used for Pubmed searching included "schizophrenia" and "blood" or "serum" or "plasma" or "CSF" or "cells." Only schizophrenia studies reporting protein or mRNA biomarkers with gene names were included. Only articles written in English and published between 1995 and April 2010 were included. Comprehensive details from these studies were recorded in an excel spreadsheet and the information compiled from each article included biomarker name (gene and protein names), type [diagnostic (D), DR, and diagnostic or drug-response (DorDR) biomarkers], assay type [e.g., immunoassay, Western blot, MS, reverse transcription-polymerase chain reaction (RT-PCR), etc.], study finding (increase or decrease in level), study design (number of patients/controls), sample source (e.g., serum, CSF, plasma, etc.), Pubmed ID (PMid), comments on each article (e.g., controls for confounding factors), journal title, and impact factor. The list of significant biomarkers extracted was subjected to *in silico* functional pathway analysis as described below.

B. Compilation of In-House Serum Biomarkers

We extracted the list of medium-high abundance serum proteins which were found to be present at significantly different levels in two independent cohorts of schizophrenia and control subjects, using the LC–MS platform (cohort 1; Levin *et al.*, 2010) and (cohort 2). We also used the list of 51 serum molecules found to be at differential levels in serum from 250 schizophrenia patients compared to 230 controls from five clinical centers using the Discovery MAP platform (RBM; http://www.rulesbasedmedicine.com/products-services/humanmap-services/humandiscoverymap/; Schwarz *et al.*, 2010). These latter molecules were incorporated into a 51-plex immunoassay panel which was used for validation of a large independent cohort (clinical center 6) of schizophrenia ($n = 577$) and control ($n = 229$) subjects. Briefly, the Discovery MAP is a multiplex immunoassay platform which simultaneously profiles 188 molecules including a range of immune, metabolic, and hormonal markers along with growth factors, tissue remodeling proteins, and cancer markers. For the purposes of this review, we analyzed a subset of demographically matched individuals from the five clinical centers (243 schizophrenia patients and 224 controls). This was carried out so that patients and control BMIs could be matched in cohort 3. Patient and control groups were matched (*T*-test and Fisher's exact test) for age, gender, BMI, disease stage (all first onset), medication status (all drug-naive or unmedicated for at least 6 weeks), time of sample collection, cumulative lifetime medication [applicable for center 3 as some patients had received previous medication: two molecules (leptin and serum amyloid P) correlating with cumulative lifetime medication were excluded]. For clinical centers 1, 3, and 4, information on BMI, smoking, and cannabis use were available and matched accordingly (Table I). For each center, raw MAP data were filtered using cutoff criteria such that molecules undetected in >60% of individuals/group (schizophrenia or control) were excluded. Significance was set at $p < 0.05$ using Student's *T*-test. The resulting molecular lists from each center were combined to produce a final nonredundant list which was consistent with our previous findings (Schwarz *et al.*, 2010).

C. *In Silico* Functional Pathway Analysis

The lists of molecules compiled from the literature and in-house datasets were subjected to functional pathway analysis using the Web-based Ingenuity Pathways Analysis (IPA) tool (Ingenuity® Systems; www.ingenuity.com). Briefly, this analysis was used to identify the most significant biological functions ("diseases and disorders," "molecular and cellular functions," and "physiological system development and function") and canonical pathways associated with the molecules. For functional analysis, right-tailed Fisher's exact test was used to calculate a

Table I
Demographic Details of Subjects that Participated in the In-House Studies (Mean ± SD).

	Control	Schizophrenia	*T*-test (*p*-value)	Fisher's exact test (*p*-value)
Clinical center 1 for MAP				
Age	30 ± 7.80	31 ± 0.10	0.59	NA
Gender (F/M)	28/31	29/42	NA	0.48
BMI	23 ± 3.60	24 ± 4.90	0.51	NA
Smoking (y/n/NA)	25/34/0	25/23/23	NA	0.34
Cannabis (y/n/NA)	31/25/3	33/22/16	NA	0.70
Clinical center 2 for MAP				
Age	27 ± 9.29	27 ± 9.39	0.98	NA
Gender (F/M)	11/35	11/35	NA	1
BMI	NA	NA	NA	NA
Smoking (y/n/NA)	NA	NA	NA	NA
Cannabis (y/n/NA)	NA	NA	NA	NA
Clinical center 3 for MAP				
Age	36 ± 11.42	38 ± 12.55	0.54	NA
Gender (F/M)	16/23	16/23	NA	1
BMI	25 ± 3.51	27 ± 5.32	0.06	NA
Smoking (y/n/NA)	11/28	18/21	NA	0.16
Cannabis (y/n/NA)	0/39	2/37	NA	0.49
Clinical center 4 for MAP				
Age	35 ± 11.10	35 ± 9.94	0.81	NA
Gender (F/M)	13/27	14/26	NA	1
BMI	24 ± 3.35	25 ± 4.56	0.36	NA
Smoking (y/n/NA)	18/22	22/18	NA	0.50
Cannabis (y/n/NA)	NA	NA	NA	NA
Clinical center 5 for MAP				
Age	27 ± 4.08	26 ± 7.46	0.68	NA
Gender (F/M)	7/33	11/36	NA	0.60
BMI	NA	NA	NA	NA
Smoking (y/n/NA)	NA	33/14	NA	NA
Cannabis (y/n/NA)	NA	23/24	NA	NA
Cohort 1 for LC–MS				
Age	28 ± 7	29.0 ± 11	0.71	NA
Gender (F/M)	15/18	7/15	NA	0.31
BMI	NA	NA	NA	NA
Smoking (y/n/NA)	2/15/16	5/11/6	NA	0.22
Cannabis (y/n/NA)	11/19/3	14/3/5	NA	0.005
Cohort 2 for LC–MS				
Age	29.1 ± 7.67	30.2 ± 7.53	0.65	NA
Gender (F/M)	8/9	10/10	NA	1
BMI	22.8 ± 3.1	23.6 ± 3.4	0.48	NA
Smoking (y/n/NA)	2/15	5/11/4	NA	0.23
Cannabis (y/n/NA)	10/7	13/5/2	NA	0.49

p-value determining the probability that the assignment of each biological function and/or disease was due to chance alone. For the canonical pathway analysis, the significance of the association between molecules in the dataset and the canonical pathway was measured in two ways: (1) the ratio of the number of molecules from the dataset associated with the pathway was divided by the total number of molecules known to map onto that pathway; (2) Fisher's exact test was used to calculate a *p*-value, determining the probability that the association between the biomarkers in the dataset and the canonical pathway was explained by chance alone.

III. Results

A. Evidence from the Literature Review

Altogether, literature searches returned 185 biomarker-reporting publications, describing a total of 365 schizophrenia biomarkers identified in peripheral tissue. Of these 365 molecules, 273 were identified in serum or plasma and classified into diagnostic ($n = 81$), DR ($n = 77$), and DorDR ($n = 115$) biomarkers. Of these 273 molecules, 103 were unique/nonredundant. Each time a molecule was reported with altered levels, it was recorded as an entry. For example, if in one publication a molecule was found to be increased in unmedicated patients and decreased following drug treatment, two entries for this molecule were recorded: one as a diagnostic biomarker increased in abundance and one as a DR biomarker decreased in abundance. Overall, 75 of these 103 molecules were involved in the inflammatory response, 37 were members of the cytokine and neurotrophic factor families, and 36 were classified as acute phase proteins (APPs). The biomarkers most frequently identified as altered in schizophrenia (reported ≥ 5 times in the literature) included 11 molecules: *BDNF* (brain-derived neurotrophic factor), *S100B* (S100 calcium-binding protein B), *PRL* (prolactin), *IL6* (interleukin 6), *IL2* (interleukin 2), *INS* (insulin), *TNF* (tumor necrosis factor), *LEP*, *IL1RN* (interleukin 1 receptor antagonist), *IL8* (interleukin 8), *IL2RA* (interleukin 2 receptor alpha; Table II).

Of the 103 biomarkers, 40 were assigned to the D group (altered in drug-naive/free patients), 29 were DR (altered in response to drug treatment), and 76 were DorDR biomarkers (biomarker category/type not stated). Of the 40 D biomarkers, 15 were found only in the D group, suggesting that these could potentially have a role in the etiology of schizophrenia. Of the 29 DR biomarkers, only 5 were uniquely assigned to this group, and of the 76 DorDR biomarkers, 54 were unique to the DorDR group. Overall, there were 15 molecules in

Table II

Summary of the Literature Findings on Blood-Based Biomarkers of Schizophrenia.

Protein name	Gene name	Diagnostic			Drug response			Diagnostic or drug response			Total
		Up	Down	Total	Up	Down	Total	Up	Down	Total	
Brain-derived neurotrophic factor	*BDNF*	–	↓9	9	↑3	↓1	4	↑4	↓6	10	23
S100 calcium-binding protein B	S100B	↑8	–	8	↑5	↓2	7	↑5	↓1	6	21
Prolactin	*PRL*	↑2	↓1	3	↑7	↓7	14	↑1	–	1	18
Interleukin 6	*IL6*	↑5	↓1	6	↑5	↓1	6	↑4	–	4	16
Interleukin 2	*IL2*	↑6	↓1	7	↑1	↓2	3	↑1	↓1	2	12
Insulin	*INS*	↑1	–	1	↑4	–	4	↑4	–	4	9
Tumor necrosis factor	*TNF*	↑3	–	3	–	↓2	2	↑3	–	3	8
Leptin	*LEP*	↑1	–	1	↑4	–	4	↑3	–	3	8
Interleukin 1 receptor antagonist	IL1RN	↑3	–	3	↑2	–	2	↑1	–	1	6
Interleukin 8	*IL8*	↑2	–	2	↑1	–	1	↑3	–	3	6
Interleukin 2 receptor, alpha	*IL2RA*	↑2	–	2	↑2	–	2	↑1	–	1	5
C-reactive protein	*CRP*	↑1	–	1	↑2	–	2	↑1	–	1	4
Nerve growth factor	*NGF*	–	↓1	1	–	↓2	2	–	↓1	1	4
Epidermal growth factor	*EGF*	–	–	–	–	–	–	↑1	↓2	3	3
Interleukin 1, beta	IL1B	↑1	–	1	–	↓1	1	↑1	–	1	3
Interleukin 4	*IL4*	–	↓1	1	–	–	–	–	↓2	2	3
Superoxide dismutase 1	*SOD1*	–	–	–	↑1	–	1	↑2	–	2	3
Adiponectin, C1Q	*ADIPOQ*	–	–	–	–	↓2	2	–	↓1	1	3
Interleukin-12 receptor	*IL2R*	–	–	–	↑2	–	2	↑1	–	1	3
Secretoglobin, family 1A, 1	SCGB1A1	–	↓2	2	↑1	–	1	–	–	–	3
Apolipoprotein D	*APOD*	↑1	↓1	2	↑1	–	1	–	–	–	3
Apolipoprotein A-I	*APOA1*	–	↓2	2	–	–	–	–	↓1	1	3
Intercellular adhesion molecule 1	*ICAM1*	–	↓1	1	–	↓1	1	↑1	–	1	3
Renal tumor antigen	*RAGE*	↑1	–	1	↑1	–	1	–	–	–	2

Insulin-like growth factor 1	*IGF1*	–	↓1	1	–	↓1	1	–	–	–	2
Albumin	*ALB*	–	↓1	1	–	–	–	–	↓1	1	2
Interferon, gamma	*IFNG*	–	↓1	1	–	–	–	↑1	–	1	2
Interleukin 10	*IL10*	–	–	–	–	–	–	↑1	↓1	2	2
Chemokine (C-C motif) ligand 11	*CCL11*	–	–	–	↑1	–	1	↑1	–	1	2
Chemokine (C-C motif) ligand 2	*CCL2*	–	–	–	–	–	–	↑2	–	2	2
Tumor necrosis factor receptor, 1A	*TNFRSF1A*	–	–	–	–	–	–	↑2	–	2	2
Von Willebrand factor	*VWF*	–	–	–	–	–	–	↑2	–	2	2
Apolipoprotein A-IV	*APOA4*	↑1	↓1	2	–	–	–	–	–	–	2
Interleukin 6 receptor	*IL6R*	↑1	–	1	–	↓1	1	–	–	–	2
Haptoglobin	*HP*	↑1	–	1	–	–	–	–	–	–	1
Alpha-2-macroglobulin	*A2M*	–	–	–	–	–	–	–	↓1	1	1
Acidphosphatase, prostate	*ACPP*	–	–	–	–	–	–	↑1	–	1	1
Alpha-2-HS-glycoprotein	*AHSG*	–	↓1	1	–	–	–	–	–	–	1
Amyloid P component, serum	*APCS*	–	–	–	–	–	–	↑1	–	1	1
Apolipoprotein A-II	*APOA2*	–	↓1	1	–	–	–	–	–	–	1
Apolipoprotein C-I	*APOC1*	–	↓1	1	–	–	–	–	–	–	1
Apolipoprotein C-III	*APOC3*	–	–	–	–	–	–	↑1	–	1	1
Apolipoprotein H (beta-2-glycoprotein I)	*APOH*	–	–	–	–	–	–	↑1	–	1	1
Beta-2-microglobulin	*B2M*	↑1	–	1	–	–	–	–	–	–	1
Complement component 3	*C3*	–	–	–	–	–	–	↑1	–	1	1
Catalase	*CAT*	–	–	–	–	–	–	↑1	–	1	1
Chemokine (C-C motif) ligand 22	*CCL22*	–	–	–	–	–	–	↑1	–	1	1
Chemokine (C-C motif) ligand 4	*CCL4*	–	–	–	–	–	–	↑1	–	1	1
Chemokine (C-C motif) ligand 5	*CCL5*	–	–	–	–	–	–	↑1	–	1	1
CD5 molecule-like	*CD5L*	–	↓1	1	–	–	–	–	–	–	1
Carcinoembryonic antigen-related cell adhesion molecule 5	*CEACAM5*	–	–	–	–	–	–	↑1	–	1	1
Complement factor B	*CFB*	↑1	–	1	–	–	–	–	–	–	1
Ceruloplasmin (ferroxidase)	*CP*	–	–	–	–	–	–	↑1	–	1	1
Colony-stimulating factor 2	*CSF2*	–	–	–	–	–	–	–	↓1	1	1

(Continued)

Table II *(Continued)*

Protein name	Gene name	Diagnostic			Drug response			Diagnostic or drug response			Total
		Up	Down	Total	Up	Down	Total	Up	Down	Total	
Chemokine (C-X-C motif) ligand 5	*CXCL5*	–	–	–	–	–	–	↑1	–	1	1
Defensin, alpha 1	*DEFA1*	↑1	–	1	–	–	–	–	–	–	1
Dipeptidyl-peptidase 4	*DPP4*	–	–	–	–	–	–	↑1	–	1	1
Endothelin 1	*EDN1*	–	–	–	–	–	–	↑1	–	1	1
Coagulation factor XIII, B polypeptide	*F13B*	–	↓1	1	–	–	–	–	–	–	1
Coagulation factor III	*F3*	–	–	–	–	–	–	↑1	–	1	1
Coagulation factor VII	F7	–	–	–	–	–	–	↑1	–	1	1
Fibrinogen alpha chain	*FGA*	–	–	–	–	–	–	↑1	–	1	1
Fibroblast growth factor 2 (basic)	*FGF2*	–	–	–	↑1	–	1	–	–	–	1
Follicle-stimulating hormone, betapolypeptide	*FSHB*	–	–	–	–	↓1	1	–	–	–	1
Growth hormone 1	*GH1*	–	–	–	–	–	–	–	↓1	1	1
Ghrelin/obestatin prepropeptide	*GHRL*	–	–	–	–	↓1	1	–	–	–	1
Glutathione peroxidase	*GPX*	–	–	–	–	–	–	↑1	–	1	1
Glutathione *S*-transferase alpha 1	*GSTA1*	–	–	–	–	–	–	↑1	–	1	1
Insulin-like growth factor-binding protein 1	*IGFBP1*	–	–	–	–	–	–	–	↓1	1	1
Immunoglobulin heavy constant mu	*IGHM*	–	↓1	1	–	–	–	–	–	–	1
Interleukin-12	*IL12*	↑1	–	1	–	–	–	–	–	–	1
Interleukin 12 A	*IL12A*	–	–	–	–	–	–	↑1	–	1	1
Interleukin 12 B	*IL12B*	–	–	–	–	–	–	–	↓1	1	1
Interleukin 13	*IL13*	–	–	–	–	–	–	–	↓1	1	1
Interleukin 15	*IL15*	–	–	–	–	–	–	↑1	–	1	1
Interleukin 16	*IL16*	–	–	–	–	–	–	↑1	–	1	1

Interleukin 18	*IL18*	–	–	–	–	–	–	↑1	–	1	1
Interleukin 1, alpha	*IL1A*	–	–	–	–	–	–	–	↓1	1	1
Interleukin 3	*IL3*	–	–	–	–	–	–	–	↓1	1	1
Inhibin, beta B	*INHBB*	–	–	–	↑1	–	1	–	–	–	1
KIT ligand	*KITLG*	–	–	–	–	–	–	↑1	–	1	1
Lymphotoxin alpha	*LTA*	–	–	–	–	–	–	–	↓1	1	1
Midkine	*MDK*	–	↓1	1	–	–	–	–	–	–	1
Matrixmetallopeptidase 2	*MMP2*	–	–	–	–	–	–	–	↓1	1	1
Matrixmetallopeptidase 9	*MMP9*	–	–	–	–	–	–	↑1	–	1	1
Mucin 16, cell surface associated	*MUC16*	–	–	–	–	–	–	↑1	–	1	1
Neural cell adhesion molecule 1	*NCAM*	–	–	–	–	–	–	↑1	–	1	1
Oxytocin, prepropeptide	*OXT*	–	–	–	–	↓1	1	–	–	–	1
Reelin	*RELN*	–	–	–	–	–	–	↑1	–	1	1
Selectin L	*SELL*	–	–	–	–	–	–	↑1	–	1	1
Serpin peptidase inhibitor, cladeA, 1	*SERPINA1*	↑1	–	1	–	–	–	↑1	–	1	2
Serpin peptidase inhibitor, cladeA, 7	*SERPINA7*	–	–	–	–	–	–	↑1	–	1	1
Serpin peptidase inhibitor, cladeE, 1	*SERPINE1*	–	–	–	–	–	–	↑1	–	1	1
Somatostatin	*SST*	–	–	–	–	–	–	↑1	–	1	1
Transferrin	*TF*	–	↓1	1	–	–	–	–	–	–	1
Transforming growth factor, beta 1	*TGFB1*	–	–	–	–	–	–	↑1	–	1	1
Thrombopoietin	*THPO*	–	–	–	–	–	–	↑1	–	1	1
TIMP metallopeptidase inhibitor 1	*TIMP1*	–	–	–	–	–	–	↑1	–	1	1
Tumor necrosis factor receptor superfamily, 14	*TNFRSF14*	–	–	–	–	–	–	↑1	–	1	1
Tumor necrosis factor receptor superfamily, 1B	*TNFRSF1B*	–	–	–	–	–	–	↑1	–	1	1
Transthyretin	*TTR*	–	↓1	1	–	–	–	–	–	–	1
Vascular cell adhesion molecule 1	*VCAM1*	–	–	–	–	–	–	↑1	–	1	1
Vascular endothelial growth factor A	*VEGFA*	–	–	–	–	–	–	↑1	–	1	1

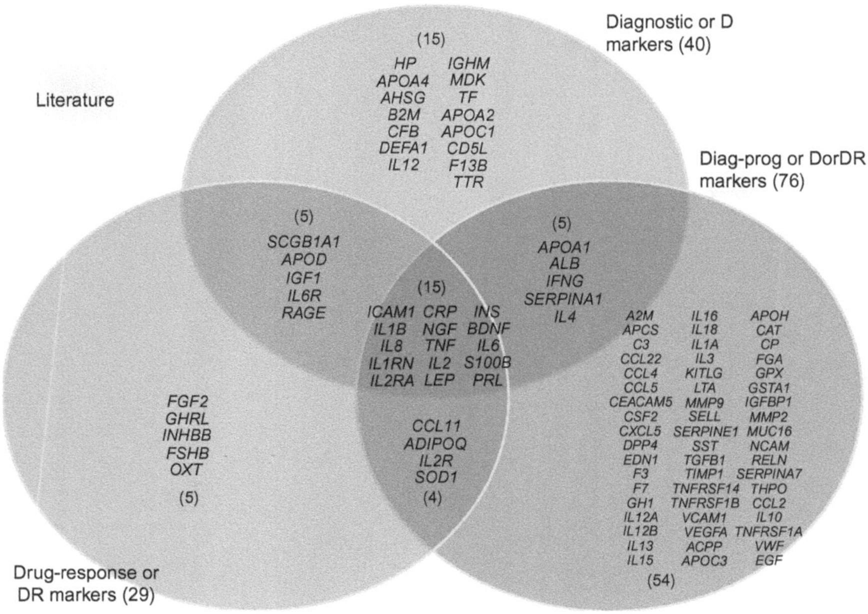

FIG. 1. Types of literature biomarkers.

common among all three biomarker categories, 5 were grouped into the D and DR groups, 5 were common between the D and DorDR groups, and 4 were common between both the DR and DorDR groups (Fig. 1).

B. Evidence from In-House Studies: Biomarkers of First-Onset Schizophrenia

The evidence from the literature for an ongoing immunological/inflammatory component in schizophrenia was consistent with the findings from in-house multicenter analysis of first-onset drug-naive/free patients. MAP analysis of serum from the five clinical centers resulted in the identification of 97 nonredundant molecules which were altered in abundance. Of these, 70 molecules were involved in inflammatory response, 27 were members of the cytokine or neurotrophic factor families, and 10 were APPs. Consistent with our previous findings (Schwarz *et al.*, 2010), 25 out of the 70 molecules were replicated in ≥ 3 centers and the directions of change were mostly consistent across the different centers (Table III). Interestingly, cortisol was found to be increased in four out of the five clinical centers. This molecule is primarily secreted by the adrenal glands under control of the hypothalamic–pituitary–adrenal (HPA) axis in response to stress or

Table III

Significantly Altered Molecules ($p < 0.05$) Identified in All the Clinical Centers (MAP Studies).

Name	Gene	Replication	Clinical Center 1	Clinical Center 2	Clinical Center 3	Clinical Center 4	Clinical Center 5
Haptoglobin	*HP*	5	↑	↑	↑	↑	↑
Glutathione *S*-transferase alpha 1	*GSTA1*	5	↑	↓	↑	↑	↑
Pancreatic polypeptide	*PPY*	4	↑	–	↑	↑	↑
Colony-stimulating factor 3 (granulocyte)	*CSF3*	4	↑	↑	–	↑	↑
Interleukin 15	*IL15*	4	–	↑	↑	↓	↑
Interleukin 7	*IL7*	4	–	↑	↑	↓	↑
Carcinoembryonic antigen-related cell adhesion molecule 5	*CEACAM5*	4	↑	↑	↑	–	↑
Interleukin 10	*IL10*	4	↑	↑	↑	–	↑
Apolipoprotein H (beta-2-glycoprotein)	*APOH*	4	↑	↑	↑	–	↑
Cortisol	–	4	–	↑	↑	↑	↑
Connective tissue growth factor	*CTGF*	3	↑	↑	↑	–	–
Resistin	*RETN*	3	↓	–	–	↑	↑
Rumor necrosis factor receptor superfamily, member 10c	*TNFRSF10C*	3	–	↓	–	↑	↑
Chemokine (C-C motif) ligand 22	*CCL22*	3	–	↑	↑	–	↑
Tumor necrosis factor receptor superfamily, member 14	*TNFRSF14*	3	–	↑	↑	–	↑
Glycoprotein hormones, alpha polypeptide	*CGA*	3	–	↑	–	↑	↑
Intercellular adhesion molecule 1	*ICAM1*	3	–	↑	↑	–	↑
Macrophage migration inhibitory factor	*MIF*	3	↑	↑	–	↑	–
Prolactin	*PRL*	3	↑	–	↑	–	↑
Vascular endothelial growth factor A	*VEGFA*	3	↓	–	↑	↑	–
Interleukin 5	*IL5*	3	↓	↓	–	↓	–
Complement component 3	*C3*	3	–	↑	↑	–	↑
Von Willebrand factor	*VWF*	3	–	–	↑	↑	↑
Interleukin 1 receptor antagonist	*IL1RN*	3	–	–	↑	↑	↑
Ferritin, heavy polypeptide 1	*FTH1*	3	–	↑	↑	↑	–
Angiotensinogen	*AGT*	2	–	–	–	↑	↓
Apolipoprotein C-III	*APOC3*	2	–	–	↓	–	↓

(Continued)

Table III *(Continued)*

Name	Gene	Replication	Clinical Center 1	Clinical Center 2	Clinical Center 3	Clinical Center 4	Clinical Center 5
Interleukin 12B	*IL12B*	2	–	–	–	↓	↑
Serpin peptidase inhibitor	*SERPINA1*	2	–	–	–	↑	↑
Amyloid P component, serum	*APCS*	2	–	–	–	↑	↑
Myeloperoxidase	*MPO*	2	–	–	–	↑	↑
Matrix metallopeptidase 7	*MMP7*	2	–	–	–	↑	↑
S100 calcium-binding protein A12	*S100A12*	2	–	–	–	↑	↑
Insulin	*INS*	2	–	↑	–	–	↑
Interleukin 3	*IL3*	2	–	–	↓	–	↑
Interleukin 13	*IL13*	2	–	–	↓	–	↑
TIMP metallopeptidase inhibitor 1	*TIMP1*	2	–	–	↑	–	↑
Chemokine (C-X-C motif) ligand 5	*CXCL5*	2	–	–	↑	–	↑
Epidermal growth factor	*EGF*	2	↓	–	–	–	↑
Chemokine (C-C motif) ligand 5	*CCL5*	2	↑	–	–	–	↑
Brain-derived neurotrophic factor	*BDNF*	2	↓	–	–	↓	–
C-reactive protein, pentraxin-related	*CRP*	2	–	↑	–	↑	–
Chemokine (C-C motif) ligand 26	*CCL26*	2	↑	↓	–	–	–
N(alpha)-acetyltransferase 15	*NAA15*	2	↑	↑	–	–	–
Fibrinogen alpha chain	*FGA*	2	↓	–	↓	–	–
Sortilin 1	*SORT1*	2	↓	–	↓	–	–
Thrombospondin 1	*THBS1*	2	↓	–	↑	–	–
Thrombopoietin	*THPO*	2	↓	–	↑	–	–
Alpha-2-macroglobulin	*A2M*	2	↑	–	↑	–	–
Interleukin 4	*IL4*	2	↑	–	↑	–	–
Platelet-derived growth factor beta polypeptide	*PDGFB*	1	–	–	–	–	↓
Growth hormone 1	*GH1*	1	–	–	–	–	↓
Secreted phosphoprotein 1	*SPP1*	1	–	–	–	–	↓
AXL receptor tyrosine kinase	*AXL*	1	–	–	–	–	↓
Fas ligand (TNF superfamily, member 6)	*FASLG*	1	–	–	–	–	↓

Cystatin C	*CST3*	1	–	–	–	–	↑
Calcitonin-related polypeptide alpha	*CALCA*	1	–	–	–	–	↑
Serpin peptidase inhibitor, cladeA	*SERPINA7*	1	–	–	–	–	↑
Protein S (alpha)	*PROS1*	1	–	–	–	–	↑
Matrix metallopeptidase 10	*MMP10*	1	–	–	–	–	↑
Chemokine (C-X-C motif) ligand 1	*CXCL1*	1	–	–	–	–	↑
Chemokine (C-C motif) ligand 20	*CCL20*	1	–	–	–	–	↑
Lipocalin 2	*LCN2*	1	–	–	–	–	↑
Alpha-2-HS-glycoprotein	*AHSG*	1	–	–	–	↓	–
Transferrin	*TF*	1	–	–	–	↓	–
Kallikrein-related peptidase 3	*KLK3*	1	–	–	–	↓	–
Advanced glycosylation end product-specific receptor	*AGER*	1	–	–	–	↓	–
Uromodulin	*UMOD*	1	–	–	–	↑	–
Creatine kinase	*CKB*	1	–	–	–	↑	–
Secretogranin II	*SCG2*	1	–	–	–	↑	–
Tenascin C	*TNC*	1	–	–	–	↑	–
Betacellulin	*BTC*	1	–	–	–	↑	–
Chemokine (C-C motif) ligand 23	*CCL23*	1	–	–	–	↑	–
Interleukin 16	*IL16*	1	–	–	–	↑	–
Endothelin 1	*EDN1*	1	–	↓	–	–	–
Fibroblast growth factor 2 (basic)	*FGF2*	1	–	↓	–	–	–
Fatty acid-binding protein 3	*FABP3*	1	–	↑	–	–	–
Platelet-derived growth factor beta polypeptide	*PDGFB*	1	–	–	↑	–	–
Serpin peptidase inhibitor, cladeE	*SERPINE1*	1	–	–	↑	–	–
Interleukin 6	*IL6*	1	–	–	↑	–	–
Interleukin 8	*IL8*	1	–	–	↑	–	–
Chemokine (C-C motif) ligand 3	*CCL3*	1	–	–	↑	–	–
Beta-2-microglobulin	*B2M*	1	–	–	↑	–	–
Interleukin 18	*IL18*	1	–	–	↑	–	–
CD40 molecule	*CD40*	1	–	–	↑	–	–
Apolipoprotein A-I	*APOA1*	1	↓	–	–	–	–
Sex hormone-binding globulin	*SHBG*	1	↓	–	–	–	–
Coagulation factor VII	*F7*	1	↓	–	–	–	–

(Continued)

Table III *(Continued)*

Name	Gene	Replication	Clinical Center 1	Clinical Center 2	Clinical Center 3	Clinical Center 4	Clinical Center 5
Chemokine (C-C motif) ligand 4	*CCL4*	1	↓	–	–	–	–
Acidphosphatase	*ACPP*	1	↓	–	–	–	–
CD40 ligand	*CD40LG*	1	↓	–	–	–	–
Angiopoietin 2	*ANGPT2*	1	↑	–	–	–	–
Chemokine (C-C motif) ligand 25	*CCL25*	1	↑	–	–	–	–
Follicle-stimulating hormone	*FSHB*	1	↑	–	–	–	–
Luteinizing hormone beta polypeptide	*LHB*	1	↑	–	–	–	–
Glutamic-oxaloacetic transaminase 1	*GOT1*	1	↑	-	–	–	–
Insulin-like growth factor-binding protein 2	*IGFBP2*	1	↑	–	–	–	–
Mucin 16, cell surface associated	*MUC16*	1	↑	–	–	–	–

to a low-blood level of glucocorticoids. Analysis of medium-high abundance proteins in the two cohorts by LC–MS revealed alterations in 40 unique/nonredundant proteins (Table IV), with 9 (23%) of these proteins participating in inflammatory responses.

C. Functional and Pathway Analysis: Overall Evidence

Functional analysis by IPA revealed that the diagnostic biomarker panels identified from the literature and in-house MAP and LC–MS studies had several biological functions, network associations, and canonical pathways in common (Tables V and VI).

1. *Biological Functions*

"Immunological disease," "inflammatory response," and "respiratory disease" were the most significant diseases and disorders associated with the literature and with the MAP data (Fig. 2). Not surprisingly, the LC–MS studies did not share the top diseases and disorders in common with the literature or MAP data as this method typically analyzes the medium-high abundance portion of the proteome (Higgs *et al.*, 2008). The proteins identified by the LC–MS studies were predominantly associated with the "cardiovascular" and "metabolic component" in schizophrenia, consistent with some of the current hypotheses on this disorder (Newcomer, 2007). This finding also highlighted the usefulness of unbiased MS profiling for identification of novel disease-associated functions. The most significant molecular and cellular functions in common between the literature and MAP studies were "cellular movement" and "cell death," and the most significant molecular and cellular functions in common between the literature and the LC–MS analysis were "lipid and molecular transport" and "small molecular biochemistry" (Fig. 2). "Antigen presentation," "carbohydrate metabolism," and "cell morphology" were significantly associated with the MAP and LC–MS studies, respectively. "Hematological system development and function" was among the most significant physiological system development and functions associated with the literature, MAP, and LC–MS studies. While "immune cell trafficking" and "tissue morphology" were shared between the literature and MAP studies, the LC–MS studies had no other physiological system functions in common (Fig. 2).

2. *Canonical Pathways*

Canonical pathway analysis of the "diagnostic biomarkers" from the literature, MAP, and LC–MS studies provided further evidence for altered immunological and/or inflammatory signaling in schizophrenia (Fig. 2). "Acute phase response (APR) signaling" was the top canonical pathway shared between all the studies (Fig. 3A). The APPs involved in APR signaling included *10 positive APPs* [*IL1RN*, *HP* (haptoglobin), *CRP* (C-reactive protein), *C3* (complement

Table IV

Summary of LC–MS Findings from the Two Patient Cohorts Combined.

Protein name	Gene name	Cohort I for LC–MS	Cohort II for LC–MS
Alpha-2-HS-glycoprotein	*AHSG*	↓	–
Apolipoprotein A-I	*APOA1*	↓	–
Apolipoprotein A-II	*APOA2*	↓	–
Apolipoprotein A-IV	*APOA4*	↓	–
Apolipoprotein C-I	*APOC1*	↓	–
Apolipoprotein D	*APOD*	↓	–
CD5 molecule-like	*CD5L*	↓	–
Immunoglobulin heavy constant mu	*IGHM*	↓	–
Transferrin	*TF*	↓	–
Coagulation factor XIII	*F13B*	↓	↓
Ubiquitin-specific peptidase 54	*USP54*	–	↓
Creatine kinase	*CKMT2*	–	↓
Growth regulation by estrogen in breast cancer 1	*GREB1*	–	↓
Putative *N*-acetylated-alpha-linked acidic dipeptidase	*FOLH1B*	–	↓
Serine/threonine kinase 38 like	*STK38L*	–	↓
Vang-like 1	*VANGL1*	–	↓
Probable ATP-dependent DNA helicase HFM1	*HFM1*	–	↓
Fibulin 1	*FBLN1*	–	↓
Polymerase (RNA) III (DNA directed) polypeptide A, 155 kDa	*POLR3A*	–	↓
Myoferlin	*MYOF*	–	↓
Actinin, alpha 2	*ACTN2*	–	↓
Peroxidasin homolog (Drosophila)	*PXDN*	–	↓
Leucine rich repeat containing 16A	*LRRC16A*	–	↓
Lumican	*LUM*	–	↓
Retinol-binding protein 4, plasma	*RBP4*	–	↓
Gelsolin	*GSN*	–	↓
Serpin peptidase inhibitor, cladeF	*SERPINF2*	–	↑
Synemin, intermediate filament protein	*SYNM*	–	↑
Neurexin 1	*NRXN1*	–	↑
Apolipoprotein L, 1	*APOL1*	–	↑
CDK5 regulatory subunit associated protein 2	*CDK5RAP2*	–	↑
Polymerase (DNA directed), theta	*POLQ*	–	↑
Collagen, type V, alpha 1	*COL5A1*	–	↑
Spermatogenesis-associated protein 21	*SPATA21*	–	↑
Microtubule-associated protein 2	*MAP2*	–	↑
RANBP2-like and GRIP domain containing 1	*RGPD1*	–	↑
Haptoglobin	*HP*	–	↑
Solute carrier family 25	*SLC25A10*	–	↑
Peroxisome proliferator-activated receptor gamma	*PPRC1*	–	↑
PDX1C-terminal inhibiting factor 1	*PCIF1*	–	↑

Table V
TOP BIOLOGICAL FUNCTIONS.

	Diagnostic			Prognostic
	First onset or chronic (literature)	First onset (RBM)	First onset (LC–MS)	First onset or chronic (literature)
Diseases and disorders	Inflammatory disease (1.01×10^{-20}–6.11×10^{-9}, $n = 28$)	Inflammatory response (1.85×10^{-46}–9.17×10^{-12}, $n = 71$)	Cardiovascular disease (1.31×10^{-4}–4.96×10^{-2}, $n = 4$)	Inflammatory response (4.91×10^{-18}–5.22×10^{-8}, $n = 22$)
	Immunological disease (2.25×10^{-20}–6.18×10^{-9}, $n = 26$)	Immunological disease (2.19×10^{-38}–6.59×10^{-12}, $n = 59$)	Genetic disorder (1.34×10^{-4}–4.96×10^{-2}, $n = 21$)	Immunological disease (1.77×10^{-17}–6.09×10^{-8}, $n = 19$)
	Inflammatory response (5.43×10^{-20}–6.18×10^{-9}, $n = 30$)	Respiratory disease (9.16×10^{-33}–8.11×10^{-12}, $n = 45$)	Neurological disease (1.34×10^{-4}–4.23×10^{-2}, $n = 19$)	Genetic disorder (2.16×10^{-17}–4.68×10^{-8}, $n = 23$)
	Respiratory disease (6.32×10^{-19}–1.70×10^{-9}, $n = 22$)	Hematological disease (9.16×10^{-33}–8.11×10^{-12}, $n = 50$)	Psychological disorders (1.34×10^{-4}–1.34×10^{-4}, $n = 8$)	Nutritional disease (4.16×10^{-17}–1.17×10^{-9}, $n = 17$)
	Organismal injury and abnormalities (2.46×10^{-18}–3.40×10^{-9}, $n = 23$)	Renal and urological disease (1.45×10^{-28}–6.52×10^{-16}, $n = 37$)	Metabolic disease (4.08×10^{-4}–4.96×10^{-2}, $n = 5$)	Endocrine system disorders (1.24×10^{-16}–4.58×10^{-8}, $n = 14$)
Molecular and cellular functions	Cellular movement (4.47×10^{-24}–6.07×10^{-8}, $n = 28$)	Cellular movement (2.10×10^{-46}–9.17×10^{-12}, $n = 71$)	Lipid metabolism (4.23×10^{-6}–4.96×10^{-2}, $n = 11$)	Cellular movement (8.14×10^{-23}–6.46×10^{-8}, $n = 23$)
	Lipid metabolism (2.87×10^{-21}–7.13×10^{-9}, $n = 25$)	Cell-to-cell signaling and interaction (1.70×10^{-45}–8.56×10^{-12}, $n = 74$)	Small molecule biochemistry (4.23×10^{-6}–4.96×10^{-2}, $n = 15$)	Cell-to-cell signaling and interaction (4.14×10^{-21}–7.42×10^{-8}, $n = 23$)
	Molecular transport (2.87×10^{-21}–6.82×10^{-9}, $n = 26$)	Cellular growth and proliferation (3.16×10^{-40}–7.61×10^{-12}, $n = 75$)	Molecular transport (4.78×10^{-6}–4.97×10^{-2}, $n = 12$)	Cell death (5.82×10^{-19}–7.27×10^{-8}, $n = 21$)

(Continued)

Table V *(Continued)*

	Diagnostic			Prognostic
	First onset or chronic (literature)	First onset (RBM)	First onset (LC–MS)	First onset or chronic (literature)
Physiological system development and function	Small molecule biochemistry (2.87×10^{-21}–7.13×10^{-9}, $n = 27$)	Antigen presentation (1.83×10^{-35}–4.15×10^{-12}, $n = 47$)	Carbohydrate metabolism (3.33×10^{-5}–4.96×10^{-2}, $n = 7$)	Cellular growth and proliferation (2.29×10^{-18}–6.16×10^{-8}, $n = 25$)
	Cell death (8.40×10^{-19}–6.82×10^{-9}, $n = 29$)	Cell death (5.46×10^{-33}–4.78×10^{-12}, $n = 71$)	Cell morphology (3.75×10^{-5}–4.71×10^{-2}, $n = 8$)	Lipid metabolism (7.28×10^{-18}–7.26×10^{-8}, $n = 20$)
	Hematological system development and function (5.54×10^{-22}–6.18×10^{-9}, $n = 31$)	Hematological system development and function (2.52×10^{-43}–9.17×10^{-12}, $n = 73$)	Organismal functions (2.36×10^{-3}–7.60×10^{-3}, $n = 3$)	Hematological system development and function (3.05×10^{-15}–6.46×10^{-8}, $n = 21$)
	Immune cell trafficking (5.54×10^{-22}–6.07×10^{-9}, $n = 29$)	Immune cell trafficking (2.52×10^{-43}–9.17×10^{-12}, $n = 66$)	Organ morphology (2.44×10^{-3}–1.77×10^{-2}, $n = 6$)	Immune cell trafficking (3.05×10^{-15}–6.46×10^{-8}, $n = 19$)
	Tissue morphology (3.35×10^{-17}–6.18×10^{-9}, $n = 26$)	Tissue development (2.28×10^{-39}–1.05×10^{-11}, $n = 60$)	Embryonic development (2.54×10^{-3}–1.88×10^{-2}, $n = 4$)	Tissue morphology (3.19×10^{-15}–7.26×10^{-8}, $n = 21$)
	Organismal survival (3.43×10^{-17}–2.78×10^{-9}, $n = 22$)	Tissue morphology (3.43×10^{-31}–1.05×10^{-11}, $n = 57$)	Endocrine system development and function (2.54×10^{-3}–2.51×10^{-2}, $n = 3$)	Behavior (1.15×10^{-13}–1.73×10^{-8}, $n = 15$)
	Lymphoid tissue structure and development (7.63×10^{-16}–1.17×10^{-9}, $n = 16$)	Cell-mediated immune response (5.73×10^{-27}–9.17×10^{-12}, $n = 34$)	Hematological system development and function (2.54×10^{-3}–4.23×10^{-2}, $n = 5$)	Digestive system development and function (1.28×10^{-13}–1.28×10^{-13}, $n = 7$)

Table VI
Top Canonical Pathways Associated with Diagnostic or Drug-Response Biomarkers Identified from the Literature or In-House Studies (MAP or LC–MS Studies).

Diagnostic			Drug response
Mixed disease stages (literature)	First onset (MAP)	First onset (LC–MS)	Mixed disease stages (literature)
Acute phase response signaling [1.34×10^{-19}, 15/173 (0.087)]	Hepatic fibrosis/hepatic stellate cell activation [2.28×10^{-19}, 18/128 (0.141)]	Acute phase response signaling [1.82×10^{-7}, 7/173 (0.04)]	Hepatic fibrosis/hepatic stellate cell activation [5.65×10^{-11}, 8/128 (0.062)]
Hepatic fibrosis/hepatic stellate cell activation [1.02×10^{-12}, 10/128 (0.078)]	Communication between Innate and adaptive immune cells [4.33×10^{-18}, 15/91 (0.165)]	LXR/RXR activation [7.03×10^{-4}, 3/76 (0.039)]	Glucocorticoid receptor signaling [5.28×10^{-10}, 9/268 (0.034)]
Role of cytokines in mediating communication between immune cells [1.06×10^{-12}, 8/52 (0.154)]	Role of cytokines in mediating communication between immune cells [7.61×10^{-16}, 12/52 (0.231)]	Coagulation system [3.75×10^{-3}, 2/36 (0.056)]	Atherosclerosis signaling [2.32×10^{-8}, 6/107 (0.056)]
Communication between innate and adaptive immune cells [3.85×10^{-11}, 8/91 (0.088)]	Acute phase response signaling [1.11×10^{-15}, 17/173 (0.098)]	Extrinsic prothrombin activation pathway [4.23×10^{-2}, 1/17 (0.059)]	Role of cytokines in mediating communication between immune cells [4.15×10^{-8}, 5/52 (0.096)]

component 3), *FGA* (fibrinogen alpha chain), *AGT* (angiotensinogen), *CFB* (complement factor B), *CP* (ceruloplasmin or ferroxidase), *FTH1* (ferritin), and *PROS1* (proteins alpha)], *5 negative APPs* [*AHSG* (alpha-2-HS-glycoprotein), *TF* (transferrin), *ALB* (albumin), *TTR* (transthyretin), and *IGF1* (insulin-like growth factor1)], and *21 other APPs* [*IL6, TNF, IL1B* (interleukin 1 beta), *IL1A* (interleukin 1 alpha), *IL8, MIF* (macrophage migration inhibitory factor), *IFN-γ* (interferon gamma), *APOA1* (apolipoprotein A-I), *APOH* (apolipoprotein H), *VWF* (von Willebrand factor), *SERPINA1* (serpin peptidase inhibitor cladeA), *A2M* (alpha-2-macroglobulin), *APCS* (amyloid P component), *APOA2* (apolipoprotein A-II), *IL18* (interleukin 18), *IL6R* (interleukin 6 receptor), *SERPINE1* (serpin peptidase inhibitor cladeE), *TNFRSF1A* (tumor necrosis factor receptor superfamily member 1A), *RBP4* (retinol-binding protein 4), *SERPINF2* (serpin peptidase inhibitor cladeF), and *TNFRSF1B* (tumor necrosis factor receptor superfamily member 1B)].

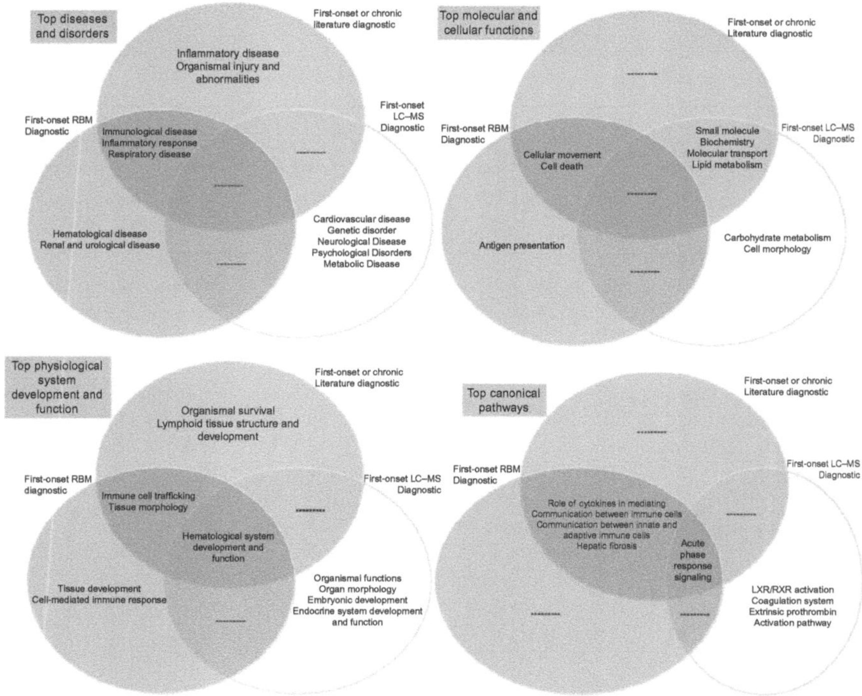

FIG. 2. Venn diagrams showing the overlaps in top biological functions and canonical pathways associated with the literature, MAP, and LC–MS studies.

Three other canonical pathways were shared between the literature and MAP studies including "hepatic fibrosis" (Fig. 3B1), "communication between innate and adaptive immune cells" (Fig. 3B2), and "role of cytokines in mediating communication between immune cells" (Fig. 3B3). The DR biomarker panel extracted from the literature was most significantly associated with "hepatic fibrosis," "glucocorticoid receptor signaling," and "atherosclerosis signaling" (Table VI).

IV. Discussion

A. Genetic, Epidemiological, and Animal Model Studies

Converging evidence from the literature and in-house serum/plasma proteome studies suggested an immunological and/or inflammatory process in first-onset schizophrenia patients (Table VII). These findings are in agreement with evidence from genetic, epidemiological, and animal model studies along with

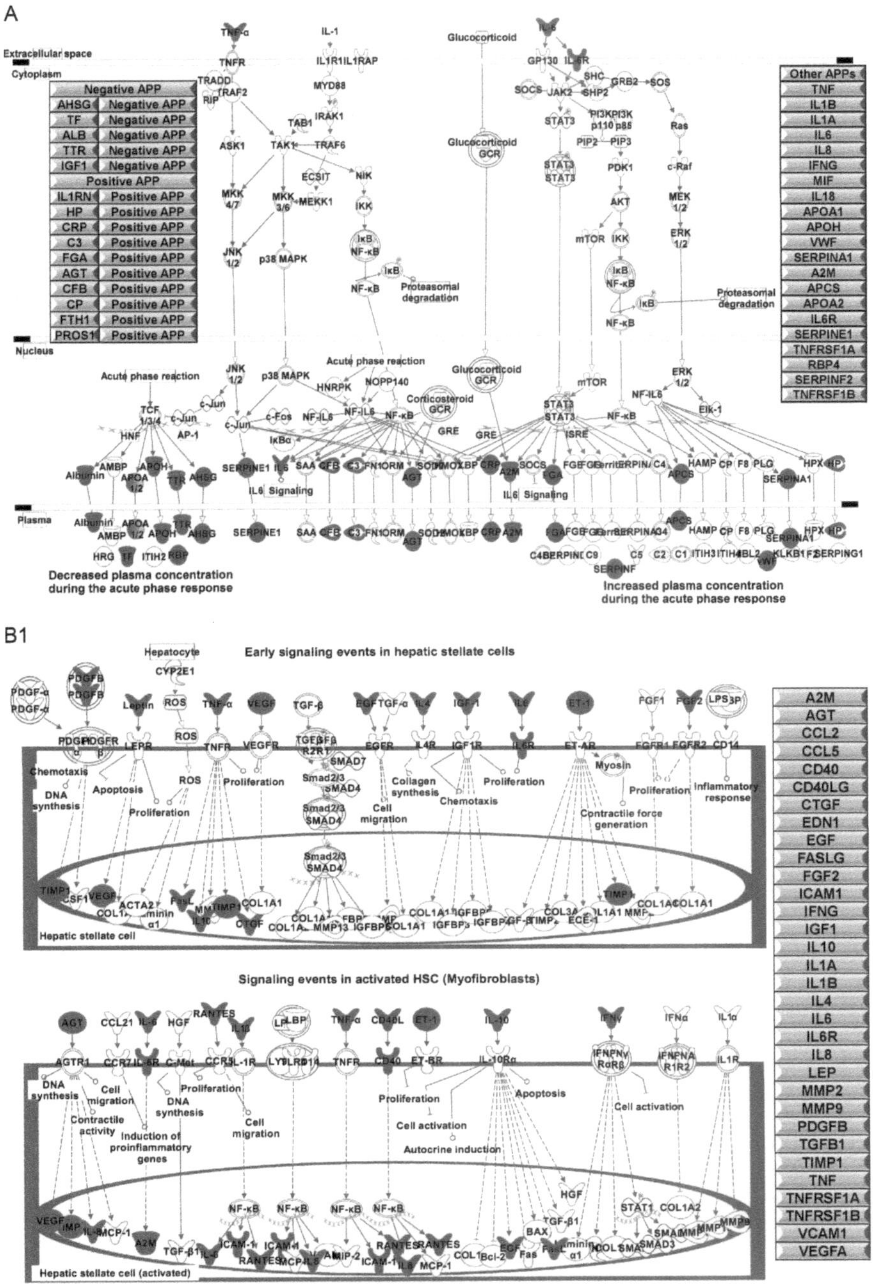

FIG. 3. (Continued)

B2

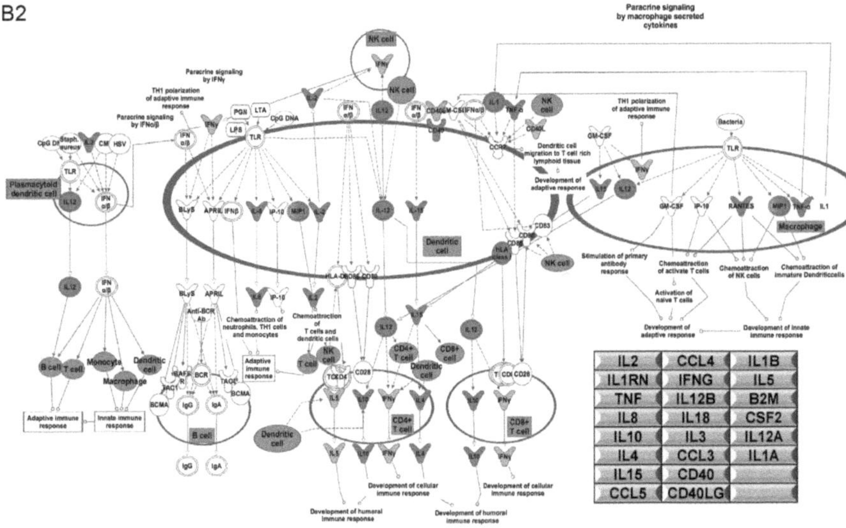

B3

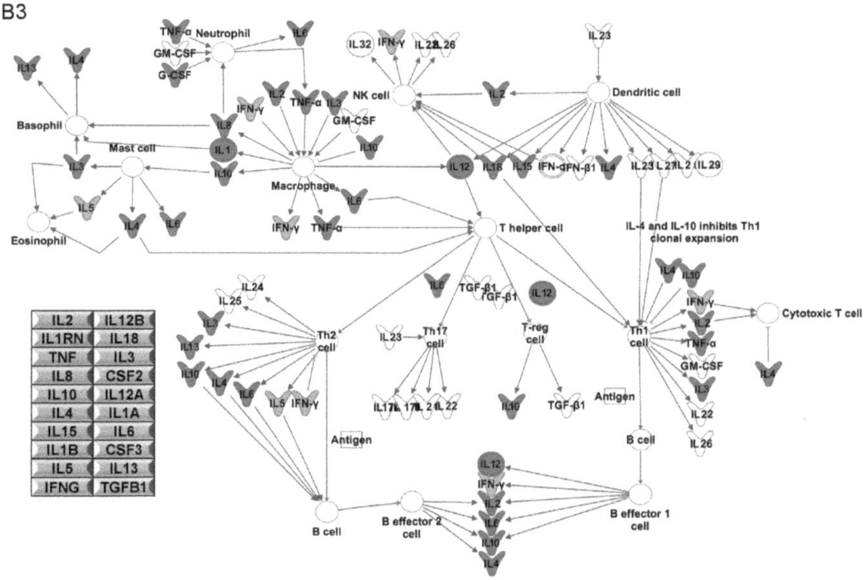

Legend:

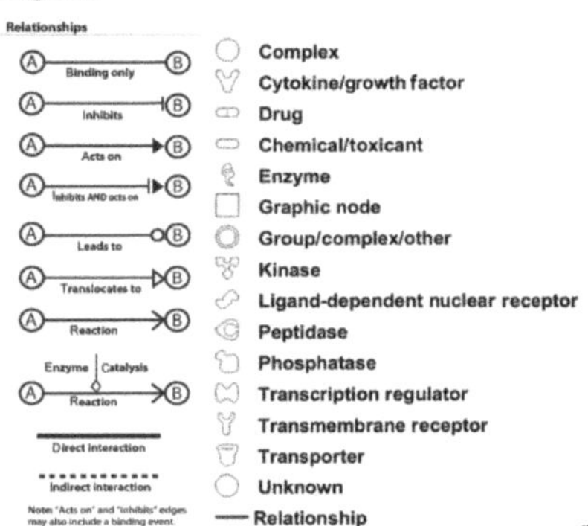

(Ingenuity® Systems, www.ingenuity.com)

aspects of the neurodevelopmental hypothesis of schizophrenia. For example, genetic studies have identified immune inflammatory response genes contributing to schizophrenia susceptibility. These include genome-wide association studies (GWAS) which revealed associations between schizophrenia and (1) single nucleotide polymorphisms (SNPs) near the colony-stimulating factor receptor 2 alpha (*CSF2RA*) gene, (2) the *CSF2RA* and interleukin 3 receptor alpha (*IL3RA*) genes (Lencz *et al.*, 2007), and (3) the major histocompatibility complex (MHC) region at 6p (Shi *et al.*, 2009; Stefansson *et al.*, 2009). The *IL1* gene complex, *IL10*, neuregulin-1, epidermal growth factor (*EGF*), and *TNF-α* are among some of those suggested through SNPs or linkage studies (Nawa *et al.*, 2000).

Epidemiological studies have suggested a role of maternal infection during prenatal life (i.e., during critical periods of central nervous system development) which can lead to increased risk of developing schizophrenia in the offspring (Brown *et al.*, 2004a; Buka *et al.*, 2001; Meyer *et al.*, 2009; Westergaard *et al.*, 1999). Examples include Finnish (Koponen *et al.*, 2004) and Brazilian (Gattaz *et al.*, 2004) epidemiological studies which demonstrated that infection of the CNS during childhood is associated with a fivefold increase in risk of developing psychosis in the offspring in later life. Maternal infection has also been discussed in the context of seasonality effects on schizophrenic births (Torrey *et al.*, 1997). Further, increased levels of maternal *IL8* in pregnancy have also been associated with a higher risk for schizophrenia in the offspring (Brown, 2006; Siegel *et al.*, 2009). The role of maternal infection has been proposed in the context of the neurodevelopmental hypothesis of schizophrenia since cytokines regulate CNS development as they act as "growth factors" of the nervous system and glial cells. Sensitization studies with proinflammatory cytokines have shown that increased production of such cytokines during the perinatal period induces long-lasting and, possibly, permanent alterations in CNS neurotransmitter systems (Siegel *et al.*, 2009).

FIG. 3. Canonical pathways most significantly associated with the diagnostic biomarkers identified from the (A) literature, MAP, and LC–MS studies, and (B1–B3) literature and MAP studies. Canonical pathways represent graphical representations of the molecular relationships between molecules. Molecules are represented as nodes, and the biological relationship between two nodes is represented as an edge (line). All edges are supported by at least one reference from the literature, from a textbook, or from canonical information stored in the Ingenuity Pathways Knowledge Base. The intensity of the node color indicates the degree of increased (red) or decreased (green) molecular levels. Nodes are displayed using various shapes that represent the functional class of the gene product. Edges are connected by various line connectors that describe the nature of the relationship between the nodes (see legend; this explanatory extract has been copied from Ingenuity® Systems; www.ingenuity.com, with permission). (A) Acute phase response signaling, (B1) hepatic fibrosis, (B2) communication between innate and adaptive immune cells, (B3) role of cytokines in mediating communication between innate and adaptive immune cells. (For interpretation of the references to color in this figure legend, the reader is referred to the Web version of this chapter.)

Table VII

Combined Findings from the Literature (References Listed in Table) and In-House Studies.

Gene name	# Times found to be altered	Total	References	Gene name	# Times found to be altered	Total	References	Gene name	# Times found to be altered	Total	References
A2M	↑2 ↓1	3	Domenici *et al.* (2010)	*FABP3*	↑1 –	1	–	*MMP2*	– ↓1	1	Domenici *et al.* (2010)
ACPP	↑1 ↓1	2	Domenici *et al.* (2010)	*FASLG*	– ↓1	1	–	*MMP7*	↑2 –	2	–
ACTN2	– ↓1	1	–	*FBLN1*	– ↓1	1	–	*MMP9*	↑1 –	1	Domenici *et al.* (2010)
ADIPOQ	– ↓3	3	Bai *et al.* (2009), Domenici *et al.* (2010), Richards *et al.* (2006)	*FGA*	↑1 ↓2	3	Domenici *et al.* (2010)	*MPO*	↑2 –	2	–
AGER	– ↓1	1	–	*FGF2*	↑1 ↓1	2	Hashimoto *et al.* (2003)	*MUC16*	↑2 –	2	Domenici *et al.* (2010)
AGT	↑1 –	1	–	*FOLH1B*	– ↓1	1	–	*MYOF*	– ↓1	1	–
AHSG	– ↓3	3	Levin *et al.* (2010)	*FSHB*	↑1 ↓1	2	Konarzewska *et al.* (2009)	*NAA15*	↑2 –	2	–
ALB	– ↓2	2	Huang (2002), Yao *et al.* (2000)	*FTH1*	↑3 –	3	–	*NCAM*	↑1 –	1	Tanaka *et al.* (2007)
ANGPT2	↑1 –	1	–	*GH1*	– ↓2	2	Domenici *et al.* (2010)	*NGF*	– ↓4	4	Jockers-Scherubl *et al.* (2006), Lee and Kim (2009), Parikh *et al.* (2003)
APCS	↑3 –	3	Domenici *et al.* (2010)	*GHRL*	– ↓1	1	Hosojima *et al.* (2006)	*NRXN1*	↑1 –	1	–
APOA1	– ↓5	5	Domenici *et al.* (2010), Levin *et al.* (2010), Yang *et al.* (2006)	*GOT1*	↑1 –	1	–	*OXT*	– ↓1	1	Keri *et al.* (2009)
APOA2	– ↓2	2	Levin *et al.* (2010)	*GPX*	↑1 –	1	Atmaca *et al.* (2005)	*PCIF1*	↑1 –	1	–
APOA4	↑1 ↓2	3	Levin *et al.* (2010), Yang *et al.* (2006)	*GREB1*	– ↓1	1	–	*PDGFB*	↑1 –	1	–
APOC1	– ↓2	2	Levin *et al.* (2010)	*GSN*	– ↓1	1	–	*POLQ*	↑1 –	1	–
APOC3	↑1 ↓2	3	Domenici *et al.* (2010)	*GSTA1*	↑5 ↓1	6	Domenici *et al.* (2010)	*POLR3A*	– ↓1	1	–
APOD	↑2 ↓2	4	Levin *et al.* (2010), Mahadik *et al.* (2002)	*HFM1*	– ↓1	1	–	*PPRC1*	↑1 –	1	–
APOH	↑5 –	5	Domenici *et al.* (2010)	*HP*	↑7 –	7	Yang *et al.* (2006)	*PPY*	↑4 –	4	–
APOL1	↑1 –	1	–	*ICAM1*	↑4 ↓2	6	Domenici *et al.* (2010), Kronig *et al.* (2005), Schwarz *et al.* (1998)	*PRL*	↑13 ↓8	21	Berwaerts *et al.* (2010), Chang *et al.* (2008), Chen *et al.* (2009), Costa *et al.* (2007), Kane *et al.* (2009),

														Kim *et al.* (2002), Kinon *et al.* (2006), Konarzewska *et al.* (2009), Kwon *et al.* (2009), Melkersson *et al.* (2001), Montgomery *et al.* (2004), Segal *et al.* (2004, 2007a, b), Wang *et al.* (2007b), Young *et al.* (2004), Zhang *et al.* (2002a)
AXL	–	↓1	1	–	*IFNG*	↑1	↓1	2	Kim *et al.* (2004), Na and Kim (2007)	*PROS1*	↑1	–	1	–
B2M	↑2	–	2	Chittiprol *et al.* (2009)	*IGF1*	–	↓2	2	Melkersson *et al.* (1999), Venkatasubramanian *et al.* (2007)	*PXDN*	–	↓1	1	–
BDNF	↑7	↓18	25	Chen da *et al.* (2009), Domenici *et al.* (2010), Gama *et al.* (2007), Grillo *et al.* (2007), Guimaraes *et al.* (2008), Huang and Hung (2009), Huang and Lee (2006), Ikeda *et al.* (2008), Jindal *et al.* (2010), Lee and Kim (2009), Palomino *et al.* (2006), Pirildar *et al.* (2004), Reis *et al.* (2008), Rizos *et al.* (2008), Tan *et al.* (2005), Toyooka *et al.* (2002), Vinogradov *et al.* (2009), Xiu *et al.* (2009), Zhang *et al.* (2008b)	*IGFBP1*	–	↓1	1	Melkersson *et al.* (2000)	*RAGE*	↑2	–	2	Steiner *et al.* (2009)
BTC	↑1	–	1	–	*IGFBP2*	↑1	–	1		*RBP4*	–	↓1	1	–
C3	↑4	–	4	Domenici *et al.* (2010)	*IL10*	↑5	↓1	6	Domenici *et al.* (2010), Maes *et al.* (2002)	*RELN*	↑1	–	1	Fatemi *et al.* (2001)
CALCA	↑1	–	1	–	*IL12*	↑1	–	1	Crespo-Facorro *et al.* (2008)	*RETN*	↑2	↓1	3	–
CAT	↑1	–	1	Atmaca *et al.* (2005)	*IL12A*	↑1	–	1	Domenici *et al.* (2010)	*RGPD1*	↑1	–	1	–
CCL11	↑2	–	2	Domenici *et al.* (2010), Teixeira *et al.* (2008)	*IL12B*	↑1	↓2	3	Domenici *et al.* (2010)	*S100A12*	↑2	–	2	–
CCL2	↑2	–	2	Domenici *et al.* (2010), Drexhage *et al.* (2008)	*IL13*	↑1	↓2	3	Domenici *et al.* (2010)	*S100B*	↑18	↓3	21	Gattaz *et al.* (2000), Lara *et al.* (2001), Ling *et al.* (2007), Pedersen *et al.* (2008), Qi *et al.* (2009), Rothermundt *et al.* (2001b, 2004, 2007), Sarandol *et al.* (2007),

(Continued)

Table VII (Continued)

Gene name	# Times found to be altered		Total	References	Gene name	# Times found to be altered		Total	References	Gene name	# Times found to be altered		Total	References
														Schmitt *et al.* (2005), Schroeter *et al.* (2003, 2009), Schroeter and Steiner (2009), Steiner *et al.* (2006, 2009, 2010), Wiesmann *et al.* (1999)
CCL20	↑1	–	1	–	*IL15*	↑4	↓1	5	Domenici *et al.* (2010)	*SCG2*	↑1	–	1	–
CCL22	↑4	–	4	Domenici *et al.* (2010)	*IL16*	↑2	–	2	Domenici *et al.* (2010)	*SCGB1A1*	↑1	↓2	3	Maes *et al.* (1996a, 1997a)
CCL23	↑1	–	1	–	*IL18*	↑2	–	2	Domenici *et al.* (2010)	*SELL*	↑1	–	1	Iwata *et al.* (2007)
CCL25	↑1	–	1	–	*IL1A*	–	↓1	1	Domenici *et al.* (2010)	*SERPINA1*	↑4	–	4	Domenici *et al.* (2010), Yang *et al.* (2006)
CCL26	↑1	–	1	–	*IL1B*	↑2	↓1	3	Song *et al.* (2009), Theodoropoulou *et al.* (2001)	*SERPINA7*	↑2	–	2	Domenici *et al.* (2010)
CCL3	↑1	–	1	–	*IL1RN*	↑9	–	9	Akiyama (1999), Maes *et al.* (1996a, 1997a), Potvin *et al.* (2008)	*SERPINE1*	↑2	–	2	Domenici *et al.* (2010)
CCL4	↑1	↓1	2	Domenici *et al.* (2010)	*IL2*	↑8	↓4	12	Ebrinc *et al.* (2002), Kim *et al.* (1998, 2000), Mahendran and Chan (2004), Na and Kim (2007), Zhang *et al.* (2002b, 2004, 2005, 2008a, 2009)	*SERPINF2*	↑1	–	1	–
CCL5	↑3	–	3	Domenici *et al.* (2010)	*IL2R*	↑3	–	3	Bresee and Rapaport (2009), Potvin *et al.* (2008), Schwarz *et al.* (1998)	*SHBG*	–	↓1	1	–
CD40	↑1	–	1	–	*IL2RA*	↑5	–	5	Akiyama (1999), Gaughran *et al.* (1998, 2002), Muller *et al.* (1997))	*SLC25A10*	↑1	–	1	–

CD40LG	–	↓1	1	–	*IL3*	↑1	↓2	3	Domenici *et al.* (2010)	*SOD1*	↑3	–	3	Atmaca *et al.* (2005), Dakhale *et al.* (2004), Gama *et al.* (2006)
CD5L	–	↓2	2	Levin *et al.* (2010)	*IL4*	↑2	↓3	5	Domenici *et al.* (2010), Na and Kim (2007), O'Brien *et al.* (2008)	*SORT1*	–	↓2	2	–
CDK5RAP2	↑1	–	1	–	*IL5*	–	↓3	3	–	*SPATA21*	↑1	–	1	–
CEACAM5	↑5	–	5	Domenici *et al.* (2010)	*IL6*	↑15	↓2	17	Akiyama (1999), Frommberger *et al.* (1997), Garcia-Miss Mdel *et al.* (2010), Lin *et al.* (1998), Maes *et al.* (1997a), Na and Kim (2007), Naudin *et al.* (1996), Potvin *et al.* (2008), Zhang *et al.* (2002b, 2005, 2008a, 2009)	*SPP1*	–	↓1	1	–
CFB	↑1	–	1	Yang *et al.* (2006)	*IL6R*	↑1	↓1	2	Maes *et al.* (1997a), Muller *et al.* (1997)	*SST*	↑1	–	1	Saiz-Ruiz *et al.* (1992)
CGA	↑3	–	3	–	*IL7*	↑3	↓1	4	–	*STK38L*	–	↓1	1	–
CKB	↑1	–	1	–	*IL8*	↑7	–	7	Brown *et al.* (2004b), Domenici *et al.* (2010), Maes *et al.* (2002), Tanaka *et al.* (2000), Zhang *et al.* (2002b)	*SYNM*	↑1	–	1	–
CKMT2	–	↓1	1	–	*INHBB*	↑1	–	1	Konarzewska *et al.* (2009)	*TF*	–	↓2	2	Levin *et al.* (2010)
COL5A1	↑1	–	1	–	*INS*	↑11	–	11	Baptista *et al.* (2007), Chen *et al.* (2008), Domenici *et al.* (2010), Fan *et al.* (2006a,b), Melkersson and Hulting (2001), Melkersson *et al.* (1999, 2000), Venkatasubramanian *et al.* (2007)	*TGFB1*	↑1	–	1	Kim *et al.* (2004)
CP	↑1	–	1	Wolf *et al.* (2006)	*KITLG*	↑1	–	1	Domenici *et al.* (2010)	*THBS1*	↑1	↓1	2	–
CRP	↑6	–	6	Baptista *et al.* (2007), Dickerson *et al.* (2007), Domenici *et al.* (2010), Fan *et al.* (2007)	*KLK3*	–	↓1	1	–	*THPO*	↑2	↓1	3	Domenici *et al.* (2010)
CSF2	–	↓1	1	Domenici *et al.* (2010)	*LCN2*	↑1	–	1	–	*TIMP1*	↑3	–	3	Domenici *et al.* (2010)
CSF3	↑4	–	4	–						*TNC*	↑1	–	1	–

(Continued)

Table VII *(Continued)*

Gene name	# Times found to be altered	Total	References	Gene name	# Times found to be altered	Total	References	Gene name	# Times found to be altered	Total	References
CST3	↑1 –	1	–	*LEP*	↑8 –	8	Atmaca *et al.* (2003), Baptista *et al.* (2007), Domenici *et al.* (2010), Hosojima *et al.* (2006), Jow *et al.* (2006), Melkersson and Hulting (2001); Melkersson *et al.* (2000), Wang *et al.* (2007a)	*TNF-α*	↑6 ↓2	8	Garcia-Miss Mdel *et al.* (2010), Monteleone *et al.* (1997), Na and Kim (2007), O'Brien *et al.* (2008), Song *et al.* (2009), Theodoropoulou *et al.* (2001)
CTGF	↑3 –	3	–	*LHB*	↑1 –	1	–	*TNFRSF10C*	↑2 ↓1	3	–
CXCL1	↑1 –	1	–	*LRRC16A*	– ↓1	1	–	*TNFRSF14*	↑4 –	4	Domenici *et al.* (2010)
CXCL5	↑3 –	3	Domenici *et al.* (2010)	*LTA*	– ↓1	1	Domenici *et al.* (2010)	*TNFRSF1A*	↑2 –	2	Coelho *et al.* (2008), Hope *et al.* (2009)
DEFA1	↑1 –	1	Craddock *et al.* (2008)	*LUM*	– ↓1	1	–	*TNFRSF1B*	↑1 –	1	Coelho *et al.* (2008)
DPP4	↑1 –	1	Maes *et al.* (1996b)	*MAP2*	↑1 –	1	–	*TTR*	– ↓1	1	Yang *et al.* (2006)
EDN1	↑1 ↓1	2	Domenici *et al.* (2010)	*MDK*	– ↓1	1	Shimizu *et al.* (2003)	*UMOD*	↑1 –	1	–
EGF	↑2 ↓3	5	Domenici *et al.* (2010), Futamura *et al.* (2002), Ikeda *et al.* (2008)	*MIF*	↑3 –	3	–	*USP54*	– ↓1	1	–
F13B	– ↓3	3	Levin *et al.* (2010)	*MMP10*	↑1 –	1	–	*VANGL1*	– ↓1	1	–
F3	↑1 –	1	Domenici *et al.* (2010)					*VCAM1*	↑1 –	1	Domenici *et al.* (2010)
F7	↑1 ↓1	2	Domenici *et al.* (2010)					*VEGFA*	↑3 ↓1	4	Domenici *et al.* (2010)
								VWF	↑5 –	5	Domenici *et al.* (2010), Hope *et al.* (2009)

The numbers indicate the number of times a molecule was found to be altered in abundance.

Maternal infection has been modeled in a number of animal studies which provided supportive evidence for the hypothesis of causality in the link between maternal infection and increased risk of schizophrenia in the offspring (reviewed in Meyer *et al.*, 2009). For example, maternal influenza infection in mice during early/middle pregnancy was found to produce schizophrenia mimicking neuropathological, behavioral, and pharmacological abnormalities in the offspring. Such alterations are likely to be attributable to the maternal immune response rather than direct viral effects on the developing fetus, as these viruses are not detected in the fetal compartments postmaternal infection. In fact, maternal exposure to cytokine-releasing agents during pregnancy was found to alter pro- and anti-inflammatory cytokine levels in the placenta, the amniotic fluid, and the fetus, and to induce critical endophenotypes modeling schizophrenia. Examples include impairments in prepulse inhibition and latent inhibition, enhanced sensitivity to dopamine-stimulating treatment with amphetamine or to NMDA-receptor blockade by dizocilpine, and working memory deficiency (Meyer *et al.*, 2009).

B. Evidence for Innate and Adaptive Immune Response Activation in Schizophrenia

Mounting evidence for activated innate and adaptive immune responses in schizophrenia has accumulated as observed from review of the current literature and in-house studies. Aberrations in such immune responses were highlighted by canonical pathway analysis which revealed pronounced alterations in "APR signaling," "hepatic fibrosis," "communication between innate and adaptive immune cells," and "role of cytokines in mediating immune response signaling." These findings agree with earlier suggestions that there is a diffuse nonspecific activation of the immunological response system and/or type-1 and type-2 immune response system activation in schizophrenia (Strous and Shoenfeld, 2006). The observation of such activation in first-onset patients further suggests that an impaired innate and adaptive immune response is already present at an early stage of the disease which prevails in later/chronic stages, as confirmed by the literature.

C. Innate Immune Response: APR Signaling and Hepatic Metabolism

The overall evidence suggests an activated innate immune response in schizophrenia. This response acts as the first line of immune response against "stressors." Its cellular component consists of activation of monocytes, macrophages, granulocytes, and natural killer (NK) cells, while its humoral component consists of the complement system and APPs which participate in APR (Strous and Shoenfeld, 2006). APR is a systemic nonspecific innate reaction to disturbances in homeostasis caused by infections or stresses (e.g., tissue injury, trauma, or surgery). Such disturbances result in release of proinflammatory cytokines and other inflammatory mediators which diffuse to the extracellular fluid compartment, circulate in the blood, and activate receptors on different targets activating a number of downstream processes (Gruys *et al.*, 2005). Many of these processes have been reported to be altered in schizophrenia including hyperactivation of the HPA axis and associated alterations in the levels of adrenocorticotrophic hormone (ACTH) and glucocorticoids (Bradley and Dinan, 2010; Kaneda *et al.*, 2002; Lammers *et al.*, 1995; Walker *et al.*, 2008), activation of the complement system (Mayilyan *et al.*, 2008), and alterations in the levels of APPs (Fig. 3A). The levels of these proteins change due to a drastic shift in the pattern of protein synthesis in the liver during the inflammatory reaction (Fig. 3B1), which results in increased and decreased hepatic mRNA levels of positive and negative APPs, respectively (Gruys *et al.*, 2005).

APR was found to be the most significantly altered canonical pathway found in common through analysis of the literature and the in-house studies. Most of the altered molecules participating in the APR were positive APPs including *IL1RN* (↑9 studies), *HP* (↑7 studies), *CRP* (↑6 studies), *C3* (↑4 studies), *FGA* (↑1 study), *AGT* (↑1 study), *CFB* (↑1 study), *CP* (↑1 study), *FTH1* (↑3 studies), and *PROS1* (↑1 study). In the in-house studies, HP was found to be consistently increased across all five clinical centers and in one of the LC–MS studies. However, this protein was altered less frequently in the literature studies which were largely dominated by the study of chronic medicated patients. This suggests a potential diagnostic role of this anti-inflammatory APP, particularly at first onset of disease. Among the negative APPs, *AHSG* (↓3 studies), *TF* (↓2 studies), *ALB* (↓2 studies), *TTR* (↓1 study), and *IGF1* (↓2 studies) were altered. Decreases in negative APPs imply a temporary increase in availability of free hormones bound to these proteins (Maes *et al.*, 1997b). Of the remaining 21 altered APPs, *IL1α* (↓1 study), *IL6* (↑15 and ↓2 studies), and *TNF-α* (↑6 and ↓2 studies) are important modulators of hepatic APR signaling, as these proteins activate hepatocytic receptors in the liver and initiate synthesis of APPs (Gruys *et al.*, 2005; Maes *et al.*, 1997b). The involvement of altered liver immunology in schizophrenia is further highlighted by changes in molecules involved in the canonical pathway "hepatic fibrosis" (Fig. 3B1). This also provides supportive evidence for the hypothesis that schizophrenia is a systemic disorder. Hepatic fibrosis is typically associated with ethanol abuse, bile acid disorders, an increase in free fatty acids, and altered glucose metabolism which has also been implicated in schizophrenia (Guest *et al.*, 2010; Steiner *et al.*, 2010). Such molecular changes initiate a cascade of proinflammatory events leading to activation of hepatic stellate cells (HSCs) which subsequently secrete a repertoire of cytokines as shown in Fig. 3B1.

IL6 and *TNF-α* are proinflammatory cytokines primarily secreted from components of the innate immune system (activated monocytes and macrophages; Siegel *et al.*, 2009). *IL6* is the chief factor regulating production of most APPs. The high frequency in which this protein has been reported to be altered in schizophrenia may therefore explain the changes in levels of APPs often observed in patients (Gabay and Kushner, 1999). *IL6* along with *IL1α* and *TNF-α* are cytokines known to induce behavioral, neuroendocrine, and metabolic changes similar to processes observed in schizophrenia. For example, *TNF-α* leads to glucagon-induced hyperglycemia and *IL1α* activates the pituitary–adrenal system (Gruys *et al.*, 2005). The link between glucocorticoids and insulin and schizophrenia is well established. Glucocorticoids enhance the stimulatory effects of cytokines on the production of APPs (Baumann *et al.*, 1987), whereas insulin decreases their effects on the production of some APPs (Campos *et al.*, 1994). The roles of *IL6* and *TNF-α* extend beyond the immune system. These molecules, along with other cytokines, are involved in a cascade of positive and negative feedback regulation, interact with hormones and neurotransmitters, and represent the key mediators of the dynamic interaction between the CNS, immune, and endocrine systems (Turck *et al.*, 2009).

The question of whether the changes in the levels of APPs observed in schizophrenia patients contribute as causative factors or simply represent a manifestation of illness remains unknown. Nevertheless, it is known that when the relevant receptors are repeatedly triggered by stimuli, APR can become chronic (Gruys *et al.*, 2005). Whether this is the case in schizophrenia patients requires further examination, although chronic infection as etiological factor has been discussed for many years (Siegel *et al.*, 2009).

D. Adaptive Immune Response: Type-1 and Type-2 Response Imbalance

Taken together, the data also pointed to an activated adaptive immune system in schizophrenia. This is a specific system involved in higher functions such as immune memory and the ability to be conditioned. Its cellular component consists of T- and B-lymphocyte cells and is mainly activated by the type-1 immune system [T-helper-1 (Th1) cells, monocytes/macrophages (M1), and other cell types] which produce activating immunotransmitters *IL2* (↑8 and ↓4 studies), *IFN-γ* (↑1 and ↓1 studies), *IL12* (↑1 study), *IL18* (↑2 studies), and *TNF-α* (↑6 and ↓2 studies; Schwarz *et al.*, 2001; Strous and Shoenfeld, 2006). The humoral component of the adaptive system is made up of specific antibodies and is predominantly activated by the type-2 system [T-helper-2 (Th2) cells and monocytes/macrophages (M2)] which produces *IL4* (↑2 and ↓3 studies), *IL5* (↓3 clinical center), *IL6* (↑15 and ↓2 studies), *IL10* (↑5 and ↓1 studies), *IL13* (↑1 and ↓1 studies), and *TGF-β* (↑1 study) (Mills *et al.*, 2000; Schwarz *et al.*, 2001; Siegel *et al.*,

2009; Strous and Shoenfeld, 2006). The cytokines produced by the Th1 and Th2 cells antagonize each other while promoting their own type of response (Schwarz *et al.*, 2001).

Several lines of evidence point to an imbalance between the type-1 and type-2 immune response in schizophrenia with an overactivation of the type-2 response and a defective type-1 response (Muller *et al.*, 2004). For example, a handful of studies have described increased antibody production (type-2 response) in schizophrenia patients suggesting the presence of an autoimmune process, observed in 20–35% patients (Siegel *et al.*, 2009). Increases in Th2 lymphocytes in the blood have also been reported (Sperner-Unterweger *et al.*, 1999). Although *IL6* (repeatedly reported to be increased) is primarily secreted from activated monocytes/macrophages [innate and adaptive responses (M2)], it is also produced upon activation of type-2 immune response and activates the type-2 response leading to antibody production. The findings that cortisol and *IL10* (Th2 cytokine) levels were consistently elevated in first-onset drug-naive/free patients (4/5 clinical centers) and the lack of replicating changes in Th1 cytokines suggest an imbalance between type-1/type-2 systems. Alterations in cortisol levels in patients have been repeatedly described and reviewed (Bradley and Dinan, 2010). Cortisol inhibits the *IFN-γ* response by acting directly on T cells or indirectly through *IL12*. Increases in the levels of this active glucocorticosteroid decrease Th1 products subsequently inducing an imbalance between Th1/Th2 cytokines and a shift to Th2 response (Pinto *et al.*, 2006). Further, cortisol also acts as a powerful stimulant of plasma *IL10* levels (Dandona *et al.*, 1999). *IL10* represents a key regulator of the immune response as it suppresses Th1-dependent cellular immunity and promotes Th2-dependent humoral immunity (Moore *et al.*, 1993). Cortisol is also stimulated by *IL6* which is induced by either stress or infectious agents. The repeated observation of absence of change in the levels of *IL6* across the clinical centers suggests that, rather than infection, stress-related glucocorticoid signaling processes (demonstrated by increased cortisol levels) might occur in first-onset patients.

E. Effects of Antipsychotic Drugs on Immune-Related Processes

Immunological effects of antipsychotic medication have been reported since the 1960s (Strous and Shoenfeld, 2006). Haloperidol and clozapine, for example, possess strong immunosuppressive properties (Leykin *et al.*, 1997) and their effects in ameliorating psychosis are thus believed to be linked to such properties. These drugs along with risperidone have also been shown to differentially affect cytokine production in patients (Cazzullo *et al.*, 1998; Strous and Shoenfeld, 2006). Other studies have shown that antipsychotic drugs are associated with normalization of the increased number of T suppressor lymphocytes following treatment in drug-naive schizophrenia patients (Masserini *et al.*, 1990). Treatment was also associated with an increase in soluble *IL2* receptors (*sIL2R*) which reflects an increase of activated *IL2* bearing T cells (Muller *et al.*, 1997). Review of the literature resulted in identification of a panel of 29 immunological biomarkers reported to be altered in response to antipsychotic medication (Table VIIIA). This highlighted their potential value as molecular predictors of DR. Seven out of these 29 DR biomarkers were APPs (*IL1RN, IGF1, IL6, TNF, IL1B, IL6R, IL8*), suggesting a role

Table VIII
DRUG-RESPONSE (DR) BIOMARKERS.

Gene name	# Times found to be altered		Gene name	# Times found to be altered	
(A) 29 DR biomarkers identified from the literature					
PRL	↑7	↓7	*APOD*	↑1	–
S100B	↑5	↓2	*CCL11*	↑1	–
IL6	↑5	↓1	*FGF2*	↑1	–
BDNF	↑3	↓1	*FSHB*	–	↓1
INS	↑4	–	*GHRL*	–	↓1
LEP	↑4	–	*ICAM1*	–	↓1
IL2	↑1	↓2	*IGF1*	–	↓1
ADIPOQ	–	↓2	*IL1B*	–	↓1
CRP	↑2	–	*IL6R*	–	↓1
IL1RN	↑2	–	*IL8*	↑1	–
IL2R	↑2	–	*INHBB*	↑1	–
IL2RA	↑2	–	*OXT*	–	↓1
NGF	–	↓2	*SCGB1A1*	↑1	–
RAGE	↑1	–	*SOD1*	↑1	–
TNF	–	↓2			

Gene	In-house clinical centers		Literature	
(B) 10 DR biomarkers shared between the literature and in-house clinical centers				
BDNF	–	↓2	↑3	↓1
CRP	↑2	–	↑2	–
FGF2	–	↓1	↑1	–
FSHB	↑1	–	–	↓1
ICAM1	↑3	–	–	↓1
IL1RN	↑3	–	↑2	–
IL6	↑1	–	↑5	↓1
IL8	↑1	–	↑1	–
INS	↑2	–	↑4	–
PRL	↑3	–	↑7	↓7

The numbers indicate the number of times a molecule was found to be altered in abundance in the literature and/or in-house studies.

of such drugs on APR signaling. Additionally, 10 out of the 29 DR biomarkers (*BDNF, CRP, FGF2, FSHB, ICAM1, IL1RN, IL6, IL8, INS,* and *PRL*) were also found to be altered in our clinical centers. Interestingly, these showed opposite directions of change (Table VIIIB), suggesting a potential role of such molecules as biomarkers for both disease etiology and response to antipsychotic medication.

F. Glucocorticoid Receptor Signaling

In addition to hepatic fibrosis, glucocorticoid receptor signaling was another canonical pathway significantly associated with the DR biomarkers. This supports the literature finding that antipsychotic drugs suppress the effects produced by glucocorticoids or stress and inhibit elements of HPA axis activity (Basta-Kaim *et al.*, 2007). Increased levels of glucocorticoids have been shown to elicit changes in serotonergic, noradrenergic, dopaminergic, and excitatory amino acid neurotransmission resembling those observed in psychiatric disorders (reviewed in Basta-Kaim *et al.*, 2007). Hyperactivity of the HPA axis has also been frequently observed in schizophrenia. Thus, normalization of HPA axis activity in schizophrenic patients has been regarded as a positive marker of successful pharmacotherapy. The HPA axis also plays an important regulatory role in glucose metabolism, blood pressure regulation, cognition, thermoregulation, satiety, and other key homeostatic functions. This may thus explain the effectiveness of glucocorticoid receptor antagonists in reversing insulin-dependent diabetes in Cushing's disease (Chu *et al.*, 2001). This also suggests a possible adjuvant therapeutic role of such glucocorticoid receptor antagonists on atypical antipsychotic drug-induced metabolic side effects (e.g., insulin resistance, type II diabetes, and weight gain; Schatzberg and Lindley, 2008). In fact, beneficial effects of such antagonists have recently been demonstrated in animal studies including reversal of olanzapine-induced weight gain and abdominal fat accumulation (Beebe *et al.*, 2006). In a recent press release by Corcept Therapeutics, reduction of olanzapine-induced weight gain has also been observed in originally nonoverweight healthy males with BMIs less than 25.

G. Type-1 and Type-2 Immune System Rebalance

Given the immunological/inflammatory process in schizophrenia, innovative anti-inflammatory therapies have been explored. The recently developed selective cyclooxygenase-2 (*COX-2*) inhibitors are examples of some of the possible options. These inhibitors elicit a type-1 and type-2 rebalancing effect in schizophrenia (Muller *et al.*, 2004) through inhibition of type-2 cytokines and induction of type-1 cytokines (Pyeon *et al.*, 2000; Stolina *et al.*, 2000). In a prospective, randomized, double-blind study, treatment with the *COX-2* inhibitor (celecoxib) was associated with improved therapeutic action of risperidone in acute exacerbation of schizophrenia (Muller *et al.*, 2002). This effect was associated with an increase of the type-1 immune response in the celecoxib treatment group (Muller *et al.*, 2004; Siegel *et al.*, 2009). The therapeutic outcome of *COX-2* inhibitors appears to be associated with duration of illness (Muller *et al.*, 2004) and has been found to be most effective in the first years of illness (Casolini *et al.*, 2002; Siegel *et al.*, 2009). This suggests an important role of hepatic function in relation to drug metabolism and hepatic immune/metabolic status and/or response to stressors. Increases in the levels of *COX-2* and prostaglandin E2 (*PGE2*, the major product of *COX-2*) have been reported in schizophrenia (Das and Khan, 1998; Kaiya *et al.*, 1989). One of the mechanisms by which *COX-2* inhibitors rebalance the type-1/type-2 immune response is by inhibition of *IL6* and *PGE2* production and induction of the type-1 immune response (Siegel *et al.*, 2009). *PGE2* not only enhances production of type-2 cytokines (*IL4, IL5, IL6*, and *IL10*) but also inhibits production of type-1 cytokines (*IFN-γ*, *IL2*, and *IL12*). Therefore, inhibition of *PGE2* synthesis has been hypothesized to have a beneficial effect in treatment of disorders with impaired T-helper cell responses (Siegel *et al.*, 2009).

H. Evidence for Type 1/Type 2 Immune Response Imbalance in the CNS

Astrocytes and microglia represent the immunological cells in the CNS. While microglial cells (derived from peripheral macrophages) primarily secrete type-1 cytokines (e.g., *IL12*), astrocytes inhibit *IL12* and *ICAM-1* production and secrete the type-2 cytokine *IL10* (Aloisi *et al.*, 2000; Xiao and Link, 1999). The imbalance in activation of microglial cells and astrocytes has been proposed to reflect the type-1/type-2 imbalance in the CNS. Since *S100β* is a marker of astrocyte activation, increases in the levels of this circulatory protein have been used as a sensitive marker of brain damage, astrocyte activation, neural death, or blood–brain-barrier dysfunction (Lara *et al.*, 2001). Therefore, the repeated finding of increased levels of *S100β* (↑18 and ↓3 studies) in schizophrenia patients may suggest overactivation of astrocytes in the CNS of schizophrenia patients.

V. Conclusion and Perspectives

Review of the literature and in-house studies revealed an extensive repertoire of candidate blood-based biomarkers in schizophrenia which were classified into diagnostic and/or DR groups. This highlights the potential approach of characterizing disorders of the central nervous system based on alterations in peripheral markers. Integration of these biomarkers proved useful for identification of a converging functional pathway associated with the disease. This revealed an ongoing immunological/inflammatory process in schizophrenia involving activation of the innate immune system as reflected by alterations in APR signaling and hepatic metabolism, and activation of the adaptive immune system response as shown by changes in the type-1 and type-2 response. The type-2 response, shown by alterations in *IL10* and *IL5*, may be associated with the disease etiology, as this was more apparent in first-onset drug-naive/free patients along with altered cortisol levels which suggested activated stress response and potential HPA axis activation in these patients. These findings are indicative of changes in peripheral systems such as immune, metabolic, and hormonal pathways which are not only in close cross talk and coregulated but are also controlled by the central nervous system. Whether or not the prevalent immunological changes observed result from persistent infection or stress requires further functional confirmation. One must also acknowledge the fact that immunological parameters are susceptible to influence, and therefore, all data should be adjusted considering potential confounding factors and interpreted with caution. Methods for sample collection, storage, preparation, and processing would also need to be standardized. There is also an increasing need to recognize the importance of clearly delineating patient subtype, disease status, and duration of illness in neuropsychiatric research. Biomarker identification in first-onset patients would be expected to be less confounded by variables such as medication, diet, duration of illness, age, and exposure to environmental stress (e.g., hospitalization). Moreover, an active immunopathology would be expected in the early stage of the disease, probably even before the manifestation of overt clinical symptoms (Rothermundt *et al.*, 2001a). This study has also demonstrated the potential of developing personalized medicine strategies through patient stratification targeting the immune, metabolic, or hormonal pathways associated with disease onset. In this way, early monotherapy targeting specific pathways or combination therapy (e.g., antipsychotics with existing antidiabetics/anti-inflammatory agents) could be potentially applied to slow disease progression and/or enhance the response to existing antipsychotic drugs.

One limitation associated with the in-house studies was that the multiplex immunoassay panel employed may be biased toward analysis of molecules associated with the immune response, metabolic, and hormonal pathways. Nevertheless, the findings from the literature and in-house LC–MS studies provided complementary supportive evidence for an altered immunological/inflammatory status in schizophrenia patients, regardless of the platforms used. It should also be noted that the immune alterations seen in schizophrenia are not specific to this disorder. For example, several cytokines and APR signaling proteins have also been found to be altered in other neuropsychiatric conditions such as bipolar disorder and major depression (Maes *et al.*, 1997b). This indicates that algorithms comprising groups of highly specific and sensitive disease biomarkers are likely to be required for delineation of complex neuropsychiatric conditions (see Chapter "Algorithm development for diagnostic biomarker assays" by Izmailov *et al.*).

Acknowledgments

This research was supported by the Stanley Medical Research Institute (SMRI) and the European Union FP7 SchizDX research program (grant reference 223427).

References

Akiyama, K. (1999). Serum levels of soluble IL-2 receptor alpha, IL-6 and IL-1 receptor antagonist in schizophrenia before and during neuroleptic administration. *Schizophr. Res.* **37**, 97–106.

Aloisi, F., Ria, F., and Adorini, L. (2000). Regulation of T-cell responses by CNS antigen-presenting cells: different roles for microglia and astrocytes. *Immunol. Today* **21**, 141–147.

Atmaca, M., Kuloglu, M., Tezcan, E., Gecici, O., and Ustundag, B. (2003). Weight gain, serum leptin and triglyceride levels in patients with schizophrenia on antipsychotic treatment with quetiapine, olanzapine and haloperidol. *Schizophr. Res.* **60**, 99–100.

Atmaca, M., Tezcan, E., Kuloglu, M., Ustundag, B., and Kirtas, O. (2005). The effect of extract of ginkgo biloba addition to olanzapine on therapeutic effect and antioxidant enzyme levels in patients with schizophrenia. *Psychiatry Clin. Neurosci.* **59**, 652–656.

Bai, Y.M., Chen, T.T., Yang, W.S., Chi, Y.C., Lin, C.C., Liou, Y.J., Wang, Y.C., Su, T.P., Chou, P., and Chen, J.Y. (2009). Association of adiponectin and metabolic syndrome among patients taking atypical antipsychotics for schizophrenia: a cohort study. *Schizophr. Res.* **111**, 1–8.

Baptista, T., Davila, A., El Fakih, Y., Uzcategui, E., Rangel, N.N., Olivares, Y., Galeazzi, T., Vargas, D., Pena, R., Marquina, D., Villarroel, V., Teneud, L., *et al.* (2007). Similar frequency of abnormal correlation between serum leptin levels and BMI before and after olanzapine treatment in schizophrenia. *Int. Clin. Psychopharmacol.* **22**, 205–211.

Basta-Kaim, A., Budziszewska, B., Jaworska-Feil, L., Leskiewicz, M., Tetich, M., Otczy, M., Kubera, M., and Lason, W. (2007). Effects of neurosteroids on glucocorticoid receptor-mediated gene transcription in LMCAT cells—a possible interaction with psychotropic drugs. *Eur. Neuropsychopharmacol.* **17**, 37–45.

Baumann, H., Richards, C., and Gauldie, J. (1987). Interaction among hepatocyte-stimulating factors, interleukin 1, and glucocorticoids for regulation of acute phase plasma proteins in human hepatoma (HepG2) cells. *J. Immunol.* **139**, 4122–4128.

Beebe, K.L., Block, T., Debattista, C., Blasey, C., and Belanoff, J.K. (2006). The efficacy of mifepristone in the reduction and prevention of olanzapine-induced weight gain in rats. *Behav. Brain Res.* **171**, 225–229.

Berwaerts, J., Cleton, A., Rossenu, S., Talluri, K., Remmerie, B., Janssens, L., Boom, S., Kramer, M., and Eerdekens, M. (2010). A comparison of serum prolactin concentrations after administration of paliperidone extended-release and risperidone tablets in patients with schizophrenia. *J. Psychopharmacol.* **24**, 1011–1018.

Bradley, A.J., and Dinan, T.G. (2010). A systematic review of hypothalamic-pituitary-adrenal axis function in schizophrenia: implications for mortality. *J. Psychopharmacol.* **24**, 91–118.

Bresee, C., and Rapaport, M.H. (2009). Persistently increased serum soluble interleukin-2 receptors in continuously ill patients with schizophrenia. *Int. J. Neuropsychopharmacol.* **12**, 861–865.

Brown, A.S. (2006). Prenatal infection as a risk factor for schizophrenia. *Schizophr. Bull.* **32**, 200–202.

Brown, A.S., Begg, M.D., Gravenstein, S., Schaefer, C.A., Wyatt, R.J., Bresnahan, M., Babulas, V.P., and Susser, E.S. (2004a). Serologic evidence of prenatal influenza in the etiology of schizophrenia. *Arch. Gen. Psychiatry* **61**, 774–780.

Brown, A.S., Hooton, J., Schaefer, C.A., Zhang, H., Petkova, E., Babulas, V., Perrin, M., Gorman, J.M., and Susser, E.S. (2004b). Elevated maternal interleukin-8 levels and risk of schizophrenia in adult offspring. *Am. J. Psychiatry* **161**, 889–895.

Buka, S.L., Tsuang, M.T., Torrey, E.F., Klebanoff, M.A., Bernstein, D., and Yolken, R.H. (2001). Maternal infections and subsequent psychosis among offspring. *Arch. Gen. Psychiatry* **58**, 1032–1037.

Campos, S.P., Wang, Y., Koj, A., and Baumann, H. (1994). Insulin cooperates with IL-1 in regulating expression of alpha 1-acid glycoprotein gene in rat hepatoma cells. *Cytokine* **6**, 485–492.

Casolini, P., Catalani, A., Zuena, A.R., and Angelucci, L. (2002). Inhibition of COX-2 reduces the age-dependent increase of hippocampal inflammatory markers, corticosterone secretion, and behavioral impairments in the rat. *J. Neurosci. Res.* **68**, 337–343.

Cazzullo, C.L., Scarone, S., Grassi, B., Vismara, C., Trabattoni, D., and Clerici, M. (1998). Cytokines production in chronic schizophrenia patients with or without paranoid behaviour. *Prog. Neuropsychopharmacol. Biol. Psychiatry* **22**, 947–957.

Chang, J.S., Ahn, Y.M., Park, H.J., Lee, K.Y., Kim, S.H., Kang, U.G., and Kim, Y.S. (2008). Aripiprazole augmentation in clozapine-treated patients with refractory schizophrenia: an 8-week, randomized, double-blind, placebo-controlled trial. *J. Clin. Psychiatry* **69**, 720–731.

Chen da, C., Wang, J., Wang, B., Yang, S.C., Zhang, C.X., Zheng, Y.L., Li, Y.L., Wang, N., Yang, K.B., Xiu, M.H., Kosten, T.R., and Zhang, X.Y. (2009). Decreased levels of serum brain-derived neurotrophic factor in drug-naive first-episode schizophrenia: relationship to clinical phenotypes. *Psychopharmacology (Berl.)* **207**, 375–380.

Chen, Q., Cai, Z.J., Mao, P.X., Zhai, Y.M., Mitchell, P.B., and Tang, Y.L. (2008). Effects of risperidone on glucose metabolism in Chinese patients with schizophrenia: a prospective study. *J. Psychiatr. Res.* **43**, 124–128.

Chen, Y.L., Cheng, T.S., and Lung, F.W. (2009). Prolactin levels in olanzapine treatment correlate with positive symptoms of schizophrenia: results from an open-label, flexible-dose study. *Prim. Care Companion J. Clin. Psychiatry* **11**, 16–20.

Chittiprol, S., Venkatasubramanian, G., Neelakantachar, N., Allha, N., Shetty, K.T., and Gangadhar, B.N. (2009). Beta2-microglobulin abnormalities in antipsychotic-naive schizophrenia: evidence for immune pathogenesis. *Brain Behav. Immun.* **23**, 189–192.

Chu, J.W., Matthias, D.F., Belanoff, J., Schatzberg, A., Hoffman, A.R., and Feldman, D. (2001). Successful long-term treatment of refractory Cushing's disease with high-dose mifepristone (RU 486). *J. Clin. Endocrinol. Metab.* **86**, 3568–3573.

Coelho, F.M., Reis, H.J., Nicolato, R., Romano-Silva, M.A., Teixeira, M.M., Bauer, M.E., and Teixeira, A.L. (2008). Increased serum levels of inflammatory markers in chronic institutionalized patients with schizophrenia. *Neuroimmunomodulation* **15**, 140–144.

Cook, I.A. (2008). Biomarkers in psychiatry: potentials, pitfalls, and pragmatics. *Prim. Psychiatry* **15**, 54–59.

Costa, A.M., de Lima, M.S., Faria, M., Filho, S.R., de Oliveira, I.R., and de Jesus Mari, J. (2007). A naturalistic, 9-month follow-up, comparing olanzapine and conventional antipsychotics on sexual function and hormonal profile for males with schizophrenia. *J. Psychopharmacol.* **21**, 165–170.

Craddock, R.M., Huang, J.T., Jackson, E., Harris, N., Torrey, E.F., Herberth, M., and Bahn, S. (2008). Increased alpha-defensins as a blood marker for schizophrenia susceptibility. *Mol. Cell. Proteomics* **7**, 1204–1213.

Crespo-Facorro, B., Carrasco-Marin, E., Perez-Iglesias, R., Pelayo-Teran, J.M., Fernandez-Prieto, L., Leyva-Cobian, F., and Vazquez-Barquero, J.L. (2008). Interleukin-12 plasma levels in drug-naive patients with a first episode of psychosis: effects of antipsychotic drugs. *Psychiatry Res.* **158**, 206–216.

Dakhale, G., Khanzode, S., Saoji, A., Khobragade, L., and Turankar, A. (2004). Oxidative damage and schizophrenia: the potential benefit by atypical antipsychotics. *Neuropsychobiology* **49**, 205–209.

Dandona, P., Aljada, A., Garg, R., and Mohanty, P. (1999). Increase in plasma interleukin-10 following hydrocortisone injection. *J. Clin. Endocrinol. Metab.* **84**, 1141–1144.

Das, I., and Khan, N.S. (1998). Increased arachidonic acid induced platelet chemiluminescence indicates cyclooxygenase overactivity in schizophrenic subjects. *Prostaglandins Leukot. Essent. Fatty Acids* **58**, 165–168.

Dickerson, F., Stallings, C., Origoni, A., Boronow, J., and Yolken, R. (2007). C-reactive protein is associated with the severity of cognitive impairment but not of psychiatric symptoms in individuals with schizophrenia. *Schizophr. Res.* **93**, 261–265.

Domenici, E., Wille, D.R., Tozzi, F., Prokopenko, I., Miller, S., McKeown, A., Brittain, C., Rujescu, D., Giegling, I., Turck, C.W., Holsboer, F., Bullmore, E.T., *et al.* (2010). Plasma protein biomarkers for depression and schizophrenia by multi analyte profiling of case-control collections. *PLoS One* **5**, e9166.

Drexhage, R.C., Padmos, R.C., de Wit, H., Versnel, M.A., Hooijkaas, H., van der Lely, A.J., van Beveren, N., deRijk, R.H., and Cohen, D. (2008). Patients with schizophrenia show raised serum levels of the pro-inflammatory chemokine CCL2: association with the metabolic syndrome in patients? *Schizophr. Res.* **102**, 352–355.

Drzyzga, L., Obuchowicz, E., Marcinowska, A., and Herman, Z.S. (2006). Cytokines in schizophrenia and the effects of antipsychotic drugs. *Brain Behav. Immun.* **20**, 532–545.

Ebrinc, S., Top, C., Oncul, O., Basoglu, C., Cavuslu, S., and Cetin, M. (2002). Serum interleukin 1 alpha and interleukin 2 levels in patients with schizophrenia. *J. Int. Med. Res.* **30**, 314–317.

Fan, X., Anderson, E.J., Copeland, P.M., Borba, C.P., Nguyen, D.D., Freudenreich, O., Goff, D.C., and Henderson, D.C. (2006a). Higher fasting serum insulin is associated with increased resting energy expenditure in nondiabetic schizophrenia patients. *Biol. Psychiatry* **60**, 1372–1377.

Fan, X., Liu, E., Pristach, C., Goff, D.C., and Henderson, D.C. (2006b). Higher fasting serum insulin levels are associated with a better psychopathology profile in acutely ill non-diabetic inpatients with schizophrenia. *Schizophr. Res.* **86**, 30–35.

Fan, X., Pristach, C., Liu, E.Y., Freudenreich, O., Henderson, D.C., and Goff, D.C. (2007). Elevated serum levels of C-reactive protein are associated with more severe psychopathology in a subgroup of patients with schizophrenia. *Psychiatry Res.* **149**, 267–271.

Fatemi, S.H., Kroll, J.L., and Stary, J.M. (2001). Altered levels of Reelin and its isoforms in schizophrenia and mood disorders. *Neuroreport* **12**, 3209–3215.

Feldcamp, L., and Wong, A. (2009). Biomarkers in schizophrenia. *In* "Biomarkers for Psychiatric Disorders", (Christoph W. Turck, Ed.), pp. 23–56. Springer, New York, NY, USA.

Frommberger, U.H., Bauer, J., Haselbauer, P., Fraulin, A., Riemann, D., and Berger, M. (1997). Interleukin-6-(IL-6) plasma levels in depression and schizophrenia: comparison between the acute state and after remission. *Eur. Arch. Psychiatry Clin. Neurosci.* **247**, 228–233.

Futamura, T., Toyooka, K., Iritani, S., Niizato, K., Nakamura, R., Tsuchiya, K., Someya, T., Kakita, A., Takahashi, H., and Nawa, H. (2002). Abnormal expression of epidermal growth factor and its receptor in the forebrain and serum of schizophrenic patients. *Mol. Psychiatry* **7**, 673–682.

Gabay, C., and Kushner, I. (1999). Acute-phase proteins and other systemic responses to inflammation. *N. Engl. J. Med.* **340**, 448–454.

Gama, C.S., Salvador, M., Andreazza, A.C., Kapczinski, F., and Silva Belmonte-de-Abreu, P. (2006). Elevated serum superoxide dismutase and thiobarbituric acid reactive substances in schizophrenia: a study of patients treated with haloperidol or clozapine. *Prog. Neuropsychopharmacol. Biol. Psychiatry* **30**, 512–515.

Gama, C.S., Andreazza, A.C., Kunz, M., Berk, M., Belmonte-de-Abreu, P.S., and Kapczinski, F. (2007). Serum levels of brain-derived neurotrophic factor in patients with schizophrenia and bipolar disorder. *Neurosci. Lett.* **420**, 45–48.

Garcia-Miss Mdel, R., Perez-Mutul, J., Lopez-Canul, B., Solis-Rodriguez, F., Puga-Machado, L., Oxte-Cabrera, A., Gurubel-Maldonado, J., and Arankowsky-Sandoval, G. (2010). Folate, homocysteine, interleukin-6, and tumor necrosis factor alfa levels, but not the methylenetetrahydrofolate reductase C677T polymorphism, are risk factors for schizophrenia. *J. Psychiatr. Res.* **44**, 441–446.

Gattaz, W.F., Lara, D.R., Elkis, H., Portela, L.V., Goncalves, C.A., Tort, A.B., Henna, J., and Souza, D.O. (2000). Decreased S100-beta protein in schizophrenia: preliminary evidence. *Schizophr. Res.* **43**, 91–95.

Gattaz, W.F., Abrahao, A.L., and Foccacia, R. (2004). Childhood meningitis, brain maturation and the risk of psychosis. *Eur. Arch. Psychiatry Clin. Neurosci.* **254**, 23–26.

Gaughran, F., O'Neill, E., Cole, M., Collins, K., Daly, R.J., and Shanahan, F. (1998). Increased soluble interleukin 2 receptor levels in schizophrenia. *Schizophr. Res.* **29**, 263–267.

Gaughran, F., O'Neill, E., Sham, P., Daly, R.J., and Shanahan, F. (2002). Soluble interleukin 2 receptor levels in families of people with schizophrenia. *Schizophr. Res.* **56**, 235–239.

Grillo, R.W., Ottoni, G.L., Leke, R., Souza, D.O., Portela, L.V., and Lara, D.R. (2007). Reduced serum BDNF levels in schizophrenic patients on clozapine or typical antipsychotics. *J. Psychiatr. Res.* **41**, 31–35.

Gruys, E., Toussaint, M.J.M., Niewold, T.A., and Koopmans, S.J. (2005). Acute phase reaction and acute phase proteins. *J. Zhejiang Univ. Sci. B* **6**, 1045–1056.

Guest, P.C., Wang, L., Harris, L.W., Burling, K., Levin, Y., Ernst, A., Wayland, M.T., Umrania, Y., Herberth, M., Koethe, D., van Beveren, J.M., Rothermundt, M., *et al.* (2010). Increased levels of circulating insulin-related peptides in first-onset, antipsychotic naive schizophrenia patients. *Mol. Psychiatry* **15**, 118–119.

Guimaraes, L.R., Jacka, F.N., Gama, C.S., Berk, M., Leitao-Azevedo, C.L., Belmonte de Abreu, M.G., Lobato, M.I., Andreazza, A.C., Ceresér, K.M., Kapczinski, F., and Belmonte-de-Abreu, P. (2008). Serum levels of brain-derived neurotrophic factor in schizophrenia on a hypocaloric diet. *Prog. Neuropsychopharmacol. Biol. Psychiatry* **32**, 1595–1598.

Hashimoto, K., Shimizu, E., Komatsu, N., Nakazato, M., Okamura, N., Watanabe, H., Kumakiri, C., Shinoda, N., Okada, S., Takei, N., and Iyo, M. (2003). Increased levels of serum basic fibroblast growth factor in schizophrenia. *Psychiatry Res.* **120**, 211–218.

Higgs, R., Knierman, M.D., Gelfanova, V., Butler, J.P., and Hale, J.E. (2008). Label-free LC-MS method for the identification of biomarkers. *In* "Clinical Proteomics: Methods and Protocols", (A. Vlahou, Ed.), Vol. 428, pp. 209–230. Humana Press, New York, NY.

Hope, S., Melle, I., Aukrust, P., Steen, N.E., Birkenaes, A.B., Lorentzen, S., Agartz, I., Ueland, T., and Andreassen, O.A. (2009). Similar immune profile in bipolar disorder and schizophrenia: selective increase in soluble tumor necrosis factor receptor I and von Willebrand factor. *Bipolar Disord.* **11**, 726–734.

Hosojima, H., Togo, T., Odawara, T., Hasegawa, K., Miura, S., Kato, Y., Kanai, A., Kase, A., Uchikado, H., and Hirayasu, Y. (2006). Early effects of olanzapine on serum levels of ghrelin, adiponectin and leptin in patients with schizophrenia. *J. Psychopharmacol.* **20**, 75–79.

Huang, T.L. (2002). Decreased serum albumin levels in Taiwanese patients with schizophrenia. *Psychiatry Clin. Neurosci.* **56**, 627–630.

Huang, T.L., and Hung, Y.Y. (2009). Lorazepam reduces the serum brain-derived neurotrophic factor level in schizophrenia patients with catatonia. *Prog. Neuropsychopharmacol. Biol. Psychiatry* **33**, 158–159.

Huang, T.L., and Lee, C.T. (2006). Associations between serum brain-derived neurotrophic factor levels and clinical phenotypes in schizophrenia patients. *J. Psychiatr. Res.* **40**, 664–668.

Ikeda, Y., Yahata, N., Ito, I., Nagano, M., Toyota, T., Yoshikawa, T., Okubo, Y., and Suzuki, H. (2008). Low serum levels of brain-derived neurotrophic factor and epidermal growth factor in patients with chronic schizophrenia. *Schizophr. Res.* **101**, 58–66.

Iwata, Y., Suzuki, K., Nakamura, K., Matsuzaki, H., Sekine, Y., Tsuchiya, K.J., Sugihara, G., Kawai, M., Minabe, Y., Takei, N., and Mori, N. (2007). Increased levels of serum soluble L-selectin in unmedicated patients with schizophrenia. *Schizophr. Res.* **89**, 154–160.

Jablensky, A. (2000). Epidemiology of schizophrenia: the global burden of disease and disability. *Eur. Arch. Psychiatry Clin. Neurosci.* **250**, 274–285.

Jindal, R.D., Pillai, A.K., Mahadik, S.P., Eklund, K., Montrose, D.M., and Keshavan, M.S. (2010). Decreased BDNF in patients with antipsychotic naive first episode schizophrenia. *Schizophr. Res.* **119**, 47–51.

Jockers-Scherubl, M.C., Rentzsch, J., Danker-Hopfe, H., Radzei, N., Schurer, F., Bahri, S., and Hellweg, R. (2006). Adequate antipsychotic treatment normalizes serum nerve growth factor concentrations in schizophrenia with and without cannabis or additional substance abuse. *Neurosci. Lett.* **400**, 262–266.

Jow, G.M., Yang, T.T., and Chen, C.L. (2006). Leptin and cholesterol levels are low in major depressive disorder, but high in schizophrenia. *J. Affect. Disord.* **90**, 21–27.

Kaiya, H., Uematsu, M., Ofuji, M., Nishida, A., Takeuchi, K., Nozaki, M., and Idaka, E. (1989). Elevated plasma prostaglandin E2 levels in schizophrenia. *J. Neural Transm.* **77**, 39–46.

Kane, J.M., Correll, C.U., Goff, D.C., Kirkpatrick, B., Marder, S.R., Vester-Blokland, E., Sun, W., Carson, W.H., Pikalov, A., and Assuncao-Talbott, S. (2009). A multicenter, randomized, double-blind, placebo-controlled, 16-week study of adjunctive aripiprazole for schizophrenia or schizoaffective disorder inadequately treated with quetiapine or risperidone monotherapy. *J. Clin. Psychiatry* **70**, 1348–1357.

Kaneda, Y., Fujii, A., and Ohmori, T. (2002). The hypothalamic-pituitary-adrenal axis in chronic schizophrenic patients long-term treated with neuroleptics. *Prog. Neuropsychopharmacol. Biol. Psychiatry* **26**, 935–938.

Keri, S., Kiss, I., and Kelemen, O. (2009). Sharing secrets: oxytocin and trust in schizophrenia. *Soc. Neurosci.* **4**, 287–293.

Kim, Y.K., Lee, M.S., and Suh, K.Y. (1998). Decreased interleukin-2 production in Korean schizophrenic patients. *Biol. Psychiatry* **43**, 701–704.

Kim, Y.K., Kim, L., and Lee, M.S. (2000). Relationships between interleukins, neurotransmitters and psychopathology in drug-free male schizophrenics. *Schizophr. Res.* **44**, 165–175.

Kim, K.S., Pae, C.U., Chae, J.H., Bahk, W.M., Jun, T.Y., Kim, D.J., and Dickson, R.A. (2002). Effects of olanzapine on prolactin levels of female patients with schizophrenia treated with risperidone. *J. Clin. Psychiatry* **63**, 408–413.

Kim, Y.K., Myint, A.M., Lee, B.H., Han, C.S., Lee, H.J., Kim, D.J., and Leonard, B.E. (2004). Th1, Th2 and Th3 cytokine alteration in schizophrenia. *Prog Neuropsychopharmacol. Biol. Psychiatry* **28**, 1129–1134.

Kinon, B.J., Ahl, J., Liu-Seifert, H., and Maguire, G.A. (2006). Improvement in hyperprolactinemia and reproductive comorbidities in patients with schizophrenia switched from conventional antipsychotics or risperidone to olanzapine. *Psychoneuroendocrinology* **31**, 577–588.

Konarzewska, B., Wolczynski, S., Szulc, A., Galinska, B., Poplawska, R., and Waszkiewicz, N. (2009). Effect of risperidone and olanzapine on reproductive hormones, psychopathology and sexual functioning in male patients with schizophrenia. *Psychoneuroendocrinology* **34**, 129–139.

Koponen, H., Rantakallio, P., Veijola, J., Jones, P., Jokelainen, J., and Isohanni, M. (2004). Childhood central nervous system infections and risk for schizophrenia. *Eur. Arch. Psychiatry Clin. Neurosci.* **254**, 9–13.

Kronig, H., Riedel, M., Schwarz, M.J., Strassnig, M., Moller, H.J., Ackenheil, M., and Muller, N. (2005). ICAM G241A polymorphism and soluble ICAM-1 serum levels: evidence for an active immune process in schizophrenia. *Neuroimmunomodulation* **12**, 54–59.

Kwon, J.S., Jang, J.H., Kang, D.H., Yoo, S.Y., Kim, Y.K., and Cho, S.J. (2009). Long-term efficacy and safety of aripiprazole in patients with schizophrenia, schizophreniform disorder, or schizoaffective disorder: 26-week prospective study. *Psychiatry Clin. Neurosci.* **63**, 73–81.

Lakhan, S.E., and Kramer, A. (2009). Schizophrenia genomics and proteomics: are we any closer to biomarker discovery? *Behav. Brain Funct.* **5**, 2.

Lakhan, S.E., Vieira, K., and Hamlat, E. (2010). Biomarkers in psychiatry: drawbacks and potential for misuse. *Int. Arch. Med.* **3**, 1.

Lammers, C.H., Garcia-Borreguero, D., Schmider, J., Gotthardt, U., Dettling, M., Holsboer, F., and Heuser, I.J. (1995). Combined dexamethasone/corticotropin-releasing hormone test in patients with schizophrenia and in normal controls: II. *Biol. Psychiatry* **38**, 803–807.

Lara, D.R., Gama, C.S., Belmonte-de-Abreu, P., Portela, L.V., Goncalves, C.A., Fonseca, M., Hauck, S., and Souza, D.O. (2001). Increased serum S100B protein in schizophrenia: a study in medication-free patients. *J. Psychiatr. Res.* **35**, 11–14.

Lee, B.H., and Kim, Y.K. (2009). Increased plasma brain-derived neurotropic factor, not nerve growth factor-Beta, in schizophrenia patients with better response to risperidone treatment. *Neuropsychobiology* **59**, 51–58.

Lencz, T., Morgan, T.V., Athanasiou, M., Dain, B., Reed, C.R., Kane, J.M., Kucherlapati, R., and Malhotra, A.K. (2007). Converging evidence for a pseudoautosomal cytokine receptor gene locus in schizophrenia. *Mol. Psychiatry* **12**, 572–580.

Levin, Y., and Bahn, S. (2010). Quantification of proteins by label-free LC-MS/MS. *Methods Mol. Biol.* **658**, 217–231.

Levin, Y., Wang, L., Schwarz, E., Koethe, D., Leweke, F.M., and Bahn, S. (2010). Global proteomic profiling reveals altered proteomic signature in schizophrenia serum. *Mol. Psychiatry* **15**, 1088–1100.

Lewis, D.A. (2000). Is there a neuropathology of schizophrenia? Recent findings converge on altered thalamic-prefrontal cortical connectivity. *Neuroscientist* **6**, 208–218.

Leykin, I., Mayer, R., and Shinitzky, M. (1997). Short and long-term immunosuppressive effects of clozapine and haloperidol. *Immunopharmacology* **37**, 75–86.

Lin, A., Kenis, G., Bignotti, S., Tura, G.J., De Jong, R., Bosmans, E., Pioli, R., Altamura, C., Scharpe, S., and Maes, M. (1998). The inflammatory response system in treatment-resistant schizophrenia: increased serum interleukin-6. *Schizophr. Res.* **32**, 9–15.

Ling, S.H., Tang, Y.L., Jiang, F., Wiste, A., Guo, S.S., Weng, Y.Z., and Yang, T.S. (2007). Plasma S-100B protein in Chinese patients with schizophrenia: comparison with healthy controls and effect of antipsychotics treatment. *J. Psychiatr. Res.* **41**, 36–42.

Maes, M., Bosmans, E., Ranjan, R., Vandoolaeghe, E., Meltzer, H.Y., De Ley, M., Berghmans, R., Stans, G., and Desnyder, R. (1996a). Lower plasma CC16, a natural anti-inflammatory protein, and increased plasma interleukin-1 receptor antagonist in schizophrenia: effects of antipsychotic drugs. *Schizophr. Res.* **21**, 39–50.

Maes, M., De Meester, I., Scharpe, S., Desnyder, R., Ranjan, R., and Meltzer, H.Y. (1996b). Alterations in plasma dipeptidyl peptidase IV enzyme activity in depression and schizophrenia: effects of antidepressants and antipsychotic drugs. *Acta Psychiatr. Scand.* **93**, 1–8.

Maes, M., Bosmans, E., Kenis, G., De Jong, R., Smith, R.S., and Meltzer, H.Y. (1997a). In vivo immunomodulatory effects of clozapine in schizophrenia. *Schizophr. Res.* **26**, 221–225.

Maes, M., Delange, J., Ranjan, R., Meltzer, H.Y., Desnyder, R., Cooremans, W., and Scharpe, S. (1997b). Acute phase proteins in schizophrenia, mania and major depression: modulation by psychotropic drugs. *Psychiatry Res.* **66**, 1–11.

Maes, M., Bocchio Chiavetto, L., Bignotti, S., Battisa Tura, G.J., Pioli, R., Boin, F., Kenis, G., Bosmans, E., de Jongh, R., and Altamura, C.A. (2002). Increased serum interleukin-8 and interleukin-10 in schizophrenic patients resistant to treatment with neuroleptics and the stimulatory effects of clozapine on serum leukemia inhibitory factor receptor. *Schizophr. Res.* **54**, 281–291.

Mahadik, S.P., Khan, M.M., Evans, D.R., and Parikh, V.V. (2002). Elevated plasma level of apolipoprotein D in schizophrenia and its treatment and outcome. *Schizophr. Res.* **58**, 55–62.

Mahendran, R., and Chan, Y.H. (2004). Interleukin-2 levels in chronic schizophrenia patients. *Ann. Acad. Med. Singapore* **33**, 320–323.

Masserini, C., Vita, A., Basile, R., Morselli, R., Boato, P., Peruzzi, C., Pugnetti, L., Ferrante, P., and Cazzullo, C.L. (1990). Lymphocyte subsets in schizophrenic disorders. Relationship with clinical, neuromorphological and treatment variables. *Schizophr. Res.* **3**, 269–275.

Mayilyan, K.R., Weinberger, D.R., and Sim, R.B. (2008). The complement system in schizophrenia. *Drug News Perspect.* **21**, 200–210.

McGrath, J.J. (2006). Variations in the incidence of schizophrenia: data versus dogma. *Schizophr. Bull.* **32**, 195–197.

McGrath, J., Saha, S., Welham, J., El Saadi, O., MacCauley, C., and Chant, D. (2004). A systematic review of the incidence of schizophrenia: the distribution of rates and the influence of sex, urbanicity, migrant status and methodology. *BMC Med.* **2**, 13.

Melkersson, K.I., and Hulting, A.L. (2001). Insulin and leptin levels in patients with schizophrenia or related psychoses—a comparison between different antipsychotic agents. *Psychopharmacology (Berl.)* **154**, 205–212.

Melkersson, K.I., Hulting, A.L., and Brismar, K.E. (1999). Different influences of classical antipsychotics and clozapine on glucose-insulin homeostasis in patients with schizophrenia or related psychoses. *J. Clin. Psychiatry* **60**, 783–791.

Melkersson, K.I., Hulting, A.L., and Brismar, K.E. (2000). Elevated levels of insulin, leptin, and blood lipids in olanzapine-treated patients with schizophrenia or related psychoses. *J. Clin. Psychiatry* **61**, 742–749.

Melkersson, K.I., Hulting, A.L., and Rane, A.J. (2001). Dose requirement and prolactin elevation of antipsychotics in male and female patients with schizophrenia or related psychoses. *Br. J. Clin. Pharmacol.* **51**, 317–324.

Meyer, U., Feldon, J., and Yee, B.K. (2009). A review of the fetal brain cytokine imbalance hypothesis of schizophrenia. *Schizophr. Bull.* **35**, 959–972.

Mills, C.D., Kincaid, K., Alt, J.M., Heilman, M.J., and Hill, A.M. (2000). M-1/M-2 macrophages and the Th1/Th2 paradigm. *J. Immunol.* **164**, 6166–6173.

Monteleone, P., Fabrazzo, M., Tortorella, A., and Maj, M. (1997). Plasma levels of interleukin-6 and tumor necrosis factor alpha in chronic schizophrenia: effects of clozapine treatment. *Psychiatry Res.* **71**, 11–17.

Montgomery, J., Winterbottom, E., Jessani, M., Kohegyi, E., Fulmer, J., Seamonds, B., and Josiassen, R.C. (2004). Prevalence of hyperprolactinemia in schizophrenia: association with typical and atypical antipsychotic treatment. *J. Clin. Psychiatry* **65**, 1491–1498.

Moore, K.W., O'Garra, A., de Waal Malefyt, R., Vieira, P., and Mosmann, T.R. (1993). Interleukin-10. *Annu. Rev. Immunol.* **11**, 165–190.

Morris, B.J., Cochran, S.M., and Pratt, J.A. (2005). PCP: from pharmacology to modelling schizophrenia. *Curr. Opin. Pharmacol.* **5**, 101–106.

Muller, N., Empl, M., Riedel, M., Schwarz, M., and Ackenheil, M. (1997). Neuroleptic treatment increases soluble IL-2 receptors and decreases soluble IL-6 receptors in schizophrenia. *Eur. Arch. Psychiatry Clin. Neurosci.* **247**, 308–313.

Muller, N., Riedel, M., Scheppach, C., Brandstatter, B., Sokullu, S., Krampe, K., Ulmschneider, M., Engel, R.R., Moller, H.J., and Schwarz, M.J. (2002). Beneficial antipsychotic effects of celecoxib add-on therapy compared to risperidone alone in schizophrenia. *Am. J. Psychiatry* **159**, 1029–1034.

Muller, N., Ulmschneider, M., Scheppach, C., Schwarz, M.J., Ackenheil, M., Moller, H.J., Gruber, R., and Riedel, M. (2004). COX-2 inhibition as a treatment approach in schizophrenia: immunological considerations and clinical effects of celecoxib add-on therapy. *Eur. Arch. Psychiatry Clin. Neurosci.* **254**, 14–22.

Müller, N., and Schwarz, M.J. (2009). Cytokines, immunity and schizophrenia with emphasis on underlying neurochemical mechanisms. *In* "The Neuroimmunological Basis of Behavior and Mental Disorders", (A. Siegel and S.S. Zalcman, Eds.), pp. 307–325. Springer, New York, NY, USA.

Na, K.S., and Kim, Y.K. (2007). Monocytic, Th1 and th2 cytokine alterations in the pathophysiology of schizophrenia. *Neuropsychobiology* **56**, 55–63.

Naudin, J., Mege, J.L., Azorin, J.M., and Dassa, D. (1996). Elevated circulating levels of IL-6 in schizophrenia. *Schizophr. Res.* **20**, 269–273.

Nawa, H., Takahashi, M., and Patterson, P.H. (2000). Cytokine and growth factor involvement in schizophrenia—support for the developmental model. *Mol. Psychiatry* **5**, 594–603.

Newcomer, J.W. (2007). Metabolic syndrome and mental illness. *Am. J. Manag. Care* **13**, S170–S177.

O'Brien, S.M., Scully, P., and Dinan, T.G. (2008). Increased tumor necrosis factor-alpha concentrations with interleukin-4 concentrations in exacerbations of schizophrenia. *Psychiatry Res.* **160**, 256–262.

Palomino, A., Vallejo-Illarramendi, A., Gonzalez-Pinto, A., Aldama, A., Gonzalez-Gomez, C., Mosquera, F., Gonzalez-Garcia, G., and Matute, C. (2006). Decreased levels of plasma BDNF in first-episode schizophrenia and bipolar disorder patients. *Schizophr. Res.* **86**, 321–322.

Parikh, V., Evans, D.R., Khan, M.M., and Mahadik, S.P. (2003). Nerve growth factor in never-medicated first-episode psychotic and medicated chronic schizophrenic patients: possible implications for treatment outcome. *Schizophr. Res.* **60**, 117–123.

Pedersen, A., Diedrich, M., Kaestner, F., Koelkebeck, K., Ohrmann, P., Ponath, G., Kipp, F., Abel, S., Siegmund, A., Suslow, T., von Eiff, C., Arolt, V., *et al.* (2008). Memory impairment correlates with increased S100B serum concentrations in patients with chronic schizophrenia. *Prog. Neuropsychopharmacol. Biol. Psychiatry* **32**, 1789–1792.

Pinto, R.A., Arredondo, S.M., Bono, M.R., Gaggero, A.A., and Diaz, P.V. (2006). T helper 1/T helper 2 cytokine imbalance in respiratory syncytial virus infection is associated with increased endogenous plasma cortisol. *Pediatrics* **117**, e878–e886.

Pirildar, S., Gonul, A.S., Taneli, F., and Akdeniz, F. (2004). Low serum levels of brain-derived neurotrophic factor in patients with schizophrenia do not elevate after antipsychotic treatment. *Prog. Neuropsychopharmacol. Biol. Psychiatry* **28**, 709–713.

Potvin, S., Stip, E., Sepehry, A.A., Gendron, A., Bah, R., and Kouassi, E. (2008). Inflammatory cytokine alterations in schizophrenia: a systematic quantitative review. *Biol. Psychiatry* **63**, 801–808.

Pyeon, D., Diaz, F.J., and Splitter, G.A. (2000). Prostaglandin E(2) increases bovine leukemia virus tax and pol mRNA levels via cyclooxygenase 2: regulation by interleukin-2, interleukin-10, and bovine leukemia virus. *J. Virol.* **74**, 5740–5745.

Qi, L.Y., Xiu, M.H., Chen da, C., Wang, F., Kosten, T.A., Kosten, T.R., and Zhang, X.Y. (2009). Increased serum S100B levels in chronic schizophrenic patients on long-term clozapine or typical antipsychotics. *Neurosci. Lett.* **462**, 113–117.

Reis, H.J., Nicolato, R., Barbosa, I.G., Teixeira do Prado, P.H., Romano-Silva, M.A., and Teixeira, A. L. (2008). Increased serum levels of brain-derived neurotrophic factor in chronic institutionalized patients with schizophrenia. *Neurosci. Lett.* **439**, 157–159.

Rice, D.P. (1999). The economic impact of schizophrenia. *J. Clin. Psychiatry* **60**(Suppl. 1), 4–6discussion 28–30.

Rice, D.P., and Miller, L.S. (1995). The economic burden of affective disorders. *Br. J. Psychiatry Suppl.* **27**, 34–42.

Richards, A.A., Hickman, I.J., Wang, A.Y., Jones, A.L., Newell, F., Mowry, B.J., Whitehead, J.P., Prins, J.B., and Macdonald, G.A. (2006). Olanzapine treatment is associated with reduced high molecular weight adiponectin in serum: a potential mechanism for olanzapine-induced insulin resistance in patients with schizophrenia. *J. Clin. Psychopharmacol.* **26**, 232–237.

Rizos, E.N., Rontos, I., Laskos, E., Arsenis, G., Michalopoulou, P.G., Vasilopoulos, D., Gournellis, R., and Lykouras, L. (2008). Investigation of serum BDNF levels in drug-naive patients with schizophrenia. *Prog. Neuropsychopharmacol. Biol. Psychiatry* **32**, 1308–1311.

Rothermundt, M., Arolt, V., and Bayer, T.A. (2001a). Review of immunological and immunopathological findings in schizophrenia. *Brain Behav. Immun.* **15**, 319–339.

Rothermundt, M., Missler, U., Arolt, V., Peters, M., Leadbeater, J., Wiesmann, M., Rudolf, S., Wandinger, K.P., and Kirchner, H. (2001b). Increased S100B blood levels in unmedicated and treated schizophrenic patients are correlated with negative symptomatology. *Mol. Psychiatry* **6**, 445–449.

Rothermundt, M., Ponath, G., Glaser, T., Hetzel, G., and Arolt, V. (2004). S100B serum levels and long-term improvement of negative symptoms in patients with schizophrenia. *Neuropsychopharmacology* **29**, 1004–1011.

Rothermundt, M., Ohrmann, P., Abel, S., Siegmund, A., Pedersen, A., Ponath, G., Suslow, T., Peters, M., Kaestner, F., Heindel, W., Arolt, V., and Pfleiderer, B. (2007). Glial cell activation in a subgroup of patients with schizophrenia indicated by increased S100B serum concentrations and elevated myo-inositol. *Prog. Neuropsychopharmacol. Biol. Psychiatry* **31**, 361–364.

Saha, S., Chant, D., Welham, J., and McGrath, J. (2005). A systematic review of the prevalence of schizophrenia. *PLoS Med.* **2**, e141.

Saiz-Ruiz, J., Carrasco, J.L., Martin, M., Manzanares, J., and Hernanz, A. (1992). Plasmatic somatostatin as a marker of positive symptoms of schizophrenia. *Prog. Neuropsychopharmacol. Biol. Psychiatry* **16**, 203–210.

Sarandol, A., Kirli, S., Akkaya, C., Altin, A., Demirci, M., and Sarandol, E. (2007). Oxidative-antioxidative systems and their relation with serum S100 B levels in patients with schizophrenia: effects of short term antipsychotic treatment. *Prog. Neuropsychopharmacol. Biol. Psychiatry* **31**, 1164–1169.

Schatzberg, A.F., and Lindley, S. (2008). Glucocorticoid antagonists in neuropsychiatric [corrected] disorders. *Eur. J. Pharmacol.* **583**, 358–364.

Schmitt, A., Bertsch, T., Henning, U., Tost, H., Klimke, A., Henn, F.A., and Falkai, P. (2005). Increased serum S100B in elderly, chronic schizophrenic patients: negative correlation with deficit symptoms. *Schizophr. Res.* **80**, 305–313.

Schroeter, M.L., and Steiner, J. (2009). Elevated serum levels of the glial marker protein S100B are not specific for schizophrenia or mood disorders. *Mol. Psychiatry* **14**, 235–237.

Schroeter, M.L., Abdul-Khaliq, H., Fruhauf, S., Hohne, R., Schick, G., Diefenbacher, A., and Blasig, I. E. (2003). Serum S100B is increased during early treatment with antipsychotics and in deficit schizophrenia. *Schizophr. Res.* **62**, 231–236.

Schroeter, M.L., Abdul-Khaliq, H., Krebs, M., Diefenbacher, A., and Blasig, I.E. (2009). Neuron-specific enolase is unaltered whereas S100B is elevated in serum of patients with schizophrenia–original research and meta-analysis. *Psychiatry Res.* **167**, 66–72.

Schwarz, M.J., Riedel, M., Gruber, R., Muller, N., and Ackenheil, M. (1998). Autoantibodies against 60-kDa heat shock protein in schizophrenia. *Eur. Arch. Psychiatry Clin. Neurosci.* **248**, 282–288.

Schwarz, M.J., Chiang, S., Muller, N., and Ackenheil, M. (2001). T-helper-1 and T-helper-2 responses in psychiatric disorders. *Brain Behav. Immun.* **15**, 340–370.

Schwarz, E., Izmailov, R., Spain, M., Barnes, A., Mapes, J.P., Guest, P.C., Rahmoune, H., Pietsch, S., Leweke, F.M., Rothermundt, M., Steiner, J., Koethe, D., *et al.* (2010). Validation of a blood-based laboratory test to aid in the confirmation of a diagnosis of schizophrenia. *Biomark Insights* **5**, 39–47.

Segal, M., Avital, A., Rojas, M., Hausvater, N., Sandbank, S., Liba, D., Moguillansky, L., Tal, I., and Weizman, A. (2004). Serum prolactin levels in unmedicated first-episode and recurrent schizophrenia patients: a possible marker for the disease's subtypes. *Psychiatry Res.* **127**, 227–235.

Segal, M., Avital, A., Berstein, S., Derevenski, A., Sandbank, S., and Weizman, A. (2007a). Prolactin and estradiol serum levels in unmedicated male paranoid schizophrenia patients. *Prog. Neuropsychopharmacol. Biol. Psychiatry* **31**, 378–382.

Segal, M., Avital, A., Derevenski, A., Berstein, S., Sandbank, S., and Weizman, A. (2007b). Prolactin serum levels in paranoid versus nonparanoid male schizophrenia patients treated with risperidone. *Int. Clin. Psychopharmacol.* **22**, 192–196.

Shi, J., Levinson, D.F., Duan, J., Sanders, A.R., Zheng, Y., Pe'er, I., Dudbridge, F., Holmans, P.A., Whittemore, A.S., Mowry, B.J., Olincy, A., Amin, F., *et al.* (2009). Common variants on chromosome 6p22.1 are associated with schizophrenia. *Nature* **460**, 753–757.

Shimizu, E., Hashimoto, K., Salama, R.H., Watanabe, H., Komatsu, N., Okamura, N., Koike, K., Shinoda, N., Nakazato, M., Kumakiri, C., Okada, S., Muramatsu, H., *et al.* (2003). Two clusters of serum midkine levels in drug-naive patients with schizophrenia. *Neurosci. Lett.* **344**, 95–98.

Singh, I., and Rose, N. (2009). Biomarkers in psychiatry. *Nature* **460**, 202–207.

Song, X.Q., Lv, L.X., Li, W.Q., Hao, Y.H., and Zhao, J.P. (2009). The interaction of nuclear factor-kappa B and cytokines is associated with schizophrenia. *Biol. Psychiatry* **65**, 481–488.

Sperner-Unterweger, B., Whitworth, A., Kemmler, G., Hilbe, W., Thaler, J., Weiss, G., and Fleischhacker, W.W. (1999). T-cell subsets in schizophrenia: a comparison between drug-naive first episode patients and chronic schizophrenic patients. *Schizophr. Res.* **38**, 61–70.

Stefansson, H., Ophoff, R.A., Steinberg, S., Andreassen, O.A., Cichon, S., Rujescu, D., Werge, T., Pietilainen, O.P., Mors, O., Mortensen, P.B., Sigurdsson, E., Gustafsson, O., *et al.* (2009). Common variants conferring risk of schizophrenia. *Nature* **460**, 744–747.

Steiner, J., Bielau, H., Bernstein, H.G., Bogerts, B., and Wunderlich, M.T. (2006). Increased cerebrospinal fluid and serum levels of S100B in first-onset schizophrenia are not related to a degenerative release of glial fibrillar acidic protein, myelin basic protein and neurone-specific enolase from glia or neurones. *J. Neurol. Neurosurg. Psychiatry* **77**, 1284–1287.

Steiner, J., Walter, M., Wunderlich, M.T., Bernstein, H.G., Panteli, B., Brauner, M., Jacobs, R., Gos, T., Rothermundt, M., and Bogerts, B. (2009). A new pathophysiological aspect of S100B in schizophrenia: potential regulation of S100B by its scavenger soluble RAGE. *Biol. Psychiatry* **65**, 1107–1110.

Steiner, J., Walter, M., Guest, P., Myint, A.M., Schiltz, K., Panteli, B., Brauner, M., Bernstein, H.G., Gos, T., Herberth, M., Schroeter, M.L., Schwarz, M.J., *et al.* (2010). Elevated S100B levels in schizophrenia are associated with insulin resistance. *Mol. Psychiatry* **15**, 3–4.

Stolina, M., Sharma, S., Lin, Y., Dohadwala, M., Gardner, B., Luo, J., Zhu, L., Kronenberg, M., Miller, P.W., Portanova, J., Lee, J.C., and Dubinett, S.M. (2000). Specific inhibition of cyclooxygenase 2 restores antitumor reactivity by altering the balance of IL-10 and IL-12 synthesis. *J. Immunol.* **164**, 361–370.

Strous, R.D., and Shoenfeld, Y. (2006). Schizophrenia, autoimmunity and immune system dysregulation: a comprehensive model updated and revisited. *J. Autoimmun.* **27**, 71–80.

Tan, Y.L., Zhou, D.F., Cao, L.Y., Zou, Y.Z., and Zhang, X.Y. (2005). Decreased BDNF in serum of patients with chronic schizophrenia on long-term treatment with antipsychotics. *Neurosci. Lett.* **382**, 27–32.

Tanaka, K.F., Shintani, F., Fujii, Y., Yagi, G., and Asai, M. (2000). Serum interleukin-18 levels are elevated in schizophrenia. *Psychiatry Res.* **96**, 75–80.

Tanaka, Y., Yoshida, S., Shimada, Y., Ueda, H., and Asai, K. (2007). Alteration in serum neural cell adhesion molecule in patients of schizophrenia. *Hum. Psychopharmacol.* **22**, 97–102.

Tandon, R., Keshavan, M.S., and Nasrallah, H.A. (2008). Schizophrenia, "just the facts" what we know in 2008. 2. Epidemiology and etiology. *Schizophr. Res.* **102**, 1–18.

Teixeira, A.L., Reis, H.J., Nicolato, R., Brito-Melo, G., Correa, H., Teixeira, M.M., and Romano-Silva, M.A. (2008). Increased serum levels of CCL11/eotaxin in schizophrenia. *Prog. Neuropsychopharmacol. Biol. Psychiatry* **32**, 710–714.

Tesch, G., Amur, S., Schousboe, J.T., Siegel, J.N., Lesko, L.J., and Bai, J.P. (2010). Successes achieved and challenges ahead in translating biomarkers into clinical applications. *AAPS J.* **12**, 243–253.

Thaker, G.K., and Carpenter, W.T. Jr. (2001). Advances in schizophrenia. *Nat. Med.* **7**, 667–671.

Theodoropoulou, S., Spanakos, G., Baxevanis, C.N., Economou, M., Gritzapis, A.D., Papamichail, M. P., and Stefanis, C.N. (2001). Cytokine serum levels, autologous mixed lymphocyte reaction and surface marker analysis in never medicated and chronically medicated schizophrenic patients. *Schizophr. Res.* **47**, 13–25.

Torrey, E.F., Miller, J., Rawlings, R., and Yolken, R.H. (1997). Seasonality of births in schizophrenia and bipolar disorder: a review of the literature. *Schizophr. Res.* **28**, 1–38.

Toyooka, K., Asama, K., Watanabe, Y., Muratake, T., Takahashi, M., Someya, T., and Nawa, H. (2002). Decreased levels of brain-derived neurotrophic factor in serum of chronic schizophrenic patients. *Psychiatry Res.* **110**, 249–257.

Venkatasubramanian, G., Chittiprol, S., Neelakantachar, N., Naveen, M.N., Thirthall, J., Gangadhar, B.N., and Shetty, K.T. (2007). Insulin and insulin-like growth factor-1 abnormalities in antipsychotic-naive schizophrenia. *Am. J. Psychiatry* **164**, 1557–1560.

Vinogradov, S., Fisher, M., Holland, C., Shelly, W., Wolkowitz, O., and Mellon, S.H. (2009). Is serum brain-derived neurotrophic factor a biomarker for cognitive enhancement in schizophrenia? *Biol. Psychiatry* **66**, 549–553.

Walker, E., Mittal, V., and Tessner, K. (2008). Stress and the hypothalamic pituitary adrenal axis in the developmental course of schizophrenia. *Annu. Rev. Clin. Psychol.* **4**, 189–216.

Wang, H.C., Yang, Y.K., Chen, P.S., Lee, I.H., Yeh, T.L., and Lu, R.B. (2007a). Increased plasma leptin in antipsychotic-naive females with schizophrenia, but not in males. *Neuropsychobiology* **56**, 213–215.

Wang, L., Yu, L., Zhang, A.P., Fang, C., Du, J., Gu, N.F., Qin, S.Y., Feng, G.Y., Li, X.W., Xing, Q.H., and He, L. (2007b). Serum prolactin levels, plasma risperidone levels, polymorphism of cytochrome P450 2D6 and clinical response in patients with schizophrenia. *J. Psychopharmacol.* **21**, 837–842.

Weickert, T.W., Goldberg, T.E., Gold, J.M., Bigelow, L.B., Egan, M.F., and Weinberger, D.R. (2000). Cognitive impairments in patients with schizophrenia displaying preserved and compromised intellect. *Arch. Gen. Psychiatry* **57**, 907–913.

Westergaard, T., Mortensen, P.B., Pedersen, C.B., Wohlfahrt, J., and Melbye, M. (1999). Exposure to prenatal and childhood infections and the risk of schizophrenia: suggestions from a study of sibship characteristics and influenza prevalence. *Arch. Gen. Psychiatry* **56**, 993–998.

Wiesmann, M., Wandinger, K.P., Missler, U., Eckhoff, D., Rothermundt, M., Arolt, V., and Kirchner, H. (1999). Elevated plasma levels of S-100b protein in schizophrenic patients. *Biol. Psychiatry* **45**, 1508–1511.

Wildenauer, D.B., Wildenauer, D.M.B., and Schwab, S.G. (2009). Dissecting the molecular causes of schizophrenia. *In* "Molecular Biology of Neuropsychiatric Disorders", (D.B. Wildenauer, Ed.), Vol. 23, pp. 51–79. Springer, Berlin, Heidelberg.

Wolf, T.L., Kotun, J., and Meador-Woodruff, J.H. (2006). Plasma copper, iron, ceruloplasmin and ferroxidase activity in schizophrenia. *Schizophr. Res.* **86**, 167–171.

Wong, A.H., and Van Tol, H.H. (2003). Schizophrenia: from phenomenology to neurobiology. *Neurosci. Biobehav. Rev.* **27**, 269–306.

Xiao, B.G., and Link, H. (1999). Is there a balance between microglia and astrocytes in regulating Th1/Th2-cell responses and neuropathologies? *Immunol. Today* **20**, 477–479.

Xiu, M.H., Hui, L., Dang, Y.F., Hou, T.D., Zhang, C.X., Zheng, Y.L., Chen da, C., Kosten, T.R., and Zhang, X.Y. (2009). Decreased serum BDNF levels in chronic institutionalized schizophrenia on long-term treatment with typical and atypical antipsychotics. *Prog. Neuropsychopharmacol. Biol. Psychiatry* **33**, 1508–1512.

Yang, Y., Wan, C., Li, H., Zhu, H., La, Y., Xi, Z., Chen, Y., Jiang, L., Feng, G., and He, L. (2006). Altered levels of acute phase proteins in the plasma of patients with schizophrenia. *Anal. Chem.* **78**, 3571–3576.

Yao, J.K., Reddy, R., and van Kammen, D.P. (2000). Abnormal age-related changes of plasma antioxidant proteins in schizophrenia. *Psychiatry Res.* **97**, 137–151.

Young, R.M., Lawford, B.R., Barnes, M., Burton, S.C., Ritchie, T., Ward, W.K., and Noble, E.P. (2004). Prolactin levels in antipsychotic treatment of patients with schizophrenia carrying the DRD2*A1 allele. *Br. J. Psychiatry* **185**, 147–151.

Zhang, X.Y., Zhou, D.F., Yuan, C.L., Zhang, P.Y., Wu, G.Y., and Shen, Y.C. (2002a). Risperidone-induced increase in serum prolactin is correlated with positive symptom improvement in chronic schizophrenia. *Psychiatry Res.* **109**, 297–302.

Zhang, X.Y., Zhou, D.F., Zhang, P.Y., Wu, G.Y., Cao, L.Y., and Shen, Y.C. (2002b). Elevated interleukin-2, interleukin-6 and interleukin-8 serum levels in neuroleptic-free schizophrenia: association with psychopathology. *Schizophr. Res.* **57**, 247–258.

Zhang, X.Y., Zhou, D.F., Cao, L.Y., Zhang, P.Y., Wu, G.Y., and Shen, Y.C. (2004). Changes in serum interleukin-2, -6, and -8 levels before and during treatment with risperidone and haloperidol: relationship to outcome in schizophrenia. *J. Clin. Psychiatry* **65**, 940–947.

Zhang, X.Y., Zhou, D.F., Cao, L.Y., Wu, G.Y., and Shen, Y.C. (2005). Cortisol and cytokines in chronic and treatment-resistant patients with schizophrenia: association with psychopathology and response to antipsychotics. *Neuropsychopharmacology* **30**, 1532–1538.

Zhang, X.Y., Cao, L.Y., Song, C., Wu, G.Y., Chen da, C., Qi, L.Y., Wang, F., Xiu, M.H., Chen, S., Zhang, Y., Lu, L., and Kosten, T.A. (2008a). Lower serum cytokine levels in smokers than nonsmokers with chronic schizophrenia on long-term treatment with antipsychotics. *Psychopharmacology (Berl.)* **201**, 383–389.

Zhang, X.Y., Zhou, D.F., Wu, G.Y., Cao, L.Y., Tan, Y.L., Haile, C.N., Li, J., Lu, L., Kosten, T.A., and Kosten, T.R. (2008b). BDNF levels and genotype are associated with antipsychotic-induced weight gain in patients with chronic schizophrenia. *Neuropsychopharmacology* **33**, 2200–2205.

Zhang, X.Y., Zhou, D.F., Qi, L.Y., Chen, S., Cao, L.Y., Chen da, C., Xiu, M.H., Wang, F., Wu, G.Y., Lu, L., Kosten, T.A., and Kosten, T.R. (2009). Superoxide dismutase and cytokines in chronic patients with schizophrenia: association with psychopathology and response to antipsychotics. *Psychopharmacology (Berl.)* **204**, 177–184.

ABNORMALITIES IN METABOLISM AND HYPOTHALAMIC–PITUITARY–ADRENAL AXIS FUNCTION IN SCHIZOPHRENIA

Paul C. Guest[1], Daniel Martins-de-Souza[1], Natacha Vanattou-Saifoudine[1], Laura W. Harris[1] and Sabine Bahn[1,2]

[1]Department of Chemical Engineering and Biotechnology, University of Cambridge, Cambridge, United Kingdom
[2]Department of Neuroscience, Erasmus Medical Centre, Rotterdam, The Netherlands

Abstract

For decades, evidence has been emerging that the pathogenesis of schizophrenia can involve perturbations in metabolic and hypothalamic–pituitary–adrenal (HPA) axis pathways. Variations in manifestation of these effects could be related to the differences in clinical symptoms between affected individuals as well as to differences in treatment response, including the finding that a high proportion of subjects fail to respond to current antipsychotic medications. Here, we review the evidence for abnormalities in metabolism and HPA axis regulation in schizophrenia. Such studies may prove critical for increasing our understanding of the multidimensional nature of psychiatric illnesses and for improving the timeliness and accuracy of diagnosis. Stratification of subjects according to molecular

INTERNATIONAL REVIEW OF
NEUROBIOLOGY, VOL. 101
DOI: 10.1016/B978-0-12-387718-5.00006-7

phenotype reflecting the disease state or trait could help to improve existing treatments through application of novel personalized medicine strategies and by the development of much-needed novel antipsychotic agents.

I. Introduction

Despite decades of research, the pathophysiology and etiology of schizophrenia are not completely understood. The main hypotheses have focused on alterations in neurotransmitter systems such as the glutamatergic and dopaminergic pathways and current antipsychotic medications mainly target these systems (Biedermann and Fleischhacker, 2009). However, schizophrenia is often associated with peripheral manifestations including dyslipidemia, hyperinsulinemia, type 2 diabetes, and inflammation (Hasnain *et al.*, 2010). Although these effects can result from currently available antipsychotic medications, they have also been observed more than 50 years previously, before the development and use of antipsychotics, and more recent evidence has emerged that some first-onset patients can show these effects prior to receiving treatment (Ryan *et al.*, 2003; Spelman *et al.*, 2007). Molecular analysis of *postmortem* brain tissue has pointed toward alterations in glucose metabolism and insulin signaling pathways (Chan *et al.*, 2010; Martins-de-Souza *et al.*, 2010), and blood-based analyzes have demonstrated hyperinsulinemia and abnormalities in secretion of other endocrine factors at first presentation of symptoms (Guest *et al.*, 2010, 2011).

There are also reports of hypothalamic–pituitary–adrenal (HPA) axis disturbance in first-onset and chronic schizophrenia subjects (Banki *et al.*, 1987; Brunelin *et al.*, 2008; Tsigos and Chrousos, 2002). The HPA axis is a major component of the diffuse neuroendocrine system governing responses to stress and maintaining whole body homeostasis through regulation of many body processes including the immune system, mood, emotion, feeding behavior, and energy storage and expenditure. These processes are controlled by releasing factors such as corticotrophin-releasing hormone (CRH) and gonadotrophin-releasing hormone in the hypothalamus; hormones such as adrenocorticorticotrophic hormone (ACTH), growth hormone, prolactin, arginine-vasopressin (AVP), and the gonadotrophins in the pituitary; corticosteroids and neurotransmitters in the adrenal glands (Engelmann *et al.*, 2004; Goeders, 2002; Schwartz *et al.*, 2000). Higher levels of CRH, AVP, ACTH, and cortisol have been reported in studies of schizophrenia and other psychiatric conditions (Banki *et al.*, 1987; Brunelin *et al.*, 2008; Tsigos and Chrousos, 2002). Functional magnetic resonance imaging (MRI) studies have also identified larger pituitary volumes in first episode schizophrenia patients and pituitaries in chronic sufferers have been shown to be smaller, which may indicate an adaptation, medication effects, or desensitization to HPA axis hyperactivity (Pariante *et al.*, 2004).

Nonetheless, such features are not observed for all subjects with schizophrenia and not all individuals with these abnormalities develop this disorder. This suggests that there may be an underlying metabolic or endocrine vulnerability in a subset of individuals which interacts with environmental or genetic factors in manifestation of schizophrenia. This may be one reason why accurate diagnosis of schizophrenia and other psychiatric disorders is difficult and why not all patients show an adequate response to existing antipsychotic medications. There has been increasing interest in identification of biomarkers that not only facilitate greater understanding of the underlying disease pathology but can also be employed in the development of early diagnostics and much-needed novel drug treatments.

Peripheral blood biomarkers are useful in clinical practice because of the ease of accessibility and minimally invasive sampling procedures, as well as the associated low costs. Here, we review the evidence for abnormalities in peripheral metabolism and HPA axis regulation in schizophrenia and other psychiatric disorders. We also take into account the likelihood that metabolic and HPA axis aspects of schizophrenia may prove critical for diagnosis, for improvement of existing treatment regimes based on patient personalized medicine strategies, and in the development of novel antipsychotic agents.

II. Peripheral Metabolic Effects

Several recent studies of schizophrenia patients have measured peripheral metabolic indices including blood glucose and insulin levels, either prior to antipsychotic treatment or in chronic antipsychotic-treated patients. Two studies (Ryan *et al.*, 2003; Spelman *et al.*, 2007) demonstrated impaired fasting glucose tolerance in first episode, antipsychotic-naive patients compared to healthy controls in small-to-medium-sized cohorts, with higher fasting levels of plasma glucose. Hyperinsulinemia and insulin resistance has also been found in antipsychotic-naive (Ryan *et al.*, 2003; Spelman *et al.*, 2007) and drug-free chronic (Arranz *et al.*, 2004; Cohn *et al.*, 2006) schizophrenia subjects. A study of schizophrenic subjects from the 1966 Northern Finland Birth Cohort identified insulin resistance in 45% of patients overall and in 33% of the unmedicated patients. However, not all findings are in agreement with this. One study found no differences in the plasma levels of glucose, insulin, or connecting (C)-peptide in a group of antipsychotic-naive patients compared to controls under nonfasting conditions (Arranz *et al.*, 2004). Another study identified impaired glucose tolerance in a group of drug-naive patients who were well matched for these parameters and another study found significant hepatic insulin resistance in schizophrenia patients compared to controls using a hyperinsulinemic clamp method (van Nimwegen *et al.*, 2008).

Other methods have also been used to study effects on metabolism in schizophrenia patients. A study using ^{1}H nuclear magnetic resonance (NMR) spectroscopy profiling identified decreased lactate and increased glucose in the cerebrospinal fluid (CSF) of antipsychotic-naive patients compared to controls (Holmes *et al.*, 2006). Interestingly, a 9-day course of antipsychotic treatment resulted in normalization of the metabolite profile in the majority of these patients. Similar deficits were found in patients prodromal for psychosis, but this did not appear to be a predictor of whether these individuals later developed schizophrenia (Huang *et al.*, 2007). Another study analyzed metabolic biomarkers in first episode patients after short-term antipsychotic treatment and found no significant differences between patients and controls in any parameter including baseline glucose, lipid metabolites, insulin levels, and glucose tolerance (Sengupta *et al.*, 2008).

Metabolic abnormalities have also been demonstrated in studies of peripheral cells obtained from first-onset antipsychotic-naive schizophrenia subjects. Microarray analysis of glucose-deprived lymphoblastoid cells showed differential expression of glucose-responsive genes in schizophrenia compared to control subjects (Martin *et al.*, 2009). Recently, we carried out proteomic profiling of stimulated peripheral blood mononuclear cells (PBMCs) and found altered expression of some glycolytic enzymes in cells from drug-naive schizophrenia patients compared to controls, suggesting an abnormality in glycolysis and potentially other related metabolic pathways (Herberth *et al.*, 2010). We also found increased levels of the glucose transporter-1 (GLUT1) and decreased levels of the insulin receptor in stimulated PBMCs from schizophrenia patients, which are used as cellular markers of insulin resistance.

Molecular studies of chronic schizophrenia patients can be confounded since routinely used antipsychotic medications have multiple side effects such as dysregulated glucose homeostasis. We carried out studies which have circumvented this problem by analysis of serum and plasma samples from first-onset, antipsychotic-naive patients (Guest *et al.*, 2010a,b). Recruitment of antipsychotic-naive patients is challenging and prohibitive since even large clinical centers can diagnose only 20–30 such patients each year and because few centers follow standard operating procedures for sample collection. To achieve adequate numbers of well-characterized first-onset antipsychotic-naive patients, we recruited subjects from four independent clinical centers over a 2-year period. Patients were diagnosed using DSM-IV criteria for schizophrenia and bipolar disorder. At the time of sample collection, schizophrenia patients were acutely psychotic and bipolar disorder subjects were euthymic. Euthymic bipolar disorder patients were chosen as these subjects often experience cognitive deficits similar to those observed in schizophrenia, and this could lead potentially to misdiagnosis (Ferrier *et al.*, 1999). Control subjects were matched to schizophrenia patients for age, gender, body mass index, and smoking.

Insulin, proinsulin, and des 31,32-proinsulin levels were measured using two-site time-resolved fluorescence assays employing different combinations of monoclonal antibodies that discriminate between the specific forms of the molecule (Sobey *et al.*, 1989). C-peptide and chromogranin A were measured using commercially available immunoassays. All of these molecules were present at significantly elevated levels in serum and plasma from schizophrenia patients (Fig. 1). The finding that these changes in insulin-related molecules occurred against a background of relatively normal glucose levels was consistent with the possibility that at least some of these patients were insulin resistant at the onset of the disease. This could have important implications since elevated insulin can have deleterious effects on brain function (Taguchi *et al.*, 2007). In contrast, no significant differences were found in the same insulin-related molecules in serum from bipolar disorder patients (Guest *et al.*, 2010a,b), providing some preliminary evidence that these molecules are not changed in all neuropsychiatric disorders. However, it will be important to carry out similar studies using blood samples from subjects with other psychiatric conditions such as major depressive disorder, anxiety disorders, and autism spectrum conditions to answer this question with greater certainty.

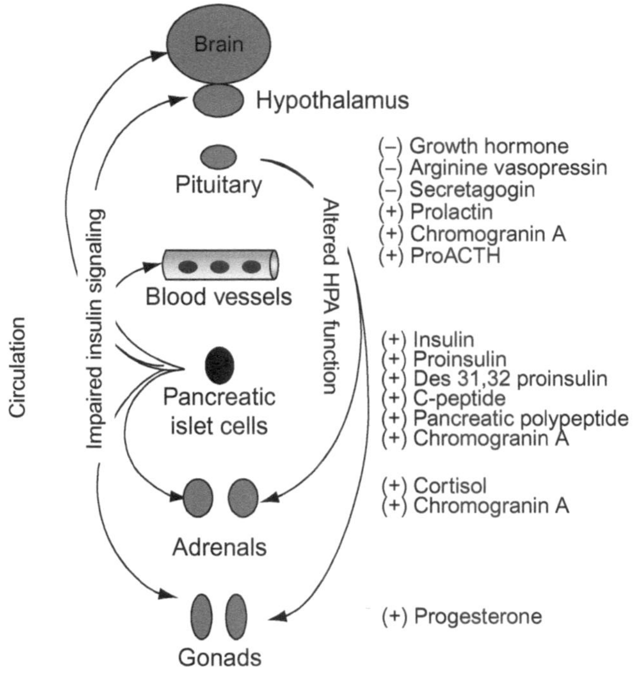

FIG. 1. Schematic diagram showing potential effects of insulin resistance on secretion of other hormones and bioactive molecules over the diffuse neuroendocrine system.

Taken together, these findings suggest that some, but not all, schizophrenia patients show signs of metabolic perturbations such as insulin resistance and altered glucose handling during the earliest phases of the disease. This leads to the possibility that increased output from pancreatic islet β cells (the insulin-producing cells) is required to maintain normal glycemia in these individuals. Through analyzes of several cohorts, we found that ~50% of the first-onset antipsychotic-naive patients had elevated insulin, proinsulin, and des 31,32 proinsulin levels. K-means clustering analysis showed that the frequency of subjects with high levels of insulin-related molecules was ~50% for schizophrenia patients (data not shown), consistent with estimates based on one epidemiological study of insulin resistance in schizophrenia patients (Timonen *et al.*, 2009). In contrast, the proportion of control subjects with high insulin levels was less than 20% (data not shown).

III. Altered Hormone Secretion in Schizophrenia

Insulin is produced by limited proteolytic cleavage of the proinsulin molecule in newly synthesized secretory granules of pancreatic islet β cells (Hutton, 1994). Residual proinsulin, the conversion intermediates, other bioactive peptides, and the proteolytic processing enzymes are cosecreted with mature insulin and C-peptide in response to secretagogue stimulation such as an elevation in blood glucose. Increased serum levels of proinsulin and the conversion intermediates have been observed in a number of pathophysiological conditions including the prodrome of type 1 diabetes, mild type 2 diabetes, and the metabolic syndrome (Creemers *et al.*, 1998). This has been attributed to such causes as an increased turnover of the secretory granule contents due to increased demands imposed by peripheral insulin resistance or by decreased functional pancreatic β cell mass (Alarcon *et al.*, 1995; Laedtke *et al.*, 2000).

Assuming that our finding of increased insulin levels in schizophrenia patients (Guest *et al.*, 2010a,b) was associated with impaired insulin signaling, we recently investigated the possibility that secretion of other hormones of pancreatic islets cells and the HPA axis is also affected in schizophrenia. In this case, we carried out multiplex immunoassay analysis of 21 hormones and hormone-related molecules using sera from 236 first and recent onset schizophrenia patients and 230 matched controls (Guest *et al.*, 2011). Similar multiplex immunoassay methods have already been used successfully in numerous clinical studies or biomarker discovery projects of diseases such as epithelial ovarian cancer, coronary artery disease, autoimmune disorders, and more recently for schizophrenia (Schwarz *et al.*, 2010; see chapter "The application of multiplexed assay systems for molecular diagnostics" by Schwarz *et al.*). In this study, analyses using the multiplex immunoassay technology revealed that the serum concentrations of insulin and chromogranin A were increased in schizophrenia subjects, consistent with our previous study. In addition, we also found elevated concentrations of pancreatic polypeptide, prolactin, progesterone, and cortisol, and decreased levels of growth hormone (Fig. 1; Guest *et al.*, 2011).

One factor that should be considered in the interpretation of these results is that many hormones are influenced by ultradian or circadian rhythms. Therefore, it is likely that the molecules measured here are coregulated as part of an oscillatory feedforward–feedback relationship between the endocrine pancreas, the pituitary, and other neuroendocrine components of the HPA and hypothalamic–pituitary–gonadal (HPG) systems (Walker *et al.*, 2010). For example, hyperinsulinemia has been linked to increased prolactin secretion (Ben-Jonathan *et al.*, 2006) and the associated insulin-induced hypoglycemia has been shown to produce a depression in the amplitude of growth hormone secretion pulses (Tannenbaum *et al.*, 1976). In addition, studies using a diet-induced insulin resistance model in rats have demonstrated hyperinsulinemia and increased progesterone levels (Akamine *et al.*, 2010). Numerous studies have found an influence of HPA and HPG axis hormones on brain regions and functions known to be involved in psychiatric and neurodegenerative disorders (Goldstein, 2006). Also, we showed previously the coregulated patterned behavior of insulin, growth hormone, leptin, and cortisol in first-onset schizophrenia patients, as determined using a targeted analyte cluster method (Cheng *et al.*, 2010). The finding of increased cortisol is consistent with that of many other researchers including a study which also found increased levels of this hormone in an antipsychotic-free cohort of schizophrenia patients (Meltzer *et al.*, 2001). However, the same study found no difference in growth hormone or prolactin levels, as we have found in our analysis.

The finding of changes in chromogranin A is intriguing as this is a precursor protein found in many neuroendocrine cell types, where it undergoes proteolytic processing to produce smaller functional peptides in a cell-dependent manner. These proteolytically processed peptides include the vasostatins I and II which have been shown to inhibit vasoconstriction in blood vessels (Helle *et al.*, 2007). This may be important as the vasodilation response is known to be altered in schizophrenia patients, as shown by differences in the niacin skin-flush response (Nilsson *et al.*, 2006). Chromogranin A processing can also result in production of another peptide termed catestatin which inhibits secretion from catecholaminergic adrenal chromaffin cells (Garcia *et al.*, 1994). This may also be relevant for schizophrenia considering the hypothesized role of catecholamines such as dopamine, serotonin, and norepinephrine in the etiology and therapeutics of the disease (Howes and Kapur, 2009). The circulating levels of chromogranin A are also elevated in type 2 diabetes (Mojiminiyi *et al.*, 2000) and conversion of the protein appears to be altered in animal models of diabetes (Guest *et al.*, 2002). Other studies have shown that chromogranin A expression is altered in the prefrontal cortex

(Iwazaki *et al.*, 2004) and CSF (Landen *et al.*, 1999) of patients with schizophrenia. Given the wide spectrum of biological activities associated with chromogranin A-derived peptides, it is possible that some of these could have important physiological consequences in the onset or development of schizophrenia.

All of the other hormones which we found to be altered in schizophrenia are known to have key roles in metabolism, development, or growth. Previous studies have shown that pancreatic polypeptide is involved in regulation of energy balance (Asakawa *et al.*, 2003), prolactin has direct effects on insulin production and insulin resistance (Ben-Jonathan *et al.*, 2006), and high levels of cortisol (Nosadini *et al.*, 1983) and progesterone (Kalkhoff, 1982) have been linked to insulin resistance. Only growth hormone showed decreased levels in schizophrenia patients. This supports other studies which have shown that subjects with growth hormone deficiency are insulin resistant (Murray and Shalet, 2005). The results of previous studies suggest that the reduced growth hormone levels may be due to schizophrenia-associated increases in dopamine or serotonin activities, which regulate the production of this hormone (Kahn *et al.*, 1992).

In separate studies, we also found alterations in expression of the neuropeptide VGF in CSF from first-onset antipsychotic-naive subjects and in *postmortem* brains from schizophrenia patients (Huang *et al.*, 2006). VGF is expressed in neurons in the central and peripheral nervous systems and in various cells of the diffuse neuroendocrine system (Levi *et al.*, 1985). It appears to be involved in regulation of energy balance and synaptic plasticity (Alder *et al.*, 2003; Hahm *et al.*, 1999). As with chromogranin A and insulin, VGF is synthesized initially as a larger precursor protein which undergoes processing to a number of bioactive peptides. Researchers have shown that administration of proVGF or synthetic peptides corresponding to the C-terminal region of proVGF may have therapeutic potential. Expression of VGF in primary spinal cord neuronal cultures has been shown to attenuate excitotoxic injury (Zhao *et al.*, 2008). In addition, central nervous system administration of a putative 22 amino acid VGF-derived peptide appears to modulate neuropathic pain (Rizzi *et al.*, 2008), increase energy expenditure, and decrease diet-induced obesity (Bartolomucci *et al.*, 2006). Interestingly, a recent study showed that exercise stimulates a neurotrophic signaling cascade which included increased expression of VGF (Hunsberger *et al.*, 2007). The authors also showed that infusion of a putative 20 amino acid VGF-derived peptide produced an antidepressant response in mice (Hunsberger *et al.*, 2007). These findings suggest that drug discovery efforts centered on VGF-derived peptides may have therapeutic potential in conditions such as chronic pain, insulin resistance, or psychiatric disorders.

Another important implication of these findings is that other proteins released from neuroendocrine secretory granules may be altered in the circulation of schizophrenia patients. Approximately 100 additional proteins have been identified in insulin secretory granules (Guest *et al.*, 1991) including the prohormone-converting proteases PC1, PC2, carboxypeptidase H, and peptidyl-amidating monooxygenase; enzymes involved in neurotransmitter biosynthesis such as glutamic acid decarboxylase and dopamine β carboxylase; and other proproteins such as 7B2, the neurokinins, and neuropeptide Y (Baekkeskov *et al.*, 1990; Gherzi *et al.*, 1994; Waeber *et al.*, 1993). Further work is warranted to determine whether any of these proteins also have a role in schizophrenia or in a subpopulation of patients at an early or late stage of the disease. Such studies may lead to additional insights into the link between central nervous system perturbations and whole body metabolic homeostatic mechanisms.

Taken together, these findings support the hypothesis that some subjects with schizophrenia show signs of metabolic conditions such as insulin resistance and HPA axis perturbations. However, it should be noted that we did not find significant changes in the levels of other molecules measured by the MetabolicMAP platform which are known to be associated with insulin resistance, such as leptin and testosterone. It will be important to determine whether any of the molecules identified here play a role in the pathophysiology of schizophrenia or if they reflect the impaired insulin signaling process as part of homeostatic mechanisms.

IV. Altered Hormone Production in Schizophrenia Pituitaries

Peripheral blood biomarkers are useful in clinical practice because of the ease of accessibility and the associated minimally invasive sampling procedures and low costs. However, initial discovery can be made using the tissue of origin for many of these biomarkers. The pituitary releases several hormones and potentially hundreds of other bioactive peptides into the peripheral circulation, rendering these ideal for translation to blood-based assays. In addition, HPA axis function is known to be perturbed in schizophrenia subjects (Ryan *et al.*, 2004). We carried out an analysis of *postmortem* pituitary tissue from schizophrenia patients to identify a schizophrenia-associated molecular fingerprint using a combination of liquid chromatography tandem mass spectrometry (LC–MS), two-dimensional difference in-gel electrophoresis (2D-DIGE), and multiplex immunoassay (Guest *et al.*, 2011). This multiplatform approach was employed to circumvent any limitations in proteomic coverage of these methods when used alone. This led to identification of differentially expressed proteins and small molecules including HPA axis-associated molecules such as cortisol, proACTH, the AVP precursor, agouti-related protein, growth hormone, prolactin, and secretagogin, and molecules associated with lipid transport and metabolism, namely the apolipoproteins A1, A2, C3, and H.

ProACTH is produced by cleavage of proopiomelanocortin (POMC) at amino acid 179 by the prohormone convertase PC1 (Brunelin *et al.*, 2008).

We found that the level of proACTH was significantly elevated in pituitaries from schizophrenia patients relative to controls. This is consistent with the finding of clinical studies which showed increased circulating levels of ACTH in schizophrenia patients (Ackland *et al.*, 1983; Walsh *et al.*, 2005). In addition, the finding of increased levels of proACTH and the concomitant elevation in cortisol supports the hypothesis that HPA axis hyperactivity may be involved in the pathophysiology of the disease (Ryan *et al.*, 2004).

AVP is stored as a larger prohormone called vasopressin–neurophysin 2–copeptin (NP2) in the posterior pituitary until the precursor and proteolyzed forms of this protein are released into the circulation by regulated exocytosis. As a consequence, reduced production of NP2 may lead to altered levels of circulating AVP. The primary role of AVP is in homeostasis and regulation of water, salt, and glucose in the blood. However, there have been studies which have also linked abnormal AVP levels to changes in mood and behavior (Heinrichs *et al.*, 2009), and to psychotic disorders (Goldman *et al.*, 1996; Mai *et al.*, 1993). In addition, AVP may affect HPA axis sensitivity since there appears to be a positive correlation between the circulating levels of AVP, ACTH, and cortisol in anti-naive schizophrenia patients (Ryan *et al.*, 2004).

The GLUT1 transporter is a transmembrane protein and therefore not likely to be released from the pituitary into the peripheral circulation, apart from causes due to cell damage. However, this molecule has been linked to abnormal glucose activity in schizophrenia brain areas such as the hippocampus, basal ganglia, and thalamus, as shown by positron emission tomography analysis (Buchsbaum *et al.*, 1986; Clark *et al.*, 2001). It is thought that acute prodromal schizophrenia symptoms can result from inadequate glucose uptake into brain cells due to deficiencies in GLUT1 and GLUT3 activities (McDermott and de Silva, 2005). Our finding of decreased GLUT1 levels in the pituitary of schizophrenia suggests that GLUT1 is a crucial component of HPA axis function and supports the case that glucose handling may be involved in the pathogenesis or development of schizophrenia.

In line with the above results, we also found reduced expression of several members of the apolipoprotein family in *postmortem* pituitaries of schizophrenia patients. These findings are consistent with those of other studies on the serum levels of these proteins (Huang *et al.*, 2008; Yang *et al.*, 2006). Apolipoprotein A1 and A2 facilitate transport of cholesterol and triglycerides in the blood (Kwiterovich, 2000), apolipoprotein C3 is thought to inhibit hepatic uptake of triglyceride-rich particles (von Eckardstein *et al.*, 1991), and apolipoprotein H appears to prevent activation of the intrinsic blood coagulation cascade by binding to phospholipids on the surface of damaged cells (Schousboe, 1985). Impairment of low-density lipoprotein oxidation and lipid transport can lead to atherosclerosis and increased risk of cardiovascular events such as myocardial infarction and stroke, which are all components of the metabolic syndrome (Alberti *et al.*, 2006). Changes in the levels of apolipoprotein A1 and C3 are already used as biomarkers for metabolic syndrome, and for prediction of cardiovascular risk (McQueen *et al.*, 2008; Onat *et al.*, 2003).

We also found decreased levels of secretagogin in schizophrenia *postmortem* pituitaries along with increased levels of the same protein in serum in first-onset antipsychotic-naive schizophrenia patients. This could be indicative of a relative depletion of this protein in the pituitary due to increased rates of secretion. The expression of secretagogin is restricted to distinct neuroendocrine cells with high levels found in the anterior pituitary, pancreatic islet cells, cerebellum, thalamus, hypothalamus, and neocortical neuronal subgroups (Gartner *et al.*, 2001). Studies of secretagogin in pancreatic β-cells suggest that it influences calcium-influx, hormone secretion, and cellular proliferation (Skovhus *et al.*, 2006). In addition, secretagogin was found to be decreased in diabetes-prone rat islets exposed to cytokines (Skovhus *et al.*, 2006). In humans, it has been used as a biomarker for neuroendocrine tumors (Birkenkamp-Demtroder *et al.*, 2005) and ischemic neuronal damage (Gartner *et al.*, 2001; Montaner *et al.*, 2008). As with secretagogin, we also found a similar decrease in prolactin levels in this study in schizophrenia *postmortem* pituitaries we have identified increased serum levels of this hormone in antipsychotic-naive schizophrenia patients in separate study (Guest *et al.*, 2011). In the case of both proteins, the fact that these effects were observed in serum from first-onset antipsychotic-naive schizophrenia subjects is suggestive of a disease-related change rather than an effect of chronic drug treatment associated with the *postmortem* samples. Moreover, detection of these biomarker candidates in an easily accessible biological fluid such as blood serum adds to the value of these findings and may lead to translation of these molecules as peripheral biomarkers for schizophrenia.

V. Evidence for Altered Insulin Signaling in Schizophrenia Brain

It is not certain whether the peripheral effects on metabolic and hormonal changes are a cause or consequence of the altered brain function seen in psychiatric disorders. However, there is at least some evidence that dysfunctions of these pathways are involved in the pathogenesis of several disorders. Decreased expression and phosphorylation of the insulin receptor and of the Akt1 and GSK3β insulin signaling proteins have been found in *postmortem* prefrontal cortex from schizophrenia patients (Zhao *et al.*, 2006). Also, reduced levels of the insulin-degrading enzyme (IDE) have been found in the dorsolateral prefrontal cortex of schizophrenia patients (Bernstein *et al.*, 2009). Interestingly, abnormalities in IDE have also been demonstrated in type 2 diabetes (Furukawa *et al.*, 2008) and Alzheimer disease (Zuo and Jia, 2009).

Hyperinsulinemia has been implicated in the pathogenesis of other neurological conditions such as Alzheimer disease. In this case, high insulin levels have been associated with aberrant phosphorylation of filamentous proteins, increased CNS inflammation, and β-amyloid plaque deposition (Convit, 2005; Craft, 2007). In addition, hippocampal volumes appear to be reduced in some type 2 diabetes patients and in insulin-resistant individuals who also have high circulating levels of this hormone (Convit, 2005). Hyperinsulinema is also known to perturb the function of various neurotransmitter systems including that of dopamine (Bello and Hajnal, 2006). It is not clear to what extent such phenomena are due to direct changes in cellular function or are secondary to effects on brain energy metabolism. However, direct effects of insulin on hippocampal synaptic plasticity have been reported (O'Malley *et al.*, 2003). Taken together, these findings suggest that abnormal insulin production or signaling can lead to perturbations of both peripheral neuroendocrine systems and the central nervous system, resulting in direct changes in cellular function and altered energy supply.

VI. Environmental Causes of Psychiatric Illness

The finding that some, but not all, schizophrenia subjects also have hyperinsulinemia, insulin resistance, or signs of HPA axis dysfunction is consistent with the hypothesis that schizophrenia is a heterogeneous disorder comprised of distinct subtypes (Seaton *et al.*, 2001). In addition, it is likely that environmental factors come to play in precipitation of the disease. Metabolic perturbations which occur during gestation, such as malnutrition, diabetes, obesity, or neuroendocrine stress, can lead to metabolic dysfunction in the offspring, which may be due to abnormalities in the development of hypothalamic connectivity (Levin, 2006). Epidemiological studies have shown that prenatal deprivation of nutrients increases the risk of schizophrenia in the offspring. The most convincing study for which full records are available concerned the Dutch Hunger Winter of 1944–1945. In this case, a widespread famine occurred due to blockade of occupied Holland in October 1944. This resulted in food shortages, particularly protein supply, in 40,000 individuals (Brown and Susser, 2008; Hoek *et al.*, 1998). A 2.7-fold increase in risk of schizophrenia spectrum disorders occurred in subjects born in December 1945. This indicated that these individuals were conceived at the peak of the famine when the protein supply was at its lowest (Hoek *et al.*, 1998). Similar findings resulted from studies of the Chinese Famine of 1959–1961 (St Clair *et al.*, 2005). In this case, detailed dietary information was not available, although birth cohorts conceived at the peak of the famine showed a 2.30-fold increased risk of schizophrenia (St Clair *et al.*, 2005).

Supporting evidence for gestational effects in the precipitation of psychiatric illness has been provided by investigation of animal models such as the *prenatal* protein deprivation rat. *Prenatal* protein deprivation has effects on the adult brain including changes in hippocampal morphology and glutamate and dopamine receptor binding (Udagawa *et al.*, 2006a,b). Another rodent model using *postnatal* protein restriction during lactation showed that the offspring had significantly lower body weights compared to controls and showed changes in circulating levels of glucose, insulin, and leptin (Udagawa *et al.*, 2006b). Hypothalamic expression of the leptin receptor; the orexigenic neuropeptides neuropeptide Y and agouti-related peptide; and the anorexigenic neuropeptides POMC and the cocaine- and amphetamine-regulated transcript were also altered (Cripps *et al.*, 2009). Other studies have shown that protein restriction during pregnancy may negatively impact normal brain development via changes in maternal lipid metabolism (Torres *et al.*, 2010). Interestingly, the behavioral abnormalities in these low protein models are not manifested until early adulthood, including effects on brain-regulated activities such as feeding, and schizophrenia-like behaviors including prepulse inhibition, apomorphine-induced stereotypy, and hyperlocomotion (Palmer *et al.*, 2004, 2008; Torres *et al.*, 2010).

The HPA axis provides a functional link between brain function and peripheral control of metabolism and growth. Many peripherally produced hormones send signals to the brain which regulate hypothalamic and pituitary responses and, likewise, hormones which are produced in the central nervous system provide regulation of peripheral organ systems. Several lines of evidence suggest that there is an abnormal HPA axis response in schizophrenia (Corcoran *et al.*, 2003; Ryan *et al.*, 2004), including elevated circulating cortisol levels and a blunted cortisol response to psychosocial stress (Brenner *et al.*, 2009). Cortisol antagonizes the effects of insulin, including those on gluconeogenesis, and chronically elevated cortisol may lead to symptoms of metabolic syndrome. Alterations in cortisol levels in schizophrenia patients have been considered a confounding factor in studies of metabolic features due to the high levels of psychosocial stress experienced by psychiatric patients. However, abnormalities in glucose tolerance have also been found independent of changes in cortisol levels (Fernandez-Egea *et al.*, 2009).

VII. Specificity of Molecular Signature for Schizophrenia

One important question is the specificity of metabolic and HPA axis effects to schizophrenia, since hyperinsulinemia and metabolic and endocrine abnormalities have been implicated in other neurological and neurodegenerative disorders such as major depressive disorder (Licinio and Wong, 2003; Rasgon and Kenna,

2005), bipolar disorder (Fagiolini *et al.*, 2005), and Alzheimer disease (Mosconi *et al.*, 2008). For example, the insulin effector protein GSK3β is thought to be a major target of lithium, the mainstay of treatment for bipolar disorder (O'Brien and Klein, 2009). Brain-derived neurotrophic factor (BDNF) levels appear to play a role in the pathophysiology of several psychiatric disorders including schizophrenia (Green *et al.*, 2010), major depressive disorder (Shimizu *et al.*, 2003), and bipolar disorder (Machado-Vieira *et al.*, 2007), and treatment with drugs such as antidepressants increases the levels of BDNF in major depressive disorder patients (Shimizu *et al.*, 2003). Changes in pituitary volume appear to be common in nonmedicated psychiatric disorder subjects including those with schizophrenia (Pariante *et al.*, 2004) and major depressive disorder (MacMaster *et al.*, 2006). Also, increased levels of cortisol have been noted in most psychiatric conditions and, at least in the case of the major depressive disorder, this increase may precede the onset of illness by several years (Goodyer *et al.*, 2000).

Recently, we reported a novel analytical approach for identifying biomarkers of schizophrenia using expression of serum analytes from first-onset, antipsychotic-naive patients (Cheng *et al.*, 2010). This method identifies molecules which exhibit coregulated behavior by analysis of the expression data in reproducing kernel spaces. This approach led to identification of a small cluster of analytes comprising only insulin, growth hormone, leptin, and cortisol that was capable of distinguishing first-onset schizophrenia subjects from controls with 80% precision. Interestingly, this same panel of metabolic markers was less strongly related to major depressive disorder and bipolar disorder. As there is a wide continuum of symptoms between many of the psychotic disorders, as well as between various manifestations of schizophrenia and schizophrenia-like conditions, further investigation of this point would be of interest for increasing our knowledge of these disorders and for improving diagnosis.

VIII. Therapeutic Implications

The finding that hyperinsulinemia may play a role in the onset of schizophrenia suggests that drugs which improve insulin signaling represent a potential novel treatment strategy. Existing antipsychotic drugs are capable of inducing metabolic side effects including insulin resistance, weight gain, and type 2 diabetes (Koller and Doraiswamy, 2002; Meyer *et al.*, 2008). Interestingly, the degree of weight gain induced by clozapine and olanzapine has been associated with improved psychopathology ratings, suggesting that these effects may be related at the biological level and that changes in these pathways may be essential for therapeutic efficacy (Czobor *et al.*, 2002; Meltzer *et al.*, 2003). In one study, alterations in body weight, blood glucose, and leptin levels were associated with improvement in positive and negative symptoms (Small *et al.*, 2004) and another study found that the metabolic abnormalities in CSF were normalized by antipsychotic treatment (Holmes *et al.*, 2006).

Therapeutic strategies which target the underlying metabolic dysfunction may provide an effective alternative to treating the traditional neurotransmitter-related endpoint of the disorder. The insulin-sensitizing agents Metformin and Rosiglitazone have been used to correct the antipsychotic-induced insulin resistance typically associated with antipsychotic treatment without compromising the psychotropic benefits (Bahtiyar *et al.*, 2007). Taken together with our finding of increased circulating levels of insulin-related peptides in patients free of antipsychotic medication suggests that the use of insulin-sensitizing agents alone might have some therapeutic benefit. Moreover, this suggests that insulin-related molecules would have utility as biomarkers not only for patient selection but also for monitoring responses or side effects of therapeutic treatment strategies. In addition, these results suggest that initiation and maintenance of treatment should include routine surveillance for clinical and/or biochemical evidence suggestive of the metabolic or syndrome or HPA axis dysfunction.

Similar strategies are already proving fruitful for treatment of memory deficits in Alzheimer's disease. Clinical trials are focusing on the use of PPAR-gamma agonist such as Rosiglitazone and Pioglitazone as an alternative therapy to enhance cognition (Sato *et al.*, 2009). One group conducted a 6-month, randomized, open-controlled trial in patients with mild Alzheimer disease accompanied with type 2 diabetes (Landreth *et al.*, 2008). They assigned patients to two groups. One group was treated with 15–30 mg Pioglitazone daily ($n = 21$) and the other was used as a control ($n = 21$). The Pioglitazone group showed improved cognition and increased regional cerebral blood flow in the parietal lobe, while the control group showed no such improvements. PPAR-gamma is required for adipocyte differentiation and fat deposition, and modulates plasma leptin levels, insulin sensitivity, and glucose homeostasis, although its role in the central nervous system is not fully understood. It was interesting that the effects of Pioglitazone were accompanied by changes in cerebral blood flow as such an effect could potentially lead to increased glucose availability in the brain.

The IDE is a target of PPAR-gamma signaling and has also been proposed as an alternative therapeutic target for Alzheimer disease (Landreth *et al.*, 2008). In addition, the development of cognitive enhancers for Alzheimer disease, based on the inhibition of insulin-regulated aminopeptidase, has been proposed (Albiston *et al.*, 2007; Chai *et al.*, 2008). This enzyme has both central and peripheral effects and again provides an example of how peripheral biomarker discovery can provide novel therapeutic strategies for the treatment of psychiatric disorders.

The adrenal steroid dehydroepiandrosterone (DHEA) has antiglucocorticoid properties that may have regulatory effects on high cortisol levels and glucocorticoid action in the brain in psychiatric patients (Strous *et al.*, 2003). Studies using DHEA augmentation in medicated schizophrenic patients found a significant improvement in negative, depressive, and anxiety symptoms (Nachshoni *et al.*, 2005). Notably, improvements in extrapyramidal side effects have also been demonstrated following DHEA treatment (Hunt *et al.*, 2000). In addition, DHEA treatment in patients with Addison's disease on standard glucocorticoid and mineralocorticoid replacement has been shown to improve self-esteem and mood (Hunt *et al.*, 2000).

IX. Special Considerations

One potential confounding factor arising in studies of psychiatric conditions is a possible difference in lifestyle factors such as diet, smoking, substance abuse, and stress between patients and controls. In particular, a higher proportion of schizophrenia patients exercise less (Brown *et al.*, 1999) and smoke more (Goff *et al.*, 2005), compared to controls. Such chronic and unhealthy lifestyle conditions could contribute to altered metabolism and HPA axis function and should therefore be controlled for as potential confounding factors. Moreover, many of these factors are known to affect brain function (Desai *et al.*, 2010). Various studies have attempted to account for this by controlling for relevant parameters such as body mass index, waist-to-hip ratio, diet, and exercise (Arranz *et al.*, 2004; Guest *et al.*, 2010a,b; Ryan *et al.*, 2003; Sengupta *et al.*, 2008; Spelman *et al.*, 2007; Venkatasubramanian *et al.*, 2007). The design of ideal clinical protocols requires consideration not only of overall levels of these factors, but also with appropriate controls for dietary intake, exercise, alcohol, and other substances prior to sampling. These considerations will prove to be critical for associating metabolic and HPA axis dysfunction with psychiatric conditions compared to those arising from potentially confounding lifestyle factors.

X. Conclusions

Converging evidence is accumulating for metabolic and hormonal components in schizophrenia, based on a wide range of methodologies. Abnormalities in glucose metabolism, insulin signaling, and the HPA axis appear to be present at the early stages of the disorder and may provide the basis for biomarker development. In turn, this could lead to improved diagnosis and patient stratification for personalized medicine strategies. Given the potential of this line of research to improve diagnosis and create alternative treatment strategies, more research is warranted. Future studies should address the metabolic and HPA axis dysfunction in subtypes of schizophrenia as well is in other neuropsychiatric conditions. In addition, longitudinal studies should be performed to investigate the status of metabolic and HPA axis factors over the course of the disorder and with antipsychotic treatment. This process will be facilitated by improved molecular technologies and rapid translation of biomarkers into clinical protocols.

Acknowledgments

This research was supported by the Stanley Medical Research Institute (SMRI) and the European Union FP7 SchizDX research program (grant reference 223427).

References

Ackland, J., Ratter, S., Bourne, G., and Rees, L.H. (1983). Pro-opiomelanocortin peptides in the human fetal pituitary. *Regul. Pept.* **6**, 51–61.

Akamine, E.H., Marcal, A.C., Camporez, J.P., Hoshida, M.S., Caperuto, L.C., Bevilacqua, E., and Carvalho, C.R. (2010). Obesity induced by high-fat diet promotes insulin resistance in the ovary. *J. Endocrinol.* **206**, 65–74.

Alarcon, C., Leahy, J.L., Schuppin, G.T., and Rhodes, C.J. (1995). Increased secretory demand rather than a defect in the proinsulin conversion mechanism causes hyperproinsulinemia in a glucose-infusion rat model of non-insulin-dependent diabetes mellitus. *J. Clin. Invest.* **95**, 1032–1039.

Alberti, K.G., Zimmet, P., and Shaw, J. (2006). Metabolic syndrome—A new world-wide definition. A consensus statement from the International Diabetes Federation. *Diabet. Med.* **23**, 469–480.

Albiston, A.L., Peck, G.R., Yeatman, H.R., Fernando, R., Ye, S., and Chai, S.Y. (2007). Therapeutic targeting of insulin-regulated aminopeptidase: heads and tails? *Pharmacol. Ther.* **116**, 417–427.

Alder, J., Thakker-Varia, S., Bangasser, D.A., Kuroiwa, M., Plummer, M.R., Shors, T.J., and Black, I.B. (2003). Brain-derived neurotrophic factor-induced gene expression reveals novel actions of VGF in hippocampal synaptic plasticity. *J. Neurosci.* **23**, 10800–10808.

Arranz, B., Rosel, P., Ramirez, N., Duenas, R., Fernandez, P., Sanchez, J.M., Navarro, M.A., and San, L. (2004). Insulin resistance and increased leptin concentrations in noncompliant schizophrenia patients but not in antipsychotic-naive first-episode schizophrenia patients. *J. Clin. Psychiatry* **65**, 1335–1342.

Asakawa, A., Inui, A., Yuzuriha, H., Ueno, N., Katsuura, G., Fujimiya, M., Fujino, M.A., Niijima, A., Meguid, M.M., and Kasuga, M. (2003). Characterization of the effects of pancreatic polypeptide in the regulation of energy balance. *Gastroenterology* **124**, 1325–1336.

Baekkeskov, S., Aanstoot, H.J., Christgau, S., Reetz, A., Solimena, M., Cascalho, M., Folli, F., Richter-Olesen, H., and De Camilli, P. (1990). Identification of the 64K autoantigen in insulin-dependent diabetes as the GABA-synthesizing enzyme glutamic acid decarboxylase. *Nature* **347**, 151–156.

Bahtiyar, G., Weiss, K., and Sacerdote, A.S. (2007). Novel endocrine disrupter effects of classic and atypical antipsychotic agents and divalproex: induction of adrenal hyperandrogenism, reversible with metformin or rosiglitazone. *Endocr. Pract.* **13**, 601–608.

Banki, C.M., Bissette, G., Arato, M., O'Connor, L., and Nemeroff, C.B. (1987). CSF corticotropin-releasing factor-like immunoreactivity in depression and schizophrenia. *Am. J. Psychiatry* **144**, 873–877.

Bartolomucci, A., La Corte, G., Possenti, R., Locatelli, V., Rigamonti, A.E., Torsello, A., Bresciani, E., Bulgarelli, I., Rizzi, R., Pavone, F., D'Amato, F.R., Severini, C., *et al.* (2006). TLQP-21, a VGF-derived peptide, increases energy expenditure and prevents the early phase of diet-induced obesity. *Proc. Natl. Acad. Sci. USA* **103**, 14584–14589.

Bello, N.T., and Hajnal, A. (2006). Alterations in blood glucose levels under hyperinsulinemia affect accumbens dopamine. *Physiol. Behav.* **88**, 138–145.

Ben-Jonathan, N., Hugo, E.R., Brandebourg, T.D., and LaPensee, C.R. (2006). Focus on prolactin as a metabolic hormone. *Trends Endocrinol. Metab.* **17**, 110–116.

Bernstein, H.G., Ernst, T., Lendeckel, U., Bukowska, A., Ansorge, S., Stauch, R., Have, S.T., Steiner, J., Dobrowolny, H., and Bogerts, B. (2009). Reduced neuronal expression of insulin-degrading enzyme in the dorsolateral prefrontal cortex of patients with haloperidol-treated, chronic schizophrenia. *J. Psychiatr. Res.* **43**, 1095–1105.

Biedermann, F., and Fleischhacker, W.W. (2009). Antipsychotics in the early stage of development. *Curr. Opin. Psychiatry* **22**, 326–330.

Birkenkamp-Demtroder, K., Wagner, L., Brandt Sorensen, F., Bording Astrup, L., Gartner, W., Scherubl, H., Heine, B., Christiansen, P., and Orntoft, T.F. (2005). Secretagogin is a novel marker for neuroendocrine differentiation. *Neuroendocrinology* **82**, 121–138.

Brenner, K., Liu, A., Laplante, D.P., Lupien, S., Pruessner, J.C., Ciampi, A., Joober, R., and King, S. (2009). Cortisol response to a psychosocial stressor in schizophrenia: blunted, delayed, or normal? *Psychoneuroendocrinology* **34**, 859–868.

Brown, A.S., and Susser, E.S. (2008). Prenatal nutritional deficiency and risk of adult schizophrenia. *Schizophr. Bull.* **34**, 1054–1063.

Brown, S., Birtwistle, J., Roe, L., and Thompson, C. (1999). The unhealthy lifestyle of people with schizophrenia. *Psychol. Med.* **29**, 697–701.

Brunelin, J., d'Amato, T., van Os, J., Cochet, A., Suaud-Chagny, M.F., and Saoud, M. (2008). Effects of acute metabolic stress on the dopaminergic and pituitary-adrenal axis activity in patients with schizophrenia, their unaffected siblings and controls. *Schizophr. Res.* **100**, 206–211.

Buchsbaum, M.S., Wu, J., DeLisi, L.E., Holcomb, H., Kessler, R., Johnson, J., King, A.C., Hazlett, E., Langston, K., and Post, R.M. (1986). Frontal cortex and basal ganglia metabolic rates assessed by positron emission tomography with [18F]2-deoxyglucose in affective illness. *J. Affect. Disord.* **10**, 137–152.

Chai, S.Y., Yeatman, H.R., Parker, M.W., Ascher, D.B., Thompson, P.E., Mulvey, H.T., and Albiston, A.L. (2008). Development of cognitive enhancers based on inhibition of insulin-regulated aminopeptidase. *BMC Neurosci.* **9**(Suppl. 2), S14.

Chan, M.K., Tsang, T.M., Harris, L.W., Guest, P.C., Holmes, E., and Bahn, S. (2010). Evidence for disease and antipsychotic medication effects in post-mortem brain from schizophrenia patients. *Mol. Psychiatry.* Oct 5. [Epub ahead of print].

Cheng, T.M., Lu, Y.E., Guest, P.C., Rahmoune, H., Harris, L.W., Wang, L., Ma, D., Stelzhammer, V., Umrania, Y., Wayland, M.T., Lio, P., and Bahn, S. (2010). Identification of targeted analyte clusters for studies of schizophrenia. *Mol. Cell. Proteomics* **9**, 510–522.

Clark, C., Kopala, L., Li, D.K., and Hurwitz, T. (2001). Regional cerebral glucose metabolism in never-medicated patients with schizophrenia. *Can. J. Psychiatry* **46**, 340–345.

Cohn, T.A., Remington, G., Zipursky, R.B., Azad, A., Connolly, P., and Wolever, T.M. (2006). Insulin resistance and adiponectin levels in drug-free patients with schizophrenia: a preliminary report. *Can. J. Psychiatry* **51**, 382–386.

Convit, A. (2005). Links between cognitive impairment in insulin resistance: an explanatory model. *Neurobiol. Aging* **26**(Suppl. 1), 31–35.

Corcoran, C., Walker, E., Huot, R., Mittal, V., Tessner, K., Kestler, L., and Malaspina, D. (2003). The stress cascade and schizophrenia: etiology and onset. *Schizophr. Bull.* **29**, 671–692.

Craft, S. (2007). Insulin resistance and Alzheimer's disease pathogenesis: potential mechanisms and implications for treatment. *Curr. Alzheimer Res.* **4**, 147–152.

Creemers, J.W., Jackson, R.S., and Hutton, J.C. (1998). Molecular and cellular regulation of prohormone processing. *Semin. Cell Dev. Biol.* **9**, 3–10.

Cripps, R.L., Martin-Gronert, M.S., Archer, Z.A., Hales, C.N., Mercer, J.G., and Ozanne, S.E. (2009). Programming of hypothalamic neuropeptide gene expression in rats by maternal dietary protein content during pregnancy and lactation. *Clin. Sci. (Lond.)* **117**, 85–93.

Czobor, P., Volavka, J., Sheitman, B., Lindenmayer, J.P., Citrome, L., McEvoy, J., Cooper, T.B., Chakos, M., and Lieberman, J.A. (2002). Antipsychotic-induced weight gain and therapeutic response: a differential association. *J. Clin. Psychopharmacol.* **22**, 244–251.

Desai, A.K., Grossberg, G.T., and Chibnall, J.T. (2010). Healthy brain aging: a road map. *Clin. Geriatr. Med.* **26**, 1–16.

Engelmann, M., Landgraf, R., and Wotjak, C.T. (2004). The hypothalamic-neurohypophysial system regulates the hypothalamic-pituitary-adrenal axis under stress: an old concept revisited. *Front. Neuroendocrinol.* **25**, 132–149.

Fagiolini, A., Frank, E., Scott, J.A., Turkin, S., and Kupfer, D.J. (2005). Metabolic syndrome in bipolar disorder: findings from the Bipolar Disorder Center for Pennsylvanians. *Bipolar Disord.* **7**, 424–430.

Fernandez-Egea, E., Bernardo, M., Donner, T., Conget, I., Parellada, E., Justicia, A., Esmatjes, E., Garcia-Rizo, C., and Kirkpatrick, B. (2009). Metabolic profile of antipsychotic-naive individuals with non-affective psychosis. *Br. J. Psychiatry* **194**, 434–438.

Ferrier, I.N., Stanton, B.R., Kelly, T.P., and Scott, J. (1999). Neuropsychological function in euthymic patients with bipolar disorder. *Br. J. Psychiatry* **175**, 246–251.

Furukawa, Y., Shimada, T., Furuta, H., Matsuno, S., Kusuyama, A., Doi, A., Nishi, M., Sasaki, H., Sanke, T., and Nanjo, K. (2008). Polymorphisms in the IDE-KIF11-HHEX gene locus are reproducibly associated with type 2 diabetes in a Japanese population. *J. Clin. Endocrinol. Metab.* **93**, 310–314.

Garcia, G.E., Gabbai, F.B., O'Connor, D.T., Dinh, T.Q., Kennedy, B., Ziegler, M.G., and Takiyyuddin, M.A. (1994). Does chromostatin influence catecholamine release or blood pressure in vivo? *Peptides* **15**, 195–197.

Gartner, W., Lang, W., Leutmetzer, F., Domanovits, H., Waldhausl, W., and Wagner, L. (2001). Cerebral expression and serum detectability of secretagogin, a recently cloned EF-hand Ca(2+)-binding protein. *Cereb. Cortex* **11**, 1161–1169.

Gherzi, R., Fehmann, H.C., Eissele, R., and Goke, B. (1994). Expression, intracellular localization, and gene transcription regulation of the secretory protein 7B2 in endocrine pancreatic cell lines and human insulinomas. *Exp. Cell Res.* **213**, 20–27.

Goeders, N.E. (2002). The HPA axis and cocaine reinforcement. *Psychoneuroendocrinology* **27**, 13–33.

Goff, D.C., Cather, C., Evins, A.E., Henderson, D.C., Freudenreich, O., Copeland, P.M., Bierer, M., Duckworth, K., and Sacks, F.M. (2005). Medical morbidity and mortality in schizophrenia: guidelines for psychiatrists. *J. Clin. Psychiatry* **66**, 183–194; quiz 147, 273–274.

Goldman, M.B., Robertson, G.L., Luchins, D.J., and Hedeker, D. (1996). The influence of polydipsia on water excretion in hyponatremic, polydipsic, schizophrenic patients. *J. Clin. Endocrinol. Metab.* **81**, 1465–1470.

Goldstein, J.M. (2006). Sex, hormones and affective arousal circuitry dysfunction in schizophrenia. *Horm. Behav.* **50**, 612–622.

Goodyer, I.M., Herbert, J., Tamplin, A., and Altham, P.M. (2000). First-episode major depression in adolescents. Affective, cognitive and endocrine characteristics of risk status and predictors of onset. *Br. J. Psychiatry* **176**, 142–149.

Green, M.J., Matheson, S.L., Shepherd, A., Weickert, C.S., and Carr, V.J. (2010). Brain-derived neurotrophic factor levels in schizophrenia: a systematic review with meta-analysis. *Mol. Psychiatry*. Aug 24. [Epub ahead of print].

Guest, P.C., Bailyes, E.M., Rutherford, N.G., and Hutton, J.C. (1991). Insulin secretory granule biogenesis. Co-ordinate regulation of the biosynthesis of the majority of constituent proteins. *Biochem. J.* **274**(Pt 1), 73–78.

Guest, P.C., Abdel-Halim, S.M., Gross, D.J., Clark, A., Poitout, V., Amaria, R., Ostenson, C.G., and Hutton, J.C. (2002). Proinsulin processing in the diabetic Goto-Kakizaki rat. *J. Endocrinol.* **175**, 637–647.

Guest, P.C., Schwarz, E., Krishnamurthy, D., Harris, L.W., Leweke, F.M., Rothermundt, M., van Beveren, N.J., Spain, M., Barnes, A., Steiner, J., Rahmoune, H., and Bahn, S. (2011). Altered levels of circulating insulin and other neuroendocrine hormones associated with the onset of schizophrenia. *Psychoneuroendocrinology.* **36**, 1062–1096.

Guest, P.C., Wang, L., Harris, L.W., Burling, K., Levin, Y., Ernst, A., Wayland, M.T., Umrania, Y., Herberth, M., Koethe, D., van Beveren, J.M., Rothermundt, M., *et al.* (2010). Increased levels of circulating insulin-related peptides in first-onset, antipsychotic naive schizophrenia patients. *Mol. Psychiatry* **15**, 118–119.

Hahm, S., Mizuno, T.M., Wu, T.J., Wisor, J.P., Priest, C.A., Kozak, C.A., Boozer, C.N., Peng, B., McEvoy, R.C., Good, P., Kelley, K.A., Takahashi, J.S., *et al.* (1999). Targeted deletion of the Vgf gene indicates that the encoded secretory peptide precursor plays a novel role in the regulation of energy balance. *Neuron* **23**, 537–548.

Hasnain, M., Fredrickson, S.K., Vieweg, W.V., and Pandurangi, A.K. (2010). Metabolic syndrome associated with schizophrenia and atypical antipsychotics. *Curr. Diab. Rep.* **10**, 209–216.

Heinrichs, M., von Dawans, B., and Domes, G. (2009). Oxytocin, vasopressin, and human social behavior. *Front. Neuroendocrinol.* **30**, 548–557.

Helle, K.B., Corti, A., Metz-Boutigue, M.H., and Tota, B. (2007). The endocrine role for chromogranin A: a prohormone for peptides with regulatory properties. *Cell. Mol. Life Sci.* **64**, 2863–2886.

Herberth, M., Koethe, D., Cheng, T.M., Krzyszton, N.D., Schoeffmann, S., Guest, P.C., Rahmoune, H., Harris, L.W., Kranaster, L., Leweke, F.M., and Bahn, S. (2010). Impaired glycolytic response in peripheral blood mononuclear cells of first-onset antipsychotic-naive schizophrenia patients. *Mol. Psychiatry* **16**, 848–859.

Hoek, H.W., Brown, A.S., and Susser, E. (1998). The Dutch famine and schizophrenia spectrum disorders. *Soc. Psychiatry Psychiatr. Epidemiol.* **33**, 373–379.

Holmes, E., Tsang, T.M., Huang, J.T., Leweke, F.M., Koethe, D., Gerth, C.W., Nolden, B.M., Gross, S., Schreiber, D., Nicholson, J.K., and Bahn, S. (2006). Metabolic profiling of CSF: evidence that early intervention may impact on disease progression and outcome in schizophrenia. *PLoS Med.* **3**, e327.

Howes, O.D., and Kapur, S. (2009). The dopamine hypothesis of schizophrenia: version III—the final common pathway. *Schizophr. Bull.* **35**, 549–562.

Huang, J.T., Leweke, F.M., Oxley, D., Wang, L., Harris, N., Koethe, D., Gerth, C.W., Nolden, B.M., Gross, S., Schreiber, D., Reed, B., and Bahn, S. (2006). Disease biomarkers in cerebrospinal fluid of patients with first-onset psychosis. *PLoS Med.* **3**, e428.

Huang, J.T., Leweke, F.M., Tsang, T.M., Koethe, D., Kranaster, L., Gerth, C.W., Gross, S., Schreiber, D., Ruhrmann, S., Schultze-Lutter, F., Klosterkotter, J., Holmes, E., *et al.* (2007). CSF metabolic and proteomic profiles in patients prodromal for psychosis. *PLoS One* **2**, e756.

Huang, J.T., Wang, L., Prabakaran, S., Wengenroth, M., Lockstone, H.E., Koethe, D., Gerth, C.W., Gross, S., Schreiber, D., Lilley, K., Wayland, M., Oxley, D., *et al.* (2008). Independent protein-profiling studies show a decrease in apolipoprotein A1 levels in schizophrenia CSF, brain and peripheral tissues. *Mol. Psychiatry* **13**, 1118–1128.

Hunsberger, J.G., Newton, S.S., Bennett, A.H., Duman, C.H., Russell, D.S., Salton, S.R., and Duman, R.S. (2007). Antidepressant actions of the exercise-regulated gene VGF. *Nat. Med.* **13**, 1476–1482.

Hunt, P.J., Gurnell, E.M., Huppert, F.A., Richards, C., Prevost, A.T., Wass, J.A., Herbert, J., and Chatterjee, V.K. (2000). Improvement in mood and fatigue after dehydroepiandrosterone replacement in Addison's disease in a randomized, double blind trial. *J. Clin. Endocrinol. Metab.* **85**, 4650–4656.

Hutton, J.C. (1994). Insulin secretory granule biogenesis and the proinsulin-processing endopeptidases. *Diabetologia* **37**(Suppl. 2), S48–S56.

Iwazaki, T., Shibata, I., Niwa, S., and Matsumoto, I. (2004). Selective reduction of chromogranin A-like immunoreactivities in the prefrontal cortex of schizophrenic subjects: a postmortem study. *Neurosci. Lett.* **367**, 293–297.

Kahn, R.S., Davidson, M., Hirschowitz, J., Stern, R.G., Davis, B.M., Gabriel, S., Moore, C., and Davis, K.L. (1992). Nocturnal growth hormone secretion in schizophrenic patients and healthy subjects. *Psychiatry Res.* **41**, 155–161.

Kalkhoff, R.K. (1982). Metabolic effects of progesterone. *Am. J. Obstet. Gynecol.* **142**, 735–738.

Koller, E.A., and Doraiswamy, P.M. (2002). Olanzapine-associated diabetes mellitus. *Pharmacotherapy* **22**, 841–852.

Kwiterovich, P.O. Jr. (2000). The metabolic pathways of high-density lipoprotein, low-density lipoprotein, and triglycerides: a current review. *Am. J. Cardiol.* **86**, 5L–10L.

Laedtke, T., Kjems, L., Porksen, N., Schmitz, O., Veldhuis, J., Kao, P.C., and Butler, P.C. (2000). Overnight inhibition of insulin secretion restores pulsatility and proinsulin/insulin ratio in type 2 diabetes. *Am. J. Physiol. Endocrinol. Metab.* **279**, E520–E528.

Landen, M., Grenfeldt, B., Davidsson, P., Stridsberg, M., Regland, B., Gottfries, C.G., and Blennow, K. (1999). Reduction of chromogranin A and B but not C in the cerebrospinal fluid in subjects with schizophrenia. *Eur. Neuropsychopharmacol.* **9**, 311–315.

Landreth, G., Jiang, Q., Mandrekar, S., and Heneka, M. (2008). PPARgamma agonists as therapeutics for the treatment of Alzheimer's disease. *Neurotherapeutics* **5**, 481–489.

Levi, A., Eldridge, J.D., and Paterson, B.M. (1985). Molecular cloning of a gene sequence regulated by nerve growth factor. *Science* **229**, 393–395.

Levin, B.E. (2006). Metabolic imprinting: critical impact of the perinatal environment on the regulation of energy homeostasis. *Philos. Trans. R. Soc. Lond. B Biol. Sci.* **361**, 1107–1121.

Licinio, J., and Wong, M.L. (2003). The interface of obesity and depression: risk factors for the metabolic syndrome. *Rev. Bras. Psiquiatr.* **25**, 196–197.

Machado-Vieira, R., Dietrich, M.O., Leke, R., Cereser, V.H., Zanatto, V., Kapczinski, F., Souza, D.O., Portela, L.V., and Gentil, V. (2007). Decreased plasma brain derived neurotrophic factor levels in unmedicated bipolar patients during manic episode. *Biol. Psychiatry* **61**, 142–144.

MacMaster, F.P., Russell, A., Mirza, Y., Keshavan, M.S., Taormina, S.P., Bhandari, R., Boyd, C., Lynch, M., Rose, M., Ivey, J., Moore, G.J., and Rosenberg, D.R. (2006). Pituitary volume in treatment-naive pediatric major depressive disorder. *Biol. Psychiatry* **60**, 862–866.

Mai, J.K., Berger, K., and Sofroniew, M.V. (1993). Morphometric evaluation of neurophysin-immunoreactivity in the human brain: pronounced inter-individual variability and evidence for altered staining patterns in schizophrenia. *J. Hirnforsch.* **34**, 133–154.

Martin, M.V., Rollins, B., Sequeira, P.A., Mesen, A., Byerley, W., Stein, R., Moon, E.A., Akil, H., Jones, E.G., Watson, S.J., Barchas, J., DeLisi, L.E., *et al.* (2009). Exon expression in lymphoblastoid cell lines from subjects with schizophrenia before and after glucose deprivation. *BMC Med. Genomics* **2**, 62.

Martins-de-Souza, D., Harris, L.W., Guest, P.C., and Bahn, S. (2010). The role of energy metabolism dysfunction and oxidative stress in schizophrenia revealed by proteomics. *Antioxid. Redox Signal.* Dec 2. [Epub ahead of print].

McDermott, E., and de Silva, P. (2005). Impaired neuronal glucose uptake in pathogenesis of schizophrenia—can GLUT 1 and GLUT 3 deficits explain imaging, post-mortem and pharmacological findings? *Med. Hypotheses* **65**, 1076–1081.

McQueen, M.J., Hawken, S., Wang, X., Ounpuu, S., Sniderman, A., Probstfield, J., Steyn, K., Sanderson, J.E., Hasani, M., Volkova, E., Kazmi, K., and Yusuf, S. (2008). Lipids, lipoproteins, and apolipoproteins as risk markers of myocardial infarction in 52 countries (the INTERHEART study): a case-control study. *Lancet* **372**, 224–233.

Meltzer, H.Y., Lee, M.A., and Jayathilake, K. (2001). The blunted plasma cortisol response to apomorphine and its relationship to treatment response in patients with schizophrenia. *Neuropsychopharmacology* **24**, 278–290.

Meltzer, H.Y., Perry, E., and Jayathilake, K. (2003). Clozapine-induced weight gain predicts improvement in psychopathology. *Schizophr. Res.* **59**, 19–27.

Meyer, J.M., Davis, V.G., Goff, D.C., McEvoy, J.P., Nasrallah, H.A., Davis, S.M., Rosenheck, R.A., Daumit, G.L., Hsiao, J., Swartz, M.S., Stroup, T.S., and Lieberman, J.A. (2008). Change in metabolic syndrome parameters with antipsychotic treatment in the CATIE Schizophrenia Trial: prospective data from phase 1. *Schizophr. Res.* **101**, 273–286.

Mojiminiyi, O.A., Abdella, N., and George, S. (2000). Evaluation of serum cystatin C and chromogranin A as markers of nephropathy in patients with type 2 diabetes mellitus. *Scand. J. Clin. Lab. Invest.* **60**, 483–489.

Montaner, J., Perea-Gainza, M., Delgado, P., Ribo, M., Chacon, P., Rosell, A., Quintana, M., Palacios, M.E., Molina, C.A., and Alvarez-Sabin, J. (2008). Etiologic diagnosis of ischemic stroke subtypes with plasma biomarkers. *Stroke* **39**, 2280–2287.

Mosconi, L., Pupi, A., and De Leon, M.J. (2008). Brain glucose hypometabolism and oxidative stress in preclinical Alzheimer's disease. *Ann. N. Y. Acad. Sci.* **1147**, 180–195.

Murray, R.D., and Shalet, S.M. (2005). Insulin sensitivity is impaired in adults with varying degrees of GH deficiency. *Clin. Endocrinol. (Oxf.)* **62**, 182–188.

Nachshoni, T., Ebert, T., Abramovitch, Y., Assael-Amir, M., Kotler, M., Maayan, R., Weizman, A., and Strous, R.D. (2005). Improvement of extrapyramidal symptoms following dehydroepiandrosterone (DHEA) administration in antipsychotic treated schizophrenia patients: a randomized, double-blind placebo controlled trial. *Schizophr. Res.* **79**, 251–256.

Nilsson, B.M., Hultman, C.M., and Wiesel, F.A. (2006). Niacin skin-flush response and electrodermal activity in patients with schizophrenia and healthy controls. *Prostaglandins Leukot. Essent. Fatty Acids* **74**, 339–346.

Nosadini, R., Del Prato, S., Tiengo, A., Valerio, A., Muggeo, M., Opocher, G., Mantero, F., Duner, E., Marescotti, C., Mollo, F., and Belloni, F. (1983). Insulin resistance in Cushing's syndrome. *J. Clin. Endocrinol. Metab.* **57**, 529–536.

O'Brien, W.T., and Klein, P.S. (2009). Validating GSK3 as an in vivo target of lithium action. *Biochem. Soc. Trans.* **37**, 1133–1138.

O'Malley, D., Shanley, L.J., and Harvey, J. (2003). Insulin inhibits rat hippocampal neurones via activation of ATP-sensitive K+ and large conductance Ca2+-activated K+ channels. *Neuropharmacology* **44**, 855–863.

Onat, A., Hergenc, G., Sansoy, V., Fobker, M., Ceyhan, K., Toprak, S., and Assmann, G. (2003). Apolipoprotein C-III, a strong discriminant of coronary risk in men and a determinant of the metabolic syndrome in both genders. *Atherosclerosis* **168**, 81–89.

Palmer, A.A., Printz, D.J., Butler, P.D., Dulawa, S.C., and Printz, M.P. (2004). Prenatal protein deprivation in rats induces changes in prepulse inhibition and NMDA receptor binding. *Brain Res.* **996**, 193–201.

Palmer, A.A., Brown, A.S., Keegan, D., Siska, L.D., Susser, E., Rotrosen, J., and Butler, P.D. (2008). Prenatal protein deprivation alters dopamine-mediated behaviors and dopaminergic and glutamatergic receptor binding. *Brain Res.* **1237**, 62–74.

Pariante, C.M., Vassilopoulou, K., Velakoulis, D., Phillips, L., Soulsby, B., Wood, S.J., Brewer, W., Smith, D.J., Dazzan, P., Yung, A.R., Zervas, I.M., Christodoulou, G.N., *et al.* (2004). Pituitary volume in psychosis. *Br. J. Psychiatry* **185**, 5–10.

Rasgon, N.L., and Kenna, H.A. (2005). Insulin resistance in depressive disorders and Alzheimer's disease: revisiting the missing link hypothesis. *Neurobiol. Aging* **26**(Suppl. 1), 103–107.

Rizzi, R., Bartolomucci, A., Moles, A., D'Amato, F., Sacerdote, P., Levi, A., La Corte, G., Ciotti, M.T., Possenti, R., and Pavone, F. (2008). The VGF-derived peptide TLQP-21: a new modulatory peptide for inflammatory pain. *Neurosci. Lett.* **441**, 129–133.

Ryan, M.C., Collins, P., and Thakore, J.H. (2003). Impaired fasting glucose tolerance in first-episode, drug-naive patients with schizophrenia. *Am. J. Psychiatry* **160**, 284–289.

Ryan, M.C., Sharifi, N., Condren, R., and Thakore, J.H. (2004). Evidence of basal pituitary-adrenal overactivity in first episode, drug naive patients with schizophrenia. *Psychoneuroendocrinology* **29**, 1065–1070.

Sato, T., Hanyu, H., Hirao, K., Kanetaka, H., Sakurai, H., and Iwamoto, T. (2009). Efficacy of PPAR-gamma agonist pioglitazone in mild Alzheimer disease. *Neurobiol. Aging* **32**, 1622–1633.

Schousboe, I. (1985). beta 2-Glycoprotein I: a plasma inhibitor of the contact activation of the intrinsic blood coagulation pathway. *Blood* **66**, 1086–1091.

Schwartz, M.W., Woods, S.C., Porte, D. Jr., Seeley, R.J., and Baskin, D.G. (2000). Central nervous system control of food intake. *Nature* **404**, 661–671.

Schwarz, E., Izmailov, R., Spain, M., Barnes, A., Mapes, J.P., Guest, P.C., Rahmoune, H., Pietsch, S., Leweke, F.M., Rothermundt, M., Steiner, J., Koethe, D., *et al.* (2010). Validation of a blood-based laboratory test to aid in the confirmation of a diagnosis of schizophrenia. *Biomark Insights* **5**, 39–47.

Seaton, B.E., Goldstein, G., and Allen, D.N. (2001). Sources of heterogeneity in schizophrenia: the role of neuropsychological functioning. *Neuropsychol. Rev.* **11**, 45–67.

Sengupta, S., Parrilla-Escobar, M.A., Klink, R., Fathalli, F., Ying Kin, N., Stip, E., Baptista, T., Malla, A., and Joober, R. (2008). Are metabolic indices different between drug-naive first-episode psychosis patients and healthy controls? *Schizophr. Res.* **102**, 329–336.

Shimizu, E., Hashimoto, K., Okamura, N., Koike, K., Komatsu, N., Kumakiri, C., Nakazato, M., Watanabe, H., Shinoda, N., Okada, S., and Iyo, M. (2003). Alterations of serum levels of brain-derived neurotrophic factor (BDNF) in depressed patients with or without antidepressants. *Biol. Psychiatry* **54**, 70–75.

Skovhus, K.V., Bergholdt, R., Erichsen, C., Sparre, T., Nerup, J., Karlsen, A.E., and Pociot, F. (2006). Identification and characterization of secretagogin promoter activity. *Scand. J. Immunol.* **64**, 639–645.

Small, J.G., Kolar, M.C., and Kellams, J.J. (2004). Quetiapine in schizophrenia: onset of action within the first week of treatment. *Curr. Med. Res. Opin.* **20**, 1017–1023.

Sobey, W.J., Beer, S.F., Carrington, C.A., Clark, P.M., Frank, B.H., Gray, I.P., Luzio, S.D., Owens, D. R., Schneider, A.E., Siddle, K., *et al.* (1989). Sensitive and specific two-site immunoradiometric assays for human insulin, proinsulin, 65-66 split and 32-33 split proinsulins. *Biochem. J.* **260**, 535–541.

Spelman, L.M., Walsh, P.I., Sharifi, N., Collins, P., and Thakore, J.H. (2007). Impaired glucose tolerance in first-episode drug-naive patients with schizophrenia. *Diabet. Med.* **24**, 481–485.

St Clair, D., Xu, M., Wang, P., Yu, Y., Fang, Y., Zhang, F., Zheng, X., Gu, N., Feng, G., Sham, P., and He, L. (2005). Rates of adult schizophrenia following prenatal exposure to the Chinese famine of 1959-1961. *JAMA* **294**, 557–562.

Strous, R.D., Maayan, R., Lapidus, R., Stryjer, R., Lustig, M., Kotler, M., and Weizman, A. (2003). Dehydroepiandrosterone augmentation in the management of negative, depressive, and anxiety symptoms in schizophrenia. *Arch. Gen. Psychiatry* **60**, 133–141.

Taguchi, A., Wartschow, L.M., and White, M.F. (2007). Brain IRS2 signaling coordinates life span and nutrient homeostasis. *Science* **317**, 369–372.

Tannenbaum, G.S., Martin, J.B., and Colle, E. (1976). Ultradian growth hormone rhythm in the rat: effects of feeding, hyperglycemia, and insulin-induced hypoglycemia. *Endocrinology* **99**, 720–727.

Timonen, M.J., Saari, K.M., Jokelainen, J.J., Meyer-Rochow, V.B., Rasanen, P.K., and Koponen, H.J. (2009). Insulin resistance and schizophrenia: results from the Northern Finland 1966 Birth Cohort. *Schizophr. Res.* **113**, 107–108.

Torres, N., Bautista, C.J., Tovar, A.R., Ordaz, G., Rodriguez-Cruz, M., Ortiz, V., Granados, O., Nathanielsz, P.W., Larrea, F., and Zambrano, E. (2010). Protein restriction during pregnancy affects maternal liver lipid metabolism and fetal brain lipid composition in the rat. *Am. J. Physiol. Endocrinol. Metab.* **298**, E270–E277.

Tsigos, C., and Chrousos, G.P. (2002). Hypothalamic-pituitary-adrenal axis, neuroendocrine factors and stress. *J. Psychosom. Res.* **53**, 865–871.

Udagawa, J., Hashimoto, R., Hioki, K., and Otani, H. (2006a). The role of leptin in the development of the cortical neuron in mouse embryos. *Brain Res.* **1120**, 74–82.

Udagawa, J., Nimura, M., and Otani, H. (2006b). Leptin affects oligodendroglial development in the mouse embryonic cerebral cortex. *Neuro Endocrinol. Lett.* **27**, 177–182.

van Nimwegen, L.J., Storosum, J.G., Blumer, R.M., Allick, G., Venema, H.W., de Haan, L., Becker, H., van Amelsvoort, T., Ackermans, M.T., Fliers, E., Serlie, M.J., and Sauerwein, H.P. (2008). Hepatic insulin resistance in antipsychotic naive schizophrenic patients: stable isotope studies of glucose metabolism. *J. Clin. Endocrinol. Metab.* **93**, 572–577.

Venkatasubramanian, G., Chittiprol, S., Neelakantachar, N., Naveen, M.N., Thirthall, J., Gangadhar, B.N., and Shetty, K.T. (2007). Insulin and insulin-like growth factor-1 abnormalities in antipsychotic-naive schizophrenia. *Am. J. Psychiatry* **164**, 1557–1560.

von Eckardstein, A., Holz, H., Sandkamp, M., Weng, W., Funke, H., and Assmann, G. (1991). Apolipoprotein C-III(Lys58——Glu). Identification of an apolipoprotein C-III variant in a family with hyperalphalipoproteinemia. *J. Clin. Invest.* **87**, 1724–1731.

Waeber, G., Thompson, N., Waeber, B., Brunner, H.R., Nicod, P., and Grouzmann, E. (1993). Neuropeptide Y expression and regulation in a differentiated rat insulin-secreting cell line. *Endocrinology* **133**, 1061–1067.

Walker, J.J., Terry, J.R., and Lightman, S.L. (2010). Origin of ultradian pulsatility in the hypothalamic-pituitary-adrenal axis. *Proc. Biol. Sci.* **277**, 1627–1633.

Walsh, P., Spelman, L., Sharifi, N., and Thakore, J.H. (2005). Male patients with paranoid schizophrenia have greater ACTH and cortisol secretion in response to metoclopramide-induced AVP release. *Psychoneuroendocrinology* **30**, 431–437.

Yang, Y., Wan, C., Li, H., Zhu, H., La, Y., Xi, Z., Chen, Y., Jiang, L., Feng, G., and He, L. (2006). Altered levels of acute phase proteins in the plasma of patients with schizophrenia. *Anal. Chem.* **78**, 3571–3576.

Zhao, Z., Ksiezak-Reding, H., Riggio, S., Haroutunian, V., and Pasinetti, G.M. (2006). Insulin receptor deficits in schizophrenia and in cellular and animal models of insulin receptor dysfunction. *Schizophr. Res.* **84**, 1–14.

Zhao, Z., Lange, D.J., Ho, L., Bonini, S., Shao, B., Salton, S.R., Thomas, S., and Pasinetti, G.M. (2008). Vgf is a novel biomarker associated with muscle weakness in amyotrophic lateral sclerosis (ALS), with a potential role in disease pathogenesis. *Int. J. Med. Sci.* **5**, 92–99.

Zuo, X., and Jia, J. (2009). Promoter polymorphisms which modulate insulin degrading enzyme expression may increase susceptibility to Alzheimer's disease. *Brain Res.* **1249**, 1–8.

IMMUNE AND NEUROIMMUNE ALTERATIONS IN MOOD DISORDERS AND SCHIZOPHRENIA

Roosmarijn C. Drexhage[1], Karin Weigelt[1], Nico van Beveren[2], Dan Cohen[3,4], Marjan A. Versnel[1], Willem A. Nolen[5] and Hemmo A. Drexhage[1]

[1]Department of Immunology, Erasmus MC, Rotterdam, The Netherlands
[2]Department of Psychiatry, Erasmus MC, Rotterdam, The Netherlands
[3]Department of Epidemiology, University Medical Center, Groningen, The Netherlands
[4]Department of Severe Mental Illness, Mental Health Care Organization, North-Holland North, Heerhugowaard, The Netherlands
[5]Department of Psychiatry, University Medical Center, University of Groningen, Groningen, The Netherlands

INTERNATIONAL REVIEW OF NEUROBIOLOGY, VOL. 101
DOI: 10.1016/B978-0-12-387718-5.00007-9

Abstract

A large number of publications over the past 20 years have indicated that immune system function is altered in schizophrenia and mood disorder patients. This chapter reviews the evidence, which suggests that a proinflammatory state of the cytokine network induces psychopathologic symptoms and may be involved in the pathogenesis and pathophysiology of these major mental illnesses. The authors also present recent data, which relates immune activation to present theories on the influence of activated immune cells in altering brain function. They also focus on the role of the environment in immune activation and on the role of the microbiome and gut flora. Increased understanding of such factors could help in the development of novel treatment strategies and improved clinical management of mental disorders.

I. Introduction

Although there have been reports of an involvement of the immune system in major mental illnesses since the first decade of the twentieth century, it has only been since the past decade of the same century before detailed studies on an immune involvement have become more numerous. These recent studies have reported aberrant levels of proinflammatory cytokines in the serum, plasma, and cerebrospinal fluid of patients with schizophrenia and major mood disorders. On the basis of these reports, it was hypothesized that a proinflammatory state of the cytokine network induces psychopathologic symptoms and is involved in the pathogenesis and pathophysiology of these major mental illnesses.

Proinflammatory cytokines are primarily produced by activated cells of the immune system such as activated endothelial cells, monocytes, monocyte-derived dendritic cells, macrophages, and T cells (for a synopsis of the different immune cells, see Fig. 1). The realization that immune cells can be involved in mental illnesses has led to the macrophage-T-cell theory of depression and schizophrenia which was proposed in 1992 and adapted in 1995 (Smith, 1992; Smith and Maes, 1995). According to this theory, chronically activated macrophages and T cells produce cytokines and inflammatory compounds, which destabilize the brain in such a way that other genetic and environmental influences are able to precipitate the signs and symptoms of schizophrenia and mania/depression (Bessis *et al.*, 2007). Indeed, receptors for inflammatory cytokines are present in various brain

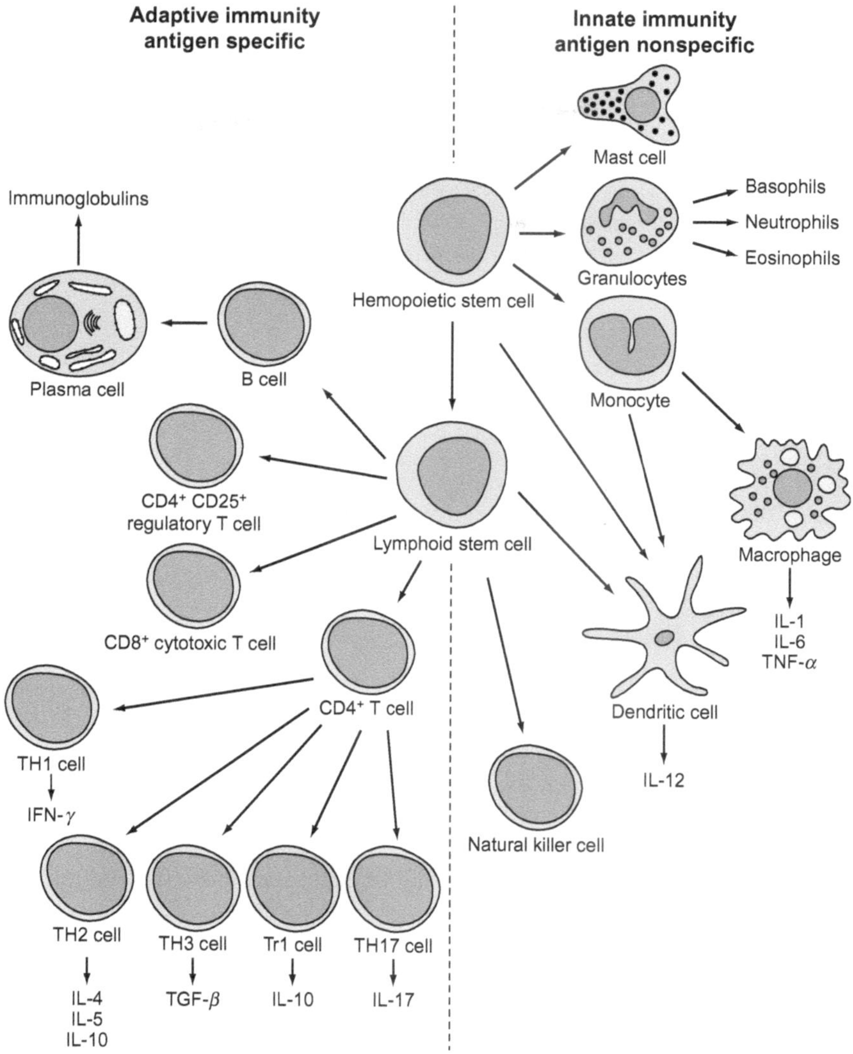

FIG. 1. The innate and adaptive immune system. Schematic diagram showing the important components of the innate and adaptive immune systems. Further information is provided in the text. IL, interleukin; TNF, tumor necrosis factor; IFN, interferon; TGF, tumor growth factor; Th, T helper cell.

nuclei (Chesnokova and Melmed, 2002), which upon triggering deregulate important neurotransmitters and neurodevelopmental systems, facilitating the development of psychiatric signs and symptoms.

In this chapter, we first give a synopsis of the literature on the reported levels of cytokines/chemokines in the serum of bipolar disorder and schizophrenia patients and relate findings to what is known on the actual numbers and activation state of immune cells in bipolar disorder and schizophrenia. Thereafter, we give some recent data of ours, summarizing the state of the art on the involvement of immune cells in major mental disorders and relate immune activation to present theories on the influence of activated immune cells in altering brain function. Finally, we discuss the role of the environment on immune activation in psychiatric disorders and focus in on the role of the microbiome and the dietary/gut flora.

II. Inflammatory Cytokines in Psychiatric Disorders—State of the Art 2010

In 2010, we reviewed the literature on cytokine networks in schizophrenia and bipolar disorder, and Table I gives a synopsis of the state of the art 2010 on the serum levels of cytokines, chemokines, and endothelial inflammatory factors in patients with schizophrenia and bipolar disorder. In summary, the data indicate that measurements of these compounds are not consistent and precise enough to reliably detect an activation of the immune system in individual or in small groups of patients. More in detail it was found that

1. The vascular inflammatory factors s-intracellular adhesion molecule (ICAM) and endothelin-1 are reduced in two out of two studies in schizophrenia, suggesting a reduced activation of the endothelial system in schizophrenia (Kronig *et al.*, 2005; Schwarz *et al.*, 2000). However, preliminary data of our group (to be published) show an increased level of s-ICAM in schizophrenia patients, while there are also unpublished data on raised levels of endothelin-1 in bipolar patients (S. Bahn, personal communication). Therefore, further studies on endothelial factors are needed before firm conclusions on the state of endothelial activation in schizophrenia and bipolar disorder can be drawn.
2. Studies on the levels of the inflammatory chemokine CCL2 are too few ($n = 2$) (Drexhage *et al.*, 2008; Teixeira *et al.*, 2008) and involve too few patients to draw firm conclusions. In our study (Drexhage *et al.*, 2008), patients with schizophrenia ($n = 145$) had higher serum levels of CCL2 as compared to healthy controls. Although schizophrenia patients with metabolic syndrome and in particular those with reduced high density lipoprotein (HDL) had the highest serum levels of CCL2, the serum levels of this protein were still higher in patients after correction for these metabolic factors (waist circumference, serum triglycerides, serum HDL cholesterol, blood pressure, fasting plasma glucose, and current adult treatment panel (ATP) III criteria; Grundy *et al.*, 2004). Nevertheless, we find that it is too early to

Table I

Synopsis of Literature Regarding Serum Levels of the Indicated Cytokines and Their Prime Sources in Schizophrenia and Bipolar Disorder.

		Schizophrenia		Bipolar disorder	
Endothelium	ICAM	↓ ↓ ↑ Kim *et al.* (2004a,b)	n.c.		
	Endothelin			↑	n.c.
Monocytes	CCL2	= ↑ Crespo-Facorro *et al.* (2008), Smith, (1992)	n.c.	= Smith and Maes (1995)	n.c.
	IL-1	= = = = = ↑ ↑ ↑ ↑ Drexhage *et al.* (2011a,b), Nikkila *et al.* (1999), Rothermundt *et al.* (1998), Smith (1992), Torres *et al.* (2009a,b), Zhang *et al.* (2005), Zorrilla *et al.* (1996)	↑	= ↑ Maino *et al.* (2007), Smith and Maes (1995)	(↑)
	IL-6	= = = = = = = = = ↑ ↑ ↑ ↑ ↑ ↑ ↑ ↑ Cazzullo *et al.* (1998), Craddock *et al.* (2007), Drexhage *et al.* (2010, 2011a,b), Henneberg *et al.* (1990), Kronfol and House (1988), Muller *et al.* (1998), Nikkila *et al.* (1999), Padmos *et al.* (2008b), Smith (1992), Sperner-Unterweger *et al.* (1999), Torres *et al.* (2009a), Weigelt *et al.* (2011), Zhang *et al.* (2005, 2009), Zorrilla *et al.* (2001)	↑	= = ↑ ↑ ↑ ↑ Cohen *et al.* (2006), Dahlman *et al.* (2005), Kamei *et al.* (2006), Maino *et al.* (2007), Smith and Maes (1995), Trayhurn *et al.* (2006)	↑
	TNF	= = = = = = ↑ ↑ ↑ ↑ ↑ ↑ ↑ Brown and Derkits (2010), Craddock *et al.* (2007), Drexhage *et al.* (2010, 2011b), Henneberg *et al.* (1990), Nikkila *et al.* (1999), Rothermundt *et al.* (1998), Smith (1992), Wang *et al.* (2010), Weigelt *et al.* (2011), Zhang *et al.* (2005), Zorrilla *et al.* (1996, 2001)	↑	= ↑ ↑ ↑ Cohen *et al.* (2006), Kamei *et al.* (2006), Maino *et al.* (2007), Trayhurn *et al.* (2006)	↑

(*Continued*)

Table I (*Continued*)

		Schizophrenia		Bipolar disorder	
T cells	IL-2	↓↓====↑↑↑↑ Craddock *et al.* (2007), Drexhage *et al.* (2011a,b), Henneberg *et al.* (1990), Kronfol and House (1988), Padmos *et al.* (2008b), Rothermundt *et al.* (1998), Torres *et al.* (2009b), Zhang *et al.* (2005), Zorrilla *et al.* (2001)	(↑)	↓↑↑ Dahlman *et al.* (2005), Kamei *et al.* (2006), Maino *et al.* (2007)	n.c.
	Th1: IL-12	=↑ Radewicz *et al.* (2000), Steiner *et al.* (2008)	n.c.	= Radewicz *et al.* (2000)	n.c.
	Th1: IFN-γ	=↑ Henneberg *et al.* (1990), Roumier *et al.* (2008)	n.c.	=↑ Chen *et al.* (2010), Kamei *et al.* (2006)	n.c.
	Th2: IL-4	↓↓== Craddock *et al.* (2007), Drexhage *et al.* (2010), Henneberg *et al.* (1990), Roumier *et al.* (2008)	n.c.	=↑↑ Chen *et al.* (2010), Kamei *et al.* (2006), Maino *et al.* (2007)	n.c.

References are indicated between brackets. ↓: downregulation, ↑: upregulation, =: normal level. The summary columns reflect our conclusions based on this literature. We concluded that there is an upregulation of IL-1, Il-6, and TNF on the basis of sufficient studies of which approximately half show upregulation and none showed a downregulation. We concluded that IL-2 is putatively upregulated in schizophrenia because there are sufficient studies of which the cumulative index shows some upregulation. n.c. means not conclusive due to low number of studies and/or conflicting outcomes.

conclude that higher levels of CCL2 are a marker of schizophrenia and further studies are needed. In bipolar disorder, we were unable to find higher serum levels of CCL2 (Padmos *et al.*, 2008a).

3. IL-1, IL-6, and TNF cytokine networks are activated in schizophrenia and bipolar disorder, be it inconsistently since only about half the studies were able to find the activation (Brietzke *et al.*, 2009; Coelho *et al.*, 2008; Drexhage *et al.*, 2008; Ebrinc *et al.*, 2002; Erbagci *et al.*, 2001; Frommberger *et al.*, 1997; Haack *et al.*, 1999; Hori *et al.*, 2007; Kaminska *et al.*, 2001; Kauer-Sant'Anna *et al.*, 2009; Kim *et al.*, 2000, 2009; Kudoh *et al.*, 2001, 2003; Lin *et al.*, 1998; Maes, 2000; Maes *et al.*, 1994; Monteleone *et al.*, 1997; Muller *et al.*, 1997; Naudin *et al.*, 1997; O'Brien *et al.*, 2006, 2008; Ortiz-Dominguez *et al.*, 2007; Padmos *et al.*, 2008a; Song *et al.*, 2009; Theodoropoulou *et al.*, 2001; Zhang *et al.*, 2002, 2004). Activation was found in general (taking schizophrenia and bipolar disorder together) in 28 out of 52 studies (none showed reduced levels), pointing to an inconsistent proinflammatory activity of the Mononuclear phagocyte system (MPS) and/or the endothelial system in patients with schizophrenia and bipolar disorder. In many of the positive studies, various confounding factors such as medication, age, gender, and lifestyle have been taken into account, which makes these observations reasonably solid.
4. The IL-2 system is probably activated in schizophrenia. Increased serum levels for the IL-2 system were found in 4 out of 10 studies (note two studies showed reduced levels) (Ebrinc *et al.*, 2002; Erbagci *et al.*, 2001; Haack *et al.*, 1999; Hori *et al.*, 2007; Kaminska *et al.*, 2001; Kim *et al.*, 2000, 2009; Muller *et al.*, 1997; Theodoropoulou *et al.*, 2001; Zhang *et al.*, 2002). Medication did not influence the levels of IL-2.

In bipolar disorder, there are fewer reports ($n = 3$) (Brietzke *et al.*, 2009; Maes *et al.*, 1994; Ortiz-Dominguez *et al.*, 2007) on the IL-2 system, but two of these studies indicate an activated T cell system particularly in periods of mania, including our study (Breunis *et al.*, 2003). In our study, sIL-2R levels were raised particularly during a manic episode, but also to a lesser extent during euthymic and depressive episodes.

5. There are too few reports and too many with contradictory results for an activation or nonbalance of the IL-12, IFN-γ, and IL-4 systems and thus insufficient indications for an activation of the $CD4^{+}$ Th1 or Th2 effector systems in schizophrenia and bipolar disorder (Brietzke *et al.*, 2009; Crespo-Facorro *et al.*, 2008; Kaminska *et al.*, 2001; Kim *et al.*, 2002, 2004a,b, 2009; O'Brien *et al.*, 2008; Ortiz-Dominguez *et al.*, 2007). Studies on the Th17 system are completely lacking.

In summary, after taking into consideration the various confounding factors as described, studies on cytokine networks only point in the direction, but are certainly not proof of an activated inflammatory response system in both schizophrenia and bipolar disorder. Of the cytokine networks, particularly the IL-1, IL-6, and TNF cytokine networks seem to be activated. There is also some evidence in the literature that the IL-2 system is activated in schizophrenia. In bipolar disorder, the IL-2 system might be activated during a manic episode. But as indicated above, the data show in particular that measurements of cytokines in plasma or serum are not consistent and precise enough to reliably detect an activation of the inflammatory systems in individual patients or in small groups of patients.

III. Circulating Immune Cells in Psychiatric Disorders—State of the Art 2010

This section examines the question of whether there are also indications of an activated inflammatory response system on the level of circulating immune cells in schizophrenia and bipolar disorder.

A. Numbers of Circulating Monocytes

There are early reports showing that the number of circulating monocytes is aberrant in schizophrenia. Rothermundt *et al.* (1998) reported a slight increase in the mean absolute and relative monocyte counts. (Zorrilla *et al.* (1996) supported these observations showing, respectively, a monocytosis and a higher number of $CD14^+$ cells (CD14 is a marker of monocytes) in nonmedicated schizophrenia patients. Another study, however, refuted this finding of monocytosis, although this was using a small sample of patients (Torres *et al.*, 2009a). In the cerebrospinal fluid of patients with schizophrenia, monocytes and macrophages show an accumulation during acute psychotic episodes (Nikkila *et al.*, 1999). For bipolar disorder, there is only one report from our group on numbers of circulating monocytes in bipolar disorder (Padmos *et al.*, 2008a) and this was focussed on numbers of mature ($CD14^+CD16^+$) and immature ($CD14^+CD16^{neg}$) circulating monocytes. We did not find abnormal numbers of these subpopulations in the circulation of bipolar patients. Recently, we confirmed the higher presence of circulating $CD14^+$ monocytes in schizophrenia (Drexhage *et al.*, 2011a), but were again unable to detect a higher number of circulating $CD14^+$ monocytes in bipolar disorder (Drexhage *et al.*, 2011b).

B. Numbers of T Cells and T Cell Subsets

Numerous studies have been performed on the number of $CD3^+$ T cells, $CD4^+$ and $CD8^+$ subsets, B cells, and NK cells in schizophrenia and bipolar disorder (Table II). In essence, it has been found that numbers of these

Table II

SYNOPSIS OF LITERATURE REGARDING PERCENTAGES OF CIRCULATING IMMUNE CELLS IN SCHIZOPHRENIA AND BIPOLAR DISORDER.

		Schizophrenia		Bipolar disorder	
Monocytes	–Number– (CD14)	(↓) ↑ ↑ Arnold *et al.* (1998), van Berckel *et al.* (2008), Wierzba-Bobrowicz *et al.* (2005)	↑	↓ = Smith and Maes (1995), van Berckel *et al.* (2008)	n.c.
T cells	CD3	↓ ↓ = = = = = ↑ Bosker *et al.* (2011), Doorduin *et al.* (2009), Liu *et al.* (2009), Miura *et al.* (2008), Oxenkrug (2010), Rothermundt *et al.* (1998), Ruhe *et al.* (2007), Wierzba-Bobrowicz *et al.* (2005)	=	= = Stefansson *et al.* (2009), Thomas *et al.* (2004)	(=)
	CD4	↓ ↓ = = = ↑ ↑ Bosker *et al.* (2011), Doorduin *et al.* (2009), Miura *et al.* (2008), Oxenkrug (2010), Rothermundt *et al.* (1998), Ruhe *et al.* (2007), van Berckel *et al.* (2008)	= ↑	= ↑ Thomas *et al.* (2004), van Berckel *et al.* (2008)	n.c.
	CD8	= = = ↑(↑) Bosker *et al.* (2011), Miura *et al.* (2008), Oxenkrug (2010), Padmos *et al.* (2009), Rothermundt *et al.* (1998)	=↑	= Thomas *et al.* (2004)	(=)
	CD45 RA	↑ ↑ Bosker *et al.* (2011), Miura *et al.* (2008)	↑		
	CD45 RO	(↓) Bosker *et al.* (2011)	n.c.		
	CD25	= ↑ ↑ Bosker *et al.* (2011), Rothermundt *et al.* (1998), Wierzba-Bobrowicz *et al.* (2005)	n.c.		
B cells	CD19	↓ = = Liu *et al.* (2009), Miura *et al.* (2008), van Berckel *et al.* (2008)	n.c.	= ↑ Thomas *et al.* (2004), van Berckel *et al.* (2008)	n.c.
NK cells	CD56	↓ = Oxenkrug (2010), van Berckel *et al.* (2008)	n.c.	= = Thomas *et al.* (2004), van Berckel *et al.* (2008)	=

References are indicated between brackets. ↓, downregulation; ↑, upregulation; =, normal level; (↓), downregulated but low number of patients. The summary columns reflect our conclusions based on this literature. We concluded that there are raised numbers of monocytes in schizophrenia because the cumulative index of three studies shows upregulation (admitted there are too few studies). We concluded that there are normal-to-putatively raised numbers of $CD8^{+}$ T cells in schizophrenia because the cumulative index of five studies shows weak upregulation to normal numbers of CD3 and CD4 T cells and the cumulative index of eight studies shows normal numbers of cells. n.c. means not conclusive due to low number of studies and/or conflicting outcomes.

lymphocyte subsets are normal (Breunis *et al.*, 2003; Cazzullo *et al.*, 1998; Craddock *et al.*, 2007; Henneberg *et al.*, 1990; Kronfol and House, 1988; Maino *et al.*, 2007; Muller *et al.*, 1998; Rothermundt *et al.*, 1998; Sperner-Unterweger *et al.*, 1999; Theodoropoulou *et al.*, 2001; Torres *et al.*, 2009a,b; Zhang *et al.*, 2005, 2009; Zorrilla *et al.*, 1996), although there might be some indication that cytotoxic $CD8^+$ T cells (particularly naive CD45RA $CD8^+$ T cells) are increased in schizophrenia (Zorrilla *et al.*, 2001).

C. Activation of Circulating Immune Cells

With regard to adhesion molecule expression, Theodoropoulou *et al.* (2001) showed that there is an increased percentage of circulating peripheral blood mononuclear cells (PBMCs; this population includes monocytes and all types of lymphocytes) expressing ICAM-1 in patients with schizophrenia, but they did not make a distinction between circulating lymphocytes and monocytes. Nevertheless their observation supports an activated state of immune cells in the circulation of patients with schizophrenia, facilitating endothelial transmigration of the cells.

The limited studies carried out on numbers of $CD25^+$ T cells (generally considered as a sign of T cell activation/proliferation) show that these cells might be increased in the circulation of schizophrenia patients and of patients with bipolar disorder (particularly in the manic phase). However, it is not known whether we are dealing here with an activation/proliferation of the T effector or the T regulatory population. To answer this question, intracellular fluorescence activated cell sorting (FACS) staining of T cells for IFN-γ (Th1), IL-4 (Th2), IL17A (Th17), and $CD25^{high}$FOXP3 (Treg) should be undertaken.

In summary, the state of the art 2009 showed only limited studies on the numbers and proinflammatory activities of immune cells in the circulation of patients with bipolar disorder and schizophrenia. Although it is difficult to draw solid conclusions from these studies, they point in the direction of a monocytosis in schizophrenia and perhaps of an activation of T cells in mania and schizophrenia (in the latter case, activation of $CD8^+$ cells). Studies on the number of circulating Th1, Th2, Th17, and Treg cells have not been carried out before 2009.

IV. Our Recent Studies on the Inflammatory State in Bipolar Disorder and Schizophrenia

Most previous studies on the proinflammatory activation of the circulating monocytes in patients with bipolar disorder and schizophrenia have been carried out through counting numbers or determining expression of inflammatory surface molecules (see above). We embarked on gene array studies to identify "gene expression fingerprints" using monocytes prepared from PBMCs of 56 bipolar disorder patients (mean age 42, range 26–61 years, 61% female, of whom one-third had active disease), 27 schizophrenia patients (mean age 27, range 17–59 years, 81% male, all with active psychosis), and the respective 40 and 32 age/gender matched healthy controls (Drexhage *et al.*, 2010).

To study the proinflammatory state of circulating $CD4^+$ T cells, we determined the percentages of IFN-γ^+ (Th1) and $IL17A^+$ (Th17) $CD4^+$ T cells using intracellular FACS staining after stimulation of the PBMCs with PMA and ionomycin. We also determined in the same FACS assay the percentages of anti-inflammatory $CD4^+CD25^{high}FOXP3^+$ natural T regulator cells and of IL-4^+ (Th2) cells (Drexhage *et al.*, 2011a,b).

Finally, we also carried out multiplex immunoassay analysis [cytometric bead array (CBA)] to measure the levels of cytokines in the serum of the patients including the innate (monocyte/macrophage) immune cytokines TNF-α, IL-1β, IL-6, and PTX3, as well as the T cell proliferation indicator sCD25 and the T effector cell cytokines IFN-γ and IL-17, and the Th2 cell cytokine IL-4. IL-10 and TGF-β were also measured in the assay. These cytokines are produced by anti-inflammatory monocytes/macrophages as well as $CD4^+CD25^{high}FOXP3^+$ natural T regulator cells (Drexhage *et al.*, 2010, 2011a,b).

A. Proinflammatory Gene Expression

First, we used whole genome Affymetrix analysis of pooled monocytes of three to four typical patients to identify mRNA transcripts of genes, which were strongly over and under expressed (Drexhage *et al.*, 2010; Padmos *et al.*, 2008a). We thus obtained 19 "bipolar-specific" transcripts and nine "schizophrenia-specific" monocyte inflammatory transcripts (Padmos *et al.*, 2008a). In addition to these inflammatory mRNA transcripts, we also tested for six additional transcripts recently described by us as overexpressed in the circulating monocytes of autoimmune diabetes patients (Padmos *et al.*, 2008b).

Using this panel of 34 genes in quantitative-PCR (Q-PCR) assays on monocytes of individual patients, we found three subclusters of coherent sets of altered mRNA levels in the circulating monocytes of bipolar and schizophrenia patients (Table III). These subclusters were characterized by different sets of transcription and/or mitogen-activated protein kinase (MAPK) regulating factors:

a. Subcluster 1A was characterized by the activating transcription factor 3 (ATF3) and the MAPK regulating factor dual specificity phosphatase 2 (DUSP2); the expression of these factors was predominantly and strongly correlated to a set of 12 inflammatory and chemotactic fingerprint genes (PDE4B, IL-6, IL-1β, TNF, TNFAIP3, BCLA2A1, PTGS2/COX2,

Table III
THE PREVALENCE OF SUBCLUSTERS IN BIPOLAR PATIENTS, SCHIZOPHRENIA PATIENTS, AND HEALTHY CONTROLS.

	Cluster 1a	Cluster 1b	Cluster 2	Cluster 2
	Dusp2 and/or ATF3 positive	MXD1 and/ or EGR3 positive	PTPN7 and/ or NAB2 positive	PTPN7 and/ or NAB2 negative
Schizophrenia	52% (14/27)[a]	41% (11/27)[a]	7% (2/27)[b]	48% (13/27)[a,b]
Healthy controls	24% (7/29)	13% (4/30)	21% (6/29)	21% (6/29)
Bipolar disorder	67% (26/39)[a]	34% (12/36)	62% (24/39)[a]	5% (2/39)[a]
Healthy controls	24% (10/41)	22% (8/36)	32% (13/41)	25% (10/40)

Positive = mRNA expression > 1standard deviation from the mean level in controls. Signature = positive/negative of the indicated the transcription factors. *P* values are obtained from Chi-squared test.
[a] $P < 0.05$ versus healthy controls.
[b] $P < 0.05$ versus bipolar disorder.

PTX3, CCL20, CXCL2, EREG, and CXCL3). Subcluster 1A was overexpressed in 67% and 52% of monocytes from bipolar and schizophrenia patients, respectively, and in only 24% of the healthy controls.

b. Subcluster 1B was characterized by the transcription factors early growth response 3 (EGR3), MAX dimerization protein, F3, and V-maf musculoaponeurotic fibrosarcoma oncogene family (MAFF). The expression of these factors was strongly associated to the inflammatory and chemotactic factors, as for DUSP2 and ATF3 above, and also, and more predominantly, to a set of two vascular pathology/metabolic syndrome factors [thrombospondin (THBS), plasminogen activator inhibitor 2 (PAI-2)]. Subcluster 1B was overexpressed in the monocytes from 41% of the schizophrenia patients and from 34% of the bipolar patients, and in only 13–22% of the healthy controls.
c. Subcluster 2 was characterized by the MAPK regulating factors HePTP, NAB2, and MAPK6. The expression of these factors was predominantly linked to a set of six motility/adhesion factors [epithelial membrane protein 1 (EMP-1), syntaxin 1A (STX1A), discs-large homologous region sequence (DHRS), chemokine C-C ligand 2 (CCL2), CCL7, and CDC42]. Transcripts in subcluster 2 were increased in the monocytes of bipolar patients (62%) and this subcluster was particularly evident in active bipolar depressive cases.

In contrast, only subcluster 2 contained transcripts (MAPK6, CCL2, CCL7, and CDC42) which were not upregulated in the monocytes of schizophrenia patients and some of these (i.e., HePTP and NAB2) were downregulated in the monocytes of 48% of the schizophrenia patients.

In summary, we showed that a transcriptomic array approach is a solid and robust assay system to detect inflammatory gene expression fingerprint patterns in monocytes of schizophrenia and bipolar disorder patients and that there were differences in monocyte fingerprint expression between schizophrenia and bipolar disorder (particularly in subcluster 1 and 2 expression). We also showed that gene expression patterns were influenced to a limited extent by the activity of disease (higher in bipolar depression than in euthymia) and by medications such as lithium and antipsychotics (both medications, in general, reduced the levels of proinflammatory mRNA transcripts), but not by age, gender, or body mass index (Drexhage *et al.*, 2010).

The fingerprints contained important transcription factors and transcription regulators. For the transcription factors early growth response 3 (EGR3) and ATF3, we have recently obtained evidence using Chromatin ImmunoPrecipitation (ChIP) analysis that these are indeed key regulators in the expression of the fingerprint genes (Weigelt *et al.*, 2011).

B. Pro- and Anti-Inflammatory $CD4^+$ T Cell Subpopulations

We extended the previous findings on the IL-2 system and on $CD25^+$ T cells (see Table I) and confirmed that the IL-2 system is activated in patients with schizophrenia and bipolar disorder: Also in the new series, we found higher serum sCD25 (=sIL-2R) levels as well as higher percentages of $CD25^+CD4^+$T cells in patients with recent onset schizophrenia and in patients with bipolar disorder, albeit only for patients younger than 40 years of age in the latter group. It is also of note that we were only able to carry out FACS analysis for 38 (mean age 41.1 years, 76% female) of the 56 patients with bipolar disorder, all of these were in a euthymic phase of their disease. At an earlier occasion, we found that the sCD25 level was highest in cases with an active mania.

We concluded from these recent data that there are clear signs of an activation/proliferation of the $CD4^+$ T cell compartment particularly in young and active cases of schizophrenia and young cases of bipolar disorder. However (as we noted in the previous section), it is not known whether we are dealing here with an activation/proliferation of the T effector or the T regulatory cell populations. For that, intracellular FACS staining of T cells for IFN-γ (Th1), IL-4 (Th2), IL17A (Th17), and FOXP3 (natural T regulator cells) should be undertaken, which we performed.

We found that the percentage of anti-inflammatory $CD4^+CD25^{high}FOXP3^+$ natural regulatory T cells was higher in the circulation of young and active recent onset schizophrenia cases and in euthymic bipolar patients of less than 40 years of age. In addition, we found for schizophrenia that those patients who had the highest global assessment of functioning (GAF) scores at discharge were the ones with the highest levels of anti-inflammatory natural T regulator cells at admission. We therefore hypothesized that high levels of $CD4^+CD25^{high}FOXP3^+$ natural regulatory

T cells are probably beneficial and counteract the proinflammatory action of the monocyte/macrophage system, since it is known that natural T regulator cells are capable of dampening down the activity of proinflammatory monocytes/macrophages. However, it must be noted that we were unable to find a significant correlation between the proinflammatory gene expression in monocytes and the number of natural T regulator cells. Therefore, this does not suggest a mutual relationship but rather separate activation mechanisms for monocytes and natural T regulator cells.

In the case of the $CD4^+$ T effector populations, we only found higher percentages of Th17 cells in our group of young and recent onset schizophrenia patients. These percentages were normal in the bipolar disorder group (also in those younger than 40 years).

C. Pro and Anti-Inflammatory Monocyte/Macrophage and T Cell Cytokines/Chemokines

With regard to the T cell cytokines, we were unable to find any abnormalities in the serum levels of IFN-γ, IL-4, IL-17, IL-10, and TGF-β from patients of both the study cohorts of recent onset schizophrenia patients and bipolar patients. We thus conclude that there probably is a balance within an activated T cell system between pro and anti-inflammatory forces.

In the case of the proinflammatory monocyte/macrophage cytokines/chemokines, we found the serum levels of IL-1β, PTX3, and CCL2 slightly (but significantly) raised in bipolar patients and in schizophrenia patients, although this was only in older patients (mean age 40 years, range 18–65 years) with a chronic form of the disease. The levels were not raised in the group of young recent onset schizophrenia patients (mean age 27, range 17–59 years). It is of note that 35% of the older schizophrenia patients suffered from metabolic syndrome (visceral obesity, hypercholesterolemia, and diabetes) and that proinflammatory cytokines were higher in patients with these comorbidities. It must, however, be noted that proinflammatory cytokines were also raised in the circulation of chronic schizophrenia patients without metabolic syndrome.

We therefore concluded that proinflammatory cytokines are to a limited extent raised in patients with bipolar disorder and schizophrenia, although this was inconsistent. This suggests that the balance within the monocyte/macrophage system tends to tip toward an active proinflammatory state in these psychiatric diseases.

With regard to the metabolic syndrome, white adipose tissue (WAT) can be seen as an endocrine organ, and in visceral obesity, WAT contains increased numbers of tissue macrophages, which are in a chronically inflamed state overproducing various cytokines including proinflammatory cytokines (in this situation also called adipokines) (Dahlman *et al.*, 2005). Apart from leptin, adiponectin, and PAI-1, CCL2 is considered an important adipokine. CCL2 promotes migration and accumulation of macrophages into WAT (Trayhurn *et al.*, 2006) and thus supports the further inflammatory state of WAT. In addition, it induces insulin resistance and therefore plays an important role as an intermediate in the development of type 2 diabetes in obese individuals (Cohen *et al.*, 2006; Weigelt *et al.*, 2011). In line with this view, the levels of CCL2 are higher in patients with visceral obesity and diabetes. Other proinflammatory cytokines produced by WAT macrophages are TNF-α and IL-6. To correct for visceral obesity and the metabolic syndrome, the body mass index, the waist–hip ratio, and levels of HDL need therefore to be taken into account in studies of proinflammatory cytokines in psychiatric patients.

D. Summary of Findings

Taken together, the data reviewed here can be interpreted as showing that the immune system in schizophrenia and bipolar disorder patients is set at a high activation set point involving both pro- as well as anti-inflammatory forces (see Fig. 2):

1. At the level of the monocyte/macrophage system overlapping but distinct gene expression patterns (mRNA transcript fingerprints) were found in the circulating monocytes of patients with schizophrenia and bipolar disorder. In the serum of bipolar disorder patients, gene transcript cluster 1 and cluster 2 were upregulated, while in schizophrenia patients, only cluster 1 gene transcripts were upregulated.

A hypothetical interaction model of the fingerprint genes is given in Fig. 3, showing cluster gene transcripts leading to a proinflammatory action (e.g., IL-1β, TNF, and CCL2) as well as cluster transcripts which give rise to an anti-inflammatory action (e.g., ATF3 and NAB2). It is thus not surprising that for the proinflammatory cytokines monocyte gene expression and circulating protein levels did not reach the same high levels: For example, in our studies, the approximate sixfold increased expression of IL-1β at the monocyte mRNA level in bipolar patients was reflected in only a twofold raised IL-1β protein level in the serum. In our studies on recent onset schizophrenia patients, the serum levels of proinflammatory cytokines were not raised at all, while there was a high gene transcript expression in circulating monocytes. Clearly, a regulation at the transcription level is operative in the "inflammatory, angry" monocytes of bipolar and schizophrenic patients to ensure a close-to-normal (but still somewhat raised) protein production. The question arises which environmental or endogenous conditions will create a failure of these "angry" monocytes to keep control over their aberrant gene transcript expressions avoiding a high actual production of the proinflammatory compounds. Psychological stresses (both acute and chronic) might be conditions which could lead to this (via adrenaline signaling). Indeed, stressors have regulating stimulatory effect on IL-1β and IL-6 protein production (see later).

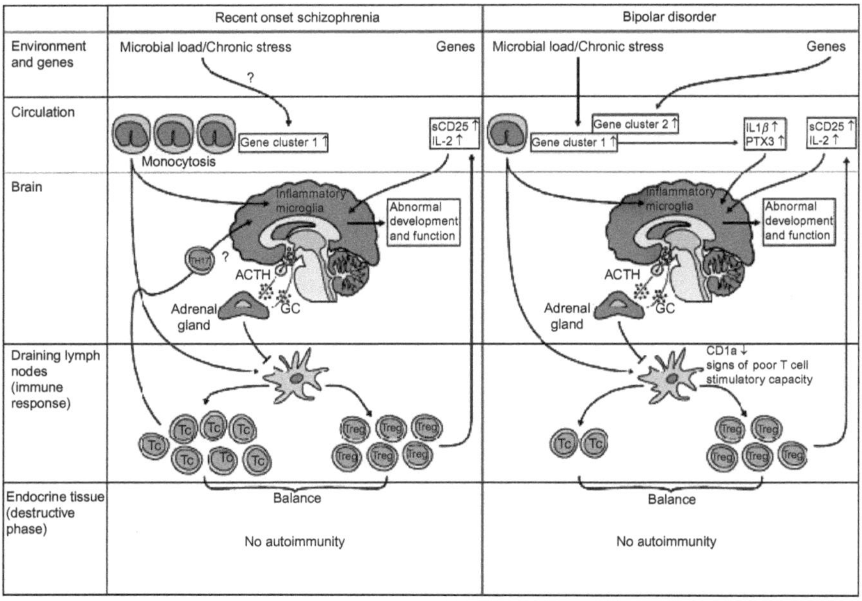

FIG. 2. Cartoon on immune mechanisms as set in recent-onset schizophrenia and bipolar disorder. Monocytes in circulation have an activated "angry" inflammatory transcriptome, which differs between bipolar disorder (clusters 1 and 2) and schizophrenia (cluster 1). Twin studies indicate that cluster 1 overexpression is determined by environmental factors, whereas genetic factors determine for a large part gene cluster 2 overexpression (see text). It can be envisaged that activated "angry" monocytes, upon arrival in the (emotional) brain as activated microglia, will display abnormal interactions with neurons and deregulate synaptic function and neuronal sprouting, and are perhaps even cytotoxic (see text). This will lead to vulnerability for psychiatric behavior. Also serum cytokines (raised to a certain extent in patients, reflecting activation of the monocyte/macrophage system) may penetrate the brain and aggravate the neuronal deregulations of the brain. It can also be envisaged that activated monocytes, when differentiated to aberrant dendritic cells in the tissues and after having traveled to the draining lymph nodes, will abnormally stimulate T cells in the secondary lymphoid tissues, such as lymph nodes, spleen, and lymphatic tissues in the mucosa of the gut and airways. There are signs of an abnormal differentiation of monocytes to dendritic cells in bipolar disorder patients and these cells have a reduced T cell stimulatory capacity (see text). With regard to expansion of T cells in secondary lymphoid tissues, such cells will recirculate and appear in the circulation. In schizophrenia patients, both Th17 cells and T regulator cells are overrepresented, whereas in bipolar patients, only T regulator cells are more numerous. As a sign of higher T cell proliferation, sCD25 is higher in the serum of both recent onset schizophrenia and bipolar disorder patients. As there is a stimulation of both the anti- and proinflammatory monocyte and T cell forces, a delicate balance is kept within the system.

2. In the case of the $CD4^{+}$T cells, we found an increase in the percentages of circulating anti-inflammatory $CD4^{+}CD25^{high}FOXP3^{+}$ natural regulatory T cells in both recent onset schizophrenia and young bipolar disorder patients. The $CD4^{+}$ T effector cells percentages of circulating Th17 cells

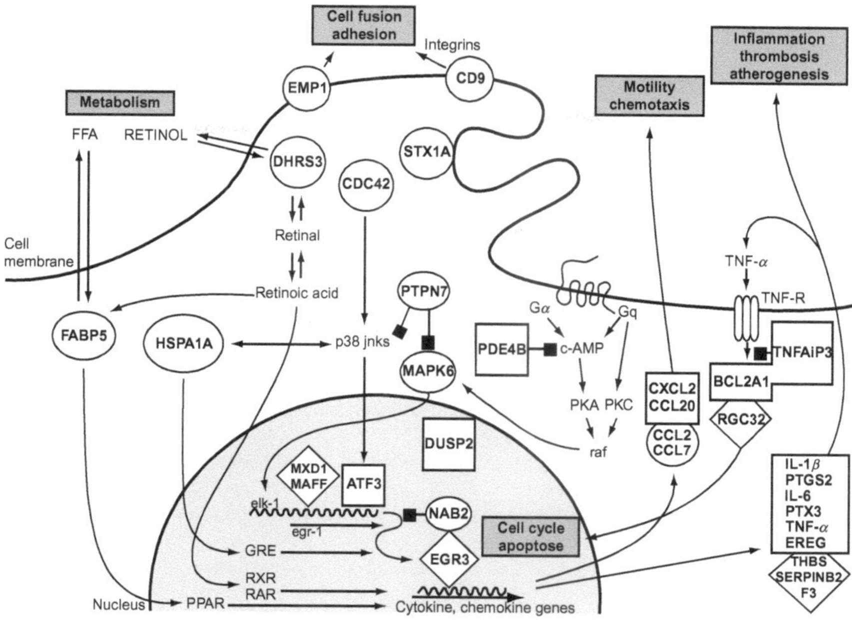

FIG. 3. Hypothetical scheme on interaction of the various fingerprint genes, shown in Table III.

Cluster 1A genes	
ATF3	Transcription factor, negative regulator of proinflammatory cytokine expression
DUSP2	Phosphatase involved in the regulation of MAPKinase activity and controlling inflammatory responses
PDE4B	Regulation of inflammatory signal transduction via the c-AMP pathway, defects in the gene elicit psychosis
IL-6	Inflammatory cytokine
IL-1β	Inflammatory cytokine
TNF	Inflammatory cytokine
PTX3	Inflammatory compound
PTGS2	Prostaglandin synthase, involved in production of the inflammatory compound PGE2
CCL20	A homeostatic and inflammatory chemokine
CXCL2	Specialized monocyte arrest chemokine promoting adherence of monocytes to endothelium
CXCL3	Specialized monocyte arrest chemokine promoting adherence of monocytes to endothelium
EREG	Epiregulin, member of the epidermal growth factor (EGF) family, an Il-6 inducing factor and a downregulating factor for inflammation
TNFAIP3	Essential negative regulator of inflammation, antiapoptotic and inducible by cytokines through NFκB
BCL2A1	Antiapoptotic factor, inducible by cytokines through NFκB
Cluster 1B genes	
EGR3	Transcription factor for early growth response, sympathetic nervous development and immune regulation
MXD1	MAD/MAD1, transcriptional repressor inhibiting cell growth of monocytic cell lines

(*Continued*)

MAFF	Transcription factor involved in cellular stress response, induced by proinflammatory cytokines
THBS	Thrombospondin 1, factor in thrombosis/atherosclerosis, low-grade inflammation, metabolic syndrome, and neuronal growth factor
F3	Thromboplastin, factor in thrombosis/atherosclerosis, low-grade inflammation and metabolic syndrome
SERPINB2	Plasminogen-activator inhibitor, antifibrinolytic factor, negative regulator of inflammatory migration
RGC32	Cell cycle progression factor, involved in diverse processes such as differentiation, inflammation, and fibrosis
Cluster 2 genes	
MAPK6	Part of the MAPKinase pathway
PTPN7	Protein tyrosine phospatase responsible for inactivation of MAPK in leukocytes
NAB2	Transcription coactivator and corepressor for EGRs
CDC42	Signaling molecule in the MAPK-pathway, factor in motility, cytoskeletal organization and chemotaxis
EMP1	Epithelial membrane protein involved in cell–cell adhesion, interactions with the extracellular membrane
CD9	Integrin-associated tetraspanin involved in cell fusion and adhesion
STX-1A	Membrane and vesicle fusion
CCL2	Monocyte-derived early chemokine to attract monocytes, activation factor for monocytes
CCL7	Similar to CCL2
DHRS3	Dehydrogenase involved in vitamin A metabolism
FABP5	Fatty acid binding protein involved in fatty acid uptake, transport and metabolism, involved in activation of the monocyte inflammatory response
HSPA1A	HSP70, protein folding, cellular stress response, chaperone of the glucocorticoid receptor, immune regulator
CCR2	The receptor for CCL2, involved in monocyte activation

were increased in recent onset schizophrenia, but not in bipolar disorder patients. These data again support a view of an existing balance between pro and anti-inflammatory forces within an activated immune system in psychiatric patients.

V. Communication of Immune System and Brain in Psychiatric Illness: The Role of Microglia

There is clear evidence from animal models that an activation of the immune system influences the brain causing an altered behavior. In our view, the best model in this respect is the "maternally induced inflammation model," which highlights several aspects of the immune to brain communication. Several studies have shown that intraperitoneal injection of a pregnant rodent at late gestational age with lipopolysaccharide (LPS) gives the pups a long-lasting activation of the immune system together with behavioral changes (Wang *et al.*, 2010). These behavioral changes include learning disabilities and reduced social behavioral performances and are often referred to as schizophrenia-like behavior. LPS treatment of the pregnant rodents is known to elevate not only levels of various proinflammatory cytokines in the serum, fetal liver, and amniotic fluid of the pregnant rodent but also that of proinflammatory TNF-α and anti-inflammatory IL-10 in the fetal brain. These cytokines are thought to be involved in the abnormal brain development and behavior of the pup. Blocking antibodies against IL-6 given to the pregnant rodents prevent the later altered behavior in the pups (Wang *et al.*, 2010).

Interestingly, there is a human correlate to this model. The relation between prenatal infections and the development of schizophrenia and mood disorders has been described for a long time. Mothers suffering from influenza in the first trimester of pregnancy have a seven times higher chance for the development of schizophrenia in their offspring, and the effect was three times higher for an infection in the second trimester (Brown and Derkits, 2010). Another study shows that mothers seropositive for herpes simplex virus (HSV)-2 in pregnancy have a two times higher chance of schizophrenia development in their offspring (Brown and Derkits, 2010). Moreover in a cohort study, IgG antibodies to Toxoplasma were two times higher in mothers who gave birth to a child with schizophrenia (Brown and Derkits, 2010). In general, these data are interpreted that the microbial pathogens cross the placenta and cause congenital brain anomalies. This has been proven for Toxoplasma, HSV, rubella, and cytomegalovirus. However, a role for immune activation, in general, should not be neglected. It has also been shown that levels of proinflammatory cytokines were higher in the serum of mothers during pregnancy, which resulted in a child who later developed a psychiatric disorder (Brown and Derkits, 2010).

There are two main pathways, which have been described by which a peripherally activated immune system can influence brain development and function:

A. Altered Inflammatory Set Point of the Brain

Proinflammatory cytokines can enter the brain via the circumventricular organs and the choroid plexus via an active transport mechanism. Although infiltrated cytokines may influence neuronal networks directly, there are no histological reports on the expression of proinflammatory cytokine networks in the brain of bipolar disorder and schizophrenia patients. There is a report on an increased ICAM-1 expression in gray and white matter of the anterior-cingulated cortex in *postmortem* brains of bipolar disorder patients. This overexpression was absent in that of schizophrenia patients. This suggests a "low-grade inflammation" of the anterior-cingulated cortex of patients with bipolar disorder (Thomas *et al.*, 2004).

However, we propose that the microglia, the "monocyte-derived macrophage/dendritic cell of the brain," is an important intermediate in the process of immune system to brain communication. A direct role of microglia was recently shown by Chen *et al*, (2010) by the finding that compulsive-like behavior in a mouse model had a hematopoietic origin: The Hoxb8 mutant mouse has mutant microglia and shows excessive grooming and hair removal; this behavior was normalized after a bone marrow transplantation of normal hematopoietic cells resulting in normal microglia (Chen *et al.*, 2010).

Also the "maternally induced mouse inflammation model" is informative for a key role of microglia in peripheral immune system–brain communication. In recent experiments carried out in our group via the EU-consortium MOODINFLAME, we found that LPS injection of pregnant mice induced a monocyte gene transcript expression profile similar to the monocyte expression profile reported here in psychiatric patients. In the mouse model, this resulted in an identical proinflammatory gene expression profile in the fetal microglia, which impacted the growth and development of neurons in the hippocampal area. In a previous study, Bessis *et al.* (Roumier *et al.*, 2008) showed that proinflammatory activated microglia alter glutamatergic synaptic function in the brain.

With regard to the number of microglial cells and their activation, histological *postmortem* reports are limited and controversial. A *postmortem* study on the brains of schizophrenia (SCZ) patients, who had committed suicide, revealed increased densities of microglia (Steiner *et al.*, 2008). Therefore, this study suggested that in the active phase of the illness microglia activation is present. Two other studies also reported on increased microglial activation in SCZ patients (Radewicz *et al.*, 2000; Wierzba-Bobrowicz *et al.*, 2005), although three studies refuted this (Arnold *et al.*, 1998; Falke *et al.*, 2000; Togo *et al.*, 2000). A potential drawback of these *postmortem* studies is that they are normally performed on older subjects and studies on brain material collected in these individuals might therefore not reflect the active disease stages which may have occurred years before death. Currently one does not need to rely on *postmortem* studies as developments in the field of positron emission tomography (PET) allow researchers to study microglia activation in real time in live patients. A PET-tracer ([11C]-PK11195) binds to the peripheral benzodiazepine receptor (PBR), which has increased expression in activated microglia (and also on circulating patient monocytes), thereby visualizing microglia and microglia activation. This technique has already successfully been applied in several patient and animal studies on Parkinson's disease and recent-onset schizophrenia (Doorduin *et al.*, 2009; van Berckel *et al.*, 2008).

B. Alterations in the Tryptophan Breakdown Pathway

Tryptophan can be broken down via two metabolic pathways:

1. The methoxyindole pathway leading to the formation of the important neurotransmitter 5-hydroxytryptamine or serotonin.
2. The kynurenine pathway leading to kynurenic acid and quinolonic acid via the production of kynurenine. Quinolonic acid has a neurotoxic effect; kynurenic acid has a neuroprotective effect. Plasma kynurenic acid and the ratio kynurenic acid/kynurenine are reduced in patients with major depressive disorder.

The rate-limiting enzyme for the second tryptophan–kynurenine pathway is indoleamine 2,3-dioxygenase (IDO) (99%). IDO activity is particularly induced at sites of inflammation in immune cells. The IDO pathway is a key regulatory element responsible for induction and maintenance of peripheral immune tolerance in normal physiological situations (Liu *et al.*, 2009). Proinflammatory cytokines, for example, those produced by macrophages and Th1 cells, induce IDO in a variety of immune cells (Oxenkrug, 2010). Inflammation and probably proinflammatory monocytes and microglia promote the second pathway, thereby depriving the first pathway of fuel and leading to a decrease in serotonin synthesis (Miura *et al.*, 2008). Serotonin deprivation is an important determinant for the development of depression. Serotonin reuptake inhibitors are commonly used in patients with depression. Also, a meta-analysis showed that acute tryptophan depletion decreases mood in healthy controls with a family history of major depressive disorder and in patients with a depression in remission, but not in a population healthy controls (Ruhe *et al.*, 2007).

Not only will this result in a deprivation of fuel for the serotonin pathway but there will also be a shift toward the formation of kynurenines, which have an apoptotic, neurotoxic, and prooxidative effect. Interestingly a higher expression of quinolate has been seen in activated microglia in the brains of suicide victims (Steiner *et al.*, 2008).

VI. The Origin of the Activated Immune System in Psychiatric Patients: Genes or Environment?

A. The Effect of Genes on the Immune Activation

Genome-wide association studies (GWASs) have been performed in the past decade in large cohorts of patients with bipolar disorder and schizophrenia with overall disappointing results. These studies revealed that it is virtually impossible to consistently find specific genes linked to the disorders and to replicate findings for individual studies. The problem is that psychiatric diseases are probably heterogeneous conditions from a pathogenesis point of view and are not generally the result of a mutation in a single or a few genes. Large meta-analyses were needed to complete the GWAS studies and presently a few genetic markers with a limited risk have been identified. Amongst these markers are the major histocompatibility complex (MHC) complex in schizophrenia (Stefansson *et al.*, 2009) and the TNF gene in major depression (Bosker *et al.*, 2011). A new approach is to find molecular pathways, which are affected in psychiatric disease. This approach is somewhat more successful and identified in schizophrenia, for instance, an involvement of the glutamate metabolism pathway and the tumor necrosis factor receptor 1 (TNFR1) pathway (Jia *et al.*, 2010). These GWAS data thus strengthen our view for a role for an activated immune system, which interacts with neurons in psychiatric disease, as discussed above, yet also suggest that the contribution of genetic polymorphisms to the activation of the monocyte/macrophage and T cell system is limited.

B. The Monocyte Inflammatory Gene Fingerprint: Environmental Effect?

Recently, Padmos *et al.* (2009) from our group carried out a case–control study using monocytes from bipolar twins to determine the contribution of genetic and environmental influences on the expression of the monocyte proinflammatory gene signature. It was found that the association of the proinflammatory monocyte gene transcript fingerprint with bipolar disorder was primarily the result of common shared environmental factors. This was in particular evident for the overexpression of cluster 1 genes, although some of the cluster 2 genes were also genetically determined. Epigenetic imprinting is the most likely mechanism via which environmental factors can create long-lasting activation set points of the immune system of psychiatric patients and via which even fetal/childhood influences can impact immune functions later in life. Chronic severe stress, such as in child abuse, has recently been shown to induce epigenetic changes to the glucocorticoid receptor gene in the brain (McGowan *et al.*, 2009). Glucocorticoid resistance is an important factor in T cell activation of psychiatric patients (Knijff *et al.*, 2006). Epigenetic modulations of important cluster 1 and 2 fingerprint genes therefore deserve further study.

C. Environmental Factors

Environmental candidates that can act as the shared common factors for immune system aberrancies and concomitant psychiatric disease are stress, diet, and infections.

1. *Stress*

The prenatal period is an interesting period to study to understand our twin data, since the environmental factors experienced by the twins *in utero* are hypothetically shared. One of the possible environmental factors that are experienced *in utero* and can influence both the immune system as well as the brain is prenatally experienced stress. The literature describes that the effect of prenatal stress on the immune system mostly leads to a reduction of immune function (Merlot *et al.*, 2008). However, a few studies report on an exaggeration of inflammatory function after prenatal stress. (Hashimoto *et al.* (2001) showed that prenatal stress in rats led to an increased fever response to LPS. In addition, Laviola *et al.* (2004) demonstrated an increase of spleen and brain frontal cortex levels of IL-1β in prenatally stressed rats. Possible mechanisms behind prenatal-stress-induced immune alterations are thought to be (1) a direct influence of maternal hormones and neurotransmitters on the ontogeny of immune cells, (2) an indirect effect via deregulation of the hypothalamic pituitary adrenal (HPA)-axis in the prenatally stressed offspring, and (3) via stress mediator induced change in placental function (Merlot *et al.*, 2008).

With regard to the effect of prenatal stress on the brain, there is increasing evidence suggesting that exposure to prenatal stress is a risk factor for psychopathology. Prenatally stressed rats, for instance, show higher emotional reactivity, higher levels of anxiety, and a depressive-like behavior (Abe *et al.*, 2007). In humans, a low birth weight is considered an index of prenatal stress, and indeed this has also been shown to be a risk factor for later development of mood symptoms (Costello *et al.*, 2007). Also, the amount of stress experienced by the mother during pregnancy was positively correlated to emotional, cognitive, and behavioral problems of the offspring (Van den Bergh *et al.*, 2005). It is suggested that stress exposure at critical time points during fetal development may (1) influence the HPA-axis (Lin *et al.*, 1998), leading to glucocorticoid resistance and hypercortisolism, (2) alter brain development, and (3) change neurotransmitter systems (Abe *et al.*, 2007; Austin *et al.*, 2005; Maccari and Morley-Fletcher, 2007; Van den Bergh *et al.*, 2005). All these events have been implicated in the pathogenesis of mood disorders (Fig. 4).

Stress experienced later in life is able to induce mood symptoms. In rats, it was shown that chronic mild stress experienced during adulthood elicited depressive-like behavior (Willner, 2005). And, in children of bipolar patients, major life events increased the risk to develop a mood disorder (Hillegers *et al.*, 2004). The impact of stress on the immune system in adulthood has also extensively been researched. It is a complex interaction in which the HPA-axis and the sympathetic nervous system play pivotal roles, especially with regards to their neuroendocrine products cortisol and catecholamines. These two main mediators of stress effects can regulate a variety of immune functions such as cytokine and chemokine production, the trafficking of immune cells and their proliferation, differentiation, and effector

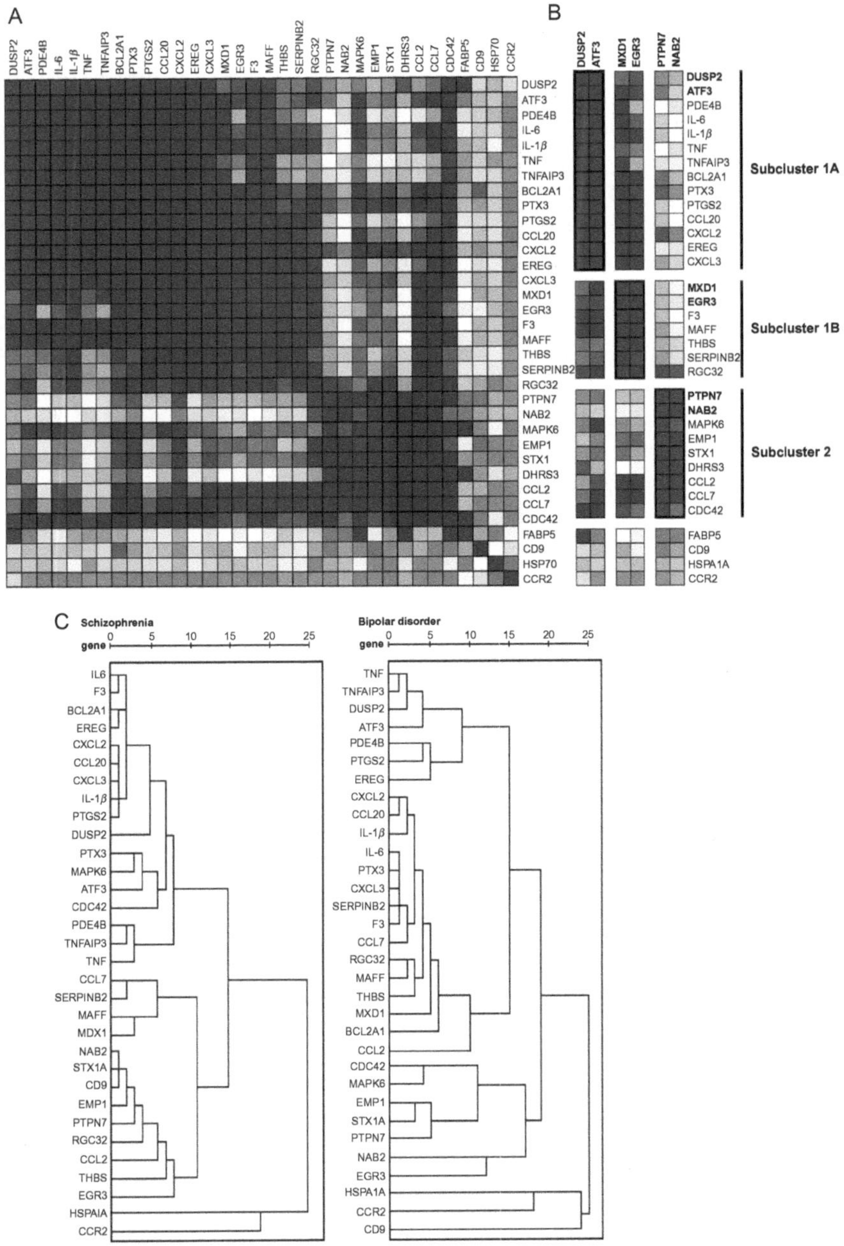

FIG. 4. Heat map of mRNA transcript correlation. The data represent Spearman's correlation coefficients, tested on relative mRNA expression in 56 bipolar disorder and 27 schizophrenic patients. Significant positive correlations ($P < 0.05$) are in a red scale (darkest red = correlations > 0.60).

functions. The final outcome is, although dependent on the quantity and quality of stress and on coping strategies, an increased susceptibility to infection and inflammatory and autoimmune diseases (Leonard, 2006; Padgett and Glaser, 2003).

2. *The Role of Omega-3 Fatty Acids*

Diet, and in particular, the consumption of omega-3 fatty acids (present in marine life and various plants) (Leaf and Weber, 1987), is a second possible shared environmental factor connecting psychiatric disease and the immune system. The immune system is known to be shaped and regulated by polyunsaturated fatty acids. Omega-6 fatty acids are recognized to potentiate inflammatory responses while omega-3 fatty acids are known to have anti-inflammatory effects. Studies reported on effects of omega-3 fatty acids on monocyte activation have been inconsistent not only with data demonstrating a reduction of IL-1, IL-6, and TNF-α production but also with many reports showing no effect on monocyte activation (Sijben and Calder, 2007).

Various epidemiological studies have reported an inverse relation between fish consumption and mood disorders (Parker *et al.*, 2006). In addition, (Tiemeier *et al.* (2003) showed that elderly subjects with depressive symptoms had lower serum levels of omega-3 fatty acids than control subjects without depressive symptoms. Further, treatment of bipolar patients with omega-3 fatty acids decreased affective symptoms and prolonged remission periods (Chiu *et al.*, 2005; Frangou *et al.*, 2006; Osher *et al.*, 2005; Stoll *et al.*, 1999). This positive effect of omega-3 fatty acids on mood disorder development is thought to be due to their effect on brain plasticity via brain-derived neurotrophic factor. In addition, omega-3 fatty acids are abundantly present in cell membranes. Changes in lipid concentrations could change structure and function of various cell membrane proteins such as receptors and enzymes. Also, the positive effects of omega-3 fatty acids could be achieved directly via the immune system.

3. *The Role of Infection*

Infections are a major environmental candidate. Next to having a clear effect on the immune system, they have also been implicated in the pathogenesis of psychiatric disorders.

Significant negative correlations are in a green scale. White = not significant. (A) Correlations between all tested mRNA transcripts. (B) Two main clusters can be seen (the cluster on the left top can be divided into subclusters 1A and 1B). Three sets of transcription factors/MAPK regulators were extracted (DUSP2/ATF3, MXD1/EGR3, and PTPN7/NAB2) and correlations to the other transcripts are shown. DUSP2/ ATF3 correlate strongest to subcluster 1A transcripts, MXD1/EGR3 correlate strongest to subcluster 1B transcripts and many subcluster 1A transcripts and PTPN7/NAB2 correlate strongest to subcluster 2 transcripts. (C) Hierarchical clustering tree, showing relationship of mRNA transcripts in schizophrenia and bipolar disorder patients. (For interpretation of the references to color in this figure legend, the reader is referred to the Web version of this chapter.)

Studies have demonstrated that being born or raised in a city is a risk factor for developing bipolar disorder. This is thought to be due to household crowding, and the consequent high exposure to infectious agents (Torrey and Yolken, 1998). Also relating bipolar disorder to infections are reports that an excess of winter and spring births resulting in increased incidence of bipolar disorder in the offspring, as these children are thought to be more prone to develop perinatal infections (Torrey *et al.*, 1997). Of the infectious agents, viruses are likely candidates (Yolken and Torrey, 1995). First, viruses are known for their neurotropism and latency. Second, viral infections can be accompanied by depressive symptoms and manic behavior. And third, the mood stabilizers lithium and valproate have been described to have antiviral effects (Amsterdam *et al.*, 1996; Witvrouw *et al.*, 1997). The viruses mentioned in the literature as being associated with bipolar disorder are HSV, CMV, and Borna virus (Hinze-Selch, 2002; Taieb *et al.*, 2001; Yolken and Torrey, 1995). However, the data are inconsistent with regard to the presence of virus specific antibodies in serum as well as with regard to detecting virus RNA in brain or in PBMCs of bipolar patients (Hinze-Selch, 2002; Taieb *et al.*, 2001; Yolken and Torrey, 1995). With regard to bacterial infections, Borrelia burgdorferi is known to be able to induce symptoms reminiscent of those seen in bipolar disorder (Fallon and Nields, 1994). Unfortunately, no systematic research has been done on the prevalence of this bacterium in bipolar disorder. *Toxoplasma gondii*, an intracellular protozoan parasite, is capable of latency and brain infiltration (Carruthers and Suzuki, 2007) and is therefore an interesting candidate to study in psychiatric diseases. In schizophrenia, many studies have reported a correlation to Toxoplasmosis infection (Torrey *et al.*, 2007), especially in cases of prenatal exposure. However, only one such study is available for bipolar disorder (to the best of our knowledge) and this reports a negative result with regard to the presence of *T. gondii* sequences in *postmortem* brains (Conejero-Goldberg *et al.*, 2003). Nevertheless, valproate does inhibit *T. gondii* development (Jones-Brando *et al.*, 2003), suggesting that more extensive research is needed to determine whether or not Toxoplasmosis plays a role of significance in bipolar disorder as has been suggested in the case of schizophrenia.

VII. Conclusions

Converging evidence is accumulating for inflammatory components in psychiatric disorders. Recent studies carried out over the past 20 years have indicated that immune system function is altered in schizophrenia and mood disorder patients. The major findings of these studies suggest that a proinflammatory state of the cytokine network can lead to psychopathologic symptoms and may be involved in the pathogenesis of mental illnesses such as schizophrenia and bipolar disorder. We presented recent data, which relates the immune activation to present theories on the influence of activated immune cells in altered brain function. We also focused on the role of the environment in immune activation and the role of the microbiome and dietary factors in the cause and potential prevention of psychiatric conditions. Increased understanding of such factors could help in the development of novel treatment strategies and improved clinical management of mental disorders. Possibly, a proinflammatory state may not be a feature of all patients, but may be a predominant feature of a subset of patients. Further research efforts should be aimed at elaborating this possibility. Indeed, such a subset of patients within the larger bipolar or schizophrenia syndromes may benefit from targeted treatment.

Acknowledgments

Studies were supported by EU-FP7-HEALTH-F2-2008-222963 "MOODINFLAME," Hersenstichting 15F07, and ZonMW-TOP 40-00812-98-08018. We thank Harm de Wit, Annemarie Wijkhuijs, and Thomas Hoogenboezem for their excellent technical assistance; Caspar Looman for statistical advice; Wendy Netten for secretarial assistance; and Sandra de Bruin for help with designing the figures.

References

Abe, H., Hidaka, N., Kawagoe, C., Odagiri, K., Watanabe, Y., Ikeda, T., Ishizuka, Y., Hashiguchi, H., Takeda, R., Nishimori, T., and Ishida, Y. (2007). Prenatal psychological stress causes higher emotionality, depression-like behavior, and elevated activity in the hypothalamo-pituitary-adrenal axis. *Neurosci. Res.* **59**(2), 145–151.

Amsterdam, J.D., Maislin, G., and Hooper, M.B. (1996). Suppression of herpes simplex virus infections with oral lithium carbonate—a possible antiviral activity. *Pharmacotherapy* **16**(6), 1070–1075.

Arnold, S.E., *et al.* (1998). Absence of neurodegeneration and neural injury in the cerebral cortex in a sample of elderly patients with schizophrenia. *Arch. Gen. Psychiatry* **55**(3), 225–232.

Austin, M.P., Leader, L.R., and Reilly, N. (2005). Prenatal stress, the hypothalamic-pituitary-adrenal axis, and fetal and infant neurobehaviour. *Early Hum. Dev.* **81**(11), 917–926.

Bessis, A., Bechade, C., Bernard, D., and Roumier, A. (2007). Microglial control of neuronal death and synaptic properties. *Glia* **55**(3), 233–238.

Bosker, F.J., *et al.* (2011). Poor replication of candidate genes for major depressive disorder using genome-wide association data. *Mol. Psychiatry* **16**, 516–532.

Breunis, M.N., Kupka, R.W., Nolen, W.A., *et al.* (2003). High numbers of circulating activated T cells and raised levels of serum IL-2 receptor in bipolar disorder. *Biol. Psychiatry* **53**(2), 157–165.

Brietzke, E., Stertz, L., Fernandes, B., *et al.* (2009). Comparison of cytokine levels in depressed, manic and euthymic patients with bipolar disorder. *J. Affect. Disord.* **116**(3), 214–217.

Brown, A.S., and Derkits, E.J. (2010). Prenatal infection and schizophrenia: a review of epidemiologic and translational studies. *Am. J. Psychiatry* **167**(3), 261–280.

Carruthers, V.B., and Suzuki, Y. (2007). Effects of Toxoplasma gondii infection on the brain. *Schizophr. Bull.* **33**(3), 745–751.

Cazzullo, C.L., Saresella, M., Roda, K., *et al.* (1998). Increased levels of CD8+ and CD4+ 45RA + lymphocytes in schizophrenic patients. *Schizophr. Res.* **31**(1), 49–55.

Chen, S.K., *et al.* (2010). Hematopoietic origin of pathological grooming in Hoxb8 mutant mice. *Cell* **141**(5), 775–785.

Chesnokova, V., and Melmed, S. (2002). Minireview: neuro-immuno-endocrine modulation of the hypothalamic-pituitary-adrenal (HPA) axis by gp130 signaling molecules. *Endocrinology* **143**(5), 1571–1574.

Chiu, C.C., Huang, S.Y., Chen, C.C., and Su, K.P. (2005). Omega-3 fatty acids are more beneficial in the depressive phase than in the manic phase in patients with bipolar I disorder. *J. Clin. Psychiatry* **66** (12), 1613–1614.

Coelho, F.M., Reis, H.J., Nicolato, R., *et al.* (2008). Increased serum levels of inflammatory markers in chronic institutionalized patients with schizophrenia. *Neuroimmunomodulation* **15**(2), 140–144.

Cohen, D., Dekker, J.J., Peen, J., and Gispen-de Wied, C.C. (2006). Prevalence of diabetes mellitus in chronic schizophrenic inpatients in relation to long-term antipsychotic treatment. *Eur. Neuropsychopharmacol.* **16**(3), 187–194.

Conejero-Goldberg, C., Torrey, E.F., and Yolken, R.H. (2003). Herpesviruses and Toxoplasma gondii in orbital frontal cortex of psychiatric patients. *Schizophr. Res.* **60**(1), 65–69.

Costello, E.J., Worthman, C., Erkanli, A., and Angold, A. (2007). Prediction from low birth weight to female adolescent depression: a test of competing hypotheses. *Arch. Gen. Psychiatry* **64**(3), 338–344.

Craddock, R.M., Lockstone, H.E., Rider, D.A., *et al.* (2007). Altered T-cell function in schizophrenia: a cellular model to investigate molecular disease mechanisms. *PLoS One* **2**(1), e692.

Crespo-Facorro, B., Carrasco-Marin, E., Perez-Iglesias, R., *et al.* (2008). Interleukin-12 plasma levels in drug-naive patients with a first episode of psychosis: effects of antipsychotic drugs. *Psychiatry Res.* **158**(2), 206–216.

Dahlman, I., Kaaman, M., Olsson, T., *et al.* (2005). A unique role of monocyte chemoattractant protein 1 among chemokines in adipose tissue of obese subjects. *J. Clin. Endocrinol. Metab.* **90**(10), 5834–5840.

Doorduin, J., *et al.* (2009). [(11)C]-DPA-713 and [(18)F]-DPA-714 as New PET tracers for TSPO: a comparison with [(11)C]-(R)-PK11195 in a rat model of herpes encephalitis. *Mol. Imaging Biol.* **11** (6), 386–398.

Drexhage, R.C., Padmos, R.C., de Wit, H., *et al.* (2008). Patients with schizophrenia show raised serum levels of the pro-inflammatory chemokine CCL2: association with the metabolic syndrome in patients? *Schizophr. Res.* **102**(1–3), 352–355.

Drexhage, R.C., van der Heul-Nieuwenhuijsen, L., Padmos, R.C., van Beveren, N., Cohen, D., Versnel, M.A., Nolen, W.A., and Drexhage, H.A. (2010). Inflammatory gene expression in monocytes of patients with schizophrenia: overlap and difference with bipolar disorder. A study in naturalistically treated patients. *Int. J. Neuropsychopharmacol.* **13**(10), 1369–1381.

Drexhage, R.C., Hoogenboezem, T.A., Cohen, D., Versnel, M.A., Nolen, W.A., van Beveren, N.J., and Drexhage, H.A. (2011a). An activated set point of T-cell and monocyte inflammatory networks in recent-onset schizophrenia patients involves both pro- and anti-inflammatory forces. *Int. J. Neuropsychopharmacol.* **24**, 1–10.

Drexhage, R.C., Hoogenboezem, T.H., Versnel, M.A., Berghout, A., Nolen, W.A., and Drexhage, H. A. (2011b). The activation of monocyte and T cell networks in patients with bipolar disorder. *Brain Behav. Immun.* **25**, 1206–1213.

Ebrinc, S., Top, C., Oncul, O., Basoglu, C., Cavuslu, S., and Cetin, M. (2002). Serum interleukin 1 alpha and interleukin 2 levels in patients with schizophrenia. *J. Int. Med. Res.* **30**(3), 314–317.

Erbagci, A.B., Herken, H., Koyluoglu, O., Yilmaz, N., and Tarakcioglu, M. (2001). Serum IL-1beta, sIL-2R, IL-6, IL-8 and TNF-alpha in schizophrenic patients, relation with symptomatology and responsiveness to risperidone treatment. *Mediators Inflamm.* **10**(3), 109–115.

Falke, E., Han, L.Y., and Arnold, S.E. (2000). Absence of neurodegeneration in the thalamus and caudate of elderly patients with schizophrenia. *Psychiatry Res.* **93**(2), 103–110.

Fallon, B.A., and Nields, J.A. (1994). Lyme disease: a neuropsychiatric illness. *Am. J. Psychiatry* **151**(11), 1571–1583.

Frangou, S., Lewis, M., and McCrone, P. (2006). Efficacy of ethyl-eicosapentaenoic acid in bipolar depression: randomised double-blind placebo-controlled study. *Br. J. Psychiatry* **188**, 46–50.

Frommberger, U.H., Bauer, J., Haselbauer, P., Fraulin, A., Riemann, D., and Berger, M. (1997). Interleukin-6-(IL-6) plasma levels in depression and schizophrenia: comparison between the acute state and after remission. *Eur. Arch. Psychiatry Clin. Neurosci.* **247**(4), 228–233.

Grundy, S.M., Brewer, H.B. Jr., Cleeman, J.I., *et al.* (2004). Definition of metabolic syndrome: report of the national heart, lung, and blood institute/American heart association conference on scientific issues related to definition. *Circulation* **109**(3), 433–438.

Haack, M., Hinze-Selch, D., Fenzel, T., *et al.* (1999). Plasma levels of cytokines and soluble cytokine receptors in psychiatric patients upon hospital admission: effects of confounding factors and diagnosis. *J. Psychiatr. Res.* **33**(5), 407–418.

Hashimoto, M., Watanabe, T., Fujioka, T., Tan, N., Yamashita, H., and Nakamura, S. (2001). Modulating effects of prenatal stress on hyperthermia induced in adult rat offspring by restraint or LPS-induced stress. *Physiol. Behav.* **73**(1–2), 125–132.

Henneberg, A., Riedl, B., Dumke, H.O., and Kornhuber, H.H. (1990). T-lymphocyte subpopulations in schizophrenic patients. *Eur. Arch. Psychiatry Neurol. Sci.* **239**(5), 283–284.

Hillegers, M.H., Burger, H., Wals, M., Reichart, C.G., Verhulst, F.C., Nolen, W.A., and Ormel, J. (2004). Impact of stressful life events, familial loading and their interaction on the onset of mood disorders: study in a high-risk cohort of adolescent offspring of parents with bipolar disorder. *Br. J. Psychiatry* **185**, 97–101.

Hinze-Selch, D. (2002). Infection, treatment and immune response in patients with bipolar disorder versus patients with major depression, schizophrenia or healthy controls. *Bipolar Disord.* **4**(Suppl. 1), 81–83.

Hori, H., Yoshimura, R., Yamada, Y., *et al.* (2007). Effects of olanzapine on plasma levels of catecholamine metabolites, cytokines, and brain-derived neurotrophic factor in schizophrenic patients. *Int. Clin. Psychopharmacol.* **22**(1), 21–27.

Jia, P., *et al.* (2010). Common variants conferring risk of schizophrenia: a pathway analysis of GWAS data. *Schizophr. Res.* **122**(1–3), 38–42.

Jones-Brando, L., Torrey, E.F., and Yolken, R. (2003). Drugs used in the treatment of schizophrenia and bipolar disorder inhibit the replication of Toxoplasma gondii. *Schizophr. Res.* **62**(3), 237–244.

Kamei, N., Tobe, K., Suzuki, R., *et al.* (2006). Overexpression of monocyte chemoattractant protein-1 in adipose tissues causes macrophage recruitment and insulin resistance. *J. Biol. Chem.* **281**(36), 26602–26614.

Kaminska, T., Wysocka, A., Marmurowska-Michalowska, H., Dubas-Slemp, H., and Kandefer-Szerszen, M. (2001). Investigation of serum cytokine levels and cytokine production in whole blood cultures of paranoid schizophrenic patients. *Arch. Immunol. Ther. Exp. (Warsz.)* **49**(6), 439–445.

Kauer-Sant'Anna, M., Kapczinski, F., Andreazza, A.C., *et al.* (2009). Brain-derived neurotrophic factor and inflammatory markers in patients with early- vs. Late-stage bipolar disorder. *Int. J. Neuropsychopharmacol.* **12**(4), 447–458.

Kim, Y.K., Kim, L., and Lee, M.S. (2000). Relationships between interleukins, neurotransmitters and psychopathology in drug-free male schizophrenics. *Schizophr. Res.* **44**(3), 165–175.

Kim, Y.K., Suh, I.B., Kim, H., *et al.* (2002). The plasma levels of interleukin-12 in schizophrenia, major depression, and bipolar mania: effects of psychotropic drugs. *Mol. Psychiatry* **7**(10), 1107–1114.

Kim, Y.K., Myint, A.M., Lee, B.H., *et al.* (2004a). T-helper types 1, 2, and 3 cytokine interactions in symptomatic manic patients. *Psychiatry Res.* **129**(3), 267–272.

Kim, Y.K., Myint, A.M., Lee, B.H., *et al.* (2004b). Th1, Th2 and Th3 cytokine alteration in schizophrenia. *Prog. Neuropsychopharmacol. Biol. Psychiatry* **28**(7), 1129–1134.

Kim, Y.K., Myint, A.M., Verkerk, R., Scharpe, S., Steinbusch, H., and Leonard, B. (2009). Cytokine changes and tryptophan metabolites in medication-naive and medication-free schizophrenic patients. *Neuropsychobiology* **59**(2), 123–129.

Knijff, E.M., *et al.* (2006). A relative resistance of T cells to dexamethasone in bipolar disorder. *Bipolar Disord.* **8**(6), 740–750.

Kronfol, Z., and House, J.D. (1988). Immune function in mania. *Biol. Psychiatry* **24**(3), 341–343.

Kronig, H., Riedel, M., Schwarz, M.J., *et al.* (2005). ICAM G241A polymorphism and soluble ICAM-1 serum levels: evidence for an active immune process in schizophrenia. *Neuroimmunomodulation* **12**(1), 54–59.

Kudoh, A., Sakai, T., Ishihara, H., and Matsuki, A. (2001). Plasma cytokine response to surgical stress in schizophrenic patients. *Clin. Exp. Immunol.* **125**(1), 89–93.

Kudoh, A., Takase, H., Takahira, Y., Katagai, H., and Takazawa, T. (2003). Postoperative confusion in schizophrenic patients is affected by interleukin-6. *J. Clin. Anesth.* **15**(6), 455–462.

Laviola, G., Rea, M., Morley-Fletcher, S., Di Carlo, S., Bacosi, A., De Simone, R., Bertini, M., and Pacifici, R. (2004). Beneficial effects of enriched environment on adolescent rats from stressed pregnancies. *Eur. J. Neurosci.* **20**(6), 1655–1664.

Leaf, A., and Weber, P.C. (1987). A new era for science in nutrition. *Am. J. Clin. Nutr.* **45**(Suppl. 5), 1048–1053.

Leonard, B.E. (2006). HPA and immune axes in stress: involvement of the serotonergic system. *Neuroimmunomodulation* **13**(5–6), 268–276.

Lin, A., Kenis, G., Bignotti, S., *et al.* (1998). The inflammatory response system in treatment-resistant schizophrenia: increased serum interleukin-6. *Schizophr. Res.* **32**(1), 9–15.

Liu, X., *et al.* (2009). Indoleamine 2,3-dioxygenase, an emerging target for anti-cancer therapy. *Curr. Cancer Drug Targets* **9**(8), 938–952.

Maccari, S., and Morley-Fletcher, S. (2007). Effects of prenatal restraint stress on the hypothalamus-pituitary-adrenal axis and related behavioural and neurobiological alterations. *Psychoneuroendocrinology* **32**(Suppl. 1), S10–S15.

Maes, M., Meltzer, H.Y., and Bosmans, E. (1994). Immune-inflammatory markers in schizophrenia: comparison to normal controls and effects of clozapine. *Acta Psychiatr. Scand.* **89**(5), 346–351.

Maes, M., Bocchio Chiavetto, L., Bignotti, S., *et al.* (2000). Effects of atypical antipsychotics on the inflammatory response system in schizophrenic patients resistant to treatment with typical neuroleptics. *Eur. Neuropsychopharmacol.* **10**(2), 119–124.

Maino, K., Gruber, R., Riedel, M., Seitz, N., Schwarz, M., and Muller, N. (2007). T- and B-lymphocytes in patients with schizophrenia in acute psychotic episode and the course of the treatment. *Psychiatry Res.* **152**(2–3), 173–180.

McGowan, P.O., *et al.* (2009). Epigenetic regulation of the glucocorticoid receptor in human brain associates with childhood abuse. *Nat. Neurosci.* **12**(3), 342–348.

Merlot, E., Couret, D., and Otten, W. (2008). Prenatal stress, fetal imprinting and immunity. *Brain Behav. Immun.* **22**(1), 42–51.

Miura, H., *et al.* (2008). A link between stress and depression: shifts in the balance between the kynurenine and serotonin pathways of tryptophan metabolism and the etiology and pathophysiology of depression. *Stress* **11**(3), 198–209.

Monteleone, P., Fabrazzo, M., Tortorella, A., and Maj, M. (1997). Plasma levels of interleukin-6 and tumor necrosis factor alpha in chronic schizophrenia: effects of clozapine treatment. *Psychiatry Res.* **71**(1), 11–17.

Muller, N., Empl, M., Riedel, M., Schwarz, M., and Ackenheil, M. (1997). Neuroleptic treatment increases soluble IL-2 receptors and decreases soluble IL-6 receptors in schizophrenia. *Eur. Arch. Psychiatry Clin. Neurosci.* **247**(6), 308–313.

Muller, N., Schlesinger, B.C., Hadjamu, M., *et al.* (1998). Increased frequency of CD8 positive gamma/delta T-lymphocytes (CD8+ gamma/delta+) in unmedicated schizophrenic patients: relation to impairment of the blood-brain barrier and HLA-DPA*02011. *Schizophr. Res.* **32**(1), 69–71.

Naudin, J., Capo, C., Giusano, B., Mege, J.L., and Azorin, J.M. (1997). A differential role for interleukin-6 and tumor necrosis factor-alpha in schizophrenia? *Schizophr. Res.* **26**(2–3), 227–233.

Nikkila, H.V., Muller, K., Ahokas, A., Miettinen, K., Rimon, R., and Andersson, L.C. (1999). Accumulation of macrophages in the CSF of schizophrenic patients during acute psychotic episodes. *Am. J. Psychiatry* **156**(11), 1725–1729.

O'Brien, S.M., Scully, P., Scott, L.V., and Dinan, T.G. (2006). Cytokine profiles in bipolar affective disorder: focus on acutely ill patients. *J. Affect. Disord.* **90**(2–3), 263–267.

O'Brien, S.M., Scully, P., and Dinan, T.G. (2008). Increased tumor necrosis factor-alpha concentrations with interleukin-4 concentrations in exacerbations of schizophrenia. *Psychiatry Res.* **160**(3), 256–262.

Ortiz-Dominguez, A., Hernandez, M.E., Berlanga, C., *et al.* (2007). Immune variations in bipolar disorder: phasic differences. *Bipolar Disord.* **9**(6), 596–602.

Osher, Y., Bersudsky, Y., and Belmaker, R.H. (2005). Omega-3 eicosapentaenoic acid in bipolar depression: report of a small open-label study. *J. Clin. Psychiatry* **66**(6), 726–729.

Oxenkrug, G.F. (2010). Metabolic syndrome, age-associated neuroendocrine disorders, and dysregulation of tryptophan-kynurenine metabolism. *Ann. N. Y. Acad. Sci.* **1199**, 1–14.

Padgett, D.A., and Glaser, R. (2003). How stress influences the immune response. *Trends Immunol.* **24**(8), 444–448.

Padmos, R.C., Hillegers, M.H., Knijff, E.M., *et al.* (2008a). A discriminating messenger RNA signature for bipolar disorder formed by an aberrant expression of inflammatory genes in monocytes. *Arch. Gen. Psychiatry* **65**(4), 395–407.

Padmos, R.C., Schloot, N.C., Beyan, H., *et al.* (2008b). Distinct monocyte gene-expression profiles in autoimmune diabetes. *Diabetes* **57**(10), 2768–2773.

Padmos, R.C., *et al.* (2009). Genetic and environmental influences on pro-inflammatory monocytes in bipolar disorder: a twin study. *Arch. Gen. Psychiatry* **66**(9), 957–965.

Parker, G., Gibson, N.A., Brotchie, H., Heruc, G., Rees, A.M., and Hadzi-Pavlovic, D. (2006). Omega-3 fatty acids and mood disorders. *Am. J. Psychiatry* **163**(6), 969–978.

Radewicz, K., *et al.* (2000). Increase in HLA-DR immunoreactive microglia in frontal and temporal cortex of chronic schizophrenics. *J. Neuropathol. Exp. Neurol.* **59**(2), 137–150.

Rothermundt, M., Arolt, V., Weitzsch, C., Eckhoff, D., and Kirchner, H. (1998). Immunological dysfunction in schizophrenia: a systematic approach. *Neuropsychobiology* **37**(4), 186–193.

Roumier, A., *et al.* (2008). Prenatal activation of microglia induces delayed impairment of glutamatergic synaptic function. *PLoS One* **3**(7), e2595.

Ruhe, H.G., Mason, N.S., and Schene, A.H. (2007). Mood is indirectly related to serotonin, norepinephrine and dopamine levels in humans: a meta-analysis of monoamine depletion studies. *Mol. Psychiatry* **12**(4), 331–359.

Schwarz, M.J., Riedel, M., Ackenheil, M., and Muller, N. (2000). Decreased levels of soluble intercellular adhesion molecule-1 (sICAM-1) in unmedicated and medicated schizophrenic patients. *Biol. Psychiatry* **47**(1), 29–33.

Sijben, J.W., and Calder, P.C. (2007). Differential immunomodulation with long-chain n-3 PUFA in health and chronic disease. *Proc. Nutr. Soc.* **66**(2), 237–259.

Smith, R.S. (1992). A comprehensive macrophage-T-lymphocyte theory of schizophrenia. *Med. Hypotheses* **39**(3), 248–257.

Smith, R.S., and Maes, M. (1995). The macrophage-T-lymphocyte theory of schizophrenia: additional evidence. *Med. Hypotheses* **45**(2), 135–141.

Song, X.Q., Lv, L.X., Li, W.Q., Hao, Y.H., and Zhao, J.P. (2009). The interaction of nuclear factor-kappa B and cytokines is associated with schizophrenia. *Biol. Psychiatry* **65**(6), 481–488.

Sperner-Unterweger, B., Whitworth, A., Kemmler, G., *et al.* (1999). T-cell subsets in schizophrenia: a comparison between drug-naive first episode patients and chronic schizophrenic patients. *Schizophr. Res.* **38**(1), 61–70.

Stefansson, H., *et al.* (2009). Common variants conferring risk of schizophrenia. *Nature* **460**(7256), 744–747.

Steiner, J., *et al.* (2008). Immunological aspects in the neurobiology of suicide: elevated microglial density in schizophrenia and depression is associated with suicide. *J. Psychiatr. Res.* **42** (2), 151–157.

Stoll, A.L., Severus, W.E., Freeman, M.P., Rueter, S., Zboyan, H.A., Diamond, E., Cress, K.K., and Marangell, L.B. (1999). Omega 3 fatty acids in bipolar disorder: a preliminary double-blind, placebo-controlled trial. *Arch. Gen. Psychiatry* **56**(5), 407–412.

Taieb, O., Baleyte, J.M., Mazet, P., and Fillet, A.M. (2001). Borna disease virus and psychiatry. *Eur. Psychiatry* **16**(1), 3–10.

Teixeira, A.L., Reis, H.J., Nicolato, R., *et al.* (2008). Increased serum levels of CCL11/eotaxin in schizophrenia. *Prog. Neuropsychopharmacol. Biol. Psychiatry* **32**(3), 710–714.

Theodoropoulou, S., Spanakos, G., Baxevanis, C.N., *et al.* (2001). Cytokine serum levels, autologous mixed lymphocyte reaction and surface marker analysis in never medicated and chronically medicated schizophrenic patients. *Schizophr. Res.* **47**(1), 13–25.

Thomas, A.J., *et al.* (2004). Elevation of cell adhesion molecule immunoreactivity in the anterior cingulate cortex in bipolar disorder. *Biol. Psychiatry* **55**(6), 652–655.

Tiemeier, H., van Tuijl, H.R., Hofman, A., Kiliaan, A.J., and Breteler, M.M. (2003). Plasma fatty acid composition and depression are associated in the elderly: the Rotterdam study. *Am. J. Clin. Nutr.* **78** (1), 40–46.

Togo, T., *et al.* (2000). Expression of CD40 in the brain of Alzheimer's disease and other neurological diseases. *Brain Res.* **885**(1), 117–121.

Torres, K.C., Souza, B.R., Miranda, D.M., *et al.* (2009a). The leukocytes expressing DARPP-32 are reduced in patients with schizophrenia and bipolar disorder. *Prog. Neuropsychopharmacol. Biol. Psychiatry* **33**(2), 214–219.

Torres, K.C., Souza, B.R., Miranda, D.M., *et al.* (2009b). Expression of neuronal calcium sensor-1 (NCS-1) is decreased in leukocytes of schizophrenia and bipolar disorder patients. *Prog. Neuropsychopharmacol. Biol. Psychiatry* **33**(2), 229–234.

Torrey, E.F., and Yolken, R.H. (1998). At issue: is household crowding a risk factor for schizophrenia and bipolar disorder? *Schizophr. Bull.* **24**(3), 321–324.

Torrey, E.F., Miller, J., Rawlings, R., and Yolken, R.H. (1997). Seasonality of births in schizophrenia and bipolar disorder: a review of the literature. *Schizophr. Res.* **28**(1), 1–38.

Torrey, E.F., Bartko, J.J., Lun, Z.R., and Yolken, R.H. (2007). Antibodies to Toxoplasma gondii in patients with schizophrenia: a meta-analysis. *Schizophr. Bull.* **33**(3), 729–736.

Trayhurn, P., Bing, C., and Wood, I.S. (2006). Adipose tissue and adipokines—energy regulation from the human perspective. *J. Nutr.* **136**(7 Suppl), 1935S–1939S.

van Berckel, B.N., *et al.* (2008). Microglia activation in recent-onset schizophrenia: a quantitative (R)-[11C]PK11195 positron emission tomography study. *Biol. Psychiatry* **64**(9), 820–822.

Van den Bergh, B.R., Mulder, E.J., Mennes, M., and Glover, V. (2005). Antenatal maternal anxiety and stress and the neurobehavioural development of the fetus and child: links and possible mechanisms. A review. *Neurosci. Biobehav. Rev.* **29**(2), 237–258.

Wang, H., *et al.* (2010). Age- and gender-dependent impairments of neurobehaviors in mice whose mothers were exposed to lipopolysaccharide during pregnancy. *Toxicol. Lett.* **192**(2), 245–251.

Weigelt, K., Carvalho, L.A., Drexhage, R.C., Wijkhuijs, A., Wit, H.D., van Beveren, N.J., Birkenhäger, T.K., Bergink, V., and Drexhage, H.A. (2011). TREM-1 and DAP12 expression in monocytes of patients with severe psychiatric disorders. EGR3, ATF3 and PU.1 as important transcription factors. *Brain Behav. Immun.* **25**, 1162–1169.

Wierzba-Bobrowicz, T., *et al.* (2005). Quantitative analysis of activated microglia, ramified and damage of processes in the frontal and temporal lobes of chronic schizophrenics. *Folia Neuropathol.* **43**(2), 81–89.

Willner, P. (2005). Chronic mild stress (CMS) revisited: consistency and behavioural-neurobiological concordance in the effects of CMS. *Neuropsychobiology* **52**(2), 90–110.

Witvrouw, M., Schmit, J.C., Van Remoortel, B., Daelemans, D., Este, J.A., Vandamme, A.M., Desmyter, J., and De Clercq, E. (1997). Cell type-dependent effect of sodium valproate on human immunodeficiency virus type 1 replication in vitro. *AIDS Res. Hum. Retroviruses* **13**(2), 187–192.

Yolken, R.H., and Torrey, E.F. (1995). Viruses, schizophrenia, and bipolar disorder. *Clin. Microbiol. Rev.* **8**(1), 131–145.

Zhang, X.Y., Zhou, D.F., Zhang, P.Y., Wu, G.Y., Cao, L.Y., and Shen, Y.C. (2002). Elevated interleukin-2, interleukin-6 and interleukin-8 serum levels in neuroleptic-free schizophrenia: association with psychopathology. *Schizophr. Res.* **57**(2–3), 247–258.

Zhang, X.Y., Zhou, D.F., Cao, L.Y., Zhang, P.Y., Wu, G.Y., and Shen, Y.C. (2004). Changes in serum interleukin-2, -6, and -8 levels before and during treatment with risperidone and haloperidol: relationship to outcome in schizophrenia. *J. Clin. Psychiatry* **65**(7), 940–947.

Zhang, X.Y., Zhou, D.F., Cao, L.Y., Wu, G.Y., and Shen, Y.C. (2005). Cortisol and cytokines in chronic and treatment-resistant patients with schizophrenia: association with psychopathology and response to antipsychotics. *Neuropsychopharmacology* **30**(8), 1532–1538.

Zhang, X.Y., Zhou, D.F., Qi, L.Y., *et al.* (2009). Superoxide dismutase and cytokines in chronic patients with schizophrenia: association with psychopathology and response to antipsychotics. *Psychopharmacology (Berl.)* **204**(1), 177–184.

Zorrilla, E.P., Cannon, T.D., Gur, R.E., and Kessler, J. (1996). Leukocytes and organ-nonspecific autoantibodies in schizophrenics and their siblings: markers of vulnerability or disease? *Biol. Psychiatry* **40**(9), 825–833.

Zorrilla, E.P., Luborsky, L., McKay, J.R., *et al.* (2001). The relationship of depression and stressors to immunological assays: a meta-analytic review. *Brain Behav. Immun.* **15**(3), 199–226.

BEHAVIORAL AND MOLECULAR BIOMARKERS IN TRANSLATIONAL ANIMAL MODELS FOR NEUROPSYCHIATRIC DISORDERS

Zoltán Sarnyai[1], Murtada Alsaif[2], Sabine Bahn[2,3], Agnes Ernst[2], Paul C. Guest[2], Eva Hradetzky[2], Wolfgang Kluge[2], Viktoria Stelzhammer[2] and Hendrik Wesseling[2]

[1]Department of Pharmacology, University of Cambridge, Cambridge, United Kingdom
[2]Department of Chemical Engineering and Biotechnology, University of Cambridge, Cambridge, United Kingdom
[3]Department of Neuroscience, Erasmus Medical Centre, Rotterdam, The Netherlands

Abstract

Modeling neuropsychiatric disorders in animals poses a significant challenge due to the subjective nature of diverse often overlapping symptoms, lack of objective biomarkers and diagnostics, and the rudimentary understanding of the pathophysiology. Successful translational research requires animal models that can inform about disease mechanisms and therapeutic targets. Here, we review behavioral and neurobiological findings from selected animal models, based on presumed etiology and risk factors, for schizophrenia, bipolar disorder, and major depressive disorder. We focus on the use of appropriate statistical tools and newly developed Research Domain Criteria (RDoC) to link biomarkers from animal models with the human disease. We argue that this approach will lead to development of only the most robust animal models for specific psychiatric

INTERNATIONAL REVIEW OF NEUROBIOLOGY, VOL. 101
DOI: 10.1016/B978-0-12-387718-5.00008-0

disorders and may ultimately lead to better understanding of the pathophysiology and identification of novel biomarkers and therapeutic targets.

I. Introduction: The Problems with Animal Models

Neuropsychiatric disorders, including schizophrenia, major depressive disorder (MDD), and bipolar disorder (BD), are among the leading causes of disability throughout the world and account for about one-third of "years lost due to disability" (YLD) among people older than 14 years (WHO, 2004). Despite the significant public health cost and personal suffering caused by psychiatric disorders, there has been frustratingly little progress made in their mechanistic understanding and in development of novel pharmacotheraupeutic drugs. Among the reasons for this lack of progress are the diverse and mostly ill-defined etiology and complexity of these disorders. However, mechanistic understanding of a human disease and the discovery of therapeutic agents cannot be achieved without good animal models. Unfortunately, most of the animal models used currently are limited in their ability to capture etiology and neurobiological mechanisms, and to predict treatment efficacy in human neuropsychological disorders. In essence, animal models must meet certain validation criteria in order to capture dimensions such as etiology, pathophysiology, symptoms, and treatment of a human disorder. These include construct, face, and predictive validity which refers to the similarities between the methods by which the model was constructed and the disease itself (Nestler and Hyman, 2010). For example, selective loss of pancreatic beta cells leading to defective insulin production can provide construct validity for a model of Type-1 diabetes. Similarly, if a disease-causing mutation is known, selective inhibition of that gene can lead to a model with good construct validity. Etiological factors, whether genetic or environmental, are still largely unknown and hotly debated in psychiatry. Therefore, it is unlikely that animal models with high construct validity can be developed on the basis of our present understanding. Face validity signifies how an animal model recreates key features of a human disease. Due to the lack of well-established and widely replicated biomarkers for psychiatric disorders, animal models can only aim to replicate certain anatomical, neurochemical, or behavioral features of the human disease, which will inherently limit their usefulness. Predictive validity refers to the utility of the model to predict effects of drugs, assuming that any similarities in effect are based on shared mechanism of action. This is less than straightforward in psychiatry as most of the drugs used therapeutically were identified by studying the mechanisms of action of drugs discovered serendipitously.

Construction of valid animal models for neuropsychiatric disorders cannot be achieved without an effective two-way translational approach. Learning more about the etiology and neurobiological mechanisms will guide the development of models with better construct validity. Identification of novel drug targets using human genetic data, along with validated biochemical, imaging, and behavioral outcome measures (efficacy biomarkers), will improve predictive validity. Similarly, once we have models with good construct validity, we will be able to identify new disease mechanisms and potential novel drug targets. Such improved animal models will also be useful to make better predictions for biomarker development to diagnose and monitor the human condition.

This chapter reviews some currently used animal models for schizophrenia, MDD, and BD from the point of view of translationally useful behavioral and molecular measures. The aim is not to provide a comprehensive review of all animal models, but rather to highlight key examples developed on the basis of different etiological theories. We intend to provide an evaluation framework that can be used in conjunction with behavioral, structural, and molecular biomarkers to facilitate translational development of valid animal models for human neuropsychiatric disorders. To achieve this, we based our presentation of the behavioral and molecular features on the framework of the recently initiated Research Domain Criteria (RDoC; Insel *et al.*, 2010) to allow better translation between animal models and the human disorders. Ultimately, overlapping matrices from animal models and human disorders will highlight behavioral, neuropathological, and molecular features that are common between the human disease and the models, leading to development of better animal models, a deeper understanding of the human condition, and the development of much-needed novel therapeutics.

II. Toward Etiological Models in Neuropsychiatry

There is a general agreement that psychiatric disorders, such as schizophrenia, MDD, and BD, are multifactorial with interacting genetic and environmental risk factors involved in the etiology (Nestler and Hyman, 2010). It has also been suggested on the basis of epidemiological studies and neurobiological findings that pre- and postnatal developmental events can contribute to the pathophysiology of mental disorders (Nestler and Hyman, 2010). Disrupted network connectivity due to abnormal functioning of neurotransmitter systems such as dopamine (DA) and glutamate may serve as a final common pathway for several psychiatric disorders (Meltzer and Stahl, 1976; Coyle, 2006a). Therefore, we review animal models that have been developed to replicate the diverse etiological factors of neuropsychiatric disorders, including genetic susceptibility, early developmental insults, environmental stressors in later life, and pharmacological modification of neurotransmitter systems. The behavioral and neurobiological phenotypes emerging from these models are viewed according to the framework proposed by the RDoC, rather than aligning them with traditional psychiatric diagnoses. The RDoC was introduced recently by the US National Institute of Mental Health to create a framework for research on pathophysiology which will ultimately inform on future disease classification schemes and diagnostic and treatment outcomes (Insel *et al.*, 2010). The main RDoC constructs, such as negative/positive effects, cognition, social processes, and arousal, have been modified to accommodate findings for the animal studies. We have also taken into consideration the different units of analysis, such as genes, molecules, circuits, and behavior.

A. Genetic Susceptibility: The Power of the Mutant Mouse

In the past 10 years, there have been massive advances in the use of genetically modified mice to study etiological mechanisms involved in psychiatric disorders. This progress has been driven by rapid developments in molecular biology techniques and has led to a better understanding of the behaviorally and neurobiologically relevant functions of genes identified through human genetic studies as risk factors. Transgenic approaches focusing on common polymorphisms, rare structural mutation, and neurotransmission hypotheses are presented in this section (Table I).

1. *Common Polymorphisms*

Several mutant mouse models for schizophrenia susceptibility genes have been created and phenotypically characterized including *NRG1*, *DISC1*, and *PRODH* (Allen *et al.*, 2008). Numerous genome-wide scan studies have identified neuregulin 1 (*NRG1*) as one of the most promising candidate genes for schizophrenia (Stefansson *et al.*, 2002). *NRG1* contributes in diverse cellular processes including neurodevelopment and regulation of synaptic plasticity through *N*-methyl-D-asparate (NMDA) receptors and glutamatergic signaling. Mice lacking any one of the several isoforms of Nrg1 show a variety of behavioral abnormalities in positive and negative effect domains and in response to novel environment or social novelty (Table I). Importantly, some of these behavioral abnormalities can be partially reversed by the atypical antipsychotic clozapine (Wang *et al.*, 2008).

Initially, the *DISC1* (*d*isrupted-*in*-*sc*hizophrenia) gene was discovered through its segregation with mental illness in a Scottish family, and this finding was replicated in a variety of populations worldwide (Hwu *et al.*, 2003, St Clair *et al.*, 1990). The functions of *DISC1* remains elusive, but a role in neurite outgrowth, cell migration, and cell signaling is proposed. Several *DISC1* mutants have been created and a natural mutation was discovered in the commonly employed 129S6 Sv/Ev strain, which exhibits a deficit in working memory (Koike *et al.*, 2006).

Table I

BEHAVIORAL PATTERNS IN ANIMAL MODELS FOR NEUROPSYCHIATRIC DISORDERS.

Animal models			Domains/constructs												
			Negative effect				Positive effect			Cognition			Social processes		Arousal
		Disease	Fear		Stress	Anhedonia	Reward seeking		Reward learning	Attention and Gating	Memory		Social interaction	Aggression	Arousal and sleep
			Anxiety	Fear cond.	HPA axis		Locomotor act.	Drug resp.			Working	Declarative			
Genetic susceptibility															
Polymorphisms															
NRG1 mutants	Nrg1(ΔEGF)$^{+/-}$	SCZ	↔ (1)	↑/↓ (2)			↑/↓ (2, 3)	↑ (1)		↔ (2)		↔ (2)	↓ (2)		
	Nrg1(ΔTM)$^{+/-}$	SCZ	↔ (4)	↓ (5)				↔ (6)		↓ (4)	↔ (5)	↔ (5)	↓ (7)	↑ (7)	
	Nrg1(ΔIG)$^{+/-}$	SCZ					↔ (8)			↓ (8)					
	Nrg1 OE	SCZ	↔ (9)				↓ (9)	↔ (9)							
DISC1 mutants	mDisc1 trunc	SCZ						↔ (10)		↔ (10)	↓ (10)				
	31L Disk	SCZ	↔ (11)					↔ (11)		↔ (11)	↓ (11)	↔ (11)	↓ (11)		
	100P Disc1	SCZ	↔ (11)				↑ (11)				↓ (11)	↔ (11)	↔ (11)		
	hDISC (DN)	SCZ	↔ (12)			↑ (14)	↑ (12)	↓ (13)	↓ (13)	↓ (14)		↓ (12)	↓ (12)	↑ (12)	
PRODH mutants	PRODH KO	SCZ	↔ (15)	↓ (15)				↓ (15)		↓ (15)	↔ (15)				
Grik2 mutant	GluR6 KO	BD	↓ (16)			↓ (16)	↑ (16)	↑ (16)	↑ (16)		↔ (16)	↔ (16)	↑ (16)	↑ (16)	
Rare structural mutations															
22q11 mutations (VCFS/ DiGeorge sy.)	Del2Aam	SCZ	↔ (17)	↓ (17)				↑ (17)		↓ (17)	↓ (17)				
	Del1Rak	SCZ		↔ (18)			↔ (18)	↔ (18)		↓ (18)					
	DelAwb	SCZ	↔ (19)	↔ (19)			↓ (19)	↔ (19)		↑ (19)					
	Df1	SCZ		↓ (20)						↓ (20)					

(Continued)

Table I (*Continued*)

Animal models			Domains/constructs												
			Negative effect				Positive effect			Cognition			Social processes		Arousal
		Disease	Fear		Stress	Anhedonia	Reward seeking		Reward learning	Attention and Gating	Memory		Social interaction	Aggression	Arousal and sleep
			Anxiety	Fear cond.	HPA axis		Locomotor act.	Drug resp.			Working	Declarative			
Neurotrasmitter-signaling hypotheses															
Dopamine	DAT KO	SCZ	↑ (21)		↔ (22)	↓ (23)	↑ (24)	↓ (24)	↓ (25)	↓ (26)	↓ (27)	↓ (28)	↔/↑ (29, 30)	↔/↑ (29, 30)	↓ Theta activity (31)
Glutamate	NR1-KD	SCZ	↓ (32)				↑ (33)	↓ (34)	↓ (35)	↓ (36, 37)	↓ (38)	↓ (38)	↓ (33)		↓ Theta activity (38)
GSK3β signaling	GSK3β[S9A]-OE	BD			↔ (39)		↑ (39)								
	clock mutant	BD	↓ (40)			↓ (40)	↑ (40)	↑ (40, 41)	↑ (40)						↓ Sleep (40)
Developmental insults															
Maternal immune activation		SCZ	↑ (42)		↔ (42)		↓ (43)	↑ (44)		↓ (42)	↓ (45)	↓ (45)	↓ (43)		
Maternal protein malnutrition		SCZ	↓ (46)					↑ (47)		↓ (48)	↓ (49)		↓ (50)		
Interruption of neurogenesis by MAM		SCZ			↑ (51)			↑ (52)		↓ (53)	↓ (54)		↓ (55)		
Neonatal ventral hippocampal lesion		SCZ	↔ (56)		↑ (56)	↓ (57)	↑ (56)	↑ (56)	↓ (57)	↓ (58)	↓ (59)	↓ (59)	↓ (60)		↑ sw EEG activity (61)
Stress and impaired adaptation															
Social isolation/isolation rearing		MDD/SCZ	↑ (62)	↓ (63)	↔/↑ (62)	↓/↔ (64, 65)	↑ (66)	↑ (66)	↑ (65)	↓ (66)		↔/↓ (62)		↑ (67)	↑ (66)

Social defeat		MDD	↑ (68)	↑ (69)	↑ (68)	↑ (70)	↓ (69)	↑ (71)	↓ (70)	↓ (69)	↔ (69)	↓ (68)		↑ sw EEG activity (72)
Chronic variable stress		MDD	↑/↔ (73, 74)	↑ (75)	↑ (75)	↑ (76)	↑ (76)	↑ (77)		↓ (78, 79)	↓ (78, 79)			
Pharmacological modifications														
Glutamate hypothesis	Phencyclidine—acute	SCZ	↓ (80)		↑ (81)	↑ (82)	↑ (83)		↓ (84)	↓ (85)	↓ (86)	↓ (87)	↑/↔ (82)	↓ (82)
	Phencyclidine—chronic	SCZ			↑ (81)	↔ (82)	↑ (88)		↓ (89)	↓ (90)	↓ (82)	↓ (91)	↑ (91)	
	Ketamine	SCZ			↑ (81)		↑ (92)		↓ (93)	↓ (94)		↓ (95)		
	MK801	SCZ			↑ (96)		↑ (97)		↓ (98)	↓ (99)		↓ (91)		
Dopamine hypothesis - Amphetamine		SCZ	↑ (100)		↑ (101)		↑ (100)		↓ (100)					

BD, bipolar disorder; DISC1, *d*isrupted-*in*-*sc*hizophrenia 1; GSK-3β, glycogen synthase kinase β; MAM, methylazoxymethanol; MDD, major depressive disorder; NREM, non-rapid eye movement sleep; NRG1, neuregulin 1; PRODH, proline dehydrogenase; SCZ, schizophrenia; sw, slow wave; VCFS, velo-cardio-facial syndrome.

(1) Duffy *et al.* (2008), (2) Ehrlichman *et al.* (2009), (3) Gerlai *et al.* (2000), (4) O'Tuathaigh *et al.* (2007), (5) Duffy *et al.* (2010), (6) van den Buuse *et al.* (2009), (7) O'Tuathaigh *et al.* (2008), (8) Rimer *et al.* (2005), (9) Deakin *et al.* (2009), (10) Koike *et al.* (2006), (11) Clapcote *et al.* (2007), (12) Pletnikov *et al.* (2008), (13) Pogorelov *et al.* (2011), (14) Hikida *et al.* (2007), (15) Gogos *et al.* (1999), (16) Shaltiel *et al.* (2008), (17) Stark *et al.* (2008), (18) Long *et al.* (2006), (19) Kimber *et al.* (1999), (20) Paylor *et al.* (2001), (21) Pogorelov *et al.* (2005), (22) Kavelaars *et al.* (2005), (23) Perona *et al.* (2008), (24) Giros *et al.* (1996), (25) Hall *et al.* (2009), (26) Ralph *et al.* (2001), (27) Li *et al.* (2009), (28) Gainetdinov *et al.* (1999), (29) Spielewoy *et al.* (2000), (30) Rodriguiz *et al.* (2004), (31) Dzirasa *et al.* (2009), (32) Halene *et al.* (2009), (33) Mohn *et al.* (1999), (34) Ramsey *et al.* (2008), (35) Wang *et al.* (2010), (36) Moy *et al.* (2006), (37) Bickel *et al.* (2008), (38) Korotkova *et al.* (2010), (39) Prickaerts *et al.* (2006), (40) Roybal *et al.* (2007), (41) McClung *et al.* (2005), (42) Romero *et al.* (2010), (43) Shi *et al.* (2003), (44) Fortier *et al.* (2004), (45) Bitanihirwe *et al.* (2010), (46) Almeida *et al.* (1996a), (47) Palmer *et al.* (2008), (48) Palmer *et al.* (2004), (49) Tonkiss and Galler (1990), (50) Almeida *et al.* (1996b), (51) Hradetzky *et al.* (submitted for publication), (52) Flagstad *et al.* (2004), (53) Moore *et al.* (2006), (54) Le Pen *et al.* (2006), (55) Flagstad *et al.* (2005), (56) Lipska *et al.* (1993), (57) Le Pen and Moreau (2002), (58) Lipska *et al.* (1995a,b), (59) Chambers *et al.* (1996), (60) Sams-Dodd *et al.* (1997), (61) Ahnaou *et al.* (2007), (62) Fone and Porkess (2008), (63) Gresack *et al.* (2010), (64) Brenes and Fornaguera (2008), (65) Lodge and Lawrence (2003), (66) Heidbreder *et al.* (2000), (67) Toth *et al.* (2011), (68) Watt *et al.* (2009), (69) Yu *et al.* (2011), (70) Rygula *et al.* (2005), (71) Kabbaj *et al.* (2001), (72) Meerlo *et al.* (2001), (73) Mineur *et al.* (2006), (74) Dagyte *et al.* (2011), (75) McGuire *et al.* (2010), (76) Willner (1997), (77) Lepsch *et al.* (2005), (78) Henningsen *et al.* (2009), (79) Song *et al.* (2006), (80) Wiley *et al.* (1992), (81) Pechnick *et al.* (1989), (82) Javitt (2007), (83) Sturgeon *et al.* (1979), (84) Geyer *et al.* (2001), (85) Nabeshima *et al.* (1985), (86) Noda *et al.* (2001), (87) Sams-Dodd (1995), (88) Yamaguchi *et al.* (1987), (89) Egerton *et al.* (2008), (90) Jentsch *et al.* (1997), (91) Seillier and Giuffrida (2009), (92) Miyamoto *et al.* (2000), (93) Becker *et al.* (2003), (94) Enomoto and Floresco (2009), (95) Becker and Grecksch (2004), (96) Baumann *et al.* (2000), (97) Zuo *et al.* (2006), (98) Uehara *et al.* (2010), (99) Paule (1994), (100) Tenn *et al.* (2005), (101) Sarnyai *et al.* (2001).

Alterations in the organization of newly formed and mature neurons and deficits in short-term plasticity may contribute to cognitive impairment. Behavioral studies suggest that *DISC1* transgenic mice produce a wide range of subtle schizophrenia-like phenotypes. Variants created by an N-ethyl-N-nitrosourea (ENU) chemical mutagenesis program showed either depression-like behavior with deficits in the forced swim test (Q31L exon) or schizophrenia-like behavior with deficits in prepulse inhibition and latent inhibition (L100P exon; Clapcote *et al.*, 2007). The Q31L-associated deficits could be ameliorated by the antidepressant bupropion, and the L100P deficits improved after antipsychotic treatment (Clapcote *et al.*, 2007).

Chromosome 6q contains the regulatory region for the human *Grik2* gene, which encodes the glutamate receptor 6 (GluR6). A specific haplotype of this region was found to be associated with BD (McQueen *et al.*, 2005). GluR6 mRNA levels were lower in the entorhinal cortex of BD patients (Beneyto *et al.*, 2007). GluR6 knockout (KO) mice showed increased exploratory behavior and hyperactivity, were more responsive to pharmacological stimulants, were more aggressive to intruders, were more dominant, and displayed hypersexual behavior. In addition, GluR6 KO mice were less anxious and took more risks and were also less immobile in the forced swim test, suggesting increased goal directed behavior. Passive avoidance learning was normal in GluR6 KO mice (Shaltiel *et al.*, 2008). Chronic lithium chloride treatment reversed the hyperactivity, aggression, and exploratory behavior in the GluR6 KO mice. These findings map quite well into the behavioral syndrome associated with BD. It should be noted that *GRIK2* abnormalities have also been linked to autism (Strutz-Seebohm *et al.*, 2006) and obsessive compulsive disorder (Sampaio *et al.*, 2010).

2. *Rare Structural Mutations*

A complementary approach to look for psychiatric disorders susceptibility genes is to study cosegregating chromosomal abnormalities, such as the microdeletion at 22q11.2 (22g11DS). This deletion causes the velocardiofacial (VCFS)/DiGeorge syndrome, which presents with congenital abnormalities affecting several tissues and organs. About 25% of the 22q11DS patients develop schizophrenia and schizoaffective disorder (Bassett *et al.*, 2005). Moreover, 22q11DS is found in 2% of patients with schizophrenia (Karayiorgou *et al.*, 1995) and in 6% of cases of childhood onset schizophrenia (Usiskin *et al.*, 1999). Importantly, genes within this region such as catechol *O*-methyl transferase (*COMT*), proline dehydrogenase (*PRODH*), zinc finger, DHHC-type containing 8 (*ZDHHC8*) and guanine nucleotide-binding protein (G protein), beta polypeptide 1-like (*GNB1L*) have been independently associated with schizophrenia. 22q11DS has been either completely or partially recreated in several mouse models (Table I). Most of these models display decreased density of dendritic spines and decreased dendritic complexity of CA1 pyramidal neurons as well as disturbance in prepulse inhibition, fear conditioning, and working memory. These models offer an interesting tool to study cognitive deficits as 22q11DS patients exhibit cognitive deficits similar to those observed in schizophrenia such as effects on attention processing and verbal working memory (Woodin *et al.*, 2001). The 22q11.2 mouse model further offers the opportunity to understand the biological basis for the increased psychosis risk associated with this genetic lesion. In addition, models of the 22q11DS are likely to capture the interactions among the affected genes (Karayiorgou *et al.*, 2010).

3. *Hypothesis-Driven Models*

The mechanism of action of drugs that provide symptomatic relief in psychiatric disorders has played a major role in the development of hypotheses to understand disease mechanisms. First generation antipsychotics act predominantly as DA receptor antagonists. Pharmacological stimulation of DA neurotransmission with indirect agonists, such as the psychostimulant amphetamine, induces psychosis-like states in susceptible individuals. These initial findings have led to formulation of the DA hypothesis of schizophrenia, which proposes that hyperactive DA neurotransmission in the mesolimbic system may underlie the psychotic features of the disease (Meltzer and Stahl, 1976). More recently, similar psychosis-inducing effects via inhibition of glutamatergic transmission, using NMDA glutamate receptor antagonists ketamine and PCP, have given rise to the glutamate hypothesis. This implicates hypoactive glutamatergic neurotransmission in development of the psychotic and cognitive symptoms of schizophrenia (Coyle, 2006a). The discovery of the effects of lithium on a key signaling molecule downstream of DA receptors, glycogen synthase kinase 3 beta (GSK3β), has led to novel hypotheses on the pathogenesis of BD (Ikonomov and Manji, 1999).

Mice lacking the DA transporter (DAT) are unable to reuptake DA released from synaptic terminals, which leads to elevated DA in the synapse. DAT-KO mice are hyperactive in novel environments, show excessive stereotypic behavior, and exhibit sleep dysregulation and dwarfism. Therefore, these mice replicate some of the features of the amphetamine model of schizophrenia (Table I). The hyperlocomotion effect can be reversed by the DA receptor antagonists haloperidol and clozapine (Spielewoy *et al.*, 2000). Based on the glutamate receptor hypofunction hypothesis of schizophrenia, a mutant with 90% reduction in NMDA receptor 1 expression was created. These NMDA receptor hypomorph mice show motor activity abnormalities and deficits in social and sexual behavior (Table I). These abnormalities can be normalized with haloperidol and clozapine, suggesting an effect on DA and glutamate systems (Mohn *et al.*, 1999).

Lithium (Li) is a frontline drug for BD and it is thought to act through GSK3β inhibition (Marmol, 2008). Therefore, Prickaerts *et al.* investigated the GSK3β [S9A]-overexpressing heterozygous mice as a model of mania. These mice displayed a number of BD-like features in the open field test, forced swim test, and acoustic startle response (Prickaerts *et al.*, 2006). In addition, the wet weight of the

GSK3β[S9A] mice was 15% lower compared to the wild type in spite of no difference in central nervous system proliferation rates as measured by [^{3}H] thymidine DNA incorporation (Prickaerts *et al.*, 2006). Adrenocorticotrophic hormone (ACTH) and corticosterone plasma levels after stress were normal in these mice (Prickaerts *et al.*, 2006), similar to the normal cortisol levels found in BD patients (O'Brien *et al.*, 2006). However, hippocampal brain-derived neurotrophic factor (BDNF) levels were significantly higher in GSK3β mice (Prickaerts *et al.*, 2006), unlike in *postmortem* brain tissue from human BD subjects where this protein was found to be lower (Grande *et al.*, 2010).

Disrupted circadian rhythm caused by or leading to less sleep often marks the beginning of a manic attack (Salvadore *et al.*, 2010). The CLOCK protein is integral to the regulation of the circadian rhythm and is regulated by GSK3β. Further, *clock* polymorphisms have been implicated in recurrence of BD episodes (Benedetti *et al.*, 2003). To this effect, *clock* mutant mice were studied and showed mania-like symptoms. These mice are hyperactive (Roybal *et al.*, 2007) and more sensitive to stimulants in the intracranial self-stimulation model (Roybal *et al.*, 2007) and to cocaine (McClung *et al.*, 2005); they show a sucrose preference and decreased anxiety, and require less sleep (Roybal *et al.*, 2007; Easton *et al.*, 2003). Chronic Li treatment reverses their hyperactivity and decreased anxiety-like behavior (Roybal *et al.*, 2007). The ventral tegmental area (VTA) is a brain reward region, and DA activity was found to be heightened in the VTA of *clock* mutant mice (McClung *et al.*, 2005). Hippocampal acetylcholine (ACh) levels are lowered in mutant *clock* mice (Sei *et al.*, 2003). Interestingly, red blood cell choline levels were lower in BD patients (Müller-Oerlinghausen *et al.*, 2002). Further, donepezil, an acetylcholinesterase inhibitor which increases synaptic Ach, has had positive effects in a small cohort of treatment-resistant BD patients (Burt *et al.*, 1999). *Clock* mutant mice express disrupted hepatic levels of Bmal1, aldolase, arginase, and catalase (Reddy *et al.*, 2006) as found in BD subjects (Kovanen *et al.*, 2010; Yanik *et al.*, 2004).

B. Developmental Insults: Setting the Stage for Life

Human *postmortem*, imaging, and epidemiological studies have provided evidence for the neurodevelopmental character of schizophrenia (Harrison, 1999). Animal models in which certain aspects of early intrauterine development is disrupted have been developed, including maternal immune activation, protein malnutrition, interruption of neurogenesis by methylazoxymethanol (MAM), and neonatal ibotenic acid lesion of ventral hippocampus (Meyer and Feldon, 2010; Lipska *et al.*, 1993; Moore *et al.*, 2006). A summary of important disease-relevant behavioral, morphological, and neurochemical abnormalities created by these models is shown in Tables I and II.

All four models reveal deficits assumed to correlate with positive, negative, and cognitive symptoms of schizophrenia and map to the major RDoC domains

Table II

Molecular and Structural Alterations in Animal Models for Neuropsychiatric Disorders.

Animal models		Disease	Neurobiological alterations		
			Transmitters/hormones	Cells	Molecules
Genetic susceptibility					
Polymorphisms					
NRG1 mutants	Nrg1(ΔEGF)$^{+/-}$	SCZ			
	Nrg1(ΔTM)$^{+/-}$	SCZ			↓ NMDA receptors; hyperphosphorylated NR2b (1, 2)
	Nrg1(ΔIG)$^{+/-}$	SCZ			
	Nrg1 OE	SCZ		Hypermyelination of small diameter axons (3)	
DISC1 mutants	mDisc1 trunc	SCZ		Abnormal DG neurons (4)	
	31L Disk	SCZ		↓ Brain volume (5)	↓ PDE4B binding and activity (5)
	100P Disc1	SCZ		↓ Brain volume (5)	↓ PDE4B binding (5)
	hDISC (DN)	SCZ		↑ Brain ventricles, ↓dendritic complexity (6)	↓ PV in CTX, ↓ DISC1, LIS1, SNAP25 (6)
PRODH mutants	PRODH KO	SCZ	↓ Glu, Asp, GABA in FC (7)		
Grik2 mutant	GluR6 KO	BD			↓ GluR5 and KA2 in HPC (8)
Rare structural mutations					
22q11 mutations (VCFS/DiGeorge sy.)	Del2Aam	SCZ		↓ Dendritic complexity in HPC (9)	Altered ox.phos. gene expression in FC/HPC (9)
				Disturbed dendritic spines, ↓ Glu synapses (10)	Altered miRNA biogenesis (10)
	Del1Rak	SCZ		Compromised neurogenesis in CTX (11)	
				Disrupted basal progenitor proliferation (11)	
	DelAwb	SCZ			
	Df1	SCZ			

(*Continued*)

Table II (*Continued*)

Animal models		Disease	Neurobiological alterations		
			Transmitters/hormones	Cells	Molecules
Neurotrasmitter-signaling hypotheses					
Dopamine	DAT KO	SCZ	↑ DA tone, ↓ PRL (12, 13)	Anterior pituitary hypoplasia (12, 13)	↓ Postsynaptic DRD1/DRD2 in STR (12, 13) ↓ ProENK-A and ↑ DYN mRNA in STR (12, 13)
Glutamate	NR1-KD	SCZ	↔ DA content in STR (14)		
GSK3β signaling	GSK3β[S9A]-OE	BD	↔ ACTH/CORT (15)	↓ Brain weight, ↔ neurogenesis in HPC (15)	↑ BDNF mRNA in HPC (15)
	clock mutant	BD	↑ DA activity in the VTA, ↓ ACh in HPC (16, 17)		Disrupted Bmal1, aldolase, arginase and catalase in liver (18)
Developmental insults					
Maternal immune activation		SCZ		↑ Brain and ventricle volume, ↓ HPC volume (19, 20) Abnormal layering, neuron morphology (19) and synaptic markers in FC and HPC (19, 20)	Altered DA, 5-HT, Glu and GABA markers in the brain (19, 21)
Maternal protein malnutrition		SCZ		↓ Brain and HPC volume (22) ↓ Cell number in HPC (22)	Altered DA, 5-HT, Glu and GABA markers in the brain (23, 24)
Interruption of neurogenesis by MAM		SCZ		↓ Brain, FC and HPC volume (25) Abnormal layering and ↓ PV interneurons in CTX and HPC (27, 28)	Altered DA, Glu and GABA markers in the brain (26, 27)

Neonatal ventral hippocampal lesion		SCZ	Altered DA and Glu release in FC (29, 30)		Altered GABAergic markers in FC (31)
Stress and impaired adaptation					
Social isolation/ isolation rearing		MDD/ SCZ	↑ CORT and TNF-α in plasma (32, 33) ↑ DA in ACB and STR, ↑ NA in HPC and CTX (37)	↓ Chandelier neurons in PFC (34) IL-1β in HPC and CTX (38)	↑ CRF-R1 mRNA in DR, ↓ BDNF mRNA in HPC (35, 36)
Social defeat		MDD	↑ CORT in serum and IL-1β in plasma (39, 40) ↓ DA in FCTX, ↑ 5-HT and NA in HPC (42)		↑ CRF mRNA in HPC, ↓ BDNF mRNA in HPC, FC and AMY (40, 41) ↑ TNF-α in splenic dendritic cells (43)
Chronic variable stress		MDD	↓ DA and 5-HT in FC, HPC and STR (44) ↑ CORT, IL-1β and TNF-α in plasma (45)		↑ CRF mRNA in PVN, ↑ TNF-α and IL-1β in HYPO (45, 46) ↔ BDNF mRNA in HPC and AMY (47)
Pharmacological modifications					
Glutamate hypothesis	Phencyclidine—acute	SCZ	↑ GABA enzymes in the brain (48) Impaired Glu metabolism (51)	Impaired cerebral glucose utilization (49)	↑ 5-HT receptors in CTX (50)
	Phencyclidine—chronic	SCZ	↑ 5-HT and 5-HIAA in STR (48) ↓ DA utilization in FC (53)		↓ LAADC mRNA expression (52)
	Ketamine	SCZ	↓ DA transmission in FC (54)	Impaired cerebral glucose utilization (55) ↑ Neurogenesis, ↓ PV interneurons (57, 58)	↑ D2 binding in HPC, ↓ Glu binding in FC (56)
	MK801	SCZ	↑ Glu in FCTX (59) ↑ Monoamines in STR (60)	Altered 2-DG uptake (55)	
Dopamine hypothesis—amphetamine		SCZ		↑ Cerebral glucose utilization (61) ↑ Oxidative stress (64)	↑ D2 dimerization, ↑ CaMKIIβ mRNA in STR (62, 63)

BD, bipolar disorder; DISC1, *d*israpted-*in*-*sc*hizophrenia 1; GSK-3β, glycogen synthase kinase β; MAM, methylazoxymethanol; MDD, major depressive disorder; NREM, non-rapid eye movement sleep; NRG1, neuregulin 1; PRODH, proline dehydrogenase; SCZ, schizophrenia; sw, slow wave; VCFS, velo-cardio-facial syndrome.

(1) Bjarnadottir *et al.* (2007), (2) Stefansson *et al.* (2002), (3) Michailov *et al.* (2004), (4) Kvajo *et al.* (2008), (5) Clapcote *et al.* (2007), (6) Pletnikov *et al.* (2008), (7) Gogos *et al.* (1999), (8) Shaltiel *et al.* (2008), (9) Mukai *et al.* (2008), (10) Stark *et al.* (2008), (11) Meechan *et al.* (2009), (12) Giros *et al.* (1996), (13) Gainetdinov *et al.* (1998), (14) Mohn *et al.* (1999), (15) Prickaerts *et al.* (2006), (16) McClung *et al.* (2005), (17) Sei *et al.* (2003), (18) Reddy *et al.* (2006), (19) Fatemi *et al.* (1999), (20) Cotter *et al.* (1995), (21) Meyer *et al.* (2008), (22) Marichich *et al.* (1979), (23) Palmer *et al.* (2004), (24) Steiger *et al.* (2003), (25) Moore *et al.* (2006), (26) Flagstad *et al.* (2004), (27) Flagstad *et al.* (2005), (28) Goto and Grace (2006), (29) Lipska *et al.* (1995), (30) Stine *et al.* (2001), (31) Tseng *et al.* (2008), (32) Sandstrom and Hart (2005), (33) Wu *et al.* (1999), (34) Bloomfield *et al.* (2008), (35) Belmaker and Agam (2008), (36) Meng *et al.* (2011), (37) Fone and Porkess (2008), (38) Pugh *et al.* (1999), (39) Marini *et al.* (2006), (40) Carobrez *et al.* (2002), (41) Pizarro *et al.* (2004), (42) Watt *et al.* (2009), (43) Powell *et al.* (2009), (44) Ahmad *et al.* (2010), (45) Duncko *et al.* (2001), (46) Grippo *et al.* (2006), (47) Allaman *et al.* (2008), (48) Hsu *et al.* (1980), (49) Tamminga *et al.* (1987), (50) Nabeshima *et al.* (1986), (51) Nishijima *et al.* (1996), (52) Buckland *et al.* (1997), (53) Jentsch *et al.* (1997), (54) Moghaddam *et al.* (1997), (55) Miyamoto *et al.* (2000), (56) Becker *et al.* (2003), (57) Keilhoff *et al.* (2004a), (58) Keilhoff *et al.* (2004b), (59) Kondziella *et al.* (2006), (60) Ali *et al.* (1994), (61) Orzi *et al.* (1983), (62) Wang *et al.* (2010), (63) Greenstein *et al.* (2007), (64) Frey *et al.* (2006).

of negative and positive effect, cognition, and social functions. Although some behavioral deficits seem to be to reversed through psychiatric medications, few antipsychotics have been tested to date (Sanberg *et al.*, 1985; Shi *et al.*, 2003).

The most reproducible neuromorphological findings in schizophrenia include reduced brain weight and hippocampal volume, and enlargement of the ventricles (Brown *et al.*, 1986; Bogerts *et al.*, 1990). Maternal malnutrition and interruption of neurogenesis reproduce those features. However, maternal immune activation mimics changes in cortical and hippocampal morphology, and in synaptic markers as seen in some schizophrenia patients (Fatemi *et al.*, 2002). Findings of disturbance in neurotransmitter systems in schizophrenia are somewhat less robust (Vollenweider *et al.*, 1998). Nevertheless, neurodevelopmental models for schizophrenia feature DA-, glutamate-, GABA-, and serotonin-related molecular and functional deficits (Table II).

C. The Stress of Life: The Cost of Impaired Adaptation

Stress is one of the most significant risk factors for development and progression of a number of psychiatric disorders (Nestler *et al.*, 2002). Several animal models involving early life stress have been used in MDD research including prenatal stress, postnatal handling, maternal separation, and social isolation at weaning (Francis *et al.*, 1996; Ladd *et al.*, 2000; Meaney, 2001). Stress can also be induced later in life by competition within a social environment. Indeed, humiliating defeats and/or entrapment are associated with a greater risk of developing depressive symptoms in women (Brown *et al.*, 1995). A relevant animal model of depression based on naturalistic stressors is social defeat. The chronic variable stress (CVS) model entails various stressors applied over long time periods, mimicking the likely situation in humans (McArthur and Borsini, 2006). In the CVS model, stressors include foot shock, cold water immersion, 48-h food and water deprivation, and milder conditions such as heat stress, cage tilt, reversal of day/night cycle, and change of cage mates (Willner, 1997).

Anhedonia, the loss of interest in and inability to experience pleasure, is one of the core symptoms of depression and has been shown in the CVS model (Rygula *et al.*, 2005; Becker *et al.*, 2008). Increased level of anxiety has also been demonstrated in the CVS and social defeat models (Fone and Porkess, 2008; Mineur *et al.*, 2006; Vidal *et al.*, 2007; Watt *et al.*, 2009). Depressed patients often show cognitive dysfunctions (Austin *et al.*, 1992), impaired visual learning and memory has also been found in all three animal models (Fone and Porkess, 2008; Henningsen *et al.*, 2009; Yu *et al.*, 2011), and impaired spatial learning can be induced by CVS (Song *et al.*, 2006).

Postmortem and magnetic-resonance imaging studies have revealed that the volume of hippocampus (MacQueen *et al.*, 2003), prefrontal cortex (Rajkowska,

2000), and amygdalae gray matter (Frodl *et al.*, 2008) is decreased in depressed patients. These morphological changes were partially reproduced in the social isolation model (Cooke *et al.*, 2000; Silva-Gomez *et al.*, 2003; Day-Wilson *et al.*, 2006), and decreased hippocampal volume has been demonstrated in the social defeat and CVS models (Becker *et al.*, 2008; Jayatissa *et al.*, 2010; Dagyte *et al.*, 2011). Decreased monoamine levels were observed in social isolation, social defeat, and CVS models, mimicking some of the early human findings which led to formulation of the monoamine hypothesis of depression (Fone and Porkess, 2008; Watt *et al.*, 2009; Ahmad *et al.*, 2010).

A significant portion of depressed patients show elevated circulating cortisol (Burke *et al.*, 2005), increased corticotrophin-releasing hormone (CRH) levels in cerebrospinal fluid (CSF), and increased levels of BDNF in hippocampus and prefrontal cortex (Karege *et al.*, 2005). Many of these findings have been demonstrated in stress-based animal models (Table II), such as elevated corticosterone (Grippo *et al.*, 2005; Sandstrom and Hart, 2005; Marini *et al.*, 2006; Becker *et al.*, 2008) which can be normalized by antidepressant treatment in the CVS model (Detanico *et al.*, 2009).

Recent studies indicate that alterations in immune system activity have been found in depressed humans, including increased serum/plasma and CSF concentrations of proinflammatory cytokines including interleukin-1(IL-1; Thomas *et al.*, 2005), IL-6 (Alesci *et al.*, 2005), and tumor necrosis factor-alpha (TNF-α; Mikova *et al.*, 2001). Social isolation, social defeat, and CVS induce a proinflammatory effect characterized by increased TNF-α and IL-1β levels in plasma, splenic dendritic cells, and hippocampus (Pugh *et al.*, 1999; Wu *et al.*, 1999; Carobrez *et al.*, 2002; Grippo *et al.*, 2005; Powell *et al.*, 2009).

D. Disrupting Communication: Pharmacological Modification

One of the most widely used approaches to develop preclinical models for psychiatric disorders is pharmacological disruption of neurotransmitter systems. The neurotransmitter hypotheses of schizophrenia are based on mechanisms of clinical efficacy of therapeutically effective drugs and early disease-related neurochemical findings in the periphery and brain (Howes and Kapur, 2009; Coyle, 2006b). Further, administration of compounds interfering with DA and glutamate transmission has been found to induce schizophrenia-like symptoms in nonschizophrenic humans, which could be reversed by antipsychotic medication (Javitt and Zukin, 1991).

The amphetamine rat model mimics certain deficits of schizophrenia closely related to positive symptoms and cognitive dysfunctions, but less so the negative symptoms (Table I). Compared to DA-based preclinical models which target mainly the positive symptoms of schizophrenia, NMDA receptor antagonists such as PCP, ketamine, and MK801 mimic a broader spectrum of symptoms including the negative symptoms and cognitive dysfunction (Table I).

III. Reverse Translation

Neuropsychiatry has gained considerable theoretical knowledge from understanding the mechanisms of action of therapeutically effective drugs. For example, identification of DA receptors as targets for early antipsychotics gave rise to the DA hypothesis of schizophrenia and the development of a series of pharmacotherapeutics with better side-effect profiles. More recently, GSK-3β has emerged as a target for lithium, leading to a novel neurotrophic signaling cascade hypothesis of BD (Shaltiel *et al.*, 2007). Therefore, comparison of molecular signatures altered by antipsychotic drug treatment in experimental animals with those obtained from analysis of human schizophrenia brain regions may lead to discovery of novel, disease-relevant pathways. Such molecular biomarker signatures can now be evaluated using contemporary proteomic and metabolomic platforms (Kaddurah-Daouk *et al.*, 2007; Martins-de-Souza *et al.*, 2010; English *et al.*, 2011; see Chapter "Proteomic technologies for biomarker studies in psychiatry: Advances and needs" by Martins-de-Souza *et al.*).

Ma *et al.* (2009) and McLoughlin *et al.* (2009) used label-free liquid chromatography tandem mass spectrometry (LC–MS^E) and ^1H NMR spectroscopy for identification of differently expressed proteins and metabolites in distinct brain regions after treatment with the typical antipsychotic, haloperidol, and the atypical antipsychotic olanzapine in rats. For simplicity, only the frontal cortex analyses are reviewed here, given the importance of this brain region in schizophrenia (Fig. 1). An overlap of 15% of the proteins and 40% of the metabolites was detected between the two drugs, potentially indicating common pathways in their mechanisms of action. However, the protein and metabolite changes unique to each antipsychotic drug potentially mirror the unique pathways and side-effect profiles evoked by each drug.

Common changes in brain metabolites after drug treatment included an increase in lactate and choline and a decrease in creatine and phosphocholine, all metabolites relevant to neuronal energy metabolism. *In silico* pathway analysis of the proteome revealed that both antipsychotics were linked with disturbances in cellular assembly and organization of the nervous system. However, the top canonical pathways were different for each drug. Haloperidol administration was associated with Hungtington's disease signaling, whereas olanzapine was linked to glycolysis/gluconeogenesis. This is in accordance with the literature

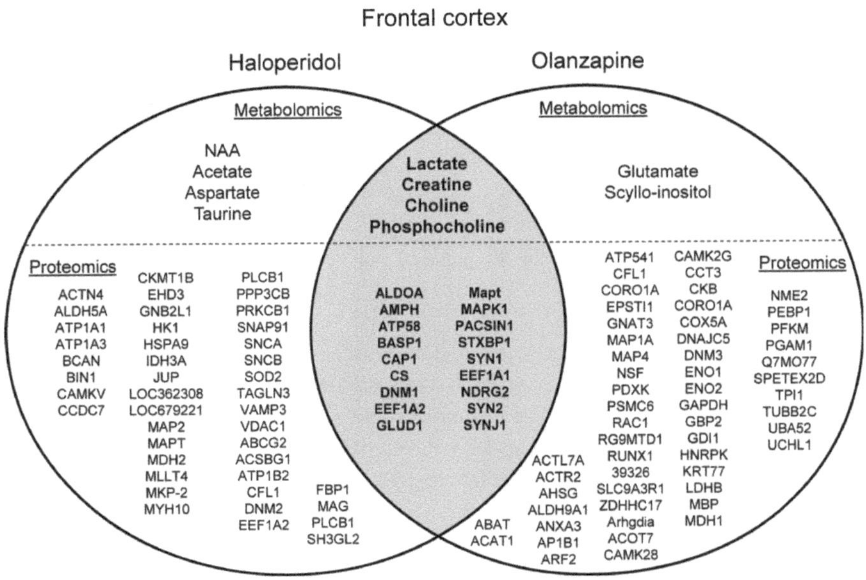

FIG. 1. Metabolomics and proteomics changes induced by chronic treatment with haloperidol and olanzapine in rat frontal cortex. Representative data from Ma *et al.* (2009) and McLoughlin *et al.* (2009) showing drug-specific and common changes.

showing that olanzapine induces more metabolic side effects than haloperidol (Deng *et al.*, 2007). Haloperidol, however, has already been used as treatment for Huntington's disease (Markianos *et al.*, 2010).

Considering the differentially expressed proteins and metabolites together, it appears that typical and atypical antipsychotics affect brain glucose metabolism and structural assembly of the presynaptic vesicle (Pellerin, 2003; Vaynman *et al.*, 2006). It is assumed that by improving neuronal activity and energy metabolism, an amelioration of psychotic symptoms can be achieved. Antidiabetic medications have been proposed on the basis of a potential role of altered energy metabolism in neuropsychiatric diseases (Laron, 2009). It has been shown that intranasal insulin improves cognition and is now a suggested therapeutic for the treatment of Alzheimer's disease (Benedict *et al.*, 2011). Such reverse translational studies using therapeutically effective drugs to identify drug-specific and class-specific biomarkers may help to pinpoint novel disease mechanisms and drug targets. Therefore, it may be worthwhile to consider the use of antidiabetic medication to improve energy metabolism pathways in the brain of schizophrenia patients to ameliorate some of the cognitive symptoms.

IV. Statistical Methods to Link Biomarkers from Animal Models with the Human Disease

A potential way of understanding the underlying molecular mechanisms of complex diseases such as schizophrenia and MDD lies in the statistical cross-comparison of data sets arising from analysis of animal models and human studies. An overlap of experimental and clinical biomarkers would lend credibility to the animal models and could potentially be used to monitor treatment effects in these models more objectively. In this context, one of the major goals is to identify changes in the molecular profiles of animal models that are reflected in the human disease. This section highlights the advantages and some of the problems arising by comparison of data within and across different species and introduces methodologies to address these difficulties.

Following data acquisition (Fig. 2A), a standard approach to assess similarity within and between different species is to find and compare lists of differentially expressed molecules using complementary univariate methods. A common statistical test to detect significantly changed molecules in proteomic or gene expression data is the nonparametric Wilcoxon rank-sum test (Fig. 2B). Then, a subsequent step is to check whether different data sets have molecules in common. There are also methods to calculate the probability that two independent lists overlap by chance alone, including Fisher's exact test (Fig. 2B) and the hypergeometric distribution or permutation tests (Fury *et al.*, 2006). In addition, *in silico* pathway analysis methods (Fig. 2B) can be applied to identify biological processes in common represented by the altered molecules in the species under comparison. Two commonly used tools to explore biological pathways are ArrayUnlock and Ingenuity Pathways Analysis (Jiménez-Marín *et al.*, 2009).

In most cases, the outcome of analyses based on lists of significantly changed molecules is rarely sufficient to assume similarity. Among the major reasons are the difficulties associated with cross-species comparison arising from the metabolic, genetic, and cellular diversities between humans and rodents. Also, statistically underpowered studies can hamper identification of specific biomarkers. Variability within a single model and across different studies might be responsible for the inability to detect significant molecular changes.

There are various statistical methods which can address this problem. Meta-analysis is a widely used approach in cross-species comparison to overcome low numbers of samples (Lu *et al.*, 2009). Statistical confidence of single clinical studies is often not sufficient enough to draw sound conclusions about the effect of treatment. Thus, meta-analysis has gained considerable popularity in medical research since it enhances the probability of detecting treatment effects by increasing the statistical power. In this way, the power is strengthened by combining different clinical studies that address similar experimental hypotheses. Also, cluster analysis (McLachlan *et al.*, 2002; Alon *et al.*, 1999) has been shown to be

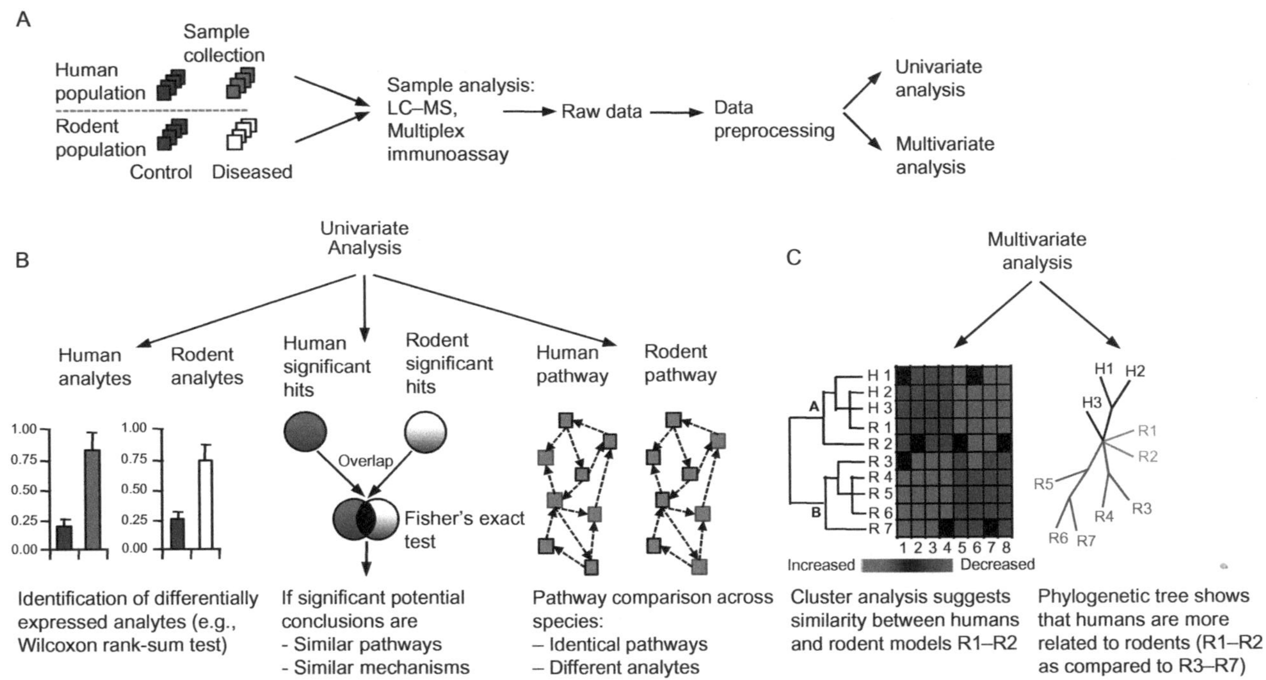

FIG. 2. Use of univariate and multivariate statistical methods to link biomarkers from animal models with the human disease. (A) Samples from both species can be analyzed using methods such as LC–MS^E or multiplexed immunoassay. (B) Univariate statistics can be used to detect similarities between the species. Differentially expressed molecules in both species can be identified using the nonparametric Wilcoxon rank-sum test and Fisher's exact test to check whether the overlap of molecules is significant by chance. Also, pathway analysis can be used to compare the effects across the different species. (C) Multivariate approach to detect similarity within and between the species. Hierarchical cluster analysis (HCA) groups animal models into clusters such that those within the same cluster are assumed to be homogenous. Cluster analysis suggests similarity between three different human cohorts (H1–H3) and the two rodent models (R1–R2). A phylogenetic tree can also be used to assess similarity. In this case, similarities between the human cohorts and R1–R2 rodent models were detected.

effective in dealing with large data sets obtained by cross-species comparisons. One advantage of using such multivariate techniques over univariate approaches (e.g., Fisher's exact test) to detect similarities within or across species is that single molecules do not necessarily have to be significantly changed to assess homogeneity. This is particularly useful when high variability within the data hampers identification of differentially expressed molecules. Clustering is based on assignment of samples or objects into distinct groups such that objects in the same group are assumed to share common features (e.g., functions or pathways), while objects in separate groups are expected to be different from each other. The distribution of objects into separate clusters is based on a distance measure which is calculated for all possible pairwise combinations between all molecules. The agglomerative hierarchical clustering approach is one of the most commonly applied procedures resulting in a tree-based hierarchical structure in which adjacent objects are considered to be more similar. To date, a number of different agglomerative algorithms are available (Meunier *et al.*, 2007). Hierarchical cluster analysis (HCA) is often applied as one of the first steps to assess both homogeneity and heterogeneity in proteomic data (Meunier *et al.*, 2007). HCA can be applied using various different distance measures and clustering algorithms but it is well known that diverse combinations might lead to different clustering results (Meunier *et al.*, 2007). Figure 2C shows how an agglomerative hierarchical clustering algorithm groups data from different species together. The outcome of the cluster analysis suggests that the human cohorts (H1–H3) are more similar to the coclustering rodent models (R1–R2) as compared to the remaining animal models (R3–R7).

V. Integration of Animal Models Using the Framework of RDoC

In this chapter, we reviewed widely used, etiologically based animal models for schizophrenia, MDD, and BD, and summarized key behavioral findings along with disease-relevant alterations in neurotransmitters, molecules, cells, and circuits. What is apparent, however, is that none of these models is able to fully replicate the human condition. This is a major problem with animal models of psychiatric disorders and is likely to be insurmountable. Therefore, it is perhaps advisable to move away from this traditional approach of trying to find animal models of certain psychiatric disorders. The ongoing debate on introduction of the new diagnostic system for mental disorders, the DSM-V, gives further impetuous to rethink the way we view and use animal models in this context.

Currently, the diagnosis of mental disorders is based on clinical observation and patients' phenomenological symptom reports (APA, 2000). It is perhaps erroneous to assume that clinical syndromes based on subjective symptoms are unique and unitary entities. Current genetic and neurobiological studies, along with the use of the same pharmacotherapeutic agents for multiple psychiatric indications, emphasize overlaps among presently established disease categories. As an attempt to overcome these problems and to align the future diagnostic system of mental disorders with the findings of modern behavioral neuroscience and neurobiology, the US National Institute of Mental Health has recently initiated a framework to guide classification of patients for research purposes and, ultimately, to help the creation of a new diagnostic system better incorporating recent neuroscience discoveries (Insel *et al.*, 2010). The RDoC framework considers specific dimensions or functions termed "constructs," such as the different behavioral domains, for example, negative, positive, and cognitive effects, and units of analysis such as behavior (Table I) and molecules, cells, and circuits (Table II). From the point of view of integrating animal models with human behavioral and neurobiological findings, the major advantage of the RDoC system over the present DSM-IV-TR and ICD-10 diagnostic systems is that it is agnostic about current disorder categories. Its intention is to generate classification stemming from behavioral neuroscience, rather than subjective patient reports. Further, RDoC is conceived as a dimensional system reflecting measurements of behavior, circuit activity, etc., spanning the range from normal to abnormal. It will also use different levels of analysis in defining certain constructs (imaging, physiological measures, and molecular biomarkers). Such a framework is better suited for integrating animal models than the traditional diagnostic categories.

Therefore, we have applied the principles of the RDoC framework to produce behavioral (Table I) and neurobiological (Table II) matrices to represent findings from diverse animal models. While the list of models is not intended to be comprehensive and we acknowledge well-known construction, interpretation, and reporting biases (e.g., models constructed with a particular purpose, data interpreted to support a rational), some interesting conclusions can be drawn. Although each of these models was developed with the intention of modeling one particular disorder, there is a considerable overlap in behavioral effects mapping into different RDoC domains (Table I). This is in line with the clinical findings of symptom overlap between major psychiatric disorders. However, the neurobiological alterations (Table II) seem to be more specific for genetic, stress-induced, or pharmacological impairments, supporting the concept that behavioral changes are downstream of molecular and cellular alterations. It is also apparent that the pattern of behavioral and neurobiological changes is more important than a single abnormality for mapping a particular model to a human disorder. This is consistent with recent knowledge emerging from the biomarker literature emphasizing altered patterns rather than individual changes of protein biomarkers (Levin *et al.*, 2010).

However, the most important conclusion of this exercise is that much more research is needed to complete the RDoC matrix with behavioral and neurobiological data from animal models. The RDoC domains/constructs should be adapted to animal research, as we have attempted here. Well-validated behavioral tests should be organized into the RDoC framework, perhaps similar to the MATRICS initiative which has identified cognition domains that are deficient in schizophrenia along with a "preclinical MATRICS" of rodent behavioral test batteries (Young *et al.*, 2009). Current and future animal models of psychiatric disorders should then be tested to fill in the RDoC matrix. These RDoC matrices can then be superimposed and overlapping patterns identified. This will lend previously unknown translational power to the system. Emerging behavioral, molecular, and circuit activation patterns from animal models can, therefore, be correlated with those from human studies to provide signatures associated with specific genetic mutations or environmental factors. Communication between the preclinical and clinical RDoC will be needed to fulfill the promise of this approach. Integration of animal models into the RDoC framework may not only remove psychiatric animal models from the ever-returning "validation crisis" but may ultimately lead to better understanding of the pathophysiology human neuropsychiatric diseases and to the identification of novel biomarkers and therapeutic targets.

Acknowledgments

This research was supported by the Stanley Medical Research Institute (SMRI), the European Union FP7 SchizDX research program (grant reference 223427), and the NEWMEDS Innovative Medicines Initiative.

References

Ahmad, A., Rasheed, N., Banu, N., and Palit, G. (2010). Alterations in monoamine levels and oxidative systems in frontal cortex, striatum, and hippocampus of the rat brain during chronic unpredictable stress. *Stress* **13**(4), 355–364.

Ahnaou, A., Nayak, S., Heylen, A., Ashton, D., and Drinkenburg, W.H. (2007). Sleep and EEG profile in neonatal hippocampal lesion model of schizophrenia. *Physiol. Behav.* **92**(3), 461–467.

Alesci, S., Martinez, P.E., Kelkar, S., Ilias, I., Ronsaville, D.S., Listwak, S.J., Ayala, A.R., Licinio, J., Gold, H.K., Kling, M.A., Chrousos, G.P., and Gold, P.W. (2005). Major depression is associated with significant diurnal elevations in plasma interleukin-6 levels, a shift of its circadian rhythm, and loss of physiological complexity in its secretion: clinical implications. *J. Clin. Endocrinol. Metab.* **90**(5), 2522–2530.

Ali, S.F., Newport, G.D., and Bracha, H.S. (1994). Phencyclidine and (+)-MK-801-induced circling preference: correlation with monoamine levels in striatum of the rat brain. *Neurotoxicol. Teratol.* **16** (4), 335–342.

Allaman, I., Papp, M., Kraftsik, R., Fiumelli, H., Magistretti, P.J., and Martin, J.L. (2008). Expression of brain-derived neurotrophic factor is not modulated by chronic mild stress in the rat hippocampus and amygdala. *Pharmacol. Rep.* **60**(6), 1001–1007.

Allen, N.C., Bagade, S., McQueen, M.B., Ioannidis, J.P., Kavvoura, F.K., Khoury, M.J., Tanzi, R.E., and Bertram, L. (2008). Systematic meta-analyses and field synopsis of genetic association studies in schizophrenia: the SzGene database. *Nat. Genet.* **40**(7), 827–834.

Almeida, S.S., Tonkiss, J., and Galler, J.R. (1996a). Prenatal protein malnutrition affects avoidance but not escape behavior in the elevated T-maze test. *Physiol. Behav.* **60**(1), 191–195.

Almeida, S.S., Tonkiss, J., and Galler, J.R. (1996b). Prenatal protein malnutrition affects the social interactions of juvenile rats. *Physiol. Behav.* **60**(1), 197–201.

Alon, U., Barkai, N., Notterman, D.A., Gish, K., Ybarra, S., Mack, D., and Levine, A.J. (1999). Broad patterns of gene expression revealed by clustering analysis of tumor and normal colon tissues probed by oligonucleotide arrays. *Proc. Natl. Acad. Sci. USA* **96**(12), 6745–6750.

American Psychiatric Association (2000). Diagnostic and Statistical Manual of Mental Disorders. 4th edn. American Psychiatric Press, Washington, DC.

Austin, M.P., Ross, M., Murray, C., O'Carroll, R.E., Ebmeier, K.P., and Goodwin, G.M. (1992). Cognitive function in major depression. *J. Affect. Disord.* **25**(1), 21–29.

Bassett, A.S., Chow, E.W., Husted, J., Weksberg, R., Caluseriu, O., Webb, G.D., and Gatzoulis, M.A. (2005). Clinical features of 78 adults with 22q11 Deletion Syndrome. *Am. J. Med. Genet. A* **138**(4), 307–313.

Baumann, M.H., Rothman, R.B., and Ali, S.F. (2000). Comparative neurobiological effects of ibogaine and MK-801 in rats. *Drug Alcohol Depend.* **59**(2), 143–151.

Becker, A., and Grecksch, G. (2004). Ketamine-induced changes in rat behaviour, a possible animal model of schizophrenia. Test of predictive validity. *Prog. Neuropsychopharmacol. Biol. Psychiatry* **28**(8), 1267–1277.

Becker, A., Peters, B., Schroeder, H., Mann, T., Huether, G., and Grecksch, G. (2003). Ketamine-induced changes in rat behaviour: a possible animal model of schizophrenia. *Prog. Neuropsychopharmacol. Biol. Psychiatry* **27**(4), 687–700.

Becker, C., Zeau, B., Rivat, C., Blugeot, A., Hamon, M., and Benoliel, J.J. (2008). Repeated social defeat-induced depression-like behavioral and biological alterations in rats, involvement of cholecystokinin. *Mol. Psychiatry* **13**(12), 1079–1092.

Belmaker, R.H., and Agam, G. (2008). Major depressive disorder. *N. Engl. J. Med.* **358**(1), 55–68.

Benedetti, F., Serretti, A., Colombo, C., Barbini, B., Lorenzi, C., Campori, E., and Smeraldi, E. (2003). Influence of CLOCK gene polymorphism on circadian mood fluctuation and illness recurrence in bipolar depression. *Am. J. Med. Genet. B Neuropsychiatr. Genet.* **123B**(1), 23–26.

Benedict, C., Frey, W.H. 2nd, Schioth, H.B., Schultes, B., Born, J., and Hallschmid, M. (2011). Intranasal insulin as a therapeutic option in the treatment of cognitive impairments. *Exp. Gerontol.* **46**(2–3), 112–115.

Beneyto, M., Kristiansen, L.V., Oni-Orisan, A., McCullumsmith, R.E., and Meador-Woodruff, J.H. (2007). Abnormal glutamate receptor expression in the medial temporal lobe in schizophrenia and mood disorders. *Neuropsychopharmacology* **32**(9), 1888–1902.

Bickel, S., Lipp, H.P., and Umbricht, D. (2008). Early auditory sensory processing deficits in mouse mutants with reduced NMDA receptor function. *Neuropsychopharmacology* **33**(7), 1680–1689.

Bitanihirwe, B.K., Weber, L., Feldon, J., and Meyer, U. (2010). Cognitive impairment following prenatal immune challenge in mice correlates with prefrontal cortical AKT1 deficiency. *Int. J. Neuropsychopharmacol.* **13**(8), 981–996.

Bjarnadottir, M., Misner, D.L., Haverfield-Gross, S., Bruun, S., Helgason, V.G., Stefansson, H., Sigmundsson, A., Firth, D.R., Nielsen, B., Stefansdottir, R., Novak, T.J., Stefansson, K., *et al.* (2007). Neuregulin1 (NRG1) signaling through Fyn modulates NMDA receptor phosphorylation, differential synaptic function in NRG1+/− knock-outs compared with wild-type mice. *J. Neurosci.* **27**(17), 4519–4529.

Bloomfield, C., French, S.J., Jones, D.N., Reavill, C., Southam, E., Cilia, J., and Totterdell, S. (2008). Chandelier cartridges in the prefrontal cortex are reduced in isolation reared rats. *Synapse* **62**(8), 628–631.

Bogerts, B., Ashtari, M., Degreef, G., Alvir, J.M., Bilder, R.M., and Lieberman, J.A. (1990). Reduced temporal limbic structure volumes on magnetic resonance images in first episode schizophrenia. *Psychiatry Res.* **35**(1), 1–13.

Brenes, J.C., and Fornaguera, J. (2008). Effects of environmental enrichment and social isolation on sucrose consumption and preference, associations with depressive-like behavior and ventral striatum dopamine. *Neurosci. Lett.* **436**(2), 278–282.

Brown, R., Colter, N., Corsellis, J.A., Crow, T.J., Frith, C.D., Jagoe, R., Johnstone, E.C., and Marsh, L. (1986). Postmortem evidence of structural brain changes in schizophrenia. Differences in brain weight, temporal horn area, and parahippocampal gyrus compared with affective disorder. *Arch. Gen. Psychiatry* **43**(1), 36–42.

Brown, G.W., Harris, T.O., and Hepworth, C. (1995). Loss, humiliation and entrapment among women developing depression, a patient and non-patient comparison. *Psychol. Med.* **25**(1), 7–21.

Buckland, P.R., Marshall, R., Watkins, P., and McGuffin, P. (1997). Does phenylethylamine have a role in schizophrenia?, LSD and PCP up-regulate aromatic L-amino acid decarboxylase mRNA levels. *Brain Res. Mol. Brain Res.* **49**(1–2), 266–270.

Burke, H.M., Davis, M.C., Otte, C., and Mohr, D.C. (2005). Depression and cortisol responses to psychological stress, a meta-analysis. *Psychoneuroendocrinology* **30**(9), 846–856.

Burt, T., Sachs, G.S., and Demopulos, C. (1999). Donepezil in treatment-resistant bipolar disorder. *Biol. Psychiatry* **45**(8), 959–964.

Carobrez, S.G., Gasparotto, O.C., Buwalda, B., and Bohus, B. (2002). Long-term consequences of social stress on corticosterone and IL-1beta levels in endotoxin-challenged rats. *Physiol. Behav.* **76**(1), 99–105.

Chambers, R.A., Moore, J., McEvoy, J.P., and Levin, E.D. (1996). Cognitive effects of neonatal hippocampal lesions in a rat model of schizophrenia. *Neuropsychopharmacology* **15**(6), 587–594.

Clapcote, S.J., Lipina, T.V., Millar, J.K., Mackie, S., Christie, S., Ogawa, F., Lerch, J.P., Trimble, K., Uchiyama, M., Sakuraba, Y., Kaneda, H., Shiroishi, T., *et al.* (2007). Behavioral phenotypes of Disc1 missense mutations in mice. *Neuron* **54**(3), 387–402.

Cooke, B.M., Chowanadisai, W., and Breedlove, S.M. (2000). Post-weaning social isolation of male rats reduces the volume of the medial amygdala and leads to deficits in adult sexual behavior. *Behav. Brain Res.* **117**(1–2), 107–113.

Cotter, D., Takei, N., Farrell, M., Sham, P., Quinn, P., Larkin, C., Oxford, J., Murray, R.M., and O'Callaghan, E. (1995). Does prenatal exposure to influenza in mice induce pyramidal cell disarray in the dorsal hippocampus? *Schizophr. Res.* **16**(3), 233–241.

Coyle, J.T. (2006a). Glutamate and schizophrenia, beyond the dopamine hypothesis. *Cell. Mol. Neurobiol.* **26**(4–6), 365–384.

Coyle, J.T. (2006b). Substance use disorders and Schizophrenia: a question of shared glutamatergic mechanisms. *Neurotox. Res.* **10**, 221–233.

Dagyte, G., Crescente, I., Postema, F., Seguin, L., Gabriel, C., Mocaer, E., Boer, J.A., and Koolhaas, J.M. (2011). Agomelatine reverses the decrease in hippocampal cell survival induced by chronic mild stress. *Behav. Brain Res.* **218**(1), 121–128.

Day-Wilson, K.M., Jones, D.N., Southam, E., Cilia, J., and Totterdell, S. (2006). Medial prefrontal cortex volume loss in rats with isolation rearing-induced deficits in prepulse inhibition of acoustic startle. *Neuroscience* **141**(3), 1113–1121.

Deakin, I.H., Law, A.J., Oliver, P.L., Schwab, M.H., Nave, K.A., Harrison, P.J., and Bannerman, D.M. (2009). Behavioural characterization of neuregulin 1 type I overexpressing transgenic mice. *Neuroreport* **20**(17), 1523–1528.

Deng, C., Weston-Green, K.L., Han, M., and Huang, X.F. (2007). Olanzapine treatment decreases the density of muscarinic M2 receptors in the dorsal vagal complex of rats. *Prog. Neuropsychopharmacol. Biol. Psychiatry* **31**(4), 915–920.

Detanico, B.C., Piato, A.L., Freitas, J.J., Lhullier, F.L., Hidalgo, M.P., Caumo, W., and Elisabetsky, E. (2009). Antidepressant-like effects of melatonin in the mouse chronic mild stress model. *Eur. J. Pharmacol.* **607**(1–3), 121–125.

Duffy, L., Cappas, E., Scimone, A., Schofield, P.R., and Karl, T. (2008). Behavioral profile of a heterozygous mutant mouse model for EGF-like domain neuregulin 1. *Behav. Neurosci.* **122**(4), 748–759.

Duffy, L., Cappas, E., Lai, D., Boucher, A.A., and Karl, T. (2010). Cognition in transmembrane domain neuregulin 1 mutant mice. *Neuroscience* **170**(3), 800–807.

Duncko, R., Kiss, A., Skultetyova, I., Rusnak, M., and Jezova, D. (2001). Corticotropin-releasing hormone mRNA levels in response to chronic mild stress rise in male but not in female rats while tyrosine hydroxylase mRNA levels decrease in both sexes. *Psychoneuroendocrinology* **26**(1), 77–89.

Dzirasa, K., Santos, L.M., Ribeiro, S., Stapleton, J., Gainetdinov, R.R., Caron, M.G., and Nicolelis, M.A. (2009). Persistent hyperdopaminergia decreases the peak frequency of hippocampal theta oscillations during quiet waking and REM sleep. *PLoS One* **4**(4), e5238.

Easton, A., Arbuzova, J., and Turek, F.W. (2003). The circadian Clock mutation increases exploratory activity and escape-seeking behavior. *Genes Brain Behav.* **2**, 11–19.

Egerton, A., Reid, L., McGregor, S., Cochran, S.M., Morris, B.J., and Pratt, J.A. (2008). Subchronic and chronic PCP treatment produces temporally distinct deficits in attentional set shifting and prepulse inhibition in rats. *Psychopharmacology (Berl.)* **198**(1), 37–49.

Ehrlichman, R.S., Luminais, S.N., White, S.L., Rudnick, N.D., Ma, N., Dow, H.C., Kreibich, A.S., Abel, T., Brodkin, E.S., Hahn, C.G., and Siegel, S.J. (2009). Neuregulin 1 transgenic mice display reduced mismatch negativity, contextual fear conditioning and social interactions. *Brain Res.* **1294**, 116–127.

English, J.A., Pennington, K., Dunn, M.J., and Cotter, D.R. (2011). The neuroproteomics of schizophrenia. *Biol. Psychiatry* **69**(2), 163–172.

Enomoto, T., and Floresco, S.B. (2009). Disruptions in spatial working memory, but not short-term memory, induced by repeated ketamine exposure. *Prog. Neuropsychopharmacol. Biol. Psychiatry* **33**(4), 668–675.

Fatemi, S.H., Emamian, E.S., Kist, D., Sidwell, R.W., Nakajima, K., Akhter, P., Shier, A., Sheikh, S., and Bailey, K. (1999). Defective corticogenesis and reduction in Reelin immunoreactivity in cortex and hippocampus of prenatally infected neonatal mice. *Mol. Psychiatry* **4**(2), 145–154.

Fatemi, S.H., Earle, J., Kanodia, R., Kist, D., Emamian, E.S., Patterson, P.H., Shi, L., and Sidwell, R. (2002). Prenatal viral infection leads to pyramidal cell atrophy and macrocephaly in adulthood, implications for genesis of autism and schizophrenia. *Cell. Mol. Neurobiol.* **22**(1), 25–33.

Flagstad, P., Mork, A., Glenthoj, B.Y., van Beek, J., Michael-Titus, A.T., and Didriksen, M. (2004). Disruption of neurogenesis on gestational day 17 in the rat causes behavioral changes relevant to positive and negative schizophrenia symptoms and alters amphetamine-induced dopamine release in nucleus accumbens. *Neuropsychopharmacology* **29**(11), 2052–2064.

Flagstad, P., Glenthoj, B.Y., and Didriksen, M. (2005). Cognitive deficits caused by late gestational disruption of neurogenesis in rats, a preclinical model of schizophrenia. *Neuropsychopharmacology* **30** (2), 250–260.

Fone, K.C., and Porkess, M.V. (2008). Behavioural and neurochemical effects of post-weaning social isolation in rodents-relevance to developmental neuropsychiatric disorders. *Neurosci. Biobehav. Rev.* **32**(6), 1087–1102.

Fortier, M.E., Joober, R., Luheshi, G.N., and Boksa, P. (2004). Maternal exposure to bacterial endotoxin during pregnancy enhances amphetamine-induced locomotion and startle responses in adult rat offspring. *J. Psychiatr. Res.* **38**(3), 335–345.

Francis, D., Diorio, J., LaPlante, P., Weaver, S., Seckl, J.R., and Meaney, M.J. (1996). The role of early environmental events in regulating neuroendocrine development. Moms, pups, stress, and glucocorticoid receptors. *Ann. N. Y. Acad. Sci.* **794**, 136–152.

Frey, B.N., Valvassori, S.S., Gomes, K.M., Martins, M.R., Dal-Pizzol, F., Kapczinski, F., *et al.* (2006). Increased oxidative stress in submitochondrial particles after chronic amphetamine exposure. *Brain Res.* **1097**(1), 224–229.

Frodl, T.S., Koutsouleris, N., Bottlender, R., Born, C., Jager, M., Scupin, I., Reiser, M., Moller, H.J., and Meisenzahl, E.M. (2008). Depression-related variation in brain morphology over 3 years, effects of stress? *Arch. Gen. Psychiatry* **65**(10), 1156–1165.

Fury, W., Batliwalla, F., Gregersen, P.K., and Li, W. (2006). Overlapping probabilities of top ranking gene lists, hypergeometric distribution, and stringency of gene selection criterion. *Conf. Proc. IEEE Eng. Med. Biol. Soc.* **1**, 5531–5534.

Gainetdinov, R.R., Jones, S.R., Fumagalli, F., Wightman, R.M., and Caron, M.G. (1998). Re-evaluation of the role of the dopamine transporter in dopamine system homeostatsis. *Brain Res. Brain Res. Rev.* **26**, 148–153.

Gainetdinov, R.R., Jones, S.R., and Caron, M.G. (1999). Functional hyperdopaminergia in dopamine transporter knock-out mice. *Biol. Psychiatry* **46**(3), 303–311Review.

Gerlai, R., Pisacane, P., and Erickson, S. (2000). Heregulin, but not ErbB2 or ErbB3 heterozygous mutant mice exhibit hyperactivity in multiple behavioural tasks. *Behav. Brain Res.* **109**(2), 219.

Geyer, M.A., Krebs-Thomson, K., Braff, D.L., and Swerdlow, N.R. (2001). Pharmacological studies of prepulse inhibition models of sensorimotor gating deficits in schizophrenia, a decade in review. *Psychopharmacology (Berl.)* **156**(2–3), 117–154.

Giros, B., Jaber, M., Jones, S.R., Wightman, R.M., and Caron, M.G. (1996). Hyperlocomotion and indifference to cocaine and amphetamine in mice lacking the dopamine transporter. *Nature* **379** (6566), 606–612.

Gogos, J.A., Satha, M., Takacs, Z., Beck, K.D., Luine, V., Lucas, L.R., Nadler, J.V., and Karayiorgou, M. (1999). The gene encoding proline dehydrogenase modulates sensorimotor gating in mice. *Nat. Genet.* **21**, 434–439.

Goto, Y., and Grace, A.A. (2006). Alterations in medial prefrontal cortical activity and plasticity in rats with disruption of cortical development. *Biol. Psychiatry* **60**(11), 1259–1267.

Grande, I., Fries, G., Kunz, M., and Kapczinski, F. (2010). The role of BDNF as a mediator of neuroplasticity in bipolar disorder. *Psychiatry Investig.* **7**(4), 243–250.

Greenstein, R., Novak, G., and Seeman, P. (2007). Amphetamine sensitization elevates CaMKIIbeta mRNA. *Synapse* **61**(10), 827–834.

Gresack, J.E., Risbrough, V.B., Scott, C.N., Coste, S., Stenzel-Poore, M., Geyer, M.A., and Powell, S.B. (2010). Isolation rearing-induced deficits in contextual fear learning do not require CRF(2) receptors. *Behav. Brain Res.* **209**(1), 80–84.

Grippo, A.J., Francis, J., Beltz, T.G., Felder, R.B., and Johnson, A.K. (2005). Neuroendocrine and cytokine profile of chronic mild stress-induced anhedonia. *Physiol. Behav.* **84**(5), 697–706.

Halene, T.B., Ehrlichman, R.S., Liang, Y., Christian, E.P., Jonak, G.J., Gur, T.L., Blendy, J.A., Dow, H. C., Brodkin, E.S., Schneider, F., Gur, R.C., and Siegel, S.J. (2009). Assessment of NMDA receptor NR1 subunit hypofunction in mice as a model for schizophrenia. *Genes Brain Behav.* **8**(7), 661–675.

Hall, F.S., Li, X.F., Randall-Thompson, J., Sora, I., Murphy, D.L., Lesch, K.P., Caron, M., and Uhl, G. R. (2009). Cocaine-conditioned locomotion in dopamine transporter, norepinephrine transporter and 5-HT transporter knockout mice. *Neuroscience* **162**(4), 870–880.

Harrison, P.J. (1999). The neuropathology of schizophrenia. A critical review of the data and their interpretation. *Brain* **122**(Pt. 4), 593–624.

Heidbreder, C.A., Weiss, I.C., Domeney, A.M., Pryce, C., Homberg, J., Hedou, G., Feldon, J., Moran, M.C., and Nelson, P. (2000). Behavioral, neurochemical and endocrinological characterization of the early social isolation syndrome. *Neuroscience* **100**(4), 749–768.

Henningsen, K., Andreasen, J.T., Bouzinova, E.V., Jayatissa, M.N., Jensen, M.S., Redrobe, J.P., and Wiborg, O. (2009). Cognitive deficits in the rat chronic mild stress model for depression, relation to anhedonic-like responses. *Behav. Brain Res.* **198**(1), 136–141.

Hikida, T., Jaaro-Peled, H., Seshadri, S., Oishi, K., Hookway, C., Kong, S., Wu, D., Xue, R., Andradé, M., Tankou, S., Mori, S., Gallagher, M., *et al.* (2007). Dominant-negative DISC1 transgenic mice display schizophrenia-associated phenotypes detected by measures translatable to humans. *Proc. Natl. Acad. Sci. USA* **104**(36), 14501–14506.

Howes, O.D., and Kapur, S. (2009). The dopamine hypothesis of schizophrenia, version III—the final common pathway. *Schizophr. Bull.* **35**(3), 549–562.

Hradetzky, E., Sanderson, T.M., Tsang, T.M., Lakics, V., Malik, N., O'Neill, M.C., Schoeffmann, S., Cheng, T., Harris, L.W., Guest, P.C., Rahmoune, H., Holmes, E., Tricklebank, M., and Bahn, S. Methylazoxymethanol acetate rat model, molecular and functional altered synaptic transmission in the hippocampus. (submitted).

Hsu, L.L., Smith, R.C., Rolsten, C., and Leelavathi, D.E. (1980). Effects of acute and chronic phencyclidine on neurotransmitter enzymes in rat brain. *Biochem. Pharmacol.* **29**(18), 2524–2526.

Hwu, H.G., Liu, C.M., Fann, C.S., Ou-Yang, W.C., and Lee, S.F. (2003). Linkage of schizophrenia with chromosome 1q loci in Taiwanese families. *Mol. Psychiatry* **8**, 445–452.

Ikonomov, O.C., and Manji, H.K. (1999). Molecular mechanisms underlying mood stabilization in manic-depressive illness, the phenotype challenge. *Am. J. Psychiatry* **156**(10), 1506–1514.

Insel, T., Cuthbert, B., Garvey, M., Heinssen, R., Pine, D.S., Quinn, K., Sanislow, C., and Wang, P. (2010). Research domain criteria (RDoC), toward a new classification framework for research on mental disorders. *Am. J. Psychiatry* **167**(7), 748–751.

Javitt, D.C. (2007). Glutamate and schizophrenia, phencyclidine, N-methyl-D-aspartate receptors, and dopamine-glutamate interactions. *Int. Rev. Neurobiol.* **78**, 69–108.

Javitt, D.C., and Zukin, S.R. (1991). Recent advances in the phencyclidine model of schizophrenia. *Am. J. Psychiatry* **148**(10), 1301–1308.

Jayatissa, M.N., Henningsen, K., Nikolajsen, G., West, M.J., and Wiborg, O. (2010). A reduced number of hippocampal granule cells does not associate with an anhedonia-like phenotype in a rat chronic mild stress model of depression. *Stress* **13**(2), 95–105.

Jentsch, J.D., Tran, A., Le, D., Youngren, K.D., and Roth, R.H. (1997). Subchronic phencyclidine administration reduces mesoprefrontal dopamine utilization and impairs prefrontal cortical-dependent cognition in the rat. *Neuropsychopharmacology* **17**(2), 92–99.

Jiménez-Marín, Á., Collado-Romero, M., Ramirez-Boo, M., Arce, C., and Garrido, J.J. (2009). Biological pathway analysis by ArrayUnlock and Ingenuity Pathway Analysis. *BMC Proc.* **3** (Suppl. 4), S6.

Kabbaj, M., Norton, C.S., Kollack-Walker, S., Watson, S.J., Robinson, T.E., and Akil, H. (2001). Social defeat alters the acquisition of cocaine self-administration in rats, role of individual differences in cocaine-taking behavior. *Psychopharmacology (Berl.)* **158**(4), 382–387.

Kaddurah-Daouk, R., McEvoy, J., Baillie, R.A., Lee, D., Yao, J.K., Doraiswamy, P.M., *et al.* (2007). Metabolomic mapping of atypical antipsychotic effects in schizophrenia. *Mol. Psychiatry* **12**(10), 934–945.

Karayiorgou, M., Morris, M.A., Morrow, B., Shprintzen, R.J., Goldberg, R., Borrow, J., Gos, A., Nestadt, G., Wolyniec, P.S., and Lasseter, V.K. (1995). Schizophrenia susceptibility associated with interstitial deletions of chromosome 22q11. *Proc. Natl. Acad. Sci. USA* **92**(17), 7612–7616.

Karayiorgou, M., Simon, T.J., and Gogos, J.A. (2010). 22q11.2 microdeletions, linking DNA structural variation to brain dysfunction and schizophrenia. *Nat. Rev. Neurosci.* **11**(6), 402–416.

Karege, F., Vaudan, G., Schwald, M., Perroud, N., and La Harpe, R. (2005). Neurotrophin levels in postmortem brains of suicide victims and the effects of antemortem diagnosis and psychotropic drugs. *Brain Res. Mol. Brain Res.* **136**(1–2), 29–37.

Kavelaars, A., Cobelens, P.M., Teunis, M.A., and Heijnen, C.J. (2005). Changes in innate and acquired immune responses in mice with targeted deletion of the dopamine transporter gene. *J. Neuroimmunol.* **161**(1–2), 162–168.

Keilhoff, G., Becker, A., Grecksch, G., Wolf, G., and Bernstein, H.G. (2004a). Repeated application of ketamine to rats induces changes in the hippocampal expression of parvalbumin, neuronal nitric oxide synthase and cFOS similar to those found in human schizophrenia. *Neuroscience* **126**(3), 591–598.

Keilhoff, G., Bernstein, H.G., Becker, A., Grecksch, G., and Wolf, G. (2004b). Increased neurogenesis in a rat ketamine model of schizophrenia. *Biol. Psychiatry* **56**(5), 317–322.

Kimber, W.L., Hsieh, P., Hirotsune, S., Yuva-Paylor, L., Sutherland, H.F., Chen, A., Ruiz-Lozano, P., Hoogstraten-Miller, S.L., Chien, K.R., Paylor, R., Scambler, P.J., and Wynshaw-Boris, A. (1999). Deletion of 150 kb in the minimal DiGeorge/velocardiofacial syndrome critical region in mouse. *Hum. Mol. Genet.* **8**(12), 2229–2237.

Koike, H., Arguello, P.A., Kvajo, M., Karayiorgou, M., and Gogos, J.A. (2006). Disc1 is mutated in the 129S6/SvEv strain and modulates working memory in mice. *Proc. Natl. Acad. Sci. USA* **103**(10), 3693–3697.

Kondziella, D., Brenner, E., Eyjolfsson, E.M., Markinhuhta, K.R., Carlsson, M.L., and Sonnewald, U. (2006). Glial-neuronal interactions are impaired in the schizophrenia model of repeated MK801 exposure. *Neuropsychopharmacology* **31**(9), 1880–1887.

Korotkova, T., Fuchs, E.C., Ponomarenko, A., von Engelhardt, J., and Monyer, H. (2010). NMDA receptor ablation on parvalbumin-positive interneurons impairs hippocampal synchrony, spatial representations, and working memory. *Neuron* **68**(3), 557–569.

Kovanen, L., Saarikoski, S., Aromaa, A., Lönnqvist, J., and Partonen, T. (2010). ARNTL (BMAL1) and NPAS2 gene variants contribute to fertility and seasonality. *PLoS One* **5**(4), e10007.

Kvajo, M., McKellar, H., Arguello, P.A., Drew, L.J., Moore, H., MacDermott, A.B., Karayiorgou, M., and Gogos, J.A. (2008). A mutation in mouse Disc1 that models a schizophrenia risk allele leads to specific alterations in neuronal architecture and cognition. *Proc. Natl. Acad. Sci. USA* **105**(19), 7076–7081.

Ladd, C.O., Huot, R.L., Thrivikraman, K.V., Nemeroff, C.B., Meaney, M.J., and Plotsky, P.M. (2000). Long-term behavioral and neuroendocrine adaptations to adverse early experience. *Prog. Brain Res.* **122**, 81–103.

Laron, Z. (2009). Insulin and the brain. *Arch. Physiol. Biochem.* **115**(2), 112–116.

Le Pen, G., and Moreau, J.L. (2002). Disruption of prepulse inhibition of startle reflex in a neurodevelopmental model of schizophrenia, reversal by clozapine, olanzapine and risperidone but not by haloperidol. *Neuropsychopharmacology* **27**(1), 1–11.

Le Pen, G., Gourevitch, R., Hazane, F., Hoareau, C., Jay, T.M., and Krebs, M.O. (2006). Peri-pubertal maturation after developmental disturbance, a model for psychosis onset in the rat. *Neuroscience* **143**(2), 395–405.

Lepsch, L.B., Gonzalo, L.A., Magro, F.J., Delucia, R., Scavone, C., and Planeta, C.S. (2005). Exposure to chronic stress increases the locomotor response to cocaine and the basal levels of corticosterone in adolescent rats. *Addict. Biol.* **10**(3), 251–256.

Levin, Y., Wang, L., Schwarz, E., Koethe, D., Leweke, F.M., and Bahn, S. (2010). Global proteomic profiling reveals altered proteomic signature in schizophrenia serum. *Mol. Psychiatry* **15**(11), 1088–1100.

Li, Q., Cheung, C., Wei, R., Hui, E.S., Feldon, J., Meyer, U., Chung, S., Chua, S.E., Sham, P.C., Wu, E. X., and McAlonan, G.M. (2009). Prenatal immune challenge is an environmental risk factor for brain and behavior change relevant to schizophrenia, evidence from MRI in a mouse model. *PLoS One* **4**(7), e6354.

Lipska, B.K., Jaskiw, G.E., and Weinberger, D.R. (1993). Postpubertal emergence of hyperresponsiveness to stress and to amphetamine after neonatal excitotoxic hippocampal damage, a potential animal model of schizophrenia. *Neuropsychopharmacology* **9**(1), 67–75.

Lipska, B.K., Chrapusta, S.J., Egan, M.F., and Weinberger, D.R. (1995a). Neonatal excitotoxic ventral hippocampal damage alters dopamine response to mild repeated stress and to chronic haloperidol. *Synapse* **20**(2), 125–130.

Lipska, B.K., Swerdlow, N.R., Geyer, M.A., Jaskiw, G.E., Braff, D.L., and Weinberger, D.R. (1995b). Neonatal excitotoxic hippocampal damage in rats causes post-pubertal changes in prepulse inhibition of startle and its disruption by apomorphine. *Psychopharmacology (Berl.)* **122**(1), 35–43.

Lodge, D.J., and Lawrence, A.J. (2003). The effect of isolation rearing on volitional ethanol consumption and central CCK/dopamine systems in Fawn-Hooded rats. *Behav. Brain Res.* **141**(2), 113–122.

Long, J.M., LaPorte, P., Merscher, S., Funke, B., Saint-Jore, B., Puech, A., Kucherlapati, R., Morrow, B. E., Skoultchi, A.I., and Wynshaw-Boris, A. (2006). Behavior of mice with mutations in the conserved region deleted in velocardiofacial/DiGeorge syndrome. *Neurogenetics* **7**(4), 247–257.

Lu, Y., Huggins, P., and Bar-Joseph, Z. (2009). Cross species analysis of microarray expression data. *Bioinformatics* **25**(12), 1476–1483.

Ma, D., Chan, M.K., Lockstone, H.E., Pietsch, S.R., Jones, D.N., Cilia, J., *et al.* (2009). Antipsychotic treatment alters protein expression associated with presynaptic function and nervous system development in rat frontal cortex. *J. Proteome Res.* **8**(7), 3284–3297.

MacQueen, G.M., Campbell, S., McEwen, B.S., Macdonald, K., Amano, S., Joffe, R.T., Nahmias, C., and Young, L.T. (2003). Course of illness, hippocampal function, and hippocampal volume in major depression. *Proc. Natl. Acad. Sci. USA* **100**(3), 1387–1392.

Marichich, E.S., Molina, V.A., and Orsingher, O.A. (1979). Persistent changes in central catecholaminergic system after recovery of perinatally undernourished rats. *J. Nutr.* **109**(6), 1045–1050.

Marini, F., Pozzato, C., Andreetta, V., Jansson, B., Arban, R., Domenici, E., and Carboni, L. (2006). Single exposure to social defeat increases corticotropin-releasing factor and glucocorticoid receptor mRNA expression in rat hippocampus. *Brain Res.* **1067**(1), 25–35.

Markianos, M., Panas, M., Kalfakis, N., Hatzimanolis, J., and Vassilopoulos, D. (2010). Neuroendocrine evidence of normal hypothalamus-pituitary dopaminergic function in Huntington's disease. *Neuro Endocrinol. Lett.* **31**(3), 359–362.

Marmol, F. (2008). Lithium, bipolar disorder and neurodegenerative diseases possible cellular mechanisms of the therapeutic effects of lithium. *Prog. Neuropsychopharmacol. Biol. Psychiatry* **32**(8), 1761–1771.

Martins-de-Souza, D., Harris, L.W., Guest, P.C., and Bahn, S. (2010). The role of energy metabolism dysfunction and oxidative stress in schizophrenia revealed by proteomics. *Antioxid. Redox Signal.* Dec 2. (Epub ahead of print).

McArthur, R., and Borsini, F. (2006). Animal models of depression in drug discovery, a historical perspective. *Pharmacol. Biochem. Behav.* **84**(3), 436–452.

McClung, C.A., Sidiropoulou, K., Vitaterna, M., Takahashi, J.S., White, F.J., Cooper, D.C., and Nestler, E.J. (2005). Regulation of dopaminergic transmission and cocaine reward by the Clock gene. *Proc. Natl. Acad. Sci. USA* **102**(26), 9377–9381.

McGuire, J., Herman, J.P., Horn, P.S., Sallee, F.R., and Sah, R. (2010). Enhanced fear recall and emotional arousal in rats recovering from chronic variable stress. *Physiol. Behav.* **101**(4), 474–482.

McLachlan, G.J., Bean, R.W., and Peel, D. (2002). A mixture model-based approach to the clustering of microarray expression data. *Bioinformatics* **18**(3), 413–422.

McLoughlin, G.A., Ma, D., Tsang, T.M., Jones, D.N., Cilia, J., Hill, M.D., *et al.* (2009). Analyzing the effects of psychotropic drugs on metabolite profiles in rat brain using 1H NMR spectroscopy. *J. Proteome Res.* **8**(4), 1943–1952.

McQueen, M.B., Devlin, B., Faraone, S.V., Nimgaonkar, V.L., Sklar, P., Smoller, J.W., Abou Jamra, R., Albus, M., Bacanu, S.-A., Baron, M., Barrett, T.B., Berrettini, W., *et al.* (2005). Combined analysis from eleven linkage studies of bipolar disorder provides strong evidence of susceptibility loci on chromosomes 6q and 8q. *Am. J. Hum. Genet.* **77**(4), 582–595.

Meaney, M.J. (2001). Maternal care, gene expression, and the transmission of individual differences in stress reactivity across generations. *Annu. Rev. Neurosci.* **24**, 1161–1192.

Meechan, D.W., Tucker, E.S., Maynard, T.M., and LaMantia, A.S. (2009). Diminished dosage of 22q11 genes disrupts neurogenesis and cortical development in a mouse model of 22q11 deletion/DiGeorge syndrome. *Proc. Natl. Acad. Sci. USA* **106**(38), 16434–16445.

Meerlo, P., de Bruin, E.A., Strijkstra, A.M., and Daan, S. (2001). A social conflict increases EEG slow-wave activity during subsequent sleep. *Physiol. Behav.* **73**(3), 331–335.

Meltzer, H.Y., and Stahl, S.M. (1976). The dopamine hypothesis of schizophrenia, a review. *Schizophr. Bull.* **2**(1), 19–76.

Meng, Q., Li, N., Han, X., Shao, F., and Wang, W. (2011). Effects of adolescent social isolation on the expression of brain-derived neurotrophic factors in the forebrain. *Eur. J. Pharmacol.* **650**(1), 229–232.

Meunier, B., Dumas, E., Piec, I., Béchet, D., Hébraud, M., and Hocquette, J. (2007). Assessment of hierarchical clustering methodologies for proteomic data mining. *J. Proteome Res.* **6**(1), 358–366.

Meyer, U., and Feldon, J. (2010). Epidemiology-driven neurodevelopmental animal models of schizophrenia. *Prog. Neurobiol.* **90**(3), 285–326.

Meyer, U., Nyffeler, M., Schwendener, S., Knuesel, I., Yee, B.K., and Feldon, J. (2008). Relative prenatal and postnatal maternal contributions to schizophrenia-related neurochemical dysfunction after in utero immune challenge. *Neuropsychopharmacology* **33**(2), 441–456.

Michailov, G.V., Sereda, M.W., Brinkmann, B.G., Fischer, T.M., Haug, B., Birchmeier, C., Role, L., Lai, C., Schwab, M.H., and Nave, K.A. (2004). Axonal neuregulin-1 regulates myelin sheath thickness. *Science* **304**, 700–703.

Mikova, O., Yakimova, R., Bosmans, E., Kenis, G., and Maes, M. (2001). Increased serum tumor necrosis factor alpha concentrations in major depression and multiple sclerosis. *Eur. Neuropsychopharmacol.* **11**(3), 203–208.

Mineur, Y.S., Belzung, C., and Crusio, W.E. (2006). Effects of unpredictable chronic mild stress on anxiety and depression-like behavior in mice. *Behav. Brain Res.* **175**(1), 43–50.

Miyamoto, S., Leipzig, J.N., Lieberman, J.A., and Duncan, G.E. (2000). Effects of ketamine, MK-801, and amphetamine on regional brain 2-deoxyglucose uptake in freely moving mice. *Neuropsychopharmacology* **22**(4), 400–412.

Moghaddam, B., Adams, B., Verma, A., and Daly, D. (1997). Activation of glutamatergic neurotransmission by ketamine, a novel step in the pathway from NMDA receptor blockade to dopaminergic and cognitive disruptions associated with the prefrontal cortex. *J. Neurosci.* **17**(8), 2921–2927.

Mohn, A.R., Gainetdinov, R.R., Caron, M.G., and Koller, B.H. (1999). Mice with reduced NMDA receptor expression display behaviors related to schizophrenia. *Cell* **98**(4), 427–436.

Moore, H., Jentsch, J.D., Ghajarnia, M., Geyer, M.A., and Grace, A.A. (2006). A neurobehavioral systems analysis of adult rats exposed to methylazoxymethanol acetate on E17, implications for the neuropathology of schizophrenia. *Biol. Psychiatry* **60**(3), 253–264.

Moy, S.S., Perez, A., Koller, B.H., and Duncan, G.E. (2006). Amphetamine-induced disruption of prepulse inhibition in mice with reduced NMDA receptor function. *Brain Res.* **1089**(1), 186–194.

Mukai, J., Dhilla, A., Drew, L.J., Stark, K.L., Cao, L., MacDermott, A.B., Karayiorgou, M., and Gogos, J.A. (2008). Palmitoylation-dependent neurodevelopmental deficits in a mouse model of 22q11 microdeletion. *Nat. Neurosci.* **11**(11), 1302–1310.

Müller-Oerlinghausen, B., Berghöfer, A., and Bauer, M. (2002). Bipolar disorder. *Lancet* **359**(9302), 241–247.

Nabeshima, T., Hiramatsu, M., Furukawa, H., and Kameyama, T. (1985). Effects of acute and chronic administrations of phencyclidine on the levels of serotonin and 5-hydroxyindoleacetic acid in discrete brain areas of mouse. *Life Sci.* **36**(10), 939–946.

Nabeshima, T., Kozawa, T., Furukawa, H., and Kameyama, T. (1986). Phencyclidine-induced retrograde amnesia in mice. *Psychopharmacology (Berl.)* **89**(3), 334–337.

Nestler, E.J., and Hyman, S.E. (2010). Animal models of neuropsychiatric disorders. *Nat. Neurosci.* **13** (10), 1161–1169.

Nestler, E.J., Barrot, M., DiLeone, R.J., Eisch, A.J., Gold, S.J., and Monteggia, L.M. (2002). Neurobiology of depression. *Neuron* **34**(1), 13–25.

Nishijima, K., Kashiwa, A., Hashimoto, A., Iwama, H., Umino, A., and Nishikawa, T. (1996). Differential effects of phencyclidine and methamphetamine on dopamine metabolism in rat frontal cortex and striatum as revealed by in vivo dialysis. *Synapse* **22**(4), 304–312.

Noda, A., Noda, Y., Kamei, H., Ichihara, K., Mamiya, T., Nagai, T., *et al.* (2001). Phencyclidine impairs latent learning in mice, interaction between glutamatergic systems and sigma(1) receptors. *Neuropsychopharmacology* **24**(4), 451–460.

O'Brien, S.M., Scully, P., Scott, L.V., and Dinan, T.G. (2006). Cytokine profiles in bipolar affective disorder: Focus on acutely ill patients. *J. Affect. Disord.* **90**(2–3), 263–267.

O'Tuathaigh, C.M., Babovic, D., O'Sullivan, G.J., Clifford, J.J., Tighe, O., Croke, D.T., Harvey, R., and Waddington, J.L. (2007). Phenotypic characterization of spatial cognition and social behavior in mice with 'knockout' of the schizophrenia risk gene neuregulin 1. *Neuroscience* **147**(1), 18–27.

O'Tuathaigh, C.M., O'Connor, A.M., O'Sullivan, G.J., Lai, D., Harvey, R., Croke, D.T., and Waddington, J.L. (2008). Disruption to social dyadic interactions but not emotional/anxiety-related behaviour in mice with heterozygous 'knockout' of the schizophrenia risk gene neuregulin-1. *Prog. Neuropsychopharmacol. Biol. Psychiatry* **32**(2), 462–466.

Orzi, F., Dow-Edwards, D., Jehle, J., Kennedy, C., and Sokoloff, L. (1983). Comparative effects of acute and chronic administration of amphetamine on local cerebral glucose utilization in the conscious rat. *J. Cereb. Blood Flow Metab.* **3**(2), 154–160.

Palmer, A.A., Printz, D.J., Butler, P.D., Dulawa, S.C., and Printz, M.P. (2004). Prenatal protein deprivation in rats induces changes in prepulse inhibition and NMDA receptor binding. *Brain Res.* **996**(2), 193–201.

Palmer, A.A., Brown, A.S., Keegan, D., Siska, L.D., Susser, E., Rotrosen, J., and Butler, P.D. (2008). Prenatal protein deprivation alters dopamine-mediated behaviors and dopaminergic and glutamatergic receptor binding. *Brain Res.* **1237**, 62–74.

Paule, M.G. (1994). Acute behavioral toxicity of MK-801 and phencyclidine, effects on rhesus monkey performance in an operant test battery. *Psychopharmacol. Bull.* **30**(4), 613–621.

Paylor, R., McIlwain, K.L., McAninch, R., Nellis, A., Yuva-Paylor, L.A., Baldini, A., and Lindsay, E.A. (2001). Mice deleted for the DiGeorge/velocardiofacial syndrome region show abnormal sensorimotor gating and learning and memory impairments. *Hum. Mol. Genet.* **10**(23), 2645–2650.

Pechnick, R.N., George, R., Poland, R.E., Hiramatsu, M., and Cho, A.K. (1989). Characterization of the effects of the acute and chronic administration of phencyclidine on the release of adrenocorticotropin, corticosterone and prolactin in the rat, evidence for the differential development of tolerance. *J. Pharmacol. Exp. Ther.* **250**(2), 534–540.

Pellerin, L. (2003). Lactate as a pivotal element in neuron-glia metabolic cooperation. *Neurochem. Int.* **43** (4–5), 331–338.

Perona, M.T., Waters, S., Hall, F.S., Sora, I., Lesch, K.P., Murphy, D.L., Caron, M., and Uhl, G.R. (2008). Animal models of depression in dopamine, serotonin, and norepinephrine transporter knockout mice, prominent effects of dopamine transporter deletions. *Behav. Pharmacol.* **19**(5–6), 566–574.

Pizarro, J.M., Lumley, L.A., Medina, W., Robison, C.L., Chang, W.E., Alagappan, A., Bah, M.J., Dawood, M.Y., Shah, J.D., Mark, B., Kendall, N., Smith, M.A., *et al.* (2004). Acute social defeat reduces neurotrophin expression in brain cortical and subcortical areas in mice. *Brain Res.* **1025**(1–2), 10–20.

Pletnikov, M.V., Ayhan, Y., Nikolskaia, O., Xu, Y., Ovanesov, M.V., Huang, H., Mori, S., Moran, T.H., and Ross, C.A. (2008). Inducible expression of mutant human DISC1 in mice is associated with brain and behavioral abnormalities reminiscent of schizophrenia. *Mol. Psychiatry* **13**(2), 173–186, 115.

Pogorelov, V.M., Rodriguiz, R.M., Insco, M.L., Caron, M.G., and Wetsel, W.C. (2005). Novelty seeking and stereotypic activation of behavior in mice with disruption of the Dat1 gene. *Neuropsychopharmacology* **30**(10), 1818–1831.

Pogorelov, V.M., Nomura, J., Kim, J., Kannan, G., Ayhan, Y., Yang, C., Taniguchi, Y., Abazyan, B., Valentine, H., Krasnova, I.N., Kamiya, A., Cadet, J.L., *et al.* (2011). Mutant DISC1 affects methamphetamine-induced sensitization and conditioned place preference, a comorbidity model. *Neuropharmacology* Feb 17. (Epub ahead of print).

Powell, N.D., Bailey, M.T., Mays, J.W., Stiner-Jones, L.M., Hanke, M.L., Padgett, D.A., and Sheridan, J.F. (2009). Repeated social defeat activates dendritic cells and enhances Toll-like receptor dependent cytokine secretion. *Brain Behav. Immun.* **23**(2), 225–231.

Prickaerts, J., Moechars, D., Cryns, K., Lenaerts, I., van Craenendonck, H., Goris, I., Daneels, G., Bouwknecht, J.A., and Steckler, T. (2006). Transgenic mice overexpressing glycogen synthase kinase 3 beta, A putative model of hyperactivity and mania. *J. Neurosci.* **26**(35), 9022–9029.

Pugh, C.R., Nguyen, K.T., Gonyea, J.L., Fleshner, M., Wakins, L.R., Maier, S.F., and Rudy, J.W. (1999). Role of interleukin-1 beta in impairment of contextual fear conditioning caused by social isolation. *Behav. Brain Res.* **106**(1–2), 109–118.

Rajkowska, G. (2000). Postmortem studies in mood disorders indicate altered numbers of neurons and glial cells. *Biol. Psychiatry* **48**(8), 766–777.

Ralph, R.J., Paulus, M.P., Fumagalli, F., Caron, M.G., and Geyer, M.A. (2001). Prepulse inhibition deficits and perseverative motor patterns in dopamine transporter knock-out mice, differential effects of D1 and D2 receptor antagonists. *J. Neurosci.* **21**(1), 305–313.

Ramsey, A.J., Laakso, A., Cyr, M., Sotnikova, T.D., Salahpour, A., Medvedev, I.O., Dykstra, L.A., Gainetdinov, R.R., and Caron, M.G. (2008). Genetic NMDA receptor deficiency disrupts acute and chronic effects of cocaine but not amphetamine. *Neuropsychopharmacology* **33**(11), 2701–2714.

Reddy, A.B., Karp, N.A., Maywood, E.S., Sage, E.A., Deery, M., O'Neill, J.S., Wong, G.K.Y., Chesham, J., Odell, M., Lilley, K.S., Kyriacou, C.P., and Hastings, M.H. (2006). Circadian orchestration of the hepatic proteome. *Curr. Biol.* **16**, 1107–1115.

Rimer, M., Barrett, D.W., Maldonado, M.A., Vock, V.M., and Gonzalez-Lima, F. (2005). Neuregulin-1 immunoglobulin-like domain mutant mice, clozapine sensitivity and impaired latent inhibition. *Neuroreport* **16**(3), 271–275.

Rodriguiz, R.M., Chu, R., Caron, M.G., and Wetsel, W.C. (2004). Aberrant responses in social interaction of dopamine transporter knockout mice. *Behav. Brain Res.* **148**(1–2), 185–198.

Romero, E., Guaza, C., Castellano, B., and Borrell, J. (2010). Ontogeny of sensorimotor gating and immune impairment induced by prenatal immune challenge in rats, implications for the etiopathology of schizophrenia. *Mol. Psychiatry* **15**(4), 372–383.

Roybal, K., Theobold, D., Graham, A., DiNieri, J.A., Russo, S.J., Krishnan, V., Chakravarty, S., Peevey, J., Oehrlein, N., Birnbaum, S., Vitaterna, M.H., Orsulak, P., *et al.* (2007). Mania-like behavior induced by disruption of CLOCK. *Proc. Natl. Acad. Sci. USA* **104**(15), 6406–6411.

Rygula, R., Abumaria, N., Flugge, G., Fuchs, E., Ruther, E., and Havemann-Reinecke, U. (2005). Anhedonia and motivational deficits in rats, impact of chronic social stress. *Behav. Brain Res.* **162**(1), 127–134.

Salvadore, G., Quiroz, J., Machado-Vieira, R., Henter, I.D., Manji, H.K., and Zarate, C.A. Jr. (2010). The neurobiology of the switch process in bipolar disorder, a review. *J. Clin. Psychiatry* **71**(11), 1488–1501.

Sampaio, A.S., Fagerness, , Crane, J., Leboyer, M., Delorme, R., Pauls, D.L., and Stewart, S.E. (2010). Association between polymorphisms in GRIK2 gene and obsessive-compulsive disorder, a family-based study. *CNS Neurosci. Ther.* **17**, 141–147.

Sams-Dodd, F. (1995). Distinct effects of d-amphetamine and phencyclidine on the social behaviour of rats. *Behav. Pharmacol.* **6**(1), 55–65.

Sams-Dodd, F., Lipska, B.K., and Weinberger, D.R. (1997). Neonatal lesions of the rat ventral hippocampus result in hyperlocomotion and deficits in social behaviour in adulthood. *Psychopharmacology (Berl.)* **132**(3), 303–310.

Sanberg, P.R., Pevsner, J., Autuono, P.G., and Coyle, J.T. (1985). Fetal methylazoxymethanol acetate-induced lesions cause reductions in dopamine receptor-mediated catalepsy and stereotypy. *Neuropharmacology* **24**(11), 1057–1062.

Sandstrom, N.J., and Hart, S.R. (2005). Isolation stress during the third postnatal week alters radial arm maze performance and corticosterone levels in adulthood. *Behav. Brain Res.* **156**(2), 289–296.

Sarnyai, Z., Shaham, Y., and Heinrichs, S.C. (2001). The role of corticotropin-releasing factor in drug addiction. *Pharmacol. Rev.* **53**(2), 209–243.

Sei, H., Sano, A., Oishi, K., Fujihara, H., Kobayashi, H., Ishida, N., and Morita, Y. (2003). Increase of hippocampal acetylcholine release at the onset of dark phase is suppressed in a mutant mice model of evening-type individuals. *Neuroscience* **117**(4), 785–789.

Seillier, A., and Giuffrida, A. (2009). Evaluation of NMDA receptor models of schizophrenia, divergences in the behavioral effects of sub-chronic PCP and MK-801. *Behav. Brain Res.* **204**(2), 410–415.

Shaltiel, G., Chen, G., and Manji, H.K. (2007). Neurotrophic signaling cascades in the pathophysiology and treatment of bipolar disorder. *Curr. Opin. Pharmacol.* **7**(1), 22–26.

Shaltiel, G., Maeng, S., Malkesman, O., Pearson, B., Schloesser, R.J., Tragon, T., Rogawski, M., Gasior, M., Luckenbaugh, D., Chen, G., and Manji, H.K. (2008). Evidence for the involvement of the kainate receptor subunit GluR6 (GRIK2) in mediating behavioral displays related to behavioral symptoms of mania. *Mol. Psychiatry* **13**(9), 858–872.

Shi, L., Fatemi, S.H., Sidwell, R.W., and Patterson, P.H. (2003). Maternal influenza infection causes marked behavioral and pharmacological changes in the offspring. *J. Neurosci.* **23**(1), 297–302.

Silva-Gomez, A.B., Rojas, D., Juarez, I., and Flores, G. (2003). Decreased dendritic spine density on prefrontal cortical and hippocampal pyramidal neurons in postweaning social isolation rats. *Brain Res.* **983**(1–2), 128–136.

Song, L., Che, W., Min-Wei, W., Murakami, Y., and Matsumoto, K. (2006). Impairment of the spatial learning and memory induced by learned helplessness and chronic mild stress. *Pharmacol. Biochem. Behav.* **83**(2), 186–193.

Spielewoy, C., Roubert, C., Hamon, M., Nosten-Bertrand, M., Betancur, C., and Giros, B. (2000). Behavioural disturbances associated with hyperdopaminergia in dopamine-transporter knockout mice. *Behav. Pharmacol.* **11**(3–4), 279–290.

St Clair, D., Blackwood, D., Muir, W., Carothers, A., Walker, M., Spowart, G., Gosden, C., and Evans, H.J. (1990). Association within a family of a balanced autosomal translocation with major mental illness. *Lancet* **336**(8706), 13–16.

Stark, K.L., Xu, B., Bagchi, A., Lai, W.S., Liu, H., Hsu, R., Wan, X., Pavlidis, P., Mills, A.A., Karayiorgou, M., and Gogos, J.A. (2008). Altered brain microRNA biogenesis contributes to phenotypic deficits in a 22q11-deletion mouse model. *Nat. Genet.* **40**(6), 751–760.

Stefansson, H., Sigurdsson, E., Steinthorsdottir, V., Bjornsdottir, S., Sigmundsson, T., Ghosh, S., Brynjolfsson, J., Gunnarsdottir, S., Ivarsson, O., Chou, T.T., Hjaltason, O., Birgisdottir, B., *et al.* (2002). Neuregulin 1 and susceptibility to schizophrenia. *Am. J. Hum. Genet.* **71**, 877–892.

Steiger, J.L., Alexander, M.J., Galler, J.R., Farb, D.H., and Russek, S.J. (2003). Effects of prenatal malnutrition on GABAA receptor alpha1, alpha3 and beta2 mRNA levels. *Neuroreport* **14**(13), 1731–1735.

Stine, C.D., Lu, W., and Wolf, M.E. (2001). Expression of AMPA receptor flip and flop mRNAs in the nucleus accumbens and prefrontal cortex after neonatal ventral hippocampal lesions. *Neuropsychopharmacology* **24**(3), 253–266.

Strutz-Seebohm, N., Korniychuk, G., Schwarz, R., Baltaev, R., Ureche, O.N., Mack, A.F., Ma, Z.L., Hollmann, M., Lang, F., and Seebohm, G. (2006). Functional significance of the kainate receptor GluR6(M836I) mutation that is linked to autism. *Cell. Physiol. Biochem.* **18**(4–5), 287–294.

Sturgeon, R.D., Fessler, R.G., and Meltzer, H.Y. (1979). Behavioral rating scales for assessing phencyclidine-induced locomotor activity, stereotyped behavior and ataxia in rats. *Eur. J. Pharmacol.* **59** (3–4), 169–179.

Tamminga, C.A., Tanimoto, K., Kuo, S., Chase, T.N., Contreras, P.C., Rice, K.C., *et al.* (1987). PCP-induced alterations in cerebral glucose utilization in rat brain, blockade by metaphit, a PCP-receptor-acylating agent. *Synapse* **1**(5), 497–504.

Tenn, C.C., Kapur, S., and Fletcher, P.J. (2005). Sensitization to amphetamine, but not phencyclidine, disrupts prepulse inhibition and latent inhibition. *Psychopharmacology (Berl.)* **180**(2), 366–376.

Thomas, A.J., Davis, S., Morris, C., Jackson, E., Harrison, R., and O'Brien, J.T. (2005). Increase in interleukin-1beta in late-life depression. *Am. J. Psychiatry* **162**(1), 175–177.

Tonkiss, J., and Galler, J.R. (1990). Prenatal protein malnutrition and working memory performance in adult rats. *Behav. Brain Res.* **40**(2), 95–107.

Toth, M., Mikics, E., Tulogdi, A., Aliczki, M., and Haller, J. (2011). Post-weaning social isolation induces abnormal forms of aggression in conjunction with increased glucocorticoid and autonomic stress responses. *Horm. Behav.* **60**, 28–36.

Tseng, K.Y., Lewis, B.L., Hashimoto, T., Sesack, S.R., Kloc, M., Lewis, D.A., and O'Donnell, P. (2008). A neonatal ventral hippocampal lesion causes functional deficits in adult prefrontal cortical interneurons. *J. Neurosci.* **28**(48), 12691–12699.

Uehara, T., Sumiyoshi, T., Seo, T., Matsuoka, T., Itoh, H., Suzuki, M., *et al.* (2010). Neonatal exposure to MK-801, an N-methyl-D-aspartate receptor antagonist, enhances methamphetamine-induced locomotion and disrupts sensorimotor gating in pre- and postpubertal rats. *Brain Res.* **1352**, 223–230.

Usiskin, S.I., Nicolson, R., Krasnewich, D.M., Yan, W., Lenane, M., Wudarsky, M., Hamburger, S.D., and Rapoport, J.L. (1999). Velocardiofacial syndrome in childhood-onset schizophrenia. *J. Am. Acad. Child Adolesc. Psychiatry* **38**(12), 1536–1543.

van den Buuse, M., Wischhof, L., Lee, R.X., Martin, S., and Karl, T. (2009). Neuregulin 1 hypomorphic mutant mice, enhanced baseline locomotor activity but normal psychotropic drug-induced hyperlocomotion and prepulse inhibition regulation. *Int. J. Neuropsychopharmacol.* **12**(10), 1383–1393.

Vaynman, S., Ying, Z., Wu, A., and Gomez-Pinilla, F. (2006). Coupling energy metabolism with a mechanism to support brain-derived neurotrophic factor-mediated synaptic plasticity. *Neuroscience* **139**(4), 1221–1234.

Vidal, J., Bie, J., Granneman, R.A., Wallinga, A.E., Koolhaas, J.M., and Buwalda, B. (2007). Social stress during adolescence in Wistar rats induces social anxiety in adulthood without affecting brain monoaminergic content and activity. *Physiol. Behav.* **92**(5), 824–830.

Vollenweider, F.X., Vollenweider-Scherpenhuyzen, M.F., Babler, A., Vogel, H., and Hell, D. (1998). Psilocybin induces schizophrenia-like psychosis in humans via a serotonin-2 agonist action. *Neuroreport* **9**(17), 3897–3902.

Wang, X.D., Su, Y.A., Guo, C.M., Yang, Y., and Si, T.M. (2008). Chronic antipsychotic drug administration alters the expression of neuregulin 1beta, ErbB2, ErbB3, and ErbB4 in the rat prefrontal cortex and hippocampus. *Int. J. Neuropsychopharmacol.* **11**(4), 553–561.

Wang, M., Pei, L., Fletcher, P.J., Kapur, S., Seeman, P., and Liu, F. (2010). Schizophrenia, amphetamine-induced sensitized state and acute amphetamine exposure all show a common alteration, increased dopamine D2 receptor dimerization. *Mol. Brain* **3**, 25.

Watt, M.J., Burke, A.R., Renner, K.J., and Forster, G.L. (2009). Adolescent male rats exposed to social defeat exhibit altered anxiety behavior and limbic monoamines as adults. *Behav. Neurosci.* **123**(3), 564–576.

Wiley, J.L., Porter, J.H., Compton, A.D., and Balster, R.L. (1992). Antipunishment effects of acute and repeated administration of phencyclidine and NPC 12626 in rats. *Life Sci.* **50**(20), 1519–1528.

Willner, P. (1997). Validity, reliability and utility of the chronic mild stress model of depression, a 10-year review and evaluation. *Psychopharmacology (Berl.)* **134**(4), 319–329.

Woodin, M., Wang, P.P., Aleman, D., McDonald-McGinn, D., Zackai, E., and Moss, E. (2001). Neuropsychological profile of children and adolescents with the 22q11.2 microdeletion. *Genet. Med.* **3**(1), 34–39.

World Health Organisation (2004). The global burden of disease, 2004 update, Geneva, World health Organization Publication.

Wu, W., Yamaura, T., Murakami, K., Ogasawara, M., Hayashi, K., Murata, J., and Saiki, I. (1999). Involvement of TNF-alpha in enhancement of invasion and metastasis of colon 26-L5 carcinoma cells in mice by social isolation stress. *Oncol. Res.* **11**(10), 461–469.

Yamaguchi, K., Nabeshima, T., Ishikawa, K., Yoshida, S., and Kameyama, T. (1987). Phencyclidine-induced head-weaving and head-twitch through interaction with 5-HT1 and 5-HT2 receptors in reserpinized rats. *Neuropharmacology* **26**(10), 1489–1497.

Yanik, M., Vural, H., Tutkun, H., Zoroğlu, S., Savaş, H., Herken, H., Koçyiğit, A., Keleş, H., and Akyol, Ö. (2004). The role of the arginine-nitric oxide pathway in the pathogenesis of bipolar affective disorder. *Eur. Arch. Psychiatry Clin. Neurosci.* **254**(1), 43–47.

Young, J.W., Powell, S.B., Risbrough, V., Marston, H.M., and Geyer, M.A. (2009). Using the MATRICS to guide development of a preclinical cognitive test battery for research in schizophrenia. *Pharmacol. Ther.* **122**(2), 150–202.

Yu, T., Guo, M., Garza, J., Rendon, S., Sun, X.L., Zhang, W., and Lu, X.Y. (2011). Cognitive and neural correlates of depression-like behaviour in socially defeated mice, an animal model of depression with cognitive dysfunction. *Int. J. Neuropsychopharmacol.* **14**(3), 303–317.

Zuo, D.Y., Zhang, Y.H., Cao, Y., Wu, C.F., Tanaka, M., and Wu, Y.L. (2006). Effect of acute and chronic MK-801 administration on extracellular glutamate and ascorbic acid release in the prefrontal cortex of freely moving mice on line with open-field behavior. *Life Sci.* **78**(19), 2172–2178.

STEM CELL MODELS FOR BIOMARKER DISCOVERY IN BRAIN DISEASE

Alan Mackay-Sim, George Mellick and Stephen Wood

National Centre for Adult Stem Cell Research, Eskitis Institute for Cell and Molecular Therapies, Griffith University, Brisbane, Queensland, Australia

Abstract

Most brain diseases arise from interactions between complex genetic and environmental risk factors. Finding biomarkers for brain diseases will require appropriate cellular models to identify dysregulated cell functions and disease-associated biochemistries. Patient-derived stem cells hold great potential as models of brain diseases. Stem cells can proliferate and can be banked, stored, and thawed for genomic, proteomic, and functional studies. Patient-derived, induced pluripotent stem cells and adult stem cells from the olfactory organ in the nose are already giving novel insights into a number of brain diseases, including Parkinson's disease and schizophrenia. Biomarker discovery may be possible from investigating disease-associated cell biologies in patient-derived stem cells.

INTERNATIONAL REVIEW OF
NEUROBIOLOGY, VOL. 101
DOI: 10.1016/B978-0-12-387718-5.00009-2

I. Introduction

The brain is the most complex organ in the human body. The intricacies of human evolution have resulted in unique species-specific anatomical and neurobiological distinctions between the human brain and the nervous systems of other higher vertebrates. As a consequence, many brain diseases and their related phenotypes (which we use here to refer essentially to phenotypes of interest to clinical neurology and psychiatry) are not only complex in their nature, but restricted to humans. This presents specific challenges for experimental scientists interested in (1) learning more about the etiology of common human brain disease, (2) developing biomarkers to further explore and monitor the nature of these conditions, and (3) discovering and testing molecular interventions targeted at modulating these phenotypes.

II. Need for Cellular Models

Most brain diseases (neurological conditions and psychiatric disorders) arise from multiple gene–environment interactions, with a minority of diseases and cases arising from mutations in single genes (Wray *et al.*, 2008). Importantly, even "monogenic" diseases differ in age of onset and penetrance in carriers of a mutation, indicating that other factors may be involved, genetic or environmental (Summers, 1996). Given that most human disease phenotypes are not observed naturally in other species, it is often difficult to directly model human brain disease in animal models, even when human gene mutations are introduced. This makes human tissue essential for understanding human brain disease. As might be expected, human *postmortem* brain tissue, both from individuals who are nominally "healthy" or those with particular brain phenotypes, "patients," continues to be critical for this work. However, human brain samples are difficult to obtain in large numbers, particularly from patients with uncommon phenotypes, and pathological experiments are often confounded by difficult-to-control artifacts resulting from the disease process itself (Sutherland *et al.*, 2009) biased sample ascertainment and the necessity to process tissue in a timely manner following death (Atz *et al.*, 2007; Marcotte *et al.*, 2003; Preece and Cairns, 2003). Moreover, experiments on *postmortem* brain tissue are largely static in nature, restricted to observations at single points in time (namely at the death of the donor), and not easily amenable to dynamic manipulation. Thus, alternative experimental tools are required. Human cell models of brain diseases may be useful in this respect as they can be derived from human sources; are available in sufficient quantity to enable manipulation and widespread investigation among the research community; and they form a logical link between human clinical studies, *postmortem* pathological investigations, and animal model systems. In this review, we will provide an overview of how cellular models are being used to advance our understanding of brain diseases, and their current and likely future roles in the development of biomarkers and novel treatments for these conditions.

III. Understanding Disease

In many brain diseases, the fundamental underlying etiology remains elusive. For example, the causes of schizophrenia (SZ) or Parkinson's disease (PD; two phenotypes of interest to the authors) remain largely unknown despite decades of tireless investigation by many groups. Population science and advances in human genetic analyses have provided clues and generated hypotheses. Epidemiology has identified risk factors associated with both diseases, while the study of rare familial forms of these conditions has revealed a small number of gene mutations that can lead to disease (Mackay-Sim *et al.*, 2004; Sutherland *et al.*, 2011). The large-scale genome-wide association studies (GWAS) have also uncovered common genetic variants with modest effects on risk, but these account for only a fraction of the suspected genetic contribution to the phenotypes (Gershon *et al.*, 2011; Nalls *et al.*, 2011). Taken together, these efforts are helping to develop more refined hypotheses for the diseases. The use of cellular models of these brain diseases is one approach that may help to build functional flesh onto the skeletons that are the emerging hypothesis generated by these efforts (Brennand *et al.*, 2011; Matigian *et al.*, 2010; Soldner *et al.*, 2009; Wang *et al.*, 2009). In particular, cell models enable the genetic candidates, emerging out of these etiological discovery efforts, to be interrogated at a molecular level. Many of the products of these gene candidates have unknown or unclear biological function and the influence of the disease-associated genetic variants on this function are also poorly understood. Moreover, the molecular interactions between identified candidate genes, proteins, and metabolites, which may be revealed through the interrogation of a cellular model, can also be further investigated to reveal networks of connections that may be more informative than individual elements in defining a disease phenotype. A goal of such work is to provide a better means to identify individuals at risk and to monitor disease progress and prognosis. In this respect, cell models of human brain diseases are providing important leads from which biomarkers may emerge. Ultimately these studies may also be used to discover novel targets for interventional strategies for therapeutic purposes.

IV. Biomarker Discovery

Biomarkers are characteristics that are objectively measured and evaluated as indicators of normal biologic processes, pathogenic processes, or pharmacologic responses to a therapeutic intervention. As such, they can provide information about a brain disease at several different levels (see Chapter "General overview: biomarkers in neuroscience research" by Filiou and Turck). For example, they may identify "at-risk" individuals, be used as a diagnostic tool, help to assess disease progression or prognosis, or assess treatment responsiveness (Mellick *et al.*, 2010). Biomarkers can have this capacity through a number of possible mechanisms. For example, they may directly reflect a physiological process involved with disease causation (casual mechanism), they may mirror the body's response to an already-established disease process (reactive mechanism), or they may reflect a process that accompanies or is coincident with the disease (independent mechanism). Biomarkers can be proteins, genes, or any biological variable that can be measured in a patient (e.g., brain images). They can derive from blood, urine, cerebrospinal fluid, saliva, and any body secretion or type of scan. Ideal biomarkers are molecular—gene, mRNA, protein, etc. These are preferred because they can be precisely defined and easily identified in different laboratories. At the regulatory level, these are necessary components for validation and clinical use. The ideal biomarker is one that is obtained with the least invasive procedure and usually one that can be obtained quickly. Patient-derived cells can be used as biomarkers, either directly (e.g., the changed morphology of red blood cells in sickle cell anemia) or indirectly (e.g., the agglutination response in blood typing). To be effective as biomarkers, cell models also need to be accessible and relevant to the disease process, with face validity, predictability, and generalizablity (Mellick *et al.*, 2010).

V. Accessible Cells for Biomarker Discovery in Brain Diseases

Understanding the cellular basis for many diseases is limited because of a lack of access to living cells affected by the disease. There are relatively few biomarkers for brain diseases that have the sensitivity and specificity to apply to individuals. Blood cells, obtained quickly and easily, make ideal sources for biomarkers. Lymphocytes and red blood cells are used routinely in hematology studies but they are limited for brain diseases by their lack of obvious tissue relevance. Blood cells are a good source of DNA for genotyping (as are cells from hair and cheek), but they are limited in number for most cellular analyses. This limitation is often overcome by transformation into lymphoblastoid cell lines but this genetic modification also changes their phenotype, further reducing their validity. Fibroblasts, another commonly used patient-derived cell type, have potential because they can be grown in large number without genetic modification. Skin fibroblasts and transformed lymphocytes have allowed the identification of differences in gene expression, protein expression, and cell functions in SZ and PD (Bowden *et al.*, 2006; Cohen *et al.*, 1987; del Hoyo *et al.*, 2010; Gavin *et al.*, 2009; Hoepken *et al.*, 2008; Mahadik *et al.*, 1991; Miyamae *et al.*, 1998; Mytilineou *et al.*, 1994; Ramchand *et al.*, 1994; Suzuki *et al.*, 2008; Vawter *et al.*, 2004; Wang *et al.*, 2009; Zhubi *et al.*, 2009), although patient-control gene expression differences in SZ can be elusive (Matigian *et al.*, 2008). These nonneural cell models may reflect systemic changes in cell biology associated with brain diseases (e.g., through genetic mutation) and may provide indirect biomarkers (through the independent mechanism defined above) associated with brain diseases, without being appropriate models of disease processes in the cells of the brain.

The utility of patient-derived cells diminishes with the cost of individual biopsy and cell manipulation. Cells requiring culture and other manipulation are likely to be less practical for individual patient assessments (compared to other biomarkers) unless the value of the data is high enough to outweigh the high financial and time costs. A more proximate use for patient-derived cells is their utility as a discovery platform for novel molecular biomarkers. In this aspect, patient-derived stem cells offer exciting prospects.

VI. Patient-Derived Stem Cells for Biomarker Discovery in Brain Diseases

Unlike other organs, taking biopsy samples of the brain is problematic, to say the least. Patient-derived stem cells have the potential to fill this void because they carry the genetic makeup of the patient and can theoretically be induced to differentiate into the cells affected in the disease (e.g., motor neurons for the study of amyotrophic lateral sclerosis or dopaminergic neurons for examining PD). In this context, patient-derived stem cells allow molecular investigations in disease-relevant cell models and thus offer great potential in the hunt for novel biomarkers (working through the casual or reactive mechanisms). Having identified such a biomarker from a cellular model, an additional challenge is encountered; there is a necessity to define biomarker levels that are sensitive and specific enough, in a clinical context, to discriminate between individuals with or without a particular phenotype or to meaningfully reflect the patients' clinical status. For example, the well-known increase in ventricular volume in SZ does not translate to the individual because variation in the normal population reduces sensitivity and increased ventricular volume occurs in many diseases and conditions, reducing specificity. Similarly, olfactory dysfunction is a sensitive marker of early PD although it lacks specificity (Haehner *et al.*, 2009; Hawkes *et al.*, 1999). Cellular models may help to translate these observational findings into molecular mechanisms. While there are few examples, to date, of patient-derived cell models leading directly to clinically relevant biomarkers measurable in easily obtainable tissues or fluids, data is emerging to suggest that this approach is starting to bear fruit.

VII. Olfactory Mucosa—An Accessible Neural Tissue for Biomarker Discovery

The olfactory mucosa is the organ of smell in the nose. The sense of smell is impaired in many brain diseases, including neurodegenerative diseases (e.g., Alzheimer's disease, PD) and neurodevelopmental disorders (e.g., SZ; Brewer *et al.*, 2003; Doty, 2009; Haehner *et al.*, 2009; Hawkes *et al.*, 1999). The olfactory system is therefore sensitive to neurological disease processes, although the mechanisms and sites of olfactory dysfunctions are not yet identified. There are suggestions that Alzheimer's disease and PD may arise from toxins entering the brain through the olfactory sensory neurons in the nose (Braak *et al.*, 2006; Hawkes *et al.*, 1999). These neurons are exposed to the external environment and sensitive to toxins, leading to their destruction. Normally, the olfactory sensory neurons are replenished through a process of neurogenesis that continues throughout the lifetime of vertebrates, including humans (Graziadei, 1973; Graziadei and Graziadei, 1979; Murrell *et al.*, 1996). Neurogenesis is made possible by "adult" stem cells that reside within the basal cells of the olfactory epithelium (Chen *et al.*, 2004; Leung *et al.*, 2007; Mackay-Sim and Kittel, 1991). The olfactory epithelium is the most superficial layer of the olfactory mucosa, separated from it by a basement membrane. The stem cells of the olfactory epithelium lie along this basement membrane and are multipotent *in situ*, able to reconstitute both the neural and nonneural elements of the olfactory mucosa (Chen *et al.*, 2004; Leung *et al.*, 2007). This regenerating neural tissue can be obtained by biopsy which is easily accessible through the external naris in human adults (Feron *et al.*, 1998). Olfactory mucosa can be grown as organ cultures and as dissociated cells (Feron *et al.*, 1998; Murrell *et al.*, 1996; Newman *et al.*, 2000; Wolozin *et al.*, 1992). When grown *in vitro*, human olfactory stem cells are multipotent, able to generate neurons, and glial cells (Murrell *et al.*, 2005; Roisen *et al.*, 2001) as well as many nonneural cell types of ectodermal, mesodermal, and endodermal origin (Murrell *et al.*, 2005, 2008, 2009).

Several brain diseases show a phenotype in the olfactory mucosa or in cells derived from it, including Alzheimer's disease, Rett syndrome, fragile X syndrome, and SZ (Abrams *et al.*, 1999; Arnold *et al.*, 2001; Feron *et al.*, 1999;

McCurdy *et al.*, 2006; Ronnett *et al.*, 2003; Wolozin *et al.*, 1993). Interestingly, no disease-associated phenotype has been identified in primary cultures of olfactory mucosa in PD (Witt *et al.*, 2009). Of particular interest in the present context, neuroblasts cultivated from patients with Alzheimer's disease showed significant alterations in their biochemical processing of the amyloid precursor polypeptide (Wolozin *et al.*, 1993) and alterations in oxidative damage (Ghanbari *et al.*, 2004), suggesting that this cell model is a relevant model and has the potential to identify biomarkers for this disease. The structure of the olfactory epithelium is disrupted in *postmortem* tissues from patients with SZ indicative of altered neurodevelopment (Arnold *et al.*, 2001), and olfactory mucosa biopsy cultures from SZ patients show a neurodevelopmental phenotype with significantly more cells in mitosis than occurs in cultures from healthy controls or from patients with bipolar disorder (Feron *et al.*, 1999; McCurdy *et al.*, 2006). These reports demonstrate the potential for the olfactory mucosa as a tool to investigate the biology of brain disease phenotypes. The presence of an accessible neural stem cell in the human olfactory mucosa has provided a new cell model for brain diseases as we recently demonstrated for SZ and PD (Matigian *et al.*, 2010).

VIII. Patient-Derived Olfactory Stem Cells as Models for Brain Diseases

Neural stem cells derived from the adult human brain or spinal cord are grown in "neurospheres." These are round, tightly packed spheroids of large numbers of cells that form when stem cells are grown *in vitro* with epidermal growth factor (EGF) and basic fibroblast growth factor (FGF2) (Rietze and Reynolds, 2006). Neurospheres contain a small number of stem cells with their progeny: proliferating neural precursors, and differentiating neurons and glia (Rietze and Reynolds, 2006). Neural stem cells from adult human olfactory mucosa are grown in similar culture conditions (Fig. 1) and, like brain neural stem cells, have been shown to be self-renewing and multipotent, the defining properties of stem cells (Delorme *et al.*, 2010; Murrell *et al.*, 2005; Roisen *et al.*, 2001). In our protocol, olfactory neurosphere (ONS) cells are dissociated and grown in standard conditions as "ONS-derived" cells. These are adhesive cultures which can be frozen, banked, thawed, and regrown in quantity for gene and protein expression analyses and functional investigations. ONS cells can be generated from large numbers of patients and controls, providing a platform for multiple studies comparing patient- and control-derived cells to identify aspects of biology that are shared by patients but different from controls (Boone *et al.*, 2010; Matigian *et al.*, 2010).

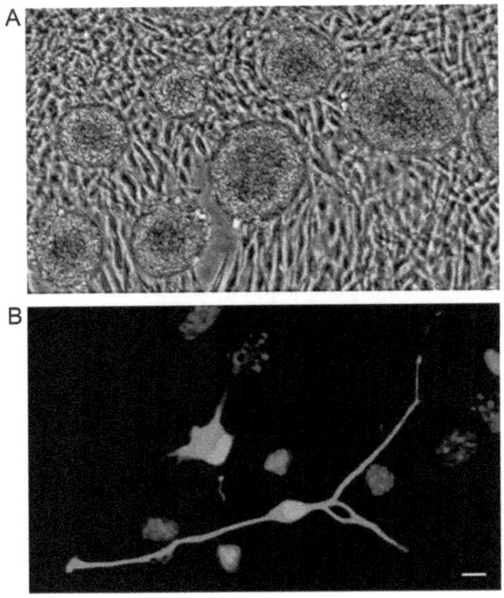

FIG. 1. Neurospheres and neurons derived from human olfactory mucosa. (A) Neurospheres, tight spherical clusters of cells, form and detach from the underlying cells when grown in serum-free medium containing epidermal growth factor and basic fibroblast growth factor. (B) When dissociated and grown in differentiating medium, neurosphere-derived cells differentiate into neurons (small cell body, long processes, immunopositive for β-tubulin III). Nuclei are stained with DAPI (blue). Bar: 30 μm in (A), 5 μm in (B). (For interpretation of the references to color in this figure legend, the reader is referred to the Web version of this chapter.)

In our initial study, we compared ONS cells from healthy controls with those from patients with either SZ or PD. We chose to contrast these cells from patients with two different brain diseases, both of which are complex heterogeneous disorders: SZ, a highly heritable neuropsychiatric, neurodevelopmental disorder (Raedler *et al.*, 1998), and sporadic PD, a neurodegenerative disease that is heritable only in about 5% of familial cases (Lesage and Brice, 2009). The guiding hypothesis is that in complex, polygenic diseases, the disease mechanisms will manifest in cell signaling pathways and genetic networks, even when single causative genes are not present. We undertook gene expression, protein expression, pathway analysis, and cell function assays that resulted in identification of 1700 genes and proteins which showed dysregulated expression in SZ cells and 500 which were dysregulated in PD cells (Matigian *et al.*, 2010). Assays of cell function identified disease-specific reductions in glutathione levels and 3-(4,5-dimethylthiazol-2-yl)-5-(3-carboxymethoxyphenyl)-2-(4-sulfophenyl)-2H-tetrazolium (MTS) metabolism (a general marker of metabolic status) in PD patient cells and increases in caspase 3 activity in SZ patient cells (Matigian *et al.*, 2010). Further assays of cellular function are underway to ascertain the effects of the many dysregulated signaling pathways specific to these cells, identified by pathway analysis of the altered gene expression profiles. In SZ cells, the significantly dysregulated pathways cluster around those affecting brain development; in PD, they show more association with mitochondrial function and oxidative stress. These cell signaling pathways are central to current theories of disease etiologies gathered from postmortem brain, genetic and epidemiological evidence gathered for each disease (Henchcliffe and Beal, 2008).

Our ONS cells model provides the first confirmation of these theories in living, patient-derived cells. We examined skin fibroblasts from the same SZ patients and found few differences in gene expression compared to controls (Matigian *et al.*, 2010), confirming our previous observation (Matigian *et al.*, 2008). Of note in SZ, cell cycle regulation and cell adhesion signaling pathways were significantly dysregulated in patient ONS cells compared to those from controls. This was notable because we previously identified cell proliferation and tissue adhesion as major differences between patient and control olfactory tissue explant cultures (Feron *et al.*, 1999; McCurdy *et al.*, 2006). Interestingly, SZ patient fibroblasts also show decreased adhesion but a slower rate of growth compared to those from controls (Mahadik *et al.*, 1991, 1994; Wang *et al.*, 2009).

Patient-derived olfactory cells have also been used to investigate a monogenic disease, familial dysautonomia caused by mutations in the *IKBKAP* gene (Boone *et al.*, 2010). Lower levels of expression of *IKBKAP* gene and protein were observed and genome-wide analysis identified altered expression in cell migration and cytoskeleton genes, with functional deficits in cell migration evident in patient cells (Boone *et al.*, 2010).

These olfactory cell models provide new tools for deeper analysis of the disease-associated phenotypes including the identification of key molecules that may be used as the starting point for biomarker discovery. For example, biomarkers may be expression levels of individual genes or proteins, or set of genes or proteins. They might be present in cells or secreted by them. As previously discussed, the most useful biomarkers identified in patient-derived stem cells in terms of practicality would be those which could be detected in blood or tissue biopsy directly, without the need for extensive cell culture or manipulation.

In summary, ONS cells have the advantages of relevance and ease of production and generation, with an ability to both self-renew and to differentiate into neurons and glial cells. Their ease of generation allows them to be used for large cohort studies in cases of sporadic diseases with unknown genetics. Their ease of banking and propagation allows them to be used for extended genomic, proteomic, and functional studies, including drug and biomarker discovery. Patient-control comparisons embrace the variability inherent in populations and will allow establishment of specificity and sensitivity of assays, including the identification of biomarkers, because the disease-associated differences will be those which are identified above the background of interindividual variability.

IX. Patient-Derived Pluripotent Stem Cells as Models for Brain Diseases

Two other types of stem cells with the capacity to differentiate into neurons are also being used as models for brain diseases. Embryonic stem (ES) cells are pluripotent cells that can differentiate into all cell types in the body including neurons and glial cells. Mouse ES cells were first isolated in 1981 (Evans and Kaufman, 1981; Martin, 1981) and human ES cells in 1998 (Thomson *et al.*, 1998). ES cells are potential models of brain diseases caused by genetic factors selected during preimplantation genetic diagnosis (Mateizel *et al.*, 2006; Urbach *et al.*, 2004; Verlinsky *et al.*, 2005). Induced pluripotent stem (iPS) cells are pluripotent cells derived from differentiated cells by introducing a few key transcription factor genes, originally demonstrated in adult mouse fibroblasts (Takahashi and Yamanaka, 2006). The expression of just four transcription factors (*Oct3/4, Sox2, c-Myc,* and *Klf4*) was sufficient to reprogram adult somatic cells back to a pluripotent ES cell-like state. Like ES cells, iPS cells display the critical ability to differentiate into all three germ layers, *in vitro* and *in vivo*. In 2007, the first human iPS cells derived from fibroblasts were reported (Takahashi *et al.*, 2007; Yu *et al.*, 2007). The implication was that accessible cells, such as dermal fibroblasts, could be collected from patients, reprogrammed to an iPS cell fate, and then redifferentiated along any lineage, so that a fibroblast could be turned into a neuron. One of the major outcomes arising from the advent of iPS cells is that they could be generated from patients to elucidate the cellular and molecular bases of disease.

Although several ES cell lines have been made from embryos with known genetic mutations, there has been little research using such cell lines to model disease processes (Urbach *et al.*, 2004). Perhaps, partly because of the ethical and legal concerns about ES cells, there has been much more interest in generating patient-derived iPS cells. Most reports have described the generation of iPS cells from cells of patients with defined genetic forms of neurological disorders, including from patients with familial forms of amyotrophic lateral sclerosis (Dimos *et al.*, 2008), muscular dystrophys, Huntington's disease, Gaucher disease, Down's syndrome (Park *et al.*, 2008), spinal muscular atrophy (Ebert *et al.*, 2009), dysautonomia (Lee *et al.*, 2009), SZ (Brennand *et al.*, 2011; Chiang *et al.*, 2011) and PD (Nguyen *et al.*, 2011; Park *et al.*, 2008). These investigations demonstrate proof-of-principle that iPS cells can be derived from patients, that these iPS cells can differentiate into neurons, and that they demonstrate altered expression of the genes and proteins of interest. As such, these publications are forerunners of an approach which will be useful for defining the molecular and cellular consequences of genetic mutations in neurons.

iPS cells have also been isolated from patients with neurological disorders of undefined genetic and/or environmental contribution (Brennand *et al.*, 2011;

Soldner *et al.*, 2009). These studies now demonstrate the feasibility of making iPS cells and, from these, neurons. While iPS cells have the potential to be generated from patients with any neurological diseases, they have not yet revealed the molecular and/or cellular basis of complex genetic and/or environmental interactions that lead to diseases such as Alzheimer's disease, PD, or SZ. The success of genetic investigations in identifying monogenetic forms of neurological diseases such as those mentioned above has provided critical mechanistic insights into the pathogenic pathways that likely underpin sporadic forms. However, it is important to recognize that these genetically loaded cases account for a minority of the disease burden. For example, the most common form of genetic PD (caused by *LRRK2* mutations) occurs in < 1% of PD patients in Australia (Huang *et al.*, 2007). Likewise, *SOD1* mutations account for less than 1% of all amyotrophic lateral sclerosis cases and it is sobering to reflect on the fact that lead molecules emerging from large drug discovery efforts based on the outcomes of *SOD1* animal studies have proven to be ostensibly unsuccessful in human amyotrophic lateral sclerosis (see Schnabel, 2008). In contrast, a major aspect of cellular models of neurological disorders is their capacity to represent, or at least account for, the wide genetic backgrounds against which most cases of the disease arise. In this aspect, iPS cells have been made from five patients with sporadic PD, from which neurons were generated (Soldner *et al.*, 2009). Interestingly, these authors did not find phenotypic differences between sporadic PD iPS and control iPS cells despite the fact that these iPS cells were derived from *LRRK2* mutation carriers (Nguyen *et al.*, 2011). In another disease of complex genetics, SZ, iPS cells have been made from two patients with a *DISC1* mutation (a familial monogenic form; Chiang *et al.*, 2011) and three patients with familial SZ of unknown genetics (Brennand *et al.*, 2011). This latter study identified reduced neurite branching and diminished neuronal connectivity and reduced glutamate receptor expression in neurons generated from iPS cells. This demonstrated that iPS cells can be useful for identifying disease-associated changes in cell biology in neurons, an apposite proof-of-principle for a cell model of a brain disease of unknown genetics.

At present, there are two primary limitations to the use of iPS cells as models of sporadic cases of neurological disorders. The first is technical. The efficiency of generating iPS cells still remains low (Hanna *et al.*, 2009). Therefore, most published studies are based on from 1 to 5 iPS clones. Given the clonal variation of iPS cells (Laurent *et al.*, 2011) derived from a single source of primary cells, it is not clear precisely what data from a single patient clone compared with a single healthy control clone represents. At present, the generation and expansion of iPS cells is not a trivial exercise, although advances in technology should overcome many present bottlenecks. Yet for complex diseases such as idiopathic PD and sporadic SZ, it might be necessary to compare hundreds of patient and control cell lines to gain insight into molecular mechanisms. Another limitation is not so much technical as inherent in the iPS derivation process. This limitation is the potential for epigenetic reprogramming of the somatic cell genome. Expression of reprogramming factors, such as *Oct3/4*, *Sox2*, *c-Myc*, and *Klf4*, reboots the pluripotent cell state in large part by stripping the epigenetic modifications of pluripotent genes which have been silenced during differentiation (Saha and Jaenisch, 2009). This is a requisite part of iPS generation. However, if epigenetic modification is a contributing factor to sporadic neurological disorders, then this vital information will be lost during the generation of the cell model. In effect, this might be throwing the baby out with the bathwater. Recent evidence suggests that iPS cells retain some memory of their original cell state, suggesting that not all epigenetic information is lost (Hu *et al.*, 2011; Kim *et al.*, 2010; Polo *et al.*, 2010; Tian *et al.*, 2011). However, the possibility remains that the epigenome of the fibroblast is not reflective of that of a neuronal cell, thereby negating a primary factor in taking a readily accessible source of cells (fibroblasts) and transforming these into a relatively inaccessible cell type (e.g., dopaminergic neurons). To answer some of these questions, we have generated iPS cells from four patients with SZ and from four controls, with multiple clones from each (unpublished findings). These cells will be induced to differentiate into neurons, with the aim of determining disease-associated differences in cell functions in these sporadic cases of SZ (i.e., nonfamilial cases).

In summary, iPS cells have the advantage of being malleable, with evidence that they can be directed to differentiate into different subtypes of neurons. iPS cells have disadvantages including poor efficiencies of production and differentiation and they are currently impractical for large cohort studies because they are time-consuming to generate, validate, and propagate. Conversely, iPS cells are being used to identify changes in neural cell biology associated with the identified mutations. This will be important for understanding the effects of disease-associated mutations on specific neurons and their circuits in the brain.

X. Advantages and Disadvantages of Current Cell Models

Fibroblasts and lymphocytes have the advantages of being accessible and minimally invasive. Their disadvantage is their validity as models for neural cells. The question remains as to whether they can show cell and molecular differences that are the same as those occurring in the brain. For biomarkers of disease, this may not matter if they show a quantifiable patient–control difference. As models of intracellular disease processes, nonneural cells may be adequate for brain diseases that arise from gene mutations affecting all cells (e.g., mitochondrial dysfunction), although the consequent cellular phenotypes may differ in cells from different tissues. Identifying disease phenotypes in specific neuronal populations, however, may require stem cell technologies to generate these from patient tissues.

This field of study is rapidly advancing and it is now possible to convert fibroblasts directly into neurons without passing through a pluripotent stage (Vierbuchen *et al.*, 2010).

It is still early days for cell models of brain diseases. The utility of current cell models varies with advantages and disadvantages associated with each cell type (Table I). For example, patient-derived cells vary in the ease with which they can be obtained and generated. Blood lymphocytes are easy to obtain but cannot be propagated without transformation into lymphoblastoid cell lines. Fibroblasts require a more invasive biopsy procedure but can be grown *in vitro* and banked. Biopsying the olfactory mucosa is still relatively more invasive but ONS cells are similar to fibroblasts in the time taken to generate these *in vitro* and are similarly robust for cell culture and banking. iPS cells are labor intensive to generate and maintain compared to other patient-derived cells, requiring many months, rather than weeks, to generate and additional months for validation. They are also more temperamental to maintain *in vitro*.

Cell models also vary in their relevance as representatives for neural tissues and their potential for reflecting neural biology. For example, ONS cells (derived from a neural tissue) but not fibroblasts (derived from skin) show gene expression differences in neurodevelopmental signaling pathways in SZ (Matigian *et al.*, 2010). However, fibroblasts can show disease-associated differences that may be associated with less tissue-specific, but systemic, changes such as cell proliferation (Wang *et al.*, 2009) and adhesion (Mahadik *et al.*, 1994). Both ONS cells and iPS cells can be differentiated into neurons but iPS cell technology is more advanced in terms of reliability and for rapid production of neurons of different classes. A potential disadvantage of iPS cells is the unknown and unpredictable effect of

Table I
UTILITY OF PATIENT-DERIVED CELLS AS DISEASE MODELS.

	Adult patient-derived	Easy to obtain	Easy to grow	Genetically modified	Proliferate *in vitro*	Potential for longitudinal monitoring	Differentiate into neurons
Nonstem cells							
Lymphocytes	✓	✓	✓			✓	
Lymphoblastoid cells	✓	✓		✓	✓	✓	
Fibroblasts	✓	✓	✓		✓	✓	
Stem cells							
ES cells					✓		✓
iPS cells	✓			✓	✓		✓
ONS cells	✓	✓	✓		✓	✓	✓

genetic reprogramming on disease-associated genes and the potential variability between clones generated from each patient. The reprogramming process is a rare stochastic event and it is not currently known whether the cells that undergo reprogramming are representative of the whole population. Resolving these issues will require investigation of multiple patient- and control-derived cell lines (Saha and Jaenisch, 2009).

XI. Future Directions: Biomarkers from Stem Cell Models

For the first time for many diseases, stem cell science will provide the ability to investigate the cellular and molecular basis of brain diseases in living neural cells from patients, including neural progenitor cells, neurons, astrocytes, and oligodendrocytes. These cell models will carry disease phenotypes, genetics, and epigenetics, and span the variability encountered across the patient population. They also provide the potential to identify novel molecular markers that distinguish patient from control cells. These may be altered levels of gene or protein expression, or altered levels of metabolites. The cellular models may also provide the keys to differential diagnosis or indicators of progress of a disease or its treatment.

The ideal next steps would be to source these markers in blood or other easily accessible tissues such as skin, hair, or cheek cells. It may be necessary to sample from more "neural" sources such as cerebrospinal fluid or olfactory mucosa. It may also be possible in the not-too-distant future that molecular biomarkers could be monitored *in vivo* using the rapidly advancing brain imaging techniques such as magnetic resonance spectroscopy (Soares and Law, 2009).

Acknowledgments

This work was supported by funding to the National Centre for Adult Stem Cell Research from the Australian Government Department of Health and Aging.

References

Abrams, M.T., Kaufmann, W.E., Rousseau, F., Oostra, B.A., Wolozin, B., Taylor, C.V., Lishaa, N., Morel, M.L., Hoogeveen, A., and Reiss, A.L. (1999). FMR1 gene expression in olfactory neuroblasts from two males with fragile X syndrome. *Am. J. Med. Genet.* **82**, 25–30.

Arnold, S.E., Han, L.Y., Moberg, P.J., Turetsky, B.I., Gur, R.E., Trojanowski, J.Q., and Hahn, C.G. (2001). Dysregulation of olfactory receptor neuron lineage in schizophrenia. *Arch. Gen. Psychiatry* **58**, 829–835.

Atz, M., Walsh, D., Cartagena, P., Li, J., Evans, S., Choudary, P., Overman, K., Stein, R., Tomita, H., Potkin, S., Myers, R., Watson, S.J., *et al.* (2007). Methodological considerations for gene expression profiling of human brain. *J. Neurosci. Methods* **163**, 295–309.

Boone, N., Loriod, B., Bergon, A., Sbai, O., Formisano-Treziny, C., Gabert, J., Khrestchatisky, M., Nguyen, C., Feron, F., Axelrod, F.B., and Ibrahim el, C. (2010). Olfactory stem cells, a new cellular model for studying molecular mechanisms underlying familial dysautonomia. *PLoS One* **5**, e15590.

Bowden, N.A., Weidenhofer, J., Scott, R.J., Schall, U., Todd, J., Michie, P.T., and Tooney, P.A. (2006). Preliminary investigation of gene expression profiles in peripheral blood lymphocytes in schizophrenia. *Schizophr. Res.* **82**, 175–183.

Braak, H., Bohl, J.R., Muller, C.M., Rub, U., de Vos, R.A., and Del Tredici, K. (2006). Stanley Fahn Lecture 2005: the staging procedure for the inclusion body pathology associated with sporadic Parkinson's disease reconsidered. *Mov. Disord.* **21**, 2042–2051.

Brennand, K.J., Simone, A., Jou, J., Gelboin-Burkhart, C., Tran, N., Sangar, S., Li, Y., Mu, Y., Chen, G., Yu, D., McCarthy, S., Sebat, J., *et al.* (2011). Modelling schizophrenia using human induced pluripotent stem cells. *Nature* **473**, 221–225.

Brewer, W.J., Wood, S.J., McGorry, P.D., Francey, S.M., Phillips, L.J., Yung, A.R., Anderson, V., Copolov, D.L., Singh, B., Velakoulis, D., and Pantelis, C. (2003). Impairment of olfactory identification ability in individuals at ultra-high risk for psychosis who later develop schizophrenia. *Am. J. Psychiatry* **160**, 1790–1794.

Chen, X., Fang, H., and Schwob, J.E. (2004). Multipotency of purified, transplanted globose basal cells in olfactory epithelium. *J. Comp. Neurol.* **469**, 457–474.

Chiang, C.H., Su, Y., Wen, Z., Yoritomo, N., Ross, C.A., Margolis, R.L., Song, H., and Ming, G.L. (2011). Integration-free induced pluripotent stem cells derived from schizophrenia patients with a DISC1 mutation. *Mol. Psychiatry* **16**, 358–360.

Cohen, M.R., Gutman, R., and McAmis, W. (1987). Cultured skin fibroblasts in schizophrenia: acute growth and susceptibility to damage. *Psychiatry Res.* **21**, 43–47.

del Hoyo, P., Garcia-Redondo, A., de Bustos, F., Molina, J.A., Sayed, Y., Alonso-Navarro, H., Caballero, L., Arenas, J., Agundez, J.A., and Jimenez-Jimenez, F.J. (2010). Oxidative stress in skin fibroblasts cultures from patients with Parkinson's disease. *BMC Neurol.* **10**, 95.

Delorme, B., Nivet, E., Gaillard, J., Haupl, T., Ringe, J., Deveze, A., Magnan, J., Sohier, J., Khrestchatisky, M., Roman, F.S., Charbord, P., Sensebe, L., *et al.* (2010). The human nose harbors a niche of olfactory ectomesenchymal stem cells displaying neurogenic and osteogenic properties. *Stem Cells Dev.* **19**, 853–866.

Dimos, J.T., Rodolfa, K.T., Niakan, K.K., Weisenthal, L.M., Mitsumoto, H., Chung, W., Croft, G.F., Saphier, G., Leibel, R., Goland, R., Wichterle, H., Henderson, C.E., *et al.* (2008). Induced pluripotent stem cells generated from patients with ALS can be differentiated into motor neurons. *Science* **321**, 1218–1221.

Doty, R.L. (2009). The olfactory system and its disorders. *Semin. Neurol.* **29**, 74–81.

Ebert, A.D., Yu, J., Rose, F.F. Jr., Mattis, V.B., Lorson, C.L., Thomson, J.A., and Svendsen, C.N. (2009). Induced pluripotent stem cells from a spinal muscular atrophy patient. *Nature* **457**, 277–280.

Evans, M.J., and Kaufman, M.H. (1981). Establishment in culture of pluripotential cells from mouse embryos. *Nature* **292**, 154–156.

Feron, F., Perry, C., McGrath, J.J., and Mackay-Sim, A. (1998). New techniques for biopsy and culture of human olfactory epithelial neurons. *Arch. Otolaryngol. Head Neck Surg.* **124**, 861–866.

Feron, F., Perry, C., Hirning, M.H., McGrath, J., and Mackay-Sim, A. (1999). Altered adhesion, proliferation and death in neural cultures from adults with schizophrenia. *Schizophr. Res.* **40**, 211–218.

Gavin, D.P., Rosen, C., Chase, K., Grayson, D.R., Tun, N., and Sharma, R.P. (2009). Dimethylated lysine 9 of histone 3 is elevated in schizophrenia and exhibits a divergent response to histone deacetylase inhibitors in lymphocyte cultures. *J. Psychiatry Neurosci.* **34**, 232–237.

Gershon, E.S., Alliey-Rodriguez, N., and Liu, C. (2011). After GWAS: searching for genetic risk for schizophrenia and bipolar disorder. *Am. J. Psychiatry* **168**, 253–256.

Ghanbari, H.A., Ghanbari, K., Harris, P.L., Jones, P.K., Kubat, Z., Castellani, R.J., Wolozin, B.L., Smith, M.A., and Perry, G. (2004). Oxidative damage in cultured human olfactory neurons from Alzheimer's disease patients. *Aging Cell* **3**, 41–44.

Graziadei, P.P. (1973). Cell dynamics in the olfactory mucosa. *Tissue Cell* **5**, 113–131.

Graziadei, G.A., and Graziadei, P.P. (1979). Neurogenesis and neuron regeneration in the olfactory system of mammals. II. Degeneration and reconstitution of the olfactory sensory neurons after axotomy. *J. Neurocytol.* **8**, 197–213.

Haehner, A., Boesveldt, S., Berendse, H.W., Mackay-Sim, A., Fleischmann, J., Silburn, P.A., Johnston, A.N., Mellick, G.D., Herting, B., Reichmann, H., and Hummel, T. (2009). Prevalence of smell loss in Parkinson's disease—a multicenter study. *Parkinsonism Relat. Disord.* **15**, 490–494.

Hanna, J., Markoulaki, S., Mitalipova, M., Cheng, A.W., Cassady, J.P., Staerk, J., Carey, B.W., Lengner, C.J., Foreman, R., Love, J., Gao, Q., Kim, J., *et al.* (2009). Metastable pluripotent states in NOD-mouse-derived ESCs. *Cell Stem Cell* **4**, 513–524.

Hawkes, C.H., Shephard, B.C., and Daniel, S.E. (1999). Is Parkinson's disease a primary olfactory disorder? *QJM* **92**, 473–480.

Henchcliffe, C., and Beal, M.F. (2008). Mitochondrial biology and oxidative stress in Parkinson disease pathogenesis. *Nat. Clin. Pract. Neurol.* **4**, 600–609.

Hoepken, H.H., Gispert, S., Azizov, M., Klinkenberg, M., Ricciardi, F., Kurz, A., Morales-Gordo, B., Bonin, M., Riess, O., Gasser, T., Kogel, D., Steinmetz, H., *et al.* (2008). Parkinson patient fibroblasts show increased alpha-synuclein expression. *Exp. Neurol.* **212**, 307–313.

Hu, Q., Friedrich, A.M., Johnson, L.V., and Clegg, D.O. (2011). Memory in induced pluripotent stem cells: reprogrammed human retinal-pigmented epithelial cells show tendency for spontaneous redifferentiation. *Stem Cells* **28**, 1981–1991.

Huang, Y., Halliday, G.M., Vandebona, H., Mellick, G.D., Mastaglia, F., Stevens, J., Kwok, J., Garlepp, M., Silburn, P.A., Horne, M.K., Kotschet, K., Venn, A., *et al.* (2007). Prevalence and clinical features of common LRRK2 mutations in Australians with Parkinson's disease. *Mov. Disord.* **22**, 982–989.

Kim, K., Doi, A., Wen, B., Ng, K., Zhao, R., Cahan, P., Kim, J., Aryee, M.J., Ji, H., Ehrlich, L.I., Yabuuchi, A., Takeuchi, A., *et al.* (2010). Epigenetic memory in induced pluripotent stem cells. *Nature* **467**, 285–290.

Laurent, L.C., Ulitsky, I., Slavin, I., Tran, H., Schork, A., Morey, R., Lynch, C., Harness, J.V., Lee, S., Barrero, M.J., Ku, S., Martynova, M., *et al.* (2011). Dynamic changes in the copy number of pluripotency and cell proliferation genes in human ESCs and iPSCs during reprogramming and time in culture. *Cell Stem Cell* **8**, 106–118.

Lee, G., Papapetrou, E.P., Kim, H., Chambers, S.M., Tomishima, M.J., Fasano, C.A., Ganat, Y.M., Menon, J., Shimizu, F., Viale, A., Tabar, V., Sadelain, M., *et al.* (2009). Modelling pathogenesis and treatment of familial dysautonomia using patient-specific iPSCs. *Nature* **461**, 402–406.

Lesage, S., and Brice, A. (2009). Parkinson's disease: from monogenic forms to genetic susceptibility factors. *Hum. Mol. Genet.* **18**, R48–R59.

Leung, C.T., Coulombe, P.A., and Reed, R.R. (2007). Contribution of olfactory neural stem cells to tissue maintenance and regeneration. *Nat. Neurosci.* **10**, 720–726.

Mackay-Sim, A., and Kittel, P. (1991). Cell dynamics in the adult mouse olfactory epithelium: a quantitative autoradiographic study. *J. Neurosci.* **11**, 979–984.

Mackay-Sim, A., Feron, F., Eyles, D., Burne, T., and McGrath, J. (2004). Schizophrenia, vitamin D, and brain development. *Int. Rev. Neurobiol.* **59**, 351–380.

Mahadik, S.P., Mukherjee, S., Laev, H., Reddy, R., and Schnur, D.B. (1991). Abnormal growth of skin fibroblasts from schizophrenic patients. *Psychiatry Res.* **37**, 309–320.

Mahadik, S.P., Mukherjee, S., Wakade, C.G., Laev, H., Reddy, R.R., and Schnur, D.B. (1994). Decreased adhesiveness and altered cellular distribution of fibronectin in fibroblasts from schizophrenic patients. *Psychiatry Res.* **53**, 87–97.

Marcotte, E.R., Srivastava, L.K., and Quirion, R. (2003). cDNA microarray and proteomic approaches in the study of brain diseases: focus on schizophrenia and Alzheimer's disease. *Pharmacol. Ther.* **100**, 63–74.

Martin, G.R. (1981). Isolation of a pluripotent cell line from early mouse embryos cultured in medium conditioned by teratocarcinoma stem cells. *Proc. Natl. Acad. Sci. USA* **78**, 7634–7638.

Mateizel, I., De Temmerman, N., Ullmann, U., Cauffman, G., Sermon, K., Van de Velde, H., De Rycke, M., Degreef, E., Devroey, P., Liebaers, I., and Van Steirteghem, A. (2006). Derivation of human embryonic stem cell lines from embryos obtained after IVF and after PGD for monogenic disorders. *Hum. Reprod.* **21**, 503–511.

Matigian, N.A., McCurdy, R.D., Feron, F., Perry, C., Smith, H., Filippich, C., McLean, D., McGrath, J., Mackay-Sim, A., Mowry, B., and Hayward, N.K. (2008). Fibroblast and lymphoblast gene expression profiles in schizophrenia: are non-neural cells informative? *PLoS One* **3**, e2412.

Matigian, N., Abrahamsen, G., Sutharsan, R., Cook, A.L., Vitale, A.M., Nouwens, A., Bellette, B., An, J., Anderson, M., Beckhouse, A.G., Bennebroek, M., Cecil, R., *et al.* (2010). Disease-specific, neurosphere-derived cells as models for brain disorders. *Dis. Model. Mech.* **3**, 785–798.

McCurdy, R.D., Feron, F., Perry, C., Chant, D.C., McLean, D., Matigian, N., Hayward, N.K., McGrath, J.J., and Mackay-Sim, A. (2006). Cell cycle alterations in biopsied olfactory neuroepithelium in schizophrenia and bipolar I disorder using cell culture and gene expression analyses. *Schizophr. Res.* **82**, 163–173.

Mellick, G.D., Silburn, P.A., Sutherland, G.T., and Siebert, G.A. (2010). Exploiting the potential of molecular profiling in Parkinson's disease: current practice and future probabilities. *Expert Rev. Mol. Diagn.* **10**, 1035–1050.

Miyamae, Y., Nakamura, Y., Kashiwagi, Y., Tanaka, T., Kudo, T., and Takeda, M. (1998). Altered adhesion efficiency and fibronectin content in fibroblasts from schizophrenic patients. *Psychiatry Clin. Neurosci.* **52**, 345–352.

Murrell, W., Bushell, G.R., Livesey, J., McGrath, J., MacDonald, K.P., Bates, P.R., and Mackay-Sim, A. (1996). Neurogenesis in adult human. *Neuroreport* **7**, 1189–1194.

Murrell, W., Feron, F., Wetzig, A., Cameron, N., Splatt, K., Bellette, B., Bianco, J., Perry, C., Lee, G., and Mackay-Sim, A. (2005). Multipotent stem cells from adult olfactory mucosa. *Dev. Dyn.* **233**, 496–515.

Murrell, W., Wetzig, A., Donnellan, M., Feron, F., Burne, T., Meedeniya, A., Kesby, J., Bianco, J., Perry, C., Silburn, P., and Mackay-Sim, A. (2008). Olfactory mucosa is a potential source for autologous stem cell therapy for Parkinson's disease. *Stem Cells* **26**, 2183–2192.

Murrell, W., Sanford, E., Anderberg, L., Cavanagh, B., and Mackay-Sim, A. (2009). Olfactory stem cells can be induced to express chondrogenic phenotype in a rat intervertebral disc injury model. *Spine J.* **9**, 585–594.

Mytilineou, C., Werner, P., Molinari, S., Di Rocco, A., Cohen, G., and Yahr, M.D. (1994). Impaired oxidative decarboxylation of pyruvate in fibroblasts from patients with Parkinson's disease. *J. Neural. Transm. Park. Dis. Dement. Sect.* **8**, 223–228.

Nalls, M.A., Plagnol, V., Hernandez, D.G., Sharma, M., Sheerin, U.M., Saad, M., Simon-Sanchez, J., Schulte, C., Lesage, S., Sveinbjornsdottir, S., Stefansson, K., Martinez, M., *et al.* (2011). Imputation of sequence variants for identification of genetic risks for Parkinson's disease: a meta-analysis of genome-wide association studies. *Lancet* **377**, 641–649.

Newman, M.P., Feron, F., and Mackay-Sim, A. (2000). Growth factor regulation of neurogenesis in adult olfactory epithelium. *Neuroscience* **99**, 343–350.

Nguyen, H.N., Byers, B., Cord, B., Shcheglovitov, A., Byrne, J., Gujar, P., Kee, K., Schule, B., Dolmetsch, R.E., Langston, W., Palmer, T.D., and Pera, R.R. (2011). LRRK2 mutant iPSC-derived DA neurons demonstrate increased susceptibility to oxidative stress. *Cell Stem Cell* **8**, 267–280.

Park, I.H., Arora, N., Huo, H., Maherali, N., Ahfeldt, T., Shimamura, A., Lensch, M.W., Cowan, C., Hochedlinger, K., and Daley, G.Q. (2008). Disease-specific induced pluripotent stem cells. *Cell* **134**, 877–886.

Polo, J.M., Liu, S., Figueroa, M.E., Kulalert, W., Eminli, S., Tan, K.Y., Apostolou, E., Stadtfeld, M., Li, Y., Shioda, T., Natesan, S., Wagers, A.J., *et al.* (2010). Cell type of origin influences the molecular and functional properties of mouse induced pluripotent stem cells. *Nat. Biotechnol.* **28**, 848–855.

Preece, P., and Cairns, N.J. (2003). Quantifying mRNA in postmortem human brain: influence of gender, age at death, postmortem interval, brain pH, agonal state and inter-lobe mRNA variance. *Brain Res. Mol. Brain Res.* **118**, 60–71.

Raedler, T.J., Knable, M.B., and Weinberger, D.R. (1998). Schizophrenia as a developmental disorder of the cerebral cortex. *Curr. Opin. Neurobiol.* **8**, 157–161.

Ramchand, C.N., Das, I., Gliddon, A., and Hirsch, S.R. (1994). Role of polyamines in the membrane pathology of schizophrenia. A study using fibroblasts from schizophrenic patients and normal controls. *Schizophr. Res.* **13**, 249–253.

Rietze, R.L., and Reynolds, B.A. (2006). Neural stem cell isolation and characterization. *Methods Enzymol.* **419**, 3–23.

Roisen, F.J., Klueber, K.M., Lu, C.L., Hatcher, L.M., Dozier, A., Shields, C.B., and Maguire, S. (2001). Adult human olfactory stem cells. *Brain Res.* **890**, 11–22.

Ronnett, G.V., Leopold, D., Cai, X., Hoffbuhr, K.C., Moses, L., Hoffman, E.P., and Naidu, S. (2003). Olfactory biopsies demonstrate a defect in neuronal development in Rett's syndrome. *Ann. Neurol.* **54**, 206–218.

Saha, K., and Jaenisch, R. (2009). Technical challenges in using human induced pluripotent stem cells to model disease. *Cell Stem Cell* **5**, 584–595.

Schnabel, J. (2008). Neuroscience: standard model. *Nature* **454**, 682–685.

Soares, D.P., and Law, M. (2009). Magnetic resonance spectroscopy of the brain: review of metabolites and clinical applications. *Clin. Radiol.* **64**, 12–21.

Soldner, F., Hockemeyer, D., Beard, C., Gao, Q., Bell, G.W., Cook, E.G., Hargus, G., Blak, A., Cooper, O., Mitalipova, M., Isacson, O., and Jaenisch, R. (2009). Parkinson's disease patient-derived induced pluripotent stem cells free of viral reprogramming factors. *Cell* **136**, 964–977.

Summers, K.M. (1996). Relationship between genotype and phenotype in monogenic diseases: relevance to polygenic diseases. *Hum. Mutat.* **7**, 283–293.

Sutherland, G.T., Matigian, N.A., Chalk, A.M., Anderson, M.J., Silburn, P.A., Mackay-Sim, A., Wells, C.A., and Mellick, G.D. (2009). A cross-study transcriptional analysis of Parkinson's disease. *PLoS One* **4**, e4955.

Sutherland, G.T., Siebert, G.A., Kril, J.J., and Mellick, G.D. (2011). Knowing me, knowing you: can knowledge of risk factors for Alzheimer's disease prove useful in understanding the pathogenesis of Parkinson's disease? *J. Alzheimers Dis.*

Suzuki, K., Nakamura, K., Iwata, Y., Sekine, Y., Kawai, M., Sugihara, G., Tsuchiya, K.J., Suda, S., Matsuzaki, H., Takei, N., Hashimoto, K., and Mori, N. (2008). Decreased expression of reelin receptor VLDLR in peripheral lymphocytes of drug-naive schizophrenic patients. *Schizophr. Res.* **98**, 148–156.

Takahashi, K., and Yamanaka, S. (2006). Induction of pluripotent stem cells from mouse embryonic and adult fibroblast cultures by defined factors. *Cell* **126**, 663–676.

Takahashi, K., Tanabe, K., Ohnuki, M., Narita, M., Ichisaka, T., Tomoda, K., and Yamanaka, S. (2007). Induction of pluripotent stem cells from adult human fibroblasts by defined factors. *Cell* **131**, 861–872.

Thomson, J.A., Itskovitz-Eldor, J., Shapiro, S.S., Waknitz, M.A., Swiergiel, J.J., Marshall, V.S., and Jones, J.M. (1998). Embryonic stem cell lines derived from human blastocysts. *Science* **282**, 1145–1147.

Tian, C., Wang, Y., Sun, L., Ma, K., and Zheng, J.C. (2011). Reprogrammed mouse astrocytes retain a "memory" of tissue origin and possess more tendencies for neuronal differentiation than reprogrammed mouse embryonic fibroblasts. *Protein Cell* **2**, 128–140.

Urbach, A., Schuldiner, M., and Benvenisty, N. (2004). Modeling for Lesch-Nyhan disease by gene targeting in human embryonic stem cells. *Stem Cells* **22**, 635–641.

Vawter, M.P., Ferran, E., Galke, B., Cooper, K., Bunney, W.E., and Byerley, W. (2004). Microarray screening of lymphocyte gene expression differences in a multiplex schizophrenia pedigree. *Schizophr. Res.* **67**, 41–52.

Verlinsky, Y., Strelchenko, N., Kukharenko, V., Rechitsky, S., Verlinsky, O., Galat, V., and Kuliev, A. (2005). Human embryonic stem cell lines with genetic disorders. *Reprod. Biomed. Online* **10**, 105–110.

Vierbuchen, T., Ostermeier, A., Pang, Z.P., Kokubu, Y., Sudhof, T.C., and Wernig, M. (2010). Direct conversion of fibroblasts to functional neurons by defined factors. *Nature* **463**, 1035–1041.

Wang, L., Lockstone, H.E., Guest, P.C., Levin, Y., Palotas, A., Pietsch, S., Schwarz, E., Rahmoune, H., Harris, L.W., Ma, D., and Bahn, S. (2009). Expression profiling of fibroblasts identifies cell cycle abnormalities in schizophrenia. *J. Proteome Res.* **9**, 521–527.

Witt, M., Bormann, K., Gudziol, V., Pehlke, K., Barth, K., Minovi, A., Hahner, A., Reichmann, H., and Hummel, T. (2009). Biopsies of olfactory epithelium in patients with Parkinson's disease. *Mov. Disord.* **24**, 906–914.

Wolozin, B., Sunderland, T., Zheng, B.B., Resau, J., Dufy, B., Barker, J., Swerdlow, R., and Coon, H. (1992). Continuous culture of neuronal cells from adult human olfactory epithelium. *J. Mol. Neurosci.* **3**, 137–146.

Wolozin, B., Lesch, P., Lebovics, R., and Sunderland, T. (1993). A.E. Bennett Research Award 1993. Olfactory neuroblasts from Alzheimer donors: studies on APP processing and cell regulation. *Biol. Psychiatry* **34**, 824–838.

Wray, N.R., Goddard, M.E., and Visscher, P.M. (2008). Prediction of individual genetic risk of complex disease. *Curr. Opin. Genet. Dev.* **18**, 257–263.

Yu, J., Vodyanik, M.A., Smuga-Otto, K., Antosiewicz-Bourget, J., Frane, J.L., Tian, S., Nie, J., Jonsdottir, G.A., Ruotti, V., Stewart, R., Slukvin, I.I., and Thomson, J.A. (2007). Induced pluripotent stem cell lines derived from human somatic cells. *Science* **318**, 1917–1920.

Zhubi, A., Veldic, M., Puri, N.V., Kadriu, B., Caruncho, H., Loza, I., Sershen, H., Lajtha, A., Smith, R.C., Guidotti, A., Davis, J.M., and Costa, E. (2009). An upregulation of DNA-methyltransferase 1 and 3a expressed in telencephalic GABAergic neurons of schizophrenia patients is also detected in peripheral blood lymphocytes. *Schizophr. Res.* **111**, 115–122.

THE APPLICATION OF MULTIPLEXED ASSAY SYSTEMS FOR MOLECULAR DIAGNOSTICS

Emanuel Schwarz[1], Nico J.M. VanBeveren[2], Paul C. Guest[1], Rauf Izmailov[3] and Sabine Bahn[1,4]

[1]Department of Chemical Engineering and Biotechnology, University of Cambridge, Cambridge, United Kingdom
[2]Department of Psychiatry, Erasmus University, Medical Centre, Rotterdam, The Netherlands
[3]Rules-Based Medicine, Inc., Austin, Texas, USA
[4]Department of Neuroscience, Erasmus Medical Centre, Rotterdam, The Netherlands

Abstract

For decades, the diagnosis of schizophrenia and other psychiatric disorders has relied on subjective assessments such as Diagnostic and Statistical Manual criteria. There is now increasing interest in the identification of altered molecular patterns in blood and other accessible body fluids that can be used to help identify, stratify, and monitor psychiatric patients. Since shorter periods of psychosis are associated with a better prognosis, an accurate molecular test may lead to early intervention and thereby improve patient outcomes. In addition, such a test would open up the possibility to stratify more accurately the disease and could represent a novel translational medicine tool, which is crucial for the discovery and development of more efficacious therapies.

INTERNATIONAL REVIEW OF
NEUROBIOLOGY, VOL. 101
DOI: 10.1016/B978-0-12-387718-5.00010-9

I. Introduction

The search for biological markers of psychiatric disorders that have applicability as clinical tools has been ongoing for several decades. There is now significant interest in the discovery of such markers, as they could be useful as objective tools to assist the diagnosis, treatment selection, and monitoring of patients. Biomarker readouts reflecting abnormal molecular processes such as the niacin skin flush response test have shown early promise for application as a schizophrenia diagnostic (Horrobin, 1980). This test is based on an attenuated flush response of schizophrenia patients after topical application of niacin to the skin. Ultimately, the low sensitivity and specificity, along with the differences between acute and multiepisode schizophrenia subjects, have prevented clinical introduction of this method (Ward *et al.*, 1998; Puri *et al.*, 2001; Smesny *et al.*, 2005).

Molecular research into psychiatric disorders such as schizophrenia, bipolar disorder, or major depressive disorder has shown that, despite the high degree of heritability, single genes only confer a small proportion of the overall risk of onset (Schwarz and Bahn, 2008). Also, studies targeted against single molecules have met with less than promising results. This is most likely due to the fact that molecules measured in peripheral body fluids or tissues often show only subtle changes in molecular levels which result in a small effect size. This is typically not sufficient for discriminating patients from healthy volunteers, or from patients with other disorders, with suitable performance. One way around this problem is to use large sample numbers for identification of statistically significant abnormalities. However, this is often challenging due to the limited availability of well-characterized high-quality samples which have been collected according to uniform standard operating procedures. In particular, the treatment state and disease stage of the subjects in question can pose potential confounding factors in molecular and other empirical investigations. For this reason, it may be most relevant to study individuals at the earliest stages of the disease before any treatment has been applied and before the disease has progressed. This is a challenging task as even large clinical centers generally collect samples from less than 30 drug-naive first-episode schizophrenia patients per year. Such confounding factors are also a particular problem for cross-disease comparisons which are essential in the development of differential diagnostic tests.

Emerging proteomic and transcriptomic platforms have facilitated the simultaneous measurement of hundreds or thousands of molecules, enabling nonhypothesis driven profiling approaches. Such profiling methods have been performed for various psychiatric disorders (Lakhan, 2006), but the resulting molecular candidates have not been validated or translated on a larger scale. Our group has recently employed an approach based on multiplexed immunoassay profiling which resulted in identification of a serum signature that could identify schizophrenia patients with an accuracy of 82% across five independent patient cohorts (Schwarz *et al.*, 2010a). Along with some components of this molecular signature that showed reproducibility across independent clinical sites, many of the molecules were found to be changed only in specific centers. The most likely interpretation of this effect is that, this is a molecular reflection of heterogeneity in a population of patients in different centers, despite the fact that these appeared to have comparable clinical presentations. Therefore, the use of multiplexed immunoassays was necessary to achieve this performance, given the greater discriminatory power afforded by multiple assays. The combined use of multiple molecular measurements can lead to greater sensitivity and specificity in the same way that a complete fingerprint aids forensic scientists in the absolute identification of an individual, whereas a partial fingerprint leads to a higher degree of uncertainty.

These issues give rise to the question of how far the construct validity of psychiatric diagnoses impacts on biomarker discovery. Currently, psychiatric disorders are diagnosed based on subjective interviews and patient history. Inevitably, the question arises of how molecular data derived from subjectively diagnosed patients can lead to the development of objective molecular diagnostics. In this chapter, we will discuss the challenge of performing molecular biomarker discovery in the case of subjects with schizophrenia and other psychiatric illnesses. The review is organized into four parts. The first part addresses the current practice and challenges of the clinical diagnosis of psychiatric disorders. The aim of this section is to illustrate the concept of diagnostic validity for psychiatric disorders and the potential impact of this on biomarker research. The second part will discuss the associated challenges and methods for the investigation of biomarkers. In particular, we will elaborate on two main types of approaches to identify disease-related molecular abnormalities when there is uncertainty regarding the validity of the clinical diagnosis. The third part of the review will present an introduction to the multiplex immunoassay approach that facilitates identification and quantitation of molecular biomarkers, and the fourth part will cover novel approaches for extending these molecular findings into the realms of identifying the associated functional consequences.

II. The Problem of Disease Heterogeneity

A. History of Clinical Diagnosis of Psychiatric Disorders

Any form of rational medicine depends on the existence of a valid method to group similar patients under a diagnosis. At present, the major psychiatric diagnostic system is the widely used Diagnostic and Statistical Manual (DSM) system

(Sadler, 2004), with the 1994 DSM-IV as the most recent edition (American Psychiatric Association, 1994), and the revised DSM-V expected in 2012.

While the influential DSM-III was introduced in 1980, the fundamental structure of this system dates back to the late nineteenth-and early twentieth-century, when Kraepelin made his influential distinction between dementia praecox and mania (Kraepelin, 1971). Kraepelins' distinction stemmed from the observation that, among the general group of relatively young previously normal functioning patients who developed alterations in behavior and mental abilities, two distinct long-term patterns could be identified. The most serious cases were those individuals who failed to show complete recovery, exhibiting lasting deficits throughout life. It was for this group that the term dementia praecox was initially used. The other group was comprised those patients who clearly improved to a state of almost complete recovery, although a remitting-relapsing course could arise. This latter group formed the manias. That such a distinction could be made with any validity has been fundamental to psychiatric classificatory systems since this time.

Dementia praecox later developed into the schizophrenia concept (Bleuler and Zinkin, 1950), whereas mania formed the basis of manic depression and depressive disorder. Finally, the distinction was incorporated into the DSM-III system, with the group of psychotic disorders incorporating schizophrenia on one side, and the broad group of mood disorders incorporating manic depression (bipolar disorder) and depressive disorders (major depressive disorder) on the other side (for an overview of the work of Kraepelin and the influence on psychiatric classification, see Braceland, 1957; Decker, 2004).

Apart from mood and psychotic disorders, three other major diagnostic categories were incorporated into the DSM system. The work of Kanner and Asperger on young children with odd, self-oriented behavior, and inability to engage in normal reciprocal social contacts gave rise to the broad group of autism spectrum disorders (Asperger, 1944; Kanner, 1968). Also, a group of anxiety disorders was recognized, partly because of the many traumatized veterans from the major twentieth century military conflicts, incorporating posttraumatic stress disorder and panic disorder (Bienvenu *et al.*, 2010). In addition, personality disorders emerged from Freud's neurosis concept (Reed, 1990).

The research scientists and clinicians responsible for developing DSM-III in the late 1970s were conscious of the fact that little was known about the biological underpinnings of psychiatric disorders. At the time, a psychological explanation existed for most disorders, usually in the form of some kind of stress-reaction process (van Praag, 1997). However, it was agreed upon that the scientific basis for psychological explanations was limited. Therefore, the DSM-III and DSM-IV systems do not take into account any hypothesized cause or process which may underlie the diagnostic categories, in order to avoid the possibility that etiological preconceptions would interfere with subsequent research (American Psychiatric

Association, 1994). In this sense, the DSM system is etiologically neutral. As a result, the DSM system defines diagnostic categories by defining cutoff criteria based on the presence or absence of symptoms, without reference to supposed associated psychological or biological processes. As such, the DSM categories do not resemble mature medical diagnostic concepts which have associated underlying biological alterations, such as structural (i.e., a bone fracture or physical trauma) or functional (i.e., diabetes) changes. An extensive overview of the origins and influence of the DSM system can be found in Sadler (2004).

The diagnostic concepts included within the DSM system have been chosen mainly because of their clinical heuristic validity. This is especially true when it comes to characterization of the treatment response. For example, schizophrenia subjects respond to dopamine D2 receptor antagonists, bipolar disorder patients respond to mood stabilizers, and major depressive disorder and some of the anxiety disorders respond to antidepressants. However, individuals with autism spectrum and personality disorders show limited response to pharmacological treatment, although some of these subjects respond to cognitive behavioral therapy. Moreover, the majority of patients seen in everyday clinical practice exhibit symptom clusters which can be associated with one or more DSM categories.

In the years following its introduction, DSM-III, and its successor DSM-IV, became the leading diagnostic system for psychiatric research (Sadler, 2004). It was intended primarily for research purposes, although it is also widely used by clinicians for patient assessment.

B. Advantages and Disadvantages of DSM Based Diagnosis

Since their introduction, DSM-III and DSM-IV have been praised for initiating a rigorous, methodologically sound approach to psychiatric diagnostics, improving the quality and quantity of psychiatric research. Indeed, the burgeoning field of psychiatric research as it is today would most likely not exist without the DSM system. Over the years, however, the DSM approach has also met with severe criticism (van Praag, 1997, 2000, 2001; Sadler, 2004). It has been argued that research into the biological determinants of abnormal behavior exacts particular standards upon psychiatric diagnosis and that the DSM system falls short in several respects. Clearly, diagnosis is the principal rate-limiting step in biological psychiatric research, and when researchers use invalid diagnostic categories, the results of the research will be either absent, clouded by inclusion noise, spurious, or, in the worst case, misleading.

The main disadvantage of the DSM diagnostic categories is that they do not reflect a true medical diagnosis, but they are arbitrary categories to a certain extent (van Praag, 1997). Through training of clinicians and psychiatrists in the use of standardized interviews, the use of DSM can lead to acceptable interrater validity. However, such increased reliability does not necessarily imply validity of the identified constructs. It has repeatedly been argued that the validity of DSM constructs is limited with respect to the underlying pathophysiological pathways, delimitation from other disorders, and in follow-up studies.

Specifically, DSM categories are notoriously heterogeneous since the system allows for several combinations of symptoms to be arranged into one category. For example, the schizophrenia concept can be generated out of 23 different combinations of symptoms and phenomena. This heterogeneity is also reflected in the debate about the validity of certain diagnostic groups within the schizophrenia spectrum such as in the case of schizoaffective disorder (Maj *et al.*, 2000).

A specific development of the past years has also contributed to the heterogeneity found in DSM-IV categories. Some researchers argue that there exists a continuum between core psychiatric symptoms and syndromes, and normal functioning (van Os *et al.*, 2000). This has lead to a situation in which the border between mental distress and mental illness is only vaguely marked. van Praag likened this situation to searching for the pathogenesis of tuberculosis but not to making a diagnostic distinction between this and the common cold (van Praag, 1997). He found this issue specifically prominent in research of depressive disorders, in which the distinction between sorrow and depression is not adequately made.

Taken together, these issues have led to the need for elucidation of the biological mechanisms of discrete disorders in biological psychiatry, including through the search for molecular biomarkers. However, one may question the construct validity of these distinguished disorders, as many of these seem to represent a variety of more or less comparable but, in many ways, dissimilar conditions. As a result, it is hard to believe that the search for particular brain dysfunctions underlying such heterogeneous diagnostic constructs stands much chance of success.

C. How Do These Issues Affect Biomarker Research?

In clinical practice, it is known that the majority of psychiatric patients do not meet the criteria of having one particular disorder, according to current definitions. Instead, subjects often show signs and symptoms associated with a multitude of disorders, or they display a patchwork from parts of different disorders. It is not clear what effect this will have on the search for molecular biomarkers. One obvious conclusion is that we should not fully rely on traditional psychiatric diagnostics, because there is some fundamental uncertainty with respect to their validity. Therefore, when trying to perform meaningful biomarker research, two distinctly different approaches to resolve inherent diagnostic problems should be taken.

The first approach has been referred to as "the narrow-to-broad" method. This involves the assumption that there is only evidence for the existence of a limited set of fundamental psychiatric syndromes, which can be identified by serious disabling core symptoms. Such fundamental syndromes might be: schizophrenia with severe negative symptoms and "first-rank symptoms" as defined by Schneider (Hoenig, 1983), bipolar mania, severe depression with autonomic dysregulation (formerly known as endogenous depression), and the classical autism syndrome as described by Kanner (1968) and Asperger (1944). These syndromes are relatively rare but can be recognized easily when encountered because of their severity and classical presentation of symptoms.

An alternative starting point may be the use of symptoms as quantified by the Positive and Negative Syndrome Scale (PANSS), which is the most widely employed symptom rating scale used in schizophrenia research (Kay *et al.*, 1987, 1988, 1989). This system is based on a set of 30 items with values ranging from 1 (absence of mental illness) to 7 (most severe clinical presentation). Similar to the DSM, PANSS has high interrater reliability (Muller and Wetzel, 1998) and has been applied frequently to assess the efficacy of treatment response (Lindenmayer *et al.*, 2007). However, PANSS scores are commonly viewed as a quantitative representation of a clinician's assessment of psychopathology and have not been directly used for the purpose of diagnosis (Mortimer, 2007). However, the correlation between rating scale scores and molecular data has often been used to underline the disease intrinsic nature of the findings (Rothermundt *et al.*, 2001). PANSS items are typically grouped into three or more sets reflecting positive and negative symptoms as well as the general psychopathology (Lindenmayer *et al.*, 1994). This grouping is based on an internal correlation structure, and factor models have been used to determine the sets of items that mostly reflect the unique features of the disease. As discussed by Mortimer, these approaches are a reflection of the heterogeneity of schizophrenia symptoms and demonstrate why single readouts such as the overall PANSS score are limited in their ability to capture the complexity of the manifestation in a given individual (Mortimer, 2007). Subscores that quantify only certain dimensions of this manifestation have already been shown to outperform categorical procedures such as DSM or the international statistical classification of diseases and related health problems (ICD) as predictors of the course of illness (van Os *et al.*, 1996).

As a first step of using these methods, one would need to identify biomarkers reliably associated with each of these syndromes. This would involve measurement of biomarkers in a large group of patients who present with less specific psychiatric symptoms, irrespective of their severity of DSM classification, and determine if the outcome could be used to classify patients according to their resemblance, if any, with either of the fundamental syndromes. It would also be important to investigate whether patients with a given biomarker profile have a prognosis or treatment response fitting one of the fundamental syndromes. This is the approach our group has taken in identifying a biomarker profile for schizophrenia, and investigating patients with atypical symptoms who later developed schizophrenia (Schwarz *et al.*, 2010a). This inherently assumes a high level of resolution and the presence of individual symptoms associated with a specific molecular phenotype. It remains to be determined whether this assumption reflects the biological reality.

The second method works the other way round and might thus be called "the broad-to-narrow" approach. This approach assumes that current diagnostic categories have limited value with respect to their underlying biological validity, whether determined by DSM, historically defined archetypical syndromes or by dimensionally defined psychological functions and should therefore not be used to steer biomarker research. Instead, the starting point should be the clinical reality that patients come into broad "problem basins" (van Praag, 2000), such as the mental and behavior disorders seen in young children, adolescents, or adults. In this approach, a first step would be to collect a large group of patients from one of the problem basins, such as adolescent-onset mental and behavioral disorders. Such a group is likely to comprise the current DSM diagnoses psychotic disorders including schizophrenia, bipolar disorder, major depressive disorder, conduct disorder, and various developing personality disorders. The next step would be to measure a multiplex biomarker fingerprint (i.e., a serum proteome profile), assuming that alterations in serum proteins will be present even though the exact nature or causes of these changes is unknown. After this, computerized clustering techniques could be used to identify clusters of patients with similar biomarker profiles. Patients belonging to a single cluster will, by definition, have the same biological profile, at least with respect to the biomarkers and tissue investigated. Such a cluster might be associated with a patient exhibiting the specific signs and symptoms of a traditional diagnostic category. However, it is more likely that a cluster will consist of mixture of patients exhibiting the signs and symptoms of a variety of DSM disorders. Therefore, it should be investigated whether the patients of a single cluster show a common, meaningful clinical characteristic such as a similar prognosis, developmental trajectory, response to treatment, or alterations in biological or psychological function.

This approach overcomes the above-mentioned assumption that symptoms have to be associated with specific molecular underpinnings. In fact, patient clustering based on molecular data alone may yield a partitioning of psychiatric patients that does not align with the current diagnostic criteria. In this way, patient clustering could ultimately lead to changes in current diagnostic criteria, in particular if patients of distinct clusters show different treatment response or side effects. The main difficulty with this approach is the requirement of large numbers of samples that are comparable from a purely analytical point of view. Such comparability requires the implementation and adherence to strict standard operating procedures for sample collection across multiple clinical sites. Also the system used to perform the molecular measurements has to be accurate, reproducible, and sufficient in throughput to cope with large numbers of samples. One method which fulfills these criteria is the multiplexed immunoassay platform utilized recently by our group (more on this below; Schwarz *et al.*, 2010a). Due to the optimization of sample dilution series, this method is capable of measuring a wide dynamic range of the most abundant molecules in serum as well as those considered to be of low abundance such as the interleukins and various hormones. The method also fulfills the high-throughput criteria, requires as little as 150 μl for measurement of approximately 200 molecules across several orders of magnitude in concentration, and yields absolute quantitation of these molecules as the output.

Obviously, a number of assumptions underlie the second approach. These include the requirement that peripheral alterations in biomarkers are present, that these biomarkers reflect underlying biological alterations common to subgroups of patients, and that a single cluster represents a characteristic of the patients that is important in the diagnostic process. One important prerequisite is the potential dynamic biological nature of psychiatric disorders. Current clinical practice considers a diagnosis valid if a given patient remains in the assigned diagnostic group for an undefined period after the diagnosis. There are several hypotheses that the underlying molecular manifestation is of dynamic nature and that time-dependent changes coincide with the exacerbation of clinical symptoms. In fact, one hypothesis which assumes that psychosis itself is toxic to the brain and causes irreversible damage is one of the major contributors to the theory that early intervention is essential for positive outcome. Similarly, symptoms observed before onset of psychosis are retrospectively considered to be characteristic for the schizophrenia prodrome. Only recently, researchers have begun to investigate such prodromal symptoms in naturalistic, prospective studies considering these as treatable disease entities to establish their predictive qualities for conversion to schizophrenia (Cornblatt *et al.*, 2001). The potential dynamic nature of the disease is of particular relevance for the development of molecular diagnostics.

As mentioned above, we recently carried out a large scale study to investigate schizophrenia specific profiles based on single sample collections during the disease course (see chapter "Algorithm development for diagnostic biomarker assays" by Izmailov and Schwarz for details). Attempts to reduce problems arising from this snapshot of a potentially dynamic disease course included factors such as recruitment of first-episode patients who were antipsychotic naïve at the time of sample collection. This is important since recruiting first-episode patients reduces the time-window relative to the onset of the disease and therefore helps to circumvent potential confounds due to treatment. Ultimately, dynamic adaptation of the molecular signal is needed for diagnostic applications to indicate the absence or decrease of an underlying pathology in patients who are receiving antipsychotic medication. This would be of use in facilitating the decision for discontinuation of treatment.

III. Multiplexed Assays are Needed to Characterize Heterogeneous Illnesses

Due to the wide spectrum of symptoms, it is likely that the molecular basis of the current classification of schizophrenia and other psychiatric disorders is not reflected in single, but in multiple molecular alterations which might differ in their importance between individuals. This is supported by several different molecular investigations which found only weak contributions of single genes to the risk of schizophrenia onset, despite the high heritability of the disorder. Molecular alterations found in more accessible, peripheral body fluids are typically low due to the large biological variability observed in asymptomatic individuals. As mentioned above, this will lead to low effect sizes, necessitating the use of higher sample numbers and, consequently, a larger number of molecular assays to obtain sufficiently high diagnostic accuracy. In our study, we identified a panel of 51 molecular assays which, when tested in a larger population, yielded a sensitivity and specificity of 82% for discriminating schizophrenia patients from controls (Schwarz *et al.*, 2010a). As stated above, the applied approach was based on multiplexed immunoassay technology that measures a prespecified set of analytes in every sample. This approach could be considered as "semihypothesis" driven since the measured molecules are derived from pathways that have high relevance for the disorder under investigation although no prior hypothesis about the alteration of individual molecules exists. Also, this method allows the measurement of candidate biomarkers that have been implicated in the scientific literature or have been found to be changed by other investigators. The measurement of molecules in a multiplex immunoassay system has the advantage that individual molecules with an effect size that is too low to lead to a biomarker may become useful when it is combined with other molecules.

As described above, the combination of multiple weak classifiers can lead to a classification rule with high performance (Freund and Schapire, 1997). Shortcomings of this method are the high cost of incorporating newly discovered molecules as novel assays into the multiplexed system and the absolute reliance on the availability suitable specific antibodies in order to achieve this. In particular, the latter requires a careful experimental setup and adherence to stringent quality control procedures to avoid shifts in measurement performance caused by changes in reagent batches. Such shifts can be detrimental for the application of a multiplexed system as a diagnostic tool. Decision rules that produce a combined output based on the measurement of the multiplexed molecules need to be trained on a reference set of samples. This inherently requires that all future samples should be comparable in their measurement relative to this reference set. This problem is exacerbated as more molecules become required for the decision rule since analytical deviations of individual measurements may add up, such that the combined classifier produces a noisy output.

A potential alternative method that is currently receiving major interest is that of selected/multiple reaction monitoring (SRM) mass spectrometry (Addona *et al.*, 2009). This method can perform absolute quantitation of proteins by measurement of targeted peptides. Important considerations are the consistency of the detectability of these peptides, in particular, when short liquid chromatography gradients are used to decrease the run time and to improve the throughput. SRM determines absolute concentrations by comparison of peptides against labeled counterparts. The technology has been applied recently to quantify protein products of somatic mutations of cancer cells and has shown promise for diagnostic applications (Wang *et al.*, 2011). Similar to immunoassays, labeled peptides can be multiplexed allowing the accurate quantitation of multiple proteins in a single sample.

IV. Multiplex Immunoassay Profiling

This section describes the procedure of the mutiplex-analyte profiling (MAP™) platform for profiling serum samples taken from patients with psychiatric disorders. The recent development and application of such multiplex immunoassay platforms allows the simultaneous measurement of tens to hundreds of molecules in individual samples. The Rules-Based Medicine (Austin, TX, USA) MAP technology has already been applied successfully in numerous clinical studies targeting diseases such as epithelial ovarian cancer (Bertenshaw *et al.*, 2008), scleroderma (Duan *et al.*, 2008), coronary artery disease (Gurbel *et al.*, 2008), myocardial infarction (Escobar and Lindsey, 2007), autoimmune disorders (Delaleu *et al.*, 2008), and sickle cell anemia (Lee *et al.*, 2007). This platform is also suitable for the development of sensitive and specific tests for use in medical practice.

The assays are basically a combination of immunoassay (reviewed elsewhere: Lowry *et al.*, 1989) and flow cytometry (reviewed elsewhere: Norman, 1980) techniques. A typical assay begins when a small volume from each sample is added to reaction wells in a plate containing the capture microspheres (Fig. 1). The microspheres are typically conjugated to antibodies and are encoded with a unique fluorescent signature that is specific to the targeted molecule. The antibody-microsphere conjugates are incubated with the samples to allow the molecules of interest sufficient time for binding. After this, a cocktail of specific, biotinylated detection reagents is added, followed by the addition of a fluorescent reporter molecule which is usually another antibody which recognizes a different epitope on the molecule of interest. Finally, the mixture is washed to remove unbound detection reagents prior to reading the reaction plate in the Luminex™ machine.

The Luminex instrument operates similar to a flow cytometer, using the principle of hydrodynamic focusing to channel the microspheres containing bound

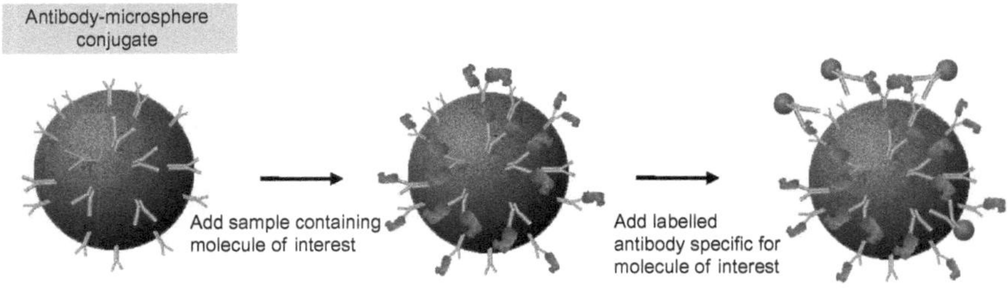

FIG. 1. Mutiple-analyte profiling scheme.

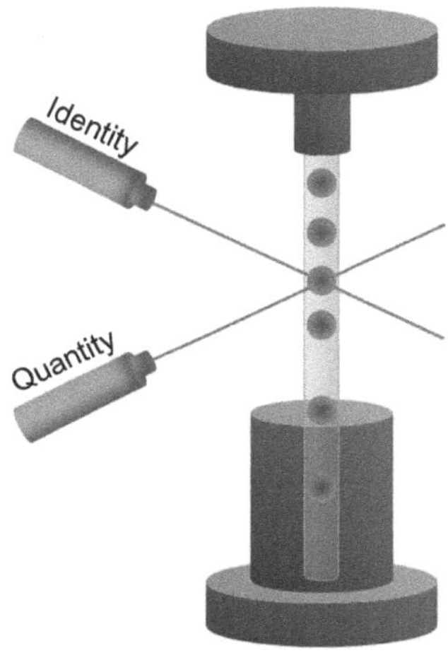

FIG. 2. Identification and quantitation of molecules using principles of flow cytometry.

molecules and assay reagents, one at a time, along a path that is analyzed by two lasers (Fig. 2). The excitation beams of the red laser measures the unique fluorescent signature of each microsphere, and the green laser determines the amount of fluorescence generated in proportion to the concentration of the molecule in the sample. Data are acquired and reported in real time, affording the ability to repeatedly measure the concentration of a given molecule in each sample.

The MAP tests incorporate specific controls for each molecule assayed within the multiplex. These controls also mimic the sample matrix, creating a realistic background for the measurements. For the majority of tests, native proteins are used as controls rather than recombinant proteins. The benefits of native controls include better assay performance over time and a greater specificity in measuring the native counterparts within the samples. In addition, low, medium, and high concentrations of controls are used to support data accuracy across the concentration range which has already been defined by testing of standards. The control values are also used to monitor assay performance longitudinally. In most cases, the intraassay coefficient of variation is less than 10%. The method also uses a set of modified Westgard rules (Westgard, 1992) (http://www.westgard.com/index.php) to evaluate the control data. This includes a set of multirule quality control decision criteria that are used to determine whether or not an assay functions as expected. This raises alerts to potential problems such as anomalies in individual control value levels, systematic problems among or within the controls, or potential unwanted trends in the data. The controls must pass all of quality control criteria to be considered valid, and only results that meet this standard are reported.

The final step in the process is the calculation and reporting of the experimentally determined molecular levels in the samples. As with most immunoassay based systems, this is achieved by plotting the readings of each sample along standard curves to derive the concentration of the target molecule. The corresponding concentration determined for that sample can then be adjusted by the appropriate dilution factor to calculate the absolute concentration of the molecule of interest. In terms of further quality control measures, one of the benefits of multiplexing is that proteins that are likely to be present in a particular sample type can be used as a point of reference. Thus, the absence of a robust signal from such a reference marker may indicate reagent or liquid handling issues.

We have recently applied this methodology in collaboration with Rules-based medicine in studies of schizophrenia (Schwarz *et al.*, 2010a), bipolar disorder (Herberth *et al.*, 2011), and Asperger syndrome (Schwarz *et al.*, 2010b). In each case, a robust multiplex signature was capable of distinguishing patients from controls with good precision. Also, these studies led to an increased biological understanding of these conditions at the molecular level. In the case of the bipolar disorder study, the identified molecules were associated with cell survival pathways (Herberth *et al.*, 2011), and the Asperger syndrome study revealed sex-specific alterations in immune pathways in males and in growth factor and hormonal systems in females (Schwarz *et al.*, 2010b). In studies of schizophrenia, we found alterations in the levels of insulin and some hormones of the hypothalamic-pitutary-adrenal-gonadal axis which is consistent with reports of insulin resistance in at least some of these subjects (Guest *et al.*, 2010, 2011; Spelman *et al.*, 2007; see chapter "Abnormalities in endocrine and metabolic function in psychiatric disorders" by Guest *et al.*).

V. Toward Functional Analysis

Up to this point, the assays highlighted in this chapter have described the multiplex assay format for measurement of the levels of proteins and small molecules in body fluids and tissues. However, this information does not necessarily lead to insights on functional consequences even if the relative levels of these molecules are found to be altered in distinct conditions, such as in psychiatric illnesses. For example, increased levels of a particular molecule could be associated with either increased or decreased activity of a given physiological pathway depending on whether it is an activator or inhibitor, a result of feed-forward or feed-back signaling, or whether it is a cause or consequence of the effect. This section describes how such information may be obtained by incorporating additional features into the multiplex assay design.

One possibility to achieve this would be to build in the ability to detect posttranslational modifications of proteins, as this could give a direct indication of function. For example, many proteins are known to undergo processing events such as proteolytic conversion, glycosylation, or phosphorylation which can lead to activation or inhibition of key pathways in a case-specific manner. The vital hormone insulin is synthesized initially as a larger inactive precursor protein called proinsulin in secretory organelles of pancreatic beta cells (Goodge and Hutton, 2000). Shortly after synthesis, this molecule is subjected to concerted proteolytic processing in the secretory pathway by prohormone convertases PC1, PC2, and carboxypeptidase H, at specific sites on the molecule containing paired basic amino acids. This process first gives rise to the conversion intermediates des 31,32 proinsulin and des 64,65 proinsulin and then to mature insulin along with the connecting (C)-peptide. Residual proinsulin and the conversion intermediates are cosecreted with insulin and the C-peptide in response to increased blood glucose levels. Increased serum levels of these molecules have been observed in a number of pathophysiological conditions including the prodrome of type 1 diabetes, mild type 2 diabetes, and the metabolic syndrome (Creemers *et al.*, 1998). Therefore, assays which can precisely measure these various forms would be of a major benefit in assessing these and potentially other disorders.

The strategy here would be to develop multiple assays using distinct sets of antibodies which discriminate between the various forms of the proinsulin-related molecules, as described by Sobey *et al.* (1989) (Fig. 3). Recently, we described the measurement of insulin, proinsulin, and des31, 32-proinsulin using these assays in an analysis of serum samples from first-onset schizophrenia subjects and controls (Guest *et al.*, 2010). This showed that all forms of the proinsulin molecule were increased in schizophrenia relative to control subjects although glucose levels were relatively unchanged. These findings suggested that hyperinsulinemia may be involved in the onset of schizophrenia, although the possibility cannot be excluded that it is a consequence of the insulin resistance known to be associated with the disorder (Ryan *et al.*, 2003).

Single targeted immunoassays have also been designed which measure glycosylation of proteins. In one study, this was achieved using a capture antibody against the peptide backbone of α-fetoprotein followed by the addition of a lectin molecule for detection of carbohydrate attachments on this protein (Kinoshita

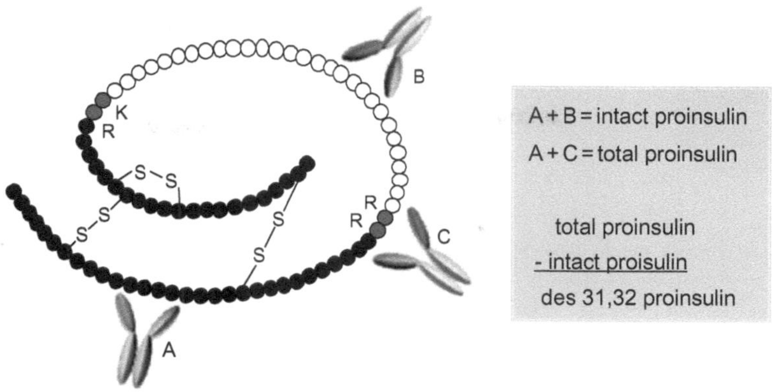

FIG. 3. Two-site immunoassays for measurement of intact proinsulin and des 31,32 proinsulin. Use of antibody A for capture and antibody B for capture recognizes intact proinsulin as this requires an intact cleave site at amino acids 31 and 32 (RR). Use of antibody A for capture and antibody C for detection recognizes all forms of proinsulin with a continuous C-peptide and will therefore recognize intact proinsulin, des 31,32 proinsulin, and des 64,65 proinsulin (total proinsulin). The levels of des 31,32 proinsulin can be calculated by subtraction of the levels obtained using antibodies A and B (intact proinsulin) from that with antibodies A and C (total proinsulin). The levels of des 64,65 (KR) proinsulin is discounted as these have been found to be negligible *in vivo* cleavage at amino acids 64 and 65 (KR) (Guest *et al.*, 2010).

et al., 1989). A similar format has been used to detect changes in glycosylation of transferrin, in this case, using an antibody against the transferrin peptide backbone and a lectin molecule which recognizes galactose residues (Pekelharing *et al.*, 1987). Likewise combinations of antibodies which target the peptide backbone and specific phosphorylated amino acid residues have been used to measure phosphorylation and dephosphorylation of proteins, a key regulatory feature in cellular signal transduction. One example of such an assay is the demonstration of changes in phosphorylation of tyrosine 1248 in the erythroblastic leukemia viral oncogene homolog 2 (ERBB2) breast cancer-related protein (Cicenas *et al.*, 2006). The general format for a phosphorylation-based immunoassay is shown below using the MAP™ bead technology (Fig. 4).

Studies of glycosylation and phosphorylation changes may be important in investigations of psychiatric and neurological disorders. Recent studies have identified N-linked glycosylation changes in the cerebrospinal fluid and serum in patients with schizophrenia (Stanta *et al.*, 2010). Reduced phosphorylation of the NMDA receptor NR1 subunit has also been observed in postmortem brain tissue from schizophrenia subjects (Emamian *et al.*, 2004). Therefore, translation of assays which can assess the states of such posttranslational modifications to the MAP™ format would be a benefit in future studies of psychiatric conditions. The use of multiplex assays in this case would lead to increased accuracy and reliability

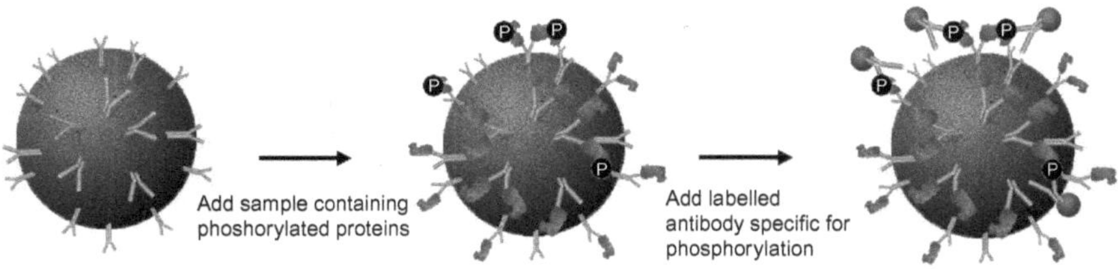

FIG. 4. Two-site immunoassay format for measurement of phosphorylation/dephosphorylation of specific proteins. (Left) The bead is coated with an antibody specific for the protein of interest (binding of both phosphorylated and nonphosphorylated forms). (Middle) Sample containing the protein of interest is added and specific binding of phosphorylated and nonphosphorylated forms of the protein occurs. (Right) An antibody specific for phosphorylation is added which detects only those proteins containing a phosphate residue. Using this format, the detection antibody could be replaced by a carbohydrate-specific molecule such as a lectin for determining the levels of glycosylation.

of the information obtained as this would allow simultaneous analysis of differentially modified forms of the same protein in the same sample. This would eliminate errors due to comparison of readings across different samples. There is also considerable scope for development of further assays based on this format for detection of other posttranslational changes such as amidation, acetylation, and sulfation of proteins.

VI. Conclusion and Outlook

In this chapter, we have discussed the clinical diagnosis of psychiatric disorders and how this impacts on patient heterogeneity and downstream biomarker discovery. Patient diagnosis remains ultimately a subjective construct with no validated biological correlates. However, multiple molecular studies have already emerged demonstrating reproducible abnormalities in these conditions. In fact, biomarker panels exist that can be used to mirror clinical decision making with high sensitivity and specificity. Such assays have started to be used in the clinic as aids in confirming the diagnosis of schizophrenia. To increase the clinical applicability of such tests, the target comparison has to be widened to include patients with other psychiatric conditions such as bipolar disorder and major depressive disorder. The overlap in symptom domains between these disorders may be an obstacle to achieve sufficient disease specificity. We have discussed several methods on how this problem can be overcome. Whether this approach starts from a narrow selection of patients with a defined set of comparable symptoms or from a broad heterogeneous group as a basis for clustering, large populations of patients will be required to increase analytical consistency. We have discussed, in particular, the two methods, multiplexed immunoassays and selective reaction monitoring, that show promise for accurate and reproducible, high-throughput profiling.

Such information will ultimately lead to a better understanding how the molecular pathophysiology aligns with the clinical manifestation of the respective disorders. This may modify current diagnostic constructs, facilitate personalized medicine strategies, and help to identify appropriate treatment approaches. Once biological pathways are identified that are specific for certain subdomains of the disorders, drug development and intervention studies can be performed in a more targeted fashion. In addition, animal models of psychiatric disorders may receive novel validity in the light of biomarkers that can be translated to clinical studies, facilitating biological substratification of patients. This, in turn, may be helpful in the development of much needed novel treatment strategies.

Acknowledgments

This research was supported by the Stanley Medical Research Institute (SMRI), the European Union FP7 SchizDX research programme (grant reference 223427), and the NEWMEDS Innovative Medicines Initiative.

References

Addona, T.A., Abbatiello, S.E., Schilling, B., Skates, S.J., Mani, D.R., Bunk, D.M., Spiegelman, C.H., Zimmerman, L.J., Ham, A.J., Keshishian, H., Hall, S.C., Allen, S., *et al.* (2009). Multi-site assessment of the precision and reproducibility of multiple reaction monitoring-based measurements of proteins in plasma. *Nat. Biotechnol.* **27**, 633–641.

American Psychiatric Association (1994). DSM-IV-TR: Diagnostic and Statistical Manual of Mental Disorders (Diagnostic & Statistical Manual of Mental Disorders) 4th revised edn. American Psychiatric Press Inc., Arlington, VA.

Asperger, H. (1944). Die "autistischen Psychopathien" im Kindesalter. *Arch. Psychiatr. Nervenkr.* **117**, 76–136.

Bertenshaw, G.P., Yip, P., Seshaiah, P., Zhao, J., Chen, T.H., Wiggins, W.S., Mapes, J.P., and Mansfield, B.C. (2008). Multianalyte profiling of serum antigens and autoimmune and infectious disease molecules to identify biomarkers dysregulated in epithelial ovarian cancer. *Cancer Epidemiol. Biomarkers Prev.* **17**, 2872–2881.

Bienvenu, O.J., Wuyek, L.A., *et al.* (2010). Anxiety disorders diagnosis: some history and controversies. *Curr. Top. Behav. Neurosci.* **2**, 3–19.

Bleuler, E., and Zinkin, J. (1950). Dementia Praecox or the Group of Schizophrenics. International Universities Press, New York.

Braceland, F.J. (1957). Kraepelin, his system and his influence. *Am. J. Psychiatry* **113**, 871–876.

Cicenas, J., Urban, P., Küng, W., Vuaroqueaux, V., Labuhn, M., Wight, E., Eppenberger, U., and Eppenberger-Castori, S. (2006). Phosphorylation of tyrosine 1248-ERBB2 measured by chemiluminescence-linked immunoassay is an independent predictor of poor prognosis in primary breast cancer patients. *Eur. J. Cancer* **42**, 636–645.

Cornblatt, B.A., Lencz, T. And, and Kane, J.M. (2001). Treatment of the schizophrenia prodrome: is it presently ethical? *Schizophr. Res.* **51**, 31–38.

Creemers, J.W., Jackson, R.S., and Hutton, J.C. (1998). Molecular and cellular regulation of prohormone processing. *Semin. Cell Dev. Biol.* **9**, 3–10.

Decker, H.S. (2004). The psychiatric works of Emil Kraepelin: a many-faceted story of modern medicine. *J. Hist. Neurosci.* **13**, 248–276.

Delaleu, N., Immervoll, H., Cornelius, J., and Jonsson, R. (2008). Biomarker profiles in serum and saliva of experimental Sjogren's syndrome: associations with specific autoimmune manifestations. *Arthritis Res. Ther.* **10**, R22.

Duan, H., Fleming, J., Pritchard, D.K., Amon, L.M., Xue, J., Arnett, H.A., Chen, G., Breen, P., Buckner, J.H., Molitor, J.A., Elkon, K.B., and Schwartz, S.M. (2008). Combined analysis of monocyte and lymphocyte messenger RNA expression with serum protein profiles in patients with scleroderma. *Arthritis Rheum.* **58**, 1465–1474.

Emamian, E.S., Karayiorgou, M., and Gogos, J.A. (2004). Decreased phosphorylation of NMDA receptor type 1 at serine 897 in brains of patients with Schizophrenia. *J. Neurosci.* **24**, 1561–1564.

Escobar, G.P., and Lindsey, M.L. (2007). Multi-analyte profiling of post-myocardial infarction plasma samples. *FASEB J.* **21**(746.11).

Freund, Y., and Schapire, R.E. (1997). A decision-theoretic generalization of on-line learning and an application to boosting. *J. Comput. Syst. Sci.* **55**, 119–139.

Goodge, K.A., and Hutton, J.C. (2000). Translational regulation of proinsulin biosynthesis and proinsulin conversion in the pancreatic beta-cell. *Semin. Cell Dev. Biol.* **11**, 235–242.

Guest, P.C., Wang, L., Harris, L.W., Burling, K., Levin, Y., Ernst, A., Wayland, M.T., Umrania, Y., Herberth, M., Koethe, D., van Beveren, J.M., Rothermundt, M., *et al.* (2010). Increased levels of circulating insulin-related peptides in first-onset, antipsychotic naive schizophrenia patients. *Mol. Psychiatry* **15**, 118–119.

Guest, P.C., Schwarz, E., Krishnamurthy, D., Harris, L.W., Leweke, F.M., Rothermundt, M., van Beveren, N.J., Spain, M., Barnes, A., Steiner, J., Rahmoune, H., and Bahn, S. (2011). Altered levels of circulating insulin and other neuroendocrine hormones associated with the onset of schizophrenia. *Psychoneuroendocrinology* **36**, 1092–1096.

Gurbel, P.A., Kreutz, R.P., Bliden, K.P., DiChiara, J., and Tantry, U.S. (2008). Biomarker analysis by fluorokine multianalyte profiling distinguishes patients requiring intervention from patients with long-term quiescent coronary artery disease: a potential approach to identify atherosclerotic disease progression. *Am. Heart J.* **155**, 56–61.

Herberth, M., Koethe, D., Levin, Y., Schwarz, E., Krzyszton, N.D., Schoeffmann, S., Ruh, H., Rahmoune, H., Kranaster, L., Schoenborn, T., Leweke, F.M., Guest, P.C., *et al.* (2011). Peripheral profiling analysis for bipolar disorder reveals markers associated with reduced cell survival. *Proteomics* **11**, 94–105.

Hoenig, J. (1983). The concept of Schizophrenia. Kraepelin-Bleuler-Schneider. *Br. J. Psychiatry* **142**, 547–556.

Horrobin, D.F. (1980). Schizophrenia: a biochemical disorder? *Biomedicine* **32**, 54–55.

Kanner, L. (1968). Autistic disturbances of affective contact. *Acta Paedopsychiatr.* **35**, 100–136.

Kay, S.R., Fiszbein, A. And, and Opler, L.A. (1987). The positive and negative syndrome scale (PANSS) for schizophrenia. *Schizophr. Bull.* **13**, 261–276.

Kay, S.R., Opler, L.A., and Lindenmayer, J.P. (1988). Reliability and validity of the positive and negative syndrome scale for schizophrenics. *Psychiatry Res.* **23**, 99–110.

Kay, S.R., Opler, L.A., and Lindenmayer, J.P. (1989). The Positive and Negative Syndrome Scale (PANSS): rationale and standardisation. *Br. J. Psychiatry Suppl.* **7**, 59–67.

Kinoshita, N., Suzuki, S., Matsuda, Y., and Taniguchi, N. (1989). Alpha-fetoprotein antibody-lectin enzyme immunoassay to characterize sugar chains for the study of liver diseases. *Clin. Chim. Acta* **179**, 143–151.

Kraepelin, E. (1971). Dementia Praecox and Paraphrenia. R. E. Krieger Pub. Co., Huntington, NY. ISBN-13: 978-1855069749.

Lakhan, S.E. (2006). Schizophrenia proteomics: biomarkers on the path to laboratory medicine? *Diagn. Pathol.* **1**, 11.

Lee, S.P., Ataga, K.I., Zayed, M., Manganello, J.M., Orringer, E.P., Phillips, D.R., and Parise, L.V. (2007). Phase I study of eptifibatide in patients with sickle cell anaemia. *Br. J. Haematol.* **139**, 612–620.

Lindenmayer, J.P., Bernstein-Hyman, R., and Grochowski, S. (1994). A new five factor model of schizophrenia. *Psychiatr. Q.* **65**, 299–322.

Lindenmayer, J.P., Khan, A., Iskander, A., Abad, M.T., and Parker, B. (2007). A randomized controlled trial of olanzapine versus haloperidol in the treatment of primary negative symptoms and neurocognitive deficits in schizophrenia. *J. Clin. Psychiatry* **68**, 368–379.

Lowry, P.J., Linton, E.A., and Hodgkinson, S.C. (1989). Analysis of peptide hormones of the hypothalamic pituitary adrenal axis using 'two-site' immunoradiometric assays. *Horm. Res.* **32**, 25–29.

Maj, M., Pirozzi, R., Formicola, A.M., Bartoli, L., and Bucci, P. (2000). Reliability and validity of the DSM-IV diagnostic category of schizoaffective disorder: preliminary data. *J. Affect. Disord.* **57**, 95–98.

Mortimer, A.M. (2007). Symptom rating scales and outcome in schizophrenia. *Br. J. Psychiatry Suppl.* **50**, s7–s14.

Muller, M.J., and Wetzel, H. (1998). Improvement of inter-rater reliability of PANSS items and subscales by a standardized rater training. *Acta Psychiatr. Scand.* **98**, 135–139.

Norman, A. (1980). Flow cytometry. *Med. Phys.* **7**, 609–615.

Pekelharing, J.M., Vissers, P., Peters, H.A., and Leijnse, B. (1987). Lectin-enzyme immunoassay of transferrin sialovariants using immobilized antitransferrin and enzyme-labeled galactose-binding lectin from *Ricinus communis*. *Anal. Biochem.* **165**, 320–326.

Puri, B.K., Easton, T., Das, I., Kidane, L., and Richardson, A.J. (2001). The niacin skin flush test in schizophrenia: a replication study. *Int. J. Clin. Pract.* **55**, 368–370.

Reed, G.S. (1990). The transference neurosis in Freud's writings. *J. Am. Psychoanal. Assoc.* **38**, 423–450.

Rothermundt, M., Missler, U., Arolt, V., Peters, M., Leadbeater, J., Wiesmann, M., Rudolf, S., Wandinger, K.P., and Kirchner, H. (2001). Increased S100B blood levels in unmedicated and treated schizophrenic patients are correlated with negative symptomatology. *Mol. Psychiatry* **6**, 445–449.

Ryan, M.C., Collins, P., and Thakore, J.H. (2003). Impaired fasting glucose tolerance in first-episode, drug-naive patients with schizophrenia. *Am. J. Psychiatry* **160**, 284–289.

Sadler, J.Z. (2004). Values and Psychiatric Diagnosis. Oxford University Press, Oxford, UK, ISBN-13: 978-0198526377.

Schwarz, E., and Bahn, S. (2008). The utility of biomarker discovery approaches for the detection of disease mechanisms in psychiatric disorders. *Br. J. Pharmacol.* **153**(Suppl. 1), S133–S136.

Schwarz, E., Izmailov, R., Spain, M., Barnes, A., Mapes, J.P., Guest, P.C., Rahmoune, H., Pietsch, S., Leweke, F.M., Rothermundt, M., Steiner, J., Koethe, D., *et al.* (2010a). Validation of a blood-based laboratory test to aid in the confirmation of a diagnosis of schizophrenia. *Biomark Insights* **5**, 39–47.

Schwarz, E., Guest, P.C., Rahmoune, H., Wang, L., Levin, Y., Ingudomnukul, E., Ruta, L., Kent, L., Spain, M., Baron-Cohen, S., and Bahn, S. (2010b). Sex-specific serum biomarker patterns in adults with Asperger's syndrome. *Mol. Psychiatry* [Epub ahead of print].

Smesny, S., Rosburg, T., Riemann, S., Baur, K., Rudolph, N., Berger, G., and Sauer, H. (2005). Impaired niacin sensitivity in acute first-episode but not in multi-episode schizophrenia. *Prostaglandins Leukot. Essent. Fatty Acids* **72**, 393–402.

Sobey, W.J., Beer, S.F., Carrington, C.A., Clark, P.M., Frank, B.H., Gray, I.P., Luzio, S.D., Owens, D. R., Schneider, A.E., Siddle, K., *et al.* (1989). Sensitive and specific two-site immunoradiometric assays for human insulin, proinsulin, 65–66 split and 32–33 split proinsulins. *Biochem. J.* **260**, 535–541.

Spelman, L.M., Walsh, P.I., Sharifi, N., Collins, P., and Thakore, J.H. (2007). Impaired glucose tolerance in first-episode drug-naive patients with schizophrenia. *Diabet. Med.* **24**, 481–485.

Stanta, J.L., Saldova, R., Struwe, W.B., Byrne, J.C., Leweke, F.M., Rothermund, M., Rahmoune, H., Levin, Y., Guest, P.C., Bahn, S., and Rudd, P.M. (2010). Identification of N-glycosylation changes in the CSF and serum in patients with schizophrenia. *J. Proteome Res.* **9**, 4476–4489.

van Os, J., Fahy, T.A., Jones, P., Harvey, I., Sham, P., Lewis, S., Bebbington, P., Toone, B., Williams, M., and Murray, R. (1996). Psychopathological syndromes in the functional psychoses: associations with course and outcome. *Psychol. Med.* **26**, 161–176.

van Os, J., Hanssen, M., Bijl, R.V., and Ravelli, A. (2000). Strauss (1969) revisited: a psychosis continuum in the general population? *Schizophr. Res.* **45**, 11–20.

van Praag, H.M. (1997). Over the mainstream: diagnostic requirements for biological psychiatric research. *Psychiatry Res.* **72**, 201–212.

van Praag, H.M. (2000). Nosologomania: a disorder of psychiatry. *World J. Biol. Psychiatry* **1**, 151–158.

Van Praag, H.M. (2001). Past expectations, present disappointments, future hopes or psychopathology as the rate-limiting step of progress in psychopharmacology. *Hum. Psychopharmacol.* **16**, 3–7.

Wang, Q., Chaerkady, R., Wu, J., Hwang, H.J., Papadopoulos, N., Kopelovich, L., Maitra, A., Matthaei, H., Eshleman, J.R., Hruban, R.H., Kinzler, K.W., Pandey, A., *et al.* (2011). Mutant proteins as cancer-specific biomarkers. *Proc. Natl. Acad. Sci. USA* **108**, 2444–2449.

Ward, P.E., Sutherland, J., Glen, E.M., and Glen, A.I. (1998). Niacin skin flush in schizophrenia: a preliminary report. *Schizophr. Res.* **29**, 269–274.

Westgard, J.O. (1992). Assuring analytical quality through process planning and quality control. *Arch. Pathol. Lab. Med.* **116**, 765–769.

ALGORITHM DEVELOPMENT FOR DIAGNOSTIC BIOMARKER ASSAYS

Rauf Izmailov[1], Paul C. Guest[2], Sabine Bahn[2,3] and Emanuel Schwarz[2]

[1]Rules-Based Medicine, Inc., Austin, Texas, USA
[2]Department of Chemical Engineering and Biotechnology, University of Cambridge, Cambridge, United Kingdom
[3]Department of Neuroscience, Erasmus Medical Centre, Rotterdam, The Netherlands

Abstract

This chapter describes the ground-breaking development of a serum-based test to help confirm the diagnosis of schizophrenia. A multiplex panel of 51 immunoassays was developed that allowed reproducible identification of schizophrenia patients compared to controls with high sensitivity and specificity. Validation of this test consisted of developing a linear support vector machine decision rule and testing its performance using cross-validation. This resulted in readjustment of the panel and algorithm to a smaller set of 40 robust assays, along with a simple procedure for maintenance and recalibration across future measurement

INTERNATIONAL REVIEW OF
NEUROBIOLOGY, VOL. 101
DOI: 10.1016/B978-0-12-387718-5.00011-0

changes associated with different reagent lots. The resulting decision rule delivered a sensitive and specific prediction for presence of schizophrenia in subjects compared to matched controls, with a receiver operating characteristic area under the curve of 88%. Performance of the recalibrated decision rule remained constant across lot changes, ensuring consistency and accuracy.

I. Introduction

We have recently described the development of a biological signature that allowed the identification of schizophrenia patients with high sensitivity and specificity (Schwarz *et al.*, 2010, 2011; see chapter "The application of multiplexed assay systems for molecular diagnostics" by Schwarz *et al.*). There is an increasing interest in the development of such objective tools that help to identify, monitor, and stratify patients with psychiatric disorders. This is particularly due to the fact that molecular tests could enable the accurate identification of schizophrenia patients early on in the disease process improving patient outcomes and reducing healthcare costs (Davies and Drummond, 1990; Hafner and Maurer, 2006; Knapp *et al.*, 2004; Wu *et al.*, 2005). Our development approach was based on the application of multiplexed immunoassays to measure proteins and small molecules in the serum of 806 schizophrenia patients and controls (Schwarz *et al.*, 2010).

The simultaneous measurement of multiple analytes is required for the investigation of complex disorders such as schizophrenia, as individual molecules have too small effect sizes to be used as accurate classifiers. Also, recent studies on other medical conditions using gene expression approaches have shown that multiplexed biomarkers can give reproducible results, which have proven useful in clinical applications (Sotiriou and Pusztais, 2009). With this in mind, we have used the Rules-Based Medicine (Austin, TX, USA) DiscoveryMAP^{TM} technology has already been applied successfully in numerous clinical studies targeting diseases such as epithelial ovarian cancer (Bertenshaw *et al.*, 2008), scleroderma (Duan *et al.*, 2008), coronary artery disease (Gurbel *et al.*, 2008), myocardial infarction (Escobar and Lindsey, 2007), autoimmune disorders (Delaleu *et al.*, 2008), and sickle cell anemia (Lee *et al.*, 2007). This platform is also suitable for the development of sensitive and specific tests for use in medical practice.

The outcome of this multicenter study was a panel of 51 molecular markers that were selected based on their reproducibility across independent patient cohorts. We constructed a support vector machine (SVM) based decision rule that utilized the entire set of molecules to discriminate schizophrenia patients from controls with an accuracy of 82%. Subsequent applications of the decision rule on new cohorts demonstrated different measurement qualities of individual assays in the 51-plex immunoassay panel. We have, therefore, implemented a procedure (1) to identify unreliably measured molecules, (2) to adjust the selection of molecular assays for the implementation of a decision rule, and (3) to adapt the decision rule based on factors causing measurement variability such as lot-to-lot changes of immunoassay reagents.

In this review, we present the entire development process of a high-dimensional diagnostic biomarker assay. First, we will recapitulate the molecular assay selection based on biological reproducibility across independent cohorts. Second, we will describe in greater detail the development of the diagnostic decision rule. Finally, we will illustrate the postdevelopment optimization of the decision rule to obtain a classifier that is stable and can be reproducibly applied as a clinical tool.

II. Methods

The present study consisted of three phases. The first phase was aimed at selection of accurate and reproducible schizophrenia biomarkers from a collection of 181 molecular assays within the Rules-Based Medicine DiscoveryMAP assay collection. Phase 1 resulted in the selection of 51 specific immunoassays to be used in assay validation. Phase 2 featured a refinement of the individual components of the multiplexed immunoassay, development of a decision rule for separating schizophrenia patients from normal controls, and validation of the decision rule using a cohort of 806 clinical samples. For biological validation of the decision rule, 480 of these samples were analyzed only during phase 2 of this study. The protocols for the study participants, clinical samples, and test methods were carried out in compliance with the Standards for Reporting of Diagnostic Accuracy (STARD) initiative (Bossuyt *et al.*, 2003). Phase 3 addressed the implementation practice of the constructed decision rule, which had to deal with variability of molecular measurements across different lots of reagents. As a result, the decision rule was further refined and recalibrated using the 40 most robust assays from the original panel of 51. In addition, a general recalibration procedure yielding only limited changes of the offset coefficient was introduced.

A. Study Participants

The subjects were recruited from the Departments of Psychiatry at the Universities of Cologne (cohort 1), Münster (cohort 2), Magdeburg (cohorts 3 and 4), Rotterdam (cohort 5), and the USA military (n = 110 bipolar disorder (BD) patients and $n = 110$ healthy controls). Cohorts used for the molecular assay selection phase were comprised of 250 first- and recent-onset schizophrenia patients and 230 healthy control subjects (Table I). Schizophrenia patients of cohort 1 ($n = 71$), 2 ($n = 46$), 4 ($n = 47$), and 5 ($n = 40$) were first onset and antipsychotic-naïve, and 32 of 46 subjects from cohort 3 had not been treated with antipsychotic medication for more than 6 weeks prior to sample collection. First onset antipsychotic-naïve patients are difficult to recruit since even large clinical facilities can only expect to diagnose about 20–30 such patients each year. To facilitate the future development of a test with differential diagnosis capability, we also carried out DiscoveryMAP analysis using samples from subjects within 30 days before their first contact with USA military psychiatric services and who later received a confirmed diagnosis of BD ($n = 110$, Table II). The cohort used to validate and implement the decision rule comprised samples from a mixture of first onset and chronic antipsychotic-treated schizophrenia ($n = 577$)

Table I
DEMOGRAPHIC DETAILS OF SUBJECTS INCLUDED IN BIOMARKER SELECTION PHASE.

Class	Cohort	1	2	3	4	5
Control	*n*	59	46	45	40	40
	M/F	31/28	35/11	27/18	33/07	26/14
	Age[a]	30 ± 8	27 ± 9	34 ± 12	27 ± 4	36 ± 11
	BMI[a]	23 ± 4	na	24 ± 4	na	24 ± 3
Schizophrenia first onset	*n*	71	46	46	47	40
	M/F	42/29	35/11	30/16	36/11	27/13
	Age[a]	31 ± 10	27 ± 9	35 ± 12	26 ± 8	35 ± 10
	BMI[a]	24 ± 5	22 ± 2	26 ± 5	na	25 ± 5

[a] Values are shown as mean ± SD.

Table II
DEMOGRAPHIC DETAILS OF PRESYMPTOMATIC BIPOLAR DISORDER AND CONTROL SUBJECTS.

Control	*n*	110
	M/F	70/40
	Age[a]	21 ± 4
Presymptomatic bipolar	*n*	110
	M/F	70/40
	Age[a]	21 ± 4

[a] Values are shown as mean ± SD.

and healthy control ($n = 229$) subjects recruited at the Universities of Cologne, Münster, and Magdeburg (Table III).

Schizophrenia was diagnosed based on the structured clinical interview according to the Diagnostic and Statistical Manual (DSM)-IV (American Psychiatric Association, 1994). Patients used for phase 1 of this study fulfilled the criteria of the paranoid subtype (DSM-IV 295.30). All diagnoses and clinical tests were performed by psychiatrists following good clinical practice guidelines. Patients whose clinical diagnosis required revision at a later stage were excluded from the study. Healthy control subjects used in phase 1 of this study were matched to the schizophrenia patients for age, gender, and social demographics and were recruited from the same economic and geographical area of the university districts. Controls with a family history of mental disease or with other medical conditions such as type II diabetes, hypertension, cardiovascular, or autoimmune diseases were excluded from the study. Presymptomatic BD

Table III
Demographic Details of 707 Subjects (51-Plex Validation).

Class	Cohort	1	2	3
Control	*n*	72	84	73
	M/F	31/40[a]	41/43	51/22
	Age[a]	31 ± 9	37 ± 14	34 ± 11
	BMI[a]	24 ± 3	na	25 ± 4
Schizophrenia first onset				
Drug naïve	*n*	132	18	56
	M/F	78/54	14/4	36/20
	Age[a]	30 ± 9	28 ± 9	37 ± 11
	BMI[a]	23 ± 4	22 ± 3	25 ± 5
Treated	n	130	71	
	M/F	73/56[a]	49/22	
	Age[a]	34 ± 12	26 ± 8	
	BMI[a]	25 ± 5	24 ± 4	
Schizophrenia chronic				
Drug free	*n*		11	
	M/F		8/3	
	Age[a]		32 ± 9	
	BMI[a]		26 ± 6	
Treated	*n*		60	
	M/F		32/28	
	Age[a]		33 ± 9	
	BMI[a]		26 ± 5	

99 follow-up samples were available from patients in cohort 2, yielding a total sample number of 806.
[a] Demographic information for one patient not available.

patients and respective controls ($n = 110$) were selected from a USA military serum bank comprising approximately 43 million sera, which facilitated matching for age, gender, ethnicity, and lifestyle.

B. Serum Samples

The medical faculty ethical committees of the respective research facilities approved the protocols of the study. Informed consent was given in writing by all participants recruited at universities, and clinical investigations were conducted according to the principles expressed in the Declaration of Helsinki. Blood samples were collected from all subjects between 8:00 and 12:00 h into S-Monovette 7.5 mL serum tubes (Sarstedt; Numbrecht, Germany). The samples were left at room temperature for 2 h to allow for blood coagulation and then centrifuged at $4000 \times g$ for 5 min. The resulting supernatants were stored at -80 °C in low-binding eppendorf tubes (Hamburg, Germany) and shipped to Rules-Based Medicine for analysis.

C. DiscoveryMAP Multiplex Immunoassay Profiling

Molecules were measured in 250 μL serum samples using the DiscoveryMAP multiplexed antigen immunoassays in the Clinical Laboratory Improved Amendments (CLIA)-certified laboratory at Rules-Based Medicine. All multiplex molecular immunoassays were calibrated individually using duplicate eight-point standard curves, and raw-intensity measurements were converted to absolute protein concentrations using proprietary software.

D. Biomarker Selection

Biomarkers were selected from the full DiscoveryMAP platform consisting of approximately 200 molecular assays. The goal of biomarker selection was to reduce the number of molecular assays under consideration to a smaller set of assays relevant for the classification task. The biomarker selection phase of the present study was aimed at identification of molecules that were altered reproducibly in schizophrenia patients compared to healthy control subjects across independent cohorts. The molecules were ranked based on the number of cohorts in which significant differences were observed using unpaired, two-tailed *t*-tests ($p < 0.05$). Molecule selection was guided by the following criteria:

i. reproducibility (including the same directional change) in three or more cohorts;
ii. high correlation (>80%) and low average measurement shift (<40%) in repeat measurements;
iii. mean experimental values distant from the least detectable dose (LDD) are more than 20-fold (LDD is defined as the average of the signal plus three standard deviations of 20 blank samples analyzed at the same time).

E. Multiplex Assay Construction

Efficient analysis of the 51 molecules required construction of new multiplexed immunoassays. This procedure was guided by optimum dilution of serum and mixing of antibodies to give the most sensitive assays. The required dilutions of serum were 1:5, 1:50, 1:200, 1:10,000, and 1:200,000. The 1:5 dilution group consisted of 31 assays which were divided into four multiplexes. For each higher dilution, only one multiplex was used, yielding a total of eight new multiplexes for the 51 assays. Once the multiplexes were created, large batches of reagents were manufactured to allow testing of approximately 7000 samples. This was important for consistency of the assays in repeat measurements. The reagents were validated using the following parameters: sensitivity, linearity, spike recovery, common serum matrix interferences, cross-reactivity, precision, correlation, freeze-thaw stability, and short-term room temperature antigen stability.

F. Decision Rule Development

After the new multiplexed immunoassays were developed, these were used to measure all 51 molecules on a combined data set of 806 subjects comprising 577 schizophrenia and 229 control subjects (from three cohorts: Universities of Cologne, Münster, and Magdeburg, Germany). The resulting 51 × 806 matrix of data was used to develop a decision rule for separating schizophrenia patients from healthy controls. There are a variety of ways of designing such decision rules. In this study, we used one of the most reliable approaches, SVM (a general description of SVM theory and its application can be found in Vapnik, 1998). Since its introduction in early 1990s, SVM technology became the default choice for classification algorithms. In particular, SVM is capable constructing nonlinear decision rules in multidimensional input spaces by using a so-called kernel function. However, in order to keep the decision rule simple in implementation, we opted for linear SVM. In other words, the kernel function is linear, and the decision rule created a separating linear hyperplane in the corresponding input space (51-dimensional, in our case). As a result, the classification algorithm comprising the decision rule was completely defined by only 52 parameters (the parameters of the separating hyperplane): one offset value and 51 coefficients corresponding to the 51 measurements of the molecules in the multiplex.

III. Results and Discussion

In this section, using the methods described in the previous section, we report the results of biomarker selection and development of classification decision rule.

A. Schizophrenia Biomarker Selection

The biomarker selection process resulted in the identification of 22 molecular assays comprised within the DiscoveryMAP assay platform, which were altered in the schizophrenia population in three or more of the clinical centers described in the previous section (Table IV, Assays (A)). Technical reproducibility was assessed by repeating the measurements ($n = 63$ subjects) approximately 3 months later. This showed an average correlation of 0.83, an average measurement shift of 29%, and an average log distance to the LDD of 1.29. In contrast, molecular assays that were not selected featured an average correlation of 0.65, measurement shift of 54%, and log distance to the LDD of 0.70.

Nine additional assays were incorporated into the 51-plex due to the known association of the targeted molecules with schizophrenia or due to the fact that we identified significant changes in these molecules in studies of schizophrenia patients using orthogonal platforms (Table IV, Assays (B)). In addition, we carried out comparative analysis of samples from 110 presymptomatic BD patients with samples from 110 matched controls and incorporated the resulting assays for significantly different molecules ($n = 20$, parametric two-tailed t-test, $p < 0.05$) in order to facilitate the future development of a test with differential diagnosis capability (Table IV, Assays (C)).

B. Decision Rule Optimization

The new multiplexed immunoassays were used to analyze sera from 806 subjects comprising 577 schizophrenia patients and 229 controls (from three cohorts: Universities of Cologne, Münster, and Magdeburg). The resulting 806 vectors of dimension 51 were used to develop a linear decision rule for separating schizophrenia patients from healthy controls.

Table IV

Selection of Molecular Assays Incorporated into the 51-Plex Test.

Assays (A)	Reproducibility	Direction of FC	Correlation > 0.8	Readoutshift <40%	Distance to LDD >20-fold	Assays (B)	Assays (C)
Cortisol	+++++	+	–	x	–	Apolipoprotein A1	Apolipoprotein B
Haptoglobin	+++++	+	+	+	+	BDNF	Apolipoprotein A2
Interleukin 10	+++++	+	+	x	x	Beta-2 Microglobulin	Apolipoprotein CI
Alpha-1 Antitrypsin	++++	+	+	+	+	Fetuin A	Cancer Antigen 125
Apolipoprotein H	++++	+	x	+	+	IgA	Calbindin
CA	++++	+	+	+	+	LH	CD5L
Complement 3	++++	+	x	+	x	MIP-1alpha	EGF-R
Ferritin	++++	+	+	+	+	Prostatic AcidPhosphatase	FSH
Interleukin 7	++++	(3/4)	x	+	+	Testosterone	IgM
Trail R3	++++	(3/4)	–	+	–		IL-6 Receptor
Betacellulin	+++	(2/3)	–	x	–		IL-11
CTGF	+++	+	–	+	–		IL-17
Endothelin 1	+++	+	–	x	x		KIM-1
ICAM 1	+++	+	+	+	+		MCP-2
MDC	+++	+	+	+	+		MMP-2
MIF	+++	+	+	+	x		PYY
Prolactin	+++	+	+	+	+		TSH
Serum Amyloid P	+++	+	+	+	+		Transferrin
Sortilin	+++	(2/3)	+	+	+		Vitronectin
TIMP 1	+++	+	+	+	+		Thrombopoietin
TNFR2	+++	+	+	+	+		
VEGF	+++	(2/3)	+	+	+		

Assays (A) selection of 22 assays was guided by (1) reproducible changes across independent cohorts of schizophrenia patients and controls (green plus; $p < 0.05$ in 3–5 cohorts), (2) consistent directional fold change ("plus" sign), (3) good correlation ($R > 0.8$), (4) a low shift (<40%) between the first and second quality control measurements, and (5) a high measurement to LDD ratio (>20:1) (x, lower; –, not tested). Assays (B) nine assays were selected from which the targeted molecules are known to be involved in schizophrenia from the scientific literature or that we have identified as being differentially expressed using orthogonal platforms. Assays (C) 20 assays were also selected for which the targeted molecules showed significant changes in bipolar disorder patients compared to controls.

For the task of separating two classes ("schizophrenia patient" from "healthy control"), the linear SVM (SVM with linear kernel) was selected. The linear SVM approach considers the data as a set of 51-dimensional vectors (corresponding to 51 measured analytes) and seeks the separating hyperplane between two classes of vectors (in our case, the class of "schizophrenia patients" and class of "healthy controls"). The hyperplane is uniquely described by a set of 51 multipliers W_1, $W_2, \ldots, W^{51}$ and one bias term B, so that the classification of any 51-dimensional vector $\{X_1, X_2, \ldots, X_{51}\}$ can be done by computing the expression

$$S = X_1 W_1 + X_2 W_2 + \cdots + X_{51} W_{51} + B$$

and comparing this with zero. If the result is positive, then the vector $\{X_1, X_2, \ldots, X_{51}\}$ is classified as an element of the first class. If the result is negative, then the vector $\{X_1, X_2, \ldots, X_{51}\}$ is classified as an element of the second class.

A perfect separation of two classes by a hyperplane is usually not possible, especially if the sample size (806 subjects in this case) is larger than the dimension of the state space (51 assays in this case). To that end, SVM allows for a certain number of misclassification errors (a vector in the first class is classified as belonging to the second class and vice versa) by making those errors count toward the overall minimization goal with a penalty parameter C. Penalty parameters can be different for both types of classification errors (one penalty C_1 for classifying the vector of the first class as belonging to the second class and another penalty C_2 for classifying the vector of the second class as belonging to the first class). Varying the ratio of penalties C_1 and C_2 yields the same effect on sensitivity and specificity of the classification SVM decision rule as moving the separation threshold in traditional statistical decision rules. For computation purposes, we operated here with two parameters of linear SVM: overall penalty parameter C and ratio F of penalties C_1 and C_2, so that $C_1 = C$ and $C_2 = C_1 F$.

Given a pair of parameters C and F, all elements of the data set are used for the training of the decision rule, and performance of the decision rule was measured using 10-fold cross-validation. For 10-fold cross-validation, the overall data set was randomly split into 10 subsets $S_1, S_2, \ldots, S_{10}$ of equal size and then the optimization run for each of the following 10 scenarios:

- The union of $S_2, S_3, \ldots, S_{10}$ is used as a training set and the set S_1 is used as a validation set.
- The union of $S_1, S_3, \ldots, S_{10}$ is used as a training set and the set S_2 is used as a validation set.
- $\vdots$
- The union of $S_1, S_2, \ldots, S_9$ is used as a training set and the set S_{10} is used as a validation set.

The measured sensitivity and specificity calculated in each of these 10 scenarios was then averaged and assumed to be the sensitivity and the specificity of the decision rule for the considered pair of parameters C and F.

This 10-fold cross-validation approach is essentially a simplified version of leave-one-out cross-validation method, in which the training would have to be done (in our scenario) 806 times, each time on the full data set with one excluded element which would be used for validation. The process of optimization would have taken significantly more time (in our case, almost 80 times longer), with little improvement in the accuracy of the designed decision rule. Another validation approach termed the holdout method would be faster and simpler than 10-fold cross-validation approach (in holdout method, one part of the data set, typically two-thirds of it, is used for training, and the remaining part is used for validation), but its accuracy would be worse than that achieved by 10-fold cross-validation.

During the optimization process, the search of optimal performance was carried out among 20,100 pairs of parameters (C, F) organized as a grid covering the following ranges:

- $\log_2 C$ was varying from -10.0 to 0.0 with step 0.1 (100 values in total)
- $\log_2 F$ was varying from -1.0 to $+1.0$ with step 0.01 (201 values in total)

We also computed the conditional probability C that a subject with a given score S is a schizophrenia patient. The conditional probability curves were calculated using the methodology developed in Vapnik, 1998. This conditional probability mapping takes as its input the diagnostic score S (already described) of the SVM classification rule and provides as its output the conditional probabilities given the score S of different classifications: probability C that the patient with the score S is "schizophrenia" and the complementary probability $1 - C$ the patient with the score S is "not schizophrenia." In other words, the decision rule states that among 100 patients each scoring the same value S on our test, $100C$ percent will be "schizophrenia," while $100(1 - C)$ percent will be "not schizophrenia."

The result of the classification decision rule is thus given by two numbers:

1. Classification "-1" or "$+1$" ("schizophrenia" if the score S is negative and "not schizophrenia" if the score S is positive).
2. Confidence level C of the classification "schizophrenia" for a given score S (among 100 patients each scoring the same value S on our test, $100C$ percent will be "schziophrenia").

With conditional probabilities, we can provide information that is more useful to the physicians who make the clinical diagnosis than is possible with traditional binary classification decision rules. In addition, traditional metrics (such as sensitivity and specificity, etc.) can now also be measured with various levels of confidence.

C. Decision Rule Performance

Two slightly different classification decision rules were constructed using linear SVM. The first decision rule (SVM-A) was optimized to discriminate schizophrenia patients from healthy controls in the combined data set of 806 subjects. This yielded a cross-validation classification accuracy of 83% (sensitivity 83%, specificity 83%, receiver operating characteristic area under the curve (ROC-AUC) 89%). Since the cohort contained multiple samples from 99 antipsychotic-treated patients, a second decision rule (SVM-B) was built using only the 707 unique samples. This decision rule yielded a similar separation between patients and controls with a cross-validation accuracy of 83% (sensitivity 83%, specificity 82%, ROC-AUC 88%).

The conditional probability curves for both SVM-A and SVM-B are shown in Fig. 1. Note that, in general, the conditional probability C of class "schizophrenia" $y = 1$ corresponding to the score $S = 0$ is not equal to 50% since the decision rules SVM-A and SVM-B were balanced to achieve approximately equal levels

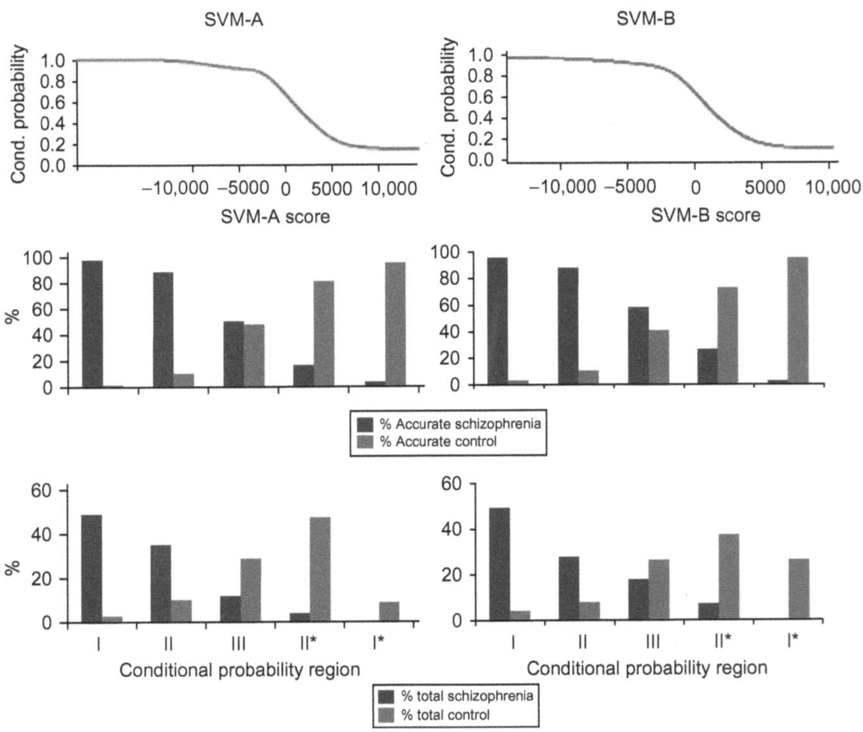

Fig. 1. Conditional probability curves for SVM-A and SVM-B decision rules.

for both sensitivity and specificity. The balancing was achieved by tilting the relative weights of the corresponding false-positive and false-negative errors (the relative weights difference is about 15% for both decision rules SVM-A and SVM-B). The conditional probability C of class "schizophrenia" $y = 1$ corresponding to the score $S = 0$ can be made equal to 50%, but for that, the decision rule has to be designed for maximizing the accuracy (overall error rate), which will have unbalanced levels of sensitivity and specificity.

We use conditional probability curve to partition the decision rule output results into five probability regions: "highly positive" Region I (corresponding to conditional probabilities 100–91%), "positive" Region II (corresponding to conditional probabilities 91–68%), "indeterminate" Region III (corresponding to conditional probabilities 68–40%), "negative" Region IV (corresponding to conditional probabilities 40–17%), and "highly negative" Region V (corresponding to conditional probabilities 17–0%).

To obtain an unbiased estimate of classification performance and a biological validation of the schizophrenia analyte signature, we determined the performance of the decision rules by testing samples which had not been used for marker selection in phase 1. Application of SVM-B ($n = 480$ subjects) yielded an overall classification accuracy of 84% (conditional probabilities are shown in Table V). We also determined the classification accuracy of the SVM-B decision rule for four main (positive and negative) regions of conditional probabilities (Table VI). This resulted in an increase in accuracy of up to 96% for schizophrenia patients and up to 97% for controls in the highest probability regions (Table VI). When

Table V
CLASSIFICATION PERFORMANCE OF SVM-B (51-PLEX DEVELOPMENT).

		SVM-B			SVM-B (480 validation samples)		
Group	Subgroup	n	Classification accuracy (%)	Conditional probability	n	Classification accuracy (%)	Conditional probability
Controls		229	83	0.69	116	77	0.61
Schizophrenia	FE drug naïve	189	77	0.86	111	78	0.85
	FE treated	201	85	0.91	173	86	0.91
	Chronic	71	96	0.94	71	96	0.94

Accuracy estimates are shown for the entire set of samples from unique patients as well as for the subset of 480 samples which were not used during phase 1 of the study. The conditional probability estimate is the median of all conditional probabilities in the respective group. FE, first episode.

Table VI

Classification Performance of SVM-B (51-Plex Development).

		SVM-B			SVM-B (480 validation samples)		
Probability region	Conditional probability	*n* patients (%)	*n* controls (%)	Classification accuracy (%)[a]	*n* patients (%)	*n* controls (%)	Classification accuracy (%)[a]
Region I	1.00–0.91	234 (49%)	9 (4%)	96	186 (51%)	0 (0%)	100
Region II	0.91–0.68	128 (27%)	17 (7%)	88	119 (32%)	21 (18%)	85
Region IV*	0.60–0.83	31 (6%)	60 (26%)	73	20 (5%)	39 (34%)	66
Region V*	0.83–0.84	2 (0.4%)	84 (37%)	97	1 (0.2%)	21 (18%)	95

Accuracy estimates are shown for the entire set of samples from unique patients as well as for the subset of 480 samples which were not used during phase 1 of the study. Individual estimates are given for four main (positive and negative) regions of conditional probabilities.

The conditional probabilities in regions marked with an asterisk reflect those determined for controls.

[a] Classification accuracies reflect the percentage of correct patient identifications in regions I and II and correct control identifications in regions IV and V.

Region III was designated as indeterminate, 17% of the total number of subjects were excluded for SVM-A, and 20% were excluded for SVM-B.

Out of the 478 total schizophrenia patients used in phase 2, 111 were suffering from a nonparanoid type of the disease (Table VII). SVM-A identified 95 (86%) of these patients correctly suggesting that the biomarker signature was present regardless of the schizophrenia subtype.

We also investigated 80 subjects with a baseline Positive and Negative Syndrome Scale (PANSS) of 66.0 ± 17.8 before and after 4–6 weeks of antipsychotic treatment. For these subjects, the treatment resulted in an overall reduction in symptoms by 13% (average reduction of 10.3 ± 17.3), as measured using the PANSS positive (18% lower (average reduction of 3.7 ± 4.1)), PANSS negative (8% lower (average reduction of 1.6 ± 4.9)), and PANSS general (12% lower (average reduction of 5.0 ± 8.5)) (Kay *et al.*, 1987). Interestingly, application of SVM-B led to correct identification of 85% of these patients at the first time-point and after the treatment period. There was an average correlation of 0.49 across all 51 molecular assays, supporting the stable identification capability of the decision rule. This suggested that schizophrenia patients in remission still feature schizophrenia-like serum profiles even after 4–6 weeks of treatment.

D. Decision Rule Refinement

Phase 3 of algorithm development started after implementation of the decision rule and replacement of the first lot of reagents with the second one. We observed that 11 molecular assays of the 51-plex panel exhibited significant lot-to-lot variations resulting in a poorer performance of the decision rule. These 11 assays targeted the following molecules:

Table VII
Subtypes of the 478 Schizophrenia Patients Investigated in Phase 2 of This Study.

DSM-IV code	Subtype	*n*
295.1	Schizophrenia, disorganized type	18
295.2	Schizophrenia, catatonic type	7
295.3	Schizophrenia, paranoid type	367
295.4	Schizophreniform disorder	26
295.6	Schizophrenia, residual type	3
295.7	Schizoaffective disorder	27
295.9	Schizophrenia, undifferentiated type	15
297.1	Delusional disorder	1
	Schizophrenia non specified	14

1. Betacellulin
2. Cancer Antigen 125
3. Calbindin
4. CTGF
5. Endothelin-1
6. IL-10
7. IL-11
8. IL-17
9. IL-7
10. MIP-1alpha
11. Thrombopoietin

Since the measurement stability of these 11 assays of the original 51-plex test was detrimental to the overall performance of the original decision rule, these were removed from the computation of the decision rule. In other words, while still remaining as components of the 51-plex test for measurement and calibration purposes, the coefficients of these assays were set to zero. As a result, it was necessary to design a revised decision rule, which would rely only on the 40 remaining assays from the original 51-plex test.

In order to design the revised decision rule, the same combined data set drawn from Universities of Cologne, Münster, and Magdeburg was tested with the second lot of reagents. Due to insufficient volume of some of the samples, the overall number of available samples decreased from 806 to 782.

The same type of decision rule optimization (search of optimal performance among 20,100 pairs of parameters (C, F) in a rectangular grid) as described in the previous section was carried out for this data set using the second lot of reagents. The performance of the diagnostic decision rules was tested using 10-fold cross-validation on the combined data set and resulted in a sensitivity of 81% and a specificity of 81%, with ROC-AUC equal to 0.88. The conditional probability curve and ROC-AUC for the refined decision rule are shown in Fig. 2.

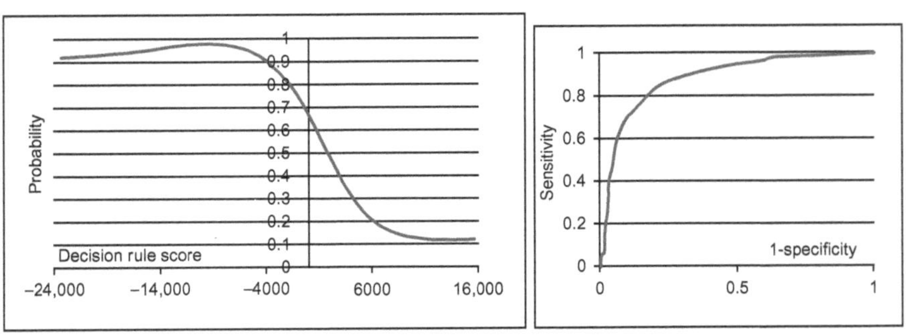

FIG. 2. Conditional probability curve and ROC for revised decision rule.

Note that the conditional probability curve does not necessarily decrease with the score S. For practical purposes, that should not be a problem, since the region of nonmonotonous behavior covers the diagnostic probabilities above 90% level.

To summarize, the performance metrics of the revised decision rule closely matched those of the original decision rule, while relying on a smaller, more rugged subset of molecular assays from the original 51-plex test.

E. Decision Rule Recalibration

As described in the previous section, the refined decision rule was designed on the second lot of reagents which caused the change of both the weight parameters and the bias term (cutoff term) that were calculated for the original decision rule. In order to avoid such sweeping changes of parameters for subsequent lots of reagents, it was decided to implement a bias adjustment procedure that only recalibrated the cutoff parameter for new lots, while retaining the values of the weight parameters the same as they were for the refined decision rule. This bias adjustment procedure is described in this section.

The key part of the bias adjustment procedure, which is executed at the time of migration from an "old" lot of reagents to a "new" one, is the "calibration pool"—a set of samples that are measured in both "old" and "new" lots. As described in the previous section, some of the samples had to be removed during the transition from the first lot to the second one due to insufficient volumes. As the migration process from one lot of reagents to another continues, the sample volumes will decrease, and thus, for maintaining a statistically meaningful calibration pool, new samples have to be added along with this migration process.

Using the calibration tool, we calculated the average measurement shift (from the "old" lot of reagents to the "new" one) for each of the 40 assays used in the refined decision rule. As a result, we obtained, for each analyte $i = 1,2, \ldots, 40$, its average shift A_i, so that any calibration pool sample that is measured X_i using the "old" lot of reagents is expected to be measured as approximately $X_i + A_i$ using the "new" one.

Since the decision rule score for the "old" lot of reagents was computed using the formula

$$S = X_1 W_1 + X_2 W_2 + \cdots + X_{40} W_{40} + B,$$

this formula, when applied to the measurements on the "new" lot, will give the value

$$S = (X_1 + A_1) W_1 + \cdots + (X_{40} + A_{40}) W_{40} + B = X_1 W_1 + \cdots + X_{40} W_{40} + B^*,$$

where

$$B^* = B + A_1 W_1 + \cdots + A_{40} W_{40}.$$

Therefore, in order to obtain the same decision rule score on the "new" lot of reagents as on the "old" one, we use the same weight parameters $W_1, \ldots, W_{40}$ of the decision rule as for the "old" lot, with replacement of only the cutoff parameter B with its recalibrated value B^*.

Note that the validity of this recalibration procedure is enabled by the linear structure of the decision rule. If we had selected a nonlinear SVM, our only recourse for changes in lots of reagents would have been the complete reoptimization of the decision rule using the recalibration pool.

IV. Conclusions

In this multicenter study, we discovered and validated a biomarker panel for schizophrenia based on biological and technical reproducibility of the molecular signature. All stages of the process, including conduction of the assays, assay selection, assay panel refinement, development, and recalibration of the decision rule, were carried out in a CLIA-certified laboratory at Rules-Based Medicine. Assay selection was based on a large number of samples collected from antipsychotic-naïve, acutely psychotic patients to facilitate relatively uniform conditions. Subjects were recruited from four independent clinical centers and samples collected according to strict standard operating procedures to maximize reliability and accuracy of the results. Biological variability arising from the collection in different geographical regions may contribute to the generality of the present findings. As the assay progresses from beta site testing into exposure to different subpopulations, the performance against present clinical classification and observed prevalence and incidence must be monitored and differences will need examination.

The implementation of the final 40-plex molecular assay decision rule was based on a cohort comprised of both untreated and treated schizophrenia patients who were either experiencing a first episode of illness or who were chronically ill (54% of patients were on current antipsychotic treatment). This collection is likely to represent more closely the patient population encountered in clinical practice. High classification performance demonstrated that the decision rule could identify schizophrenia patients with high accuracy irrespective of the disease duration or treatment state. Interestingly, the biomarker signal was still apparent in subjects even after 4–6 weeks of successful treatment with antipsychotic medication. This suggests that the 40-plex test is robust for identification of subjects with schizophrenia at different stages of the schizophrenia disease process. Further work is required for the development of a biomarker panel aiding in the monitoring of patient responses to treatment.

In summary, the present findings demonstrate the applicability of a rapid and noninvasive test to confirm the presence of schizophrenia. This first attempt to develop a molecular test with clinical utility for the diagnosis of schizophrenia was focused on the distinction of schizophrenia patients against healthy controls. For this application, we have developed a 40-plex assay panel and an optimized decision rule with a sensitivity and specificity of 81%. We have also described the procedure of decision rule adjustment for subsequent lots of reagents: the already-computed weight coefficients of the revised decision rule will remain fixed, while the bias term will change accordingly to capture average shifts in measurement values, observed over calibration sets of samples. We anticipate that the 40-plex assay panel will result in the future development of a differential diagnostic test that can be used to distinguish among various neuropsychiatric disorders such as schizophrenia, BD, and major depressive disorder. Therefore, the next stage toward clinical translation is to conduct a large scale clinical validation study using samples from diverse psychiatric patient populations and settings in a series of prospective studies with the Rules-Based Medicine assay platform.

Acknowledgments

This study was instigated and supported by Rules-Based Medicine, Psynova Neurotech Ltd, and the Stanley Medical Research Institute (SMRI). We want to thank Anke Dudeck, Jeanette Schadow, Dr Wolfgang Jordan, Dr Bernd Hahndorf, Dr Florian Kästner, Dr Anya Pedersen, Dr Ansgar Siegmund, Dr Katja Kölkebeck, Torsten Schoenborn, Dr Christoph W. Gerth, Dr Christian Mauss, Dr Brit M. Nolden, Dr M. A. Neatby, Sandra Pietsch, and Christin Oheim for their participation in sample characterization and collection. We would like to thank Dr Fuller Torrey for his support and suggestions. We would also like to thank Michael G. Walker, Ph.D. for suggestions concerning data analysis. Most of all, thanks to all patients and healthy volunteers for their selfless donation of samples used in this study

Disclaimer: 1. The views expressed are those of the authors and should not be construed to represent the positions of the United States of America Department of the Army or Department of Defense. 2. None of the authors have any associations, financial or otherwise, that may present a conflict of interest. 3. This effort was funded by the Stanley Medical Research Institute, Bethesda, MD, and the United States of America Department of the Army.

References

American Psychiatric Association (1994). DSM-IV-TR: Diagnostic and Statistical Manual of Mental Disorders (Diagnostic & Statistical Manual of Mental Disorders). 4th revised edn. American Psychiatric Press Inc. Arlington, VA. ISBN-13:978-0890420256.

Bertenshaw, G.P., Yip, P., Seshaiah, P., Zhao, J., Chen, T.H., Wiggins, W.S., Mapes, J.P., and Mansfield, B.C. (2008). Multianalyte profiling of serum antigens and autoimmune and infectious disease molecules to identify biomarkers dysregulated in epithelial ovarian cancer. *Cancer Epidemiol. Biomarkers Prev.* **17**, 2872–2881.

Bossuyt, P.M., Reitsma, J.B., Bruns, D.E., Gatsonis, C.A., Glasziou, P.P., Irwig, L.M., Lijmer, J.G., Moher, D., Rennie, D., and de Vet, H.C., Standards for Reporting of Diagnostic Accuracy (2003). Towards complete and accurate reporting of studies of diagnostic accuracy: the STARD initiative. *Clin. Chem.* **49**, 1–6.

Davies, L.M., and Drummond, M.F. (1990). The economic burden of schizophrenia. *Psychiatric Bull.* **14**, 522–525.

Delaleu, N., Immervoll, H., Cornelius, J., and Jonsson, R. (2008). Biomarker profiles in serum and saliva of experimental Sjogren's syndrome: associations with specific autoimmune manifestations. *Arthritis Res. Ther.* **10**, R22.

Duan, H., Fleming, J., Pritchard, D.K., Amon, L.M., Xue, J., Arnett, H.A., Chen, G., Breen, P., Buckner, J.H., Molitor, J.A., Elkon, K.B., and Schwartz, S.M. (2008). Combined analysis of monocyte and lymphocyte messenger RNA expression with serum protein profiles in patients with scleroderma. *Arthritis Rheum.* **58**, 1465–1474.

Escobar, G.P., and Lindsey, M.L. (2007). Multi-analyte profiling of post-myocardial infarction plasma samples. *FASEB J.* **21**(746.11).

Gurbel, P.A., Kreutz, R.P., Bliden, K.P., DiChiara, J., and Tantry, U.S. (2008). Biomarker analysis by fluorokine multianalyte profiling distinguishes patients requiring intervention from patients with long-term quiescent coronary artery disease: a potential approach to identify atherosclerotic disease progression. *Am. Heart J.* **155**, 56–61.

Hafner, H., and Maurer, K. (2006). Early detection of schizophrenia: current evidence and future perspectives. *World Psychiatry* **5**, 130–138.

Kay, S.R., Fiszbein, A., and Opler, L.A. (1987). The positive and negative syndrome scale (PANSS) for schizophrenia. *Schizophr. Bull.* **13**, 261–276.

Knapp, M., Mangalore, R., and Simon, J. (2004). The global costs of schizophrenia. *Schizophr. Bull.* **30**, 279–293.

Lee, S.P., Ataga, K.I., Zayed, M., Manganello, J.M., Orringer, E.P., Phillips, D.R., and Parise, L.V. (2007). Phase I study of eptifibatide in patients with sickle cell anaemia. *Br. J. Haematol.* **139**, 612–620.

Schwarz, E., Izmailov, R., Spain, M., Barnes, A., Mapes, J.P., Guest, P.C., Rahmoune, H., Pietsch, S., Leweke, F.M., Rothermundt, M., Steiner, J., Koethe, D., *et al.* (2010). Validation of a blood-based laboratory test to aid in the confirmation of a diagnosis of schizophrenia. *Biomark Insights* **5**, 39–47.

Schwarz, E., Guest, P.C., Rahmoune, H., Harris, L.W., Wang, L., Leweke, F.M., Rothermundt, M., Bogerts, B., Koethe, D., Kranaster, L., Ohrmann, P., Suslow, T., *et al.* (2011). Identification of a biological signature for schizophrenia in serum. *Mol. Psychiatry* [Epub ahead of print].

Sotiriou, C., and Pusztais, L. (2009). Gene-expression signatures in breast cancer. *N. Engl. J. Med.* **360**, 790–800.

Vapnik, V. (1998). Statistical Learning Theory (Adaptive and Learning Systems for Signal Processing, Communications and Control Series). John Wiley & Sons, New York, NY. ISBN-13:978-0471030034.

Wu, E.Q., Birnbaum, H.G., Shi, L., Ball, D.E., Kessler, R.C., Moulis, M., and Aggarwal, J. (2005). The economic burden of schizophrenia in the United States in 2002. *J. Clin. Psychiatry* **66**, 1122–1129.

CHALLENGES OF INTRODUCING NEW BIOMARKER PRODUCTS FOR NEUROPSYCHIATRIC DISORDERS INTO THE MARKET

Sabine Bahn[1,2], Richard Noll[3], Anthony Barnes[4], Emanuel Schwarz[1] and Paul C. Guest[1]

[1]Department of Chemical Engineering and Biotechnology, University of Cambridge, Cambridge, United Kingdom
[2]Department of Neuroscience, Erasmus Medical Centre, Rotterdam, The Netherlands
[3]DeSales University, Center Valley, Pennsylvania, USA
[4]Rules Based Medicine, Austin, Texas, USA

Abstract

There are many challenges associated with the discovery and development of serum-based biomarkers for psychiatric disorders such as schizophrenia. Here, we review these challenges from the point of view of psychiatrists, general practitioners, the regulatory agencies, and biomarker scientists. There is a general opinion in psychiatric medicine that improvements over the current subjective tests are essential. Despite this, there is a reluctance to accept that peripheral molecules can do the

INTERNATIONAL REVIEW OF NEUROBIOLOGY, VOL. 101
DOI: 10.1016/B978-0-12-387718-5.00012-2

job any better. In addition, psychiatrists find it difficult to accept that peripheral molecules, such as those found in blood, can reflect what is happening in the brain. However, the regulatory health authorities now consider biomarkers as important for the future of drug development and have called for efforts to modernize methods, tools, and techniques for the purpose of developing more efficient and safer drugs. We also describe here the development of the first ever molecular blood test for schizophrenia, and its reception in the market place, as a case in point.

I. Introduction

Studies attempting to identify molecular biomarkers for psychiatric disorders have been ongoing for many years. There is a significant interest in the discovery of such biomarkers by molecular scientists, psychiatrists, general practitioners, and the regulatory authorities. It is anticipated that these molecules could be used as empirical tools to assist the diagnosis, treatment, and monitoring of patients (see Chapter "General overview: Biomarkers in neuroscience research" by Filiou and Turck). Currently, psychiatric disorders are diagnosed based on subjective interviews and patient descriptions. It is expected that biomarkers which are associated with the disease state or with the mechanism of action of psychiatric medications will lead to improved diagnosis and potentially pave the way for more effective treatment of patients (Fig. 1).

The development of biomarkers and the implications of using these in diagnostics and clinical trials are constantly moving forward. This has led to the need for establishing standard operating procedures to meet the regulatory demands. Regulatory health authorities such as the Food and Drug Administration (FDA), consider biomarkers important in the pharmaceutical industry. The FDA has now called for efforts to modernize methods, tools, and techniques for the purpose of delivering more efficacious safer drugs (Ovens, 2006; Marson, 2007).

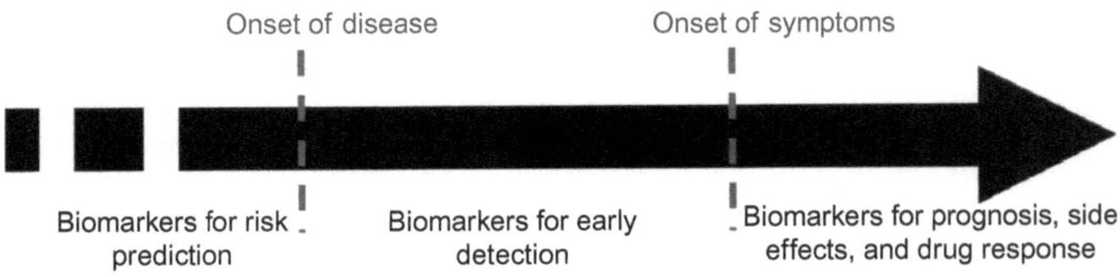

FIG. 1. Potential stages of biomarker use for neuropsychiatric disorders.

This requires that molecules must achieve the status of validated biomarkers to be used in regulatory decisions for clinical trials. The FDA has now established three types of biomarkers: (1) exploratory biomarkers, (2) probable valid biomarkers, and (3) known valid biomarkers (Goodsaid and Frueh, 2007). For the first class, there must be scientific evidence for proof of concept. The second class requires that biomarkers can be measured in an analytical test system with strict performance characteristics and that there is established scientific evidence that explains the significance of the results. The third class requires replication of the results at different sites, for cross-validation purposes.

However, this is not a simple task. The identification of biomarkers for psychiatric disorder diagnostics is challenging due to the poor understanding and delineation of these conditions using the current subjective methods, the overlap of symptoms across different disorders, and marked heterogeneity across human subjects. However, emerging proteomic platforms have facilitated the identification of biomarkers by simultaneous measurement of hundreds or thousands of molecules in nonhypothesis-driven profiling studies. Our group has recently employed an approach based on multiplexed immunoassay profiling which resulted in identification of a serum signature that could identify schizophrenia patients with an accuracy of 82% across five independent patient cohorts (Schwarz *et al.*, 2010).

In this chapter, we discuss the challenge of bringing a molecular test for neuropsychiatric disorders to the market. The first part discusses the general problem of introducing the new paradigm of molecular biomarkers into the conventionally nonmolecular field of psychiatry. The second part discusses the associated challenges and methods for the identification of biomarkers. In particular, we elaborate on the potential uses of molecular biomarkers in the field of psychiatric disorders, particularly for improved clinical classification and management of patients and as a means of facilitating the process within the pharmaceutical industry for improved drug discovery.

II. Biomarker Blood Tests for Diagnosis and Management of Mental Disorders

Recently, Thomas Laughren from the FDA published a "Festschrift" outlining the vision for psychiatric drug development over the next 50 years (Laughren, 2010). In this report, he discusses the applicability of the FDA's "Critical Path Initiative" for modernizing psychiatric drug and product development, highlighting new opportunities in the fields of biomarkers, clinical trial design, bioinformatics, and other areas.

A major problem in the field of psychiatric drug development is the lack of truly novel compounds and breakthrough concepts. Most drug candidates that have been approved over the past decade are modifications of similar parent drugs or "me-too" drugs, which are members of the same drug class only with minor modifications (DiMasi and Faden, 2011). For both antipsychotic and antidepressant medications, only 30–40% of patients achieve remission with initial treatment, emphasizing the pressing need for more efficacious drugs (Lieberman *et al.*, 2005; Rush *et al.*, 2006). A further problem is the high failure rate in psychopharmacological clinical trials, which costs the pharmaceutical industry billions of dollars in losses and most recently has led to several companies leaving the psychiatric drug discovery field altogether, including the industry giants GlaxoSmithKline and AstraZeneca. There is agreement that there is a fundamental lack of understanding of the biological abnormalities associated with severe mental illnesses, which are still defined by vague symptomatic descriptions that do not address the etiological heterogeneity of these conditions.

The FDA and pharmaceutical and biotechnology companies believe that biomarkers might help in the development of better and more efficacious drugs. A biomarker is defined as "measurable characteristics that reflect physiological, pharmacological, or disease processes" on the European Medicines Agency Web site (http://www.emea.europa.eu). Similarly, it is defined as an "indicator signaling an event or condition in a biological system or sample and giving a measure of exposure, effect, or susceptibility." In the case of biomonitoring, a biomarker is the presence of any substance or a change in any biological structure or process that can be measured as a result of exposure (Biomonitoringinfo.org/glossary). Table I summarizes the potential benefits of biomarker applications in psychiatric conditions.

To date only few biomarker tests have entered the field of psychiatric disorders. For decades, the niacin skin-flush response test (Kashshai and Mate, 1961; Vaddadi, 1981) was in sporadic use for diagnosis of schizophrenia (Wilson and Douglass, 1986). In this test, some schizophrenia patients show a reduced skin flush in response to topical application of niacin. More recently, examples of genomic markers have emerged that indicate different activities in P450-metabolizing enzymes such as various liver enzyme variants (CYP2C9, CYP2B6, CYP2C19, and CYP2D6), which can result in toxicities in patient subpopulations treated with antipsychotic medications (Fleeman *et al.*, 2010). Encouraging results are also emerging that various polymorphisms in serotonergic transporter and receptor genes are associated with response to selective serotonin reuptake inhibitor (SSRI)-based antidepressants (Kato and Serretti, 2010).

There are only a few molecular tests that predict pharmacodynamic response and these are mainly restricted to the oncology field. Perhaps the best example of this is *Her2* gene expression in breast cancer cells. This cell surface receptor can be blocked by an antibody-based therapeutic called Herceptin™ (Trastuzumab; Desmedt *et al.*, 2009; Table II).

Table I
List of Clinical Utilities for Disease Biomarkers of Psychiatric Disorders.

1. Early diagnosis and treatment decisions
   - Earlier treatment leads to better patient outcomes and healthcare cost savings
   - Differentiation of a specific psychiatric disorder from other potentially confounding conditions and aid clinical treatment decisions
2. Personalized or stratified medicine approaches
   - Predicting responders to specific therapeutic interventions (e.g., enabling decision of which antipsychotic for which patient)
   - Predicting which patients will develop specific side effects (such as weight gain or insulin resistance with olanzapine or agranulocytosis with clozapine)
3. Patient monitoring
   - Testing for "normalization" of biomarker signature with treatment (efficacy surrogate)
   - Testing for reappearance of signature on recurrence of psychosis
   - Testing for medication compliance (are patients staying on their medications?)
   - Testing to see if patients experience side effects (safety biomarker screen)
4. Development of disease modification strategies
   - Determine if biomarker signatures can predict who will develop schizophrenia in prodromal or high-risk populations and allow prophylactic treatments to be tested efficiently

Table II
List of Protein/Gene Expression Assay Tests for a Range of Indications.

1. *HER2* expression and breast cancer (Desmedt *et al.*, 2009)
2. Human chorionic gonadotropin (HCG) protein and pregnancy (Spadoni *et al.*, 1964)
3. Triple test/Quad test for Down's syndrome (Ball *et al.*, 2007)
   - Estriol, β-HCG, αfetoprotein—70% sensitivity, 95% specificity
   - Estriol, β-HCG, αfetoprotein, inhibin—81% sensitivity and 95% specificity
4. Genomic Health OncoDx test for breast cancer (Cronin *et al.*, 2007)
   - expression of 21 genes for prediction of best treatment approach
5. Glycosylated hemoglobin (HbA1) and serum fructosamine in type II diabetes (Howey *et al.*, 1989)

III. Current Dilemmas in Psychiatric Diagnosis

Currently used diagnostic classification systems for psychiatric disorder, such as the Diagnostic and Statistical Manual of Mental Disorders IV (DSM-IV; Schaffer, 1996) and the International Classification of Disease 10 (ICD-10; http://www.who.int/classifications/icd/en/), are known to have a certain degree of reliability. Therefore, a patient presenting with the same symptoms in one hospital is likely to be given the same diagnosis in another hospital if the same classification system is used. However, there is an apparent increase in the prevalence of schizophrenia when ICD-10 criteria are used for diagnosis, compared with the use of DSM-IV (Cheniaux *et al.*, 2009). This may be because DSM-IV and ICD-10 have developed to include a modern compendium of mental disorders that can be reliably diagnosed based on signs and symptoms, but have not been validated (Spitzer *et al.*, 1980; Pierre, 2008; Keller *et al.*, 2011). It is not likely that specific symptoms are linked to a defined natural disease entity. It is well known that patients with neurological, traumatic, infectious, and metabolic disorders can present with symptoms indistinguishable to symptoms of schizophrenia (Yolken *et al.*, 2009; Lovatt *et al.*, 2010; Scaglia, 2010). In addition, some subjects are known to have feigned symptoms of schizophrenia and other mental disorders (Bagby *et al.*, 1997) for reasons such as gaining access to disability payments, social housing, and other benefits.

Most psychiatrists agree that the current construct of schizophrenia is an umbrella term for a complex chimera of etiologies that happens to present with similar symptoms, in the same way that most acute infectious disorders present with fever (Tsuang, 1975). Misdiagnosis is thus a common occurrence in psychiatric practice. For example, Gonzalez-Pinto *et al.* (1998) found that 31% of bipolar patients were diagnosed with schizophrenia. Follette and Houts (1996) went further in their criticism challenging the fundamental assumptions or theoretical underpinnings of current classifications systems. They pointed out that there is no method to validate current diagnostic concepts with externally validated measures which are independent of the concept itself.

A further potential factor for misdiagnosis and inconsistency is that clinicians do not usually use classification systems to establish a psychiatric diagnosis. Instead, they mostly apply heuristic unstructured interviews. This means their diagnosis may be based on experience and personal views, instead of matching the guidelines or criteria of the diagnostic system. This can be associated with systematic errors in judgment based on misconception and experience, which may rely on selective memory. There has also been a failure to address the problem of false positives in diagnoses of mental disorders (Wakefield, 2010). A study published by Strakowski *et al.* (2003) investigated the influence of ethnicity on patient diagnosis. This study found that clinicians tend to overdiagnose schizophrenia in African Americans. The bias was removed when examiners were provided with ethnicity-blinded transcripts of otherwise identical patient interviews.

IV. The Potential for Biomarker-Based Diagnostic Tests in Psychiatry

Psychiatry must ultimately develop a diagnostic system based on the pathology or correlated with the biological disease state to move forward. This has been the case for most areas of medicine. Above, we mentioned the development of

VeriPsych™, a blood test recently launched by Rules Based Medicine Incorporated and Psynova Neurotech Limited, in collaboration with our laboratory. VeriPsych™ is the first biomarker blood test with a diagnostic application that has entered clinical practice to date. The problems that we have faced, and continue to face, in the development of molecular and diagnostic tools relate to the fact that the patients we aim to diagnose have already been selected and classified on the basis of DSM-IV criteria. As outlined above, the validity of the DSM-IV diagnosis is sometimes questionable since most of the effort has been geared toward improving reliability.

Therefore, in our initial studies, we selected patients systematically with regard to psychopathology and disease stage. We focused on paranoid schizophrenia patients who were in the first episode of illness and who were mostly drug naive. We excluded any subjects with comorbidities, and we matched patients to control subjects with similar socioeconomical backgrounds and education status (Schwarz *et al.*, 2010, 2011a,b). Patients were followed up over several years to establish disease and symptom stability over time. At the time of sample collection, patients were assessed using the positive and negative syndrome scale (PANSS), DSM-IV, and various other rating scales. Healthy control subjects were assessed using the Structured Interview for Diagnosis for DSM-IV (SCID). We initially included cohorts from three independent German clinical centers, which used identical standard operating procedures for sample acquisition and we evaluated interrater reliability across the centers.

This systematic experimental approach allowed us to establish the expression levels of approximately 200 proteins and small molecules in serum samples from all subjects, which were then used to select those molecules that correlate most significantly with the diagnosis of paranoid schizophrenia. In parallel, we also examined serum samples from bipolar disorder, major depressive disorder, and Asperger syndrome subjects and were able to establish a schizophrenia-specific signature of 34 molecules (Schwarz *et al.*, 2011a,b). A further 17 molecules were added to the panel which correlated more specifically with a diagnosis of bipolar disorder and major depressive disorder. This was important for developing a differential diagnostic for schizophrenia. Subsequently, we found that less well characterized and controlled samples from drug-treated, chronic, and more broadly diagnosed schizophrenia patients (three patient cohorts from Europe and the United States) also displayed a significant change in some components of the molecular signature.

This does not mean that all of these molecules are changing in all schizophrenia patients, or even that they are changing in the same direction. The specificity was achieved using an algorithm comprising multiple molecules. This algorithm was trained to identify molecules associated with disease status, and then we were able to discern stable molecular differences in the patient populations. Through studies of first-onset patients, we have discovered that most show differences in the levels of insulin-related molecules (Guest *et al.*, 2010), other hormones of the diffuse neuroendocrine system (Guest *et al.*, 2011), inflammatory factors (Steiner *et al.*, 2010; Schwarz *et al.*, 2010), and molecules associated with endothelial cell function (Schwarz *et al.*, 2011a,b). Determination of which of these factors are predominant in a subject might be useful for patient stratification purposes prior to antipsychotic treatment.

V. Why Is Early Diagnosis So Important?

The concordance rate for identical twins to develop schizophrenia lies between 11% and 69%, according to different studies (Torrey, 1992; McGue, 1992; Tsuang, 2000). Such twin studies provide indisputable evidence that a genetic component and predisposition to develop schizophrenia exists. However, it also means that, even when such a predisposition exists (as in identical twins), an individual will not necessarily develop schizophrenia. Environmental and other nongenetic factors appear to play a more important role in most patients. A range of environmental factors could affect brain function in subjects with psychiatric illnesses. These could include pregnancy and delivery complications, such as intrauterine hypoxia, infections, and malnutrition (Dauncey and Bicknell, 1999; Schlotz and Phillips, 2009). There are also nonbiological factors which could precipitate the onset of mental illness, including psychosocial stressors such as experiencing natural disasters, loss of a family member or close friend, residence in a poor or dangerous area, or experiencing a dysfunctional family life (Koenig *et al.*, 2002). It is likely that such environmental factors can interact with genetic components in a negative manner in the development of psychiatric conditions.

This means that disease prevention or minimization might be possible if discrete environmental risk-factors can be determined. Above, we indicated the occurrence of metabolic abnormalities such as insulin resistance in a proportion of schizophrenia patients. We and other researchers have estimated this proportion to be 20–50% of in first-onset subjects (Ryan *et al.*, 2003; Spelman *et al.*, 2007; van Nimwegen *et al.*, 2008; see Chapter "Abnormalities in metabolism and hypothalamic-pituitary-adrenal axis function in schizophrenia" by Guest *et al.*). In addition, several researchers have indicated the presence of circulating inflammatory and immune response factors in first-onset schizophrenia patients (Szulc *et al.*, 2001; Riedel *et al.*, 2005; van Venrooij *et al.*, 2010). In a preliminary study, we have shown that various markers relating to these subgroups can be identified in patients even prior to disease onset (Schwarz *et al.*, in press). This study analyzed sera obtained from U.S. military personnel approximately 30 days before the onset of symptoms. An important hypothesis to test will be whether or not disease conversion can be prevented in at-risk individuals. There is an extensive literature outlining the importance of early intervention in individuals either prodromal or with a high risk of developing schizophrenia (Agius *et al.*, 2007; Salokangas and McGlashan, 2008; Yap, 2010).

A delay in diagnosis can have devastating and/or irreversible consequences for patient lives, including dropping out of education, turning to substance abuse, social alienation from family and friends due to paranoid cognitions and behavior, increased accidents, self harm, and harm to others. In addition, misdiagnosis can lead to ineffective or even harmful treatment. A misdiagnosis of bipolar disorder as schizophrenia, for example, has been found to be associated with increased risk of suicide attempts (Thomas, 2004); longer hospitalization times (Gonzalez-Pinto *et al.*, 1998); and major psychological, legal, and financial problems (Hirschfeld, 2001). In addition to consequences for patients, misdiagnosis has a number of socioeconomic consequences including high medical costs, absence from work, and negative effects on family and relationships (Post, 2005). Early and correct diagnosis in turn can improve disease progression and outcome substantially (Table III).

VI. Historical Perspective—The Blood of the "Insane"

Human blood has been regarded as an eloquent source of information concerning systemic illness and health since ancient times. Blood, of course, was one of the four classic "humors." With the rise of experimental techniques in physiology and medicine in the mid-1800s, serological investigations of persons

Table III
Estimated Benefits of Early Diagnosis and Treatment of Schizophrenia.

Intervention: Early symptom phase	
Premature deaths	↓4%
Hospitalizations	↓12–25%
Inpatient days	↓64%
Relapse	↓30–70%
Time to remission	↓74%
Poorer outcomes	↓27%
Duration of untreated state	↓88 weeks
Intervention: prodromal phase	
Incidence	↓90%
Conversion to psychosis	↓75%

The data are based on current diagnosis and compared to the established disease phase (Davies and Drummond, 1990; Knapp *et al.*, 2004; Wu *et al.*, 2005; Hafner and Maurer, 2006).

with psychiatric disorders attempted to answer two fundamental questions: (a) were there observable physical characteristics in blood serum that could be used to differentially diagnose the ill from the well and (b) is "insanity" simply one continuum of a "unitary psychosis" or, instead, could these "blood characters" (serum biomarkers) confirm an array of discrete natural disease entities that could be specified in the laboratory and sorted into clinical categories? The history of serological investigations in psychiatry over the past 150 years comprised four main phases (see Noll, 2006, for a fuller account):

1. The corpuscular richness paradigm (1854)

   The first microscopic investigations of blood cell morphology in the insane were conducted in 1854 in a Scottish asylum by W. Lauder Lindsey, a physician whose later writings on other topics influenced Charles Darwin. Using a low-powered microscope to examine and count the numbers of different observable blood cells in samples from his patients and his staff, his pioneering study rendered a negative conclusion: "insanity and the different types and phases thereof are not characterized by a particular morbid state of the blood." Macphail (1884) reviewed subsequent studies and tentatively concluded that there was an overall "deficiency of corpuscular richness of the blood in the first stages of insanity." Such morphological blood analyses of insane persons continued into the first two decades of the twentieth century.

2. The metabolic paradigm (ca. 1895)

   As the field of endocrinology emerged from physiology in the 1890s, blood became a primary medium to detect and measure "inner secretions" (after 1905, "hormones") not only from "secreting organs," particularly the ductless glands (thyroid, adrenals, pituitary), but also from those organs with ducts (liver, pancreas, and kidneys). This emerging new endocrinological paradigm was immediately seized upon by the first "modern" biological psychiatrists seeking a new orientation for laboratory studies of the insane. If an over- or underproduction of inner secretions could produce physical diseases such as myxoedema, cretinism, or diabetes (the "internal secretion" suspected in the 1890s was identified as "insulin" in 1921), then the same might be true of psychiatric disorders. From this paradigm arose the lifelong suspicion of German psychiatrist Emil Kraepelin (1856–1926), whom historians regard as the central figure in the history of modern psychiatry, that the severe psychotic disorder he described as "dementia praecox" (in 1896) was the result of an *ongoing systemic (whole body) disease process of a metabolic (autotoxic) nature, affecting the cerebral cortex in a final and decisive stage that led to chronic mental "deterioration," "defect," and "weakness"* (Noll, 2011). In 1908, Swiss psychiatrist Eugen Bleuler (1857–1939) proposed "schizophrenia" as an expanded version of this psychotic disorder, and with a more favorable prognosis, though only Bleuler's term and not his original disease concept survived into our time. For more than a century, changes in pituitary function and related hormonal abnormalities in various psychiatric disorders have been established and remain a strong biological finding today. Endocrinology also provided a direct analogical bridge which led to the discovery of the first neurotransmitter (neurohormone), acetylcholine, by German biochemist Otto Loewi in 1921.

3. The immunoserodiagnostic paradigm (1906)

The development of the Wasserman reaction test for neurosyphillis in 1906 was arguably the first great breakthrough in biological psychiatry. It was the first diagnostic blood test for a discrete form of insanity (general paralysis of the insane) commonly observed in asylums. The discovery inspired other immunological investigations. In 1909, two German psychiatrists injected cobra venom into patients with dementia praecox and manic-depressive insanity and reported that all of the former, and only some of the latter, evidenced blood serum reactions to the toxin, and healthy people did not. However, the so-called Much-Holzmann psycho-reaction could not be replicated and was quickly refuted. A much more influential immunoserodiagnostic test developed by the prominent Swiss biochemist Emil Abderhalden (1877–1950), the "defensive ferments reaction test," was found by the German psychiatrist August Fauser (1856–1938) to differentially diagnose dementia praecox and manic-depressive insanity from healthy subjects in a series of studies. From December 1912 to perhaps as late as 1920, many in the international scientific community believed a valid blood test for madness had been found. However, beginning in 1914, a series of studies were published that could not verify the existence of Abderhalden's defensive ferments (*Abwehrfermente*), and not only the test but also its usefulness to psychiatrists were cast into doubt. In the past century, changes in immune function and inflammation have been linked to various psychiatric disorders through many lines of evidence (see Chapter "Immune and neuro-immune alterations in mood disorders and schizophrenia" by Drexhage *et al.*).

4. The medical genomics (2005)/postgenomics (2010) paradigm

Twenty-first-century blood tests have targeted both the genome and the proteome. However, attempts to specify most medical diseases (especially neuropsychiatric disorders) at the level of the genome have been disappointing. The hereditary component of some (mostly severe) psychiatric disorders has now been established beyond doubt. However, there is a lack of agreement as to how strong the genetic and environmental contributions are and how they interact to precipitate the onset of severe mental illness. For example, despite 20 years of profound and expensive efforts, no single gene or combination of genes has been identified that substantially increase the probability of developing schizophrenia. In 2009, results from the largest Genome Wide Association Studies

(GWAS) were published as three papers in *Nature* alongside a meta-analysis of these three studies (Purcell *et al.*, 2009; Shi *et al.*, 2009; Stefansson *et al.*, 2009). None of the papers alone identified any genetic marker that was significantly associated with schizophrenia. The meta-analysis, however, implicated three chromosomal regions, most significantly the short arm of chromosome 6, which is the location of the major histocompatibility complex (MHC) genes.

VII. Biomarkers: Not Quite Living up to the Promise?

The field of clinical proteomics has raised high hopes generated by reports on potential biomarkers. In most cases, these could not be substantiated in validation studies or in clinical trials. Potential reasons for the failure to link biomarkers into clinical studies include deficiencies in design and analysis, the problem that drug targets and biomarkers may not be causal to the disease but rather a result of the disease process, a lack of congruence in the animal models of the disease with the human condition, or the enrolment of patients in clinical trials who are too advanced in disease stage to show any response to potential therapeutics (Flood *et al.*, 2011). In the case of Alzheimer's disease, a consensus has now been reached for testing drug candidates in the earlier stages of the disease (Aisen *et al.*, 2011).

The suggestion that biomarker research has not lived up to the initial hype has been demonstrated by the fact that multiple "breakthrough" biomarkers have been publicized but have not reached the market place. Apart from some biomarkers in the field of cancer research, most have not been validated and have faded from the spotlight. Major cancer biomarkers that have received FDA approval over the past few decades include prostate-specific antigen for prostate cancer (Kuriyama *et al.*, 1980), carcinoembryonic antigen (CA)-125 for ovarian cancer (Klug *et al.*, 1984), and CA-19-9 for pancreatic cancer (Satake *et al.*, 1985). However, most of these biomarkers are used mainly for monitoring treatment response and are not suitable for early diagnosis with the exception of prostate-specific antigen (Gjertson and Albertsen, 2011).

The Human Proteome Organization (HUPO) emerged from the Human Genome Project as a means of understanding gene function. HUPO has developed several initiatives targeted at overcoming the problem of reproducibility, as a major problem has been the fact that most proteomic researchers have been unable to reproduce their data. These initiatives are focused on plasma, liver, brain, disease glycomics/proteomics, disease biomarkers, mouse disease models, model organisms, kidney/urine, cardiovascular disease, stem cells, the Human Antibody Initiative, and the Proteomics Standards Initiative (http://hupo.org/). Each of these initiatives is based in one country and includes subprojects involving international partner laboratories. The rationale for this structure is that most of the reproducibility problem is likely to be due to several sources of variability including the biological sample, sample handling, study design, technical and user-related differences. However, the first proteomics-based test approved by the FDA, the OVA1 ovarian cancer triage test (Quest Diagnostics), demonstrates that this is possible (Zhang and Chan, 2010).

VIII. Biomarkers: What Are the Issues?

A. Regulatory

The development of biomarkers and the implications of using these in diagnostics and clinical trials are constantly moving forward. The development and use of biomarkers trigger the need to establish standard operating procedures to meet the regulatory demands. Regulatory health authorities, in particular, the FDA, now consider biomarkers as important for the future of drug development. In the Critical Path Initiative, initiated by a FDA white paper (Ovens, 2006), the FDA called for efforts to modernize methods, tools, and techniques for the purpose of developing more efficient and safer drugs (Marson, 2007). The regulatory aspects of biomarkers were first described in a guidance associated with the Pharmacogenomic Data Submission by the FDA (Dieterle *et al.*, 2010; Guidance for Industry, Pharmacogenomic Data Submissions, March 2005, U.S. Department of Health and Human Services, FDA, Center for Drug Evaluation and Research, Center for Biologics Evaluation and Research, Center for Devices and Radiological Health. http://www.fda.gov/OHRMS/DOCKETS/98fr/2003d-0497-gdl0002.pdf). Molecules need the status of validated biomarkers to be used in regulatory decision making, such as those regarding dose regimen in a clinical trial and selection of patients. Currently, only traditional and well-established tests have been used for regulatory decision making, such as serum creatinine levels to monitor kidney function and a fasting glucose tolerance test combined with insulin and glucose measurements or glucose clamping to establish insulin sensitivity (Truglia *et al.*, 1985; Khristov, 1986). According to the FDA guidance for pharmaceutical companies on pharmacogenomic data, there is discrimination between three types of biomarkers: (1) exploratory biomarkers, (2) probable valid biomarkers, and (3) known valid biomarkers (Goodsaid and Frueh, 2007). For exploratory biomarkers, evidence for a scientific proof of concept must exist from in-house experiments or literature. The transition of a biomarker to the status of a probable valid biomarker implies that it can be measured in an analytical test system with well-established performance characteristic and that there is an established scientific framework or body of evidence that explains the biological, pharmacological, toxicological, or clinical significance of the results. A biomarker will achieve the status of a known valid biomarker, given replication of the results at different sites, laboratories, or agencies in cross-validation experiments (Fig. 2).

A first proposal of the biomarker qualification process was on rat kidney safety biomarkers as part of a Cooperative Research and Development Agreement (CRADA; http://www.usgs.gov/tech-transfer/what-crada.html) between the FDA and Novartis (Basel, Switzerland). These biomarkers became part of a cross-validation study to achieve the status of known valid biomarkers as part of the Predictive Safety Testing Consortium (PSTC; Ovens, 2006; Marson, 2007). The PSTC was founded by the FDA to act as a liaison between the FDA, pharmaceutical companies, and academia in order to qualify biomarkers for preclinical and clinical use. It now includes the FDA, several pharmaceutical companies, and the European Agency for the Evaluation of Medicinal Products (EMEA) as members. Recently, the nephrotoxicity subgroup of the PSTC identified seven kidney safety biomarkers for limited use in preclinical and clinical drug development (Dieterle *et al.*, 2010; Ozer *et al.*, 2010). A Rat KidneyMAP™ (http://www.rulesbasedmedicine.com/products-services/rodentmap-services/ rat-kidneymap/) has now been developed as a multiplexed immunoassay by Rules Based Medicine, Inc. (www.rulesbasedmedicine.com), in conjunction with the PSTC for detection of early signs of renal damage, which is a common problem in drug development programs (Swain *et al.*, 2011). This serves as an example that the process of biomarker standardization should lead to a better understanding of how the qualification process works and helps to set requirements necessary to evaluate the performance of biomarkers for specific applications.

Another useful strategy for biomarker qualification is through their codevelopment with drugs (Goodsaid and Frueh, 2006). This necessitates that the use of these biomarkers is limited to the applications involving the corresponding drug. This strategy was first described in a draft guidance issued by the FDA [U.S. Department of Health and Human Services, FDA (2005) Drug-Diagnostic Co-Development Concept Paper (http://www.fda.gov/cder/genomics/pharmacoconceptfn.pdf)]. The idea of this draft is that increased knowledge about the biology of a biomarker and a robust association between the biomarker

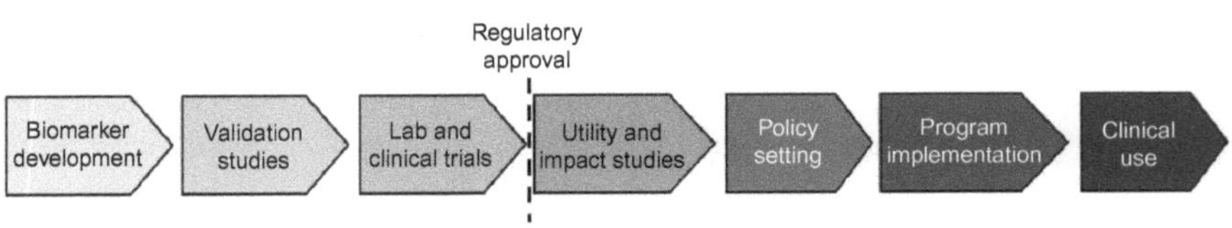

FIG. 2. From laboratory bench to bedside—the biomarker development process.

signal and the clinical result will lead to a more efficient and less risky development process. Therefore, inclusion of biomarkers into clinical development will only be achieved by a rigorous scientific approach, including standardized sample collection, analysis, and data processing. Early interaction with the appropriate regulatory agencies is advised to ensure that studies are designed and biomarker tests are carried out appropriately. One of the best examples of a codeveloped biomarker and drug combination is that of the HER2/neu subtype of epidermal growth factor receptor (EGFR) with the humanized monoclonal antibody trastuzumab (Herceptin™; Taube *et al.*, 2009), as mentioned above. In this case, breast cancer subjects who overexpress the HER2 subtype of EGFR are more likely to respond to Herceptin™ treatment (Pegram *et al.*, 1998). The example of HER2/Herceptin™ and other biomarker/drug combinations involving the use of scientifically and analytically validated biomarkers and rationally designed hypothesis-testing may lead to a paradigm shift in clinical trials, as appears to be the case in the field of cancer therapeutics (Tan *et al.*, 2009).

The European health authorities have also initiated activities to support the development and implementation of biomarkers through agencies such as the Innovative Medicines Initiative (Kamel *et al.*, 2008; Hunter, 2008). The Innovative Medicines Initiative is a partnership between the European Commission and the pharmaceutical industry that aims to promote more efficient discovery and development of medicines by supporting research into the drug development process. One of the main objectives is the discovery of translational biomarkers, including for psychiatric conditions such as schizophrenia and autism spectrum disorders. The European Commission contributes one billion euros to the program and this amount is matched by in kind contributions consisting mostly of research activities worth an equivalent amount from member companies of the European Federation of Pharmaceutical Industries and Associations (EFPIA).

B. Technologies

In the case of genetic linkage studies for psychiatric disorders, there have been many linked loci which were claimed and withdrawn, many association studies published and not confirmed, and many new and different chromosomal regions implicated for the same disorders. Thus, a lack of reproducibility in other linkage and association studies has generated considerable doubt from scientists that only a concerted effort will be able to rectify. Without this, it is unlikely that such knowledge will be translated into the clinic (Bondy, 2011). These results have led to skepticism from clinicians, scientists, and regulatory agencies, which will make the introduction of valid biomarkers into clinical diagnostics or the drug discovery industry even more difficult (Kopec *et al.*, 2005).

This is due in part to the lack of a sensible pipeline connecting marker discovery with technologies for validation and translation to a platform that offers accuracy and ease of use in a clinical setting (Plymoth and Hainaut, 2011). The multiplex immunoassay format (MAP^{TM}) employed by Rules Based Medicine, Inc., is an example of a technology that is positioned perfectly to fulfill these criteria (Schwarz *et al.*, 2010, 2011a,b; see Chapter "The application of multiplexed assay systems for molecular diagnostics" by Schwarz *et al*). Due to the typical low effect size associated with single molecular readouts, the diagnosis of complex disorders often requires analysis of multiple molecules and implementation of an algorithm that combines multiple measurements into one diagnostic output, as mentioned above (see Chapter "Algorithm development for diagnostic biomarker assays" by Izmailov *et al.*).

The applicability of assay systems for such diagnostic purposes in the United States is regulated by the Clinical Laboratory Improved Amendments (CLIA) (US Food and Drug Administration—CLIA—http://www.fda.gov/MedicalDevices/DeviceRegulationandGuidance/IVDRegulatoryAssistance/ucm124105.htm). These federal regulatory standards govern any tests performed in a clinical laboratory on human specimen for the purpose of diagnosis, prevention, treatment, or assessment of health. Commercially available tests that are marketed under CLIA are categorized by the FDA into three groups depending on their potential risk for human health. This categorization considers the required knowledge, training, materials, and judgment to carry out the tests and other factors such as operational, maintenance, and quality control procedures. Tests can be issued a waiver if they are accurate, basic, exclude erroneous interpretation, and pose no risk to human health if interpreted incorrectly.

Rules Based Medicine develops, manufactures, and validates their multiplexed assays to clinical laboratory standards and they operate a CLIA-certified laboratory that can support Good Laboratory Practice (GLP) studies (Swanson, 2002). This requires the repeated demonstration of specific performance characteristics of the assays which include accuracy, precision, sensitivity, and specificity. These factors are now considered absolute requirements since variability in biomarker measurements can be affected by biological (e.g., gender, age, disease), environmental (e.g., diet, toxicity, circadian rhythms), sample collection (e.g., collection procedure, preservatives, storage, transport), and analytical (e.g., assay precision, specificity, data analysis) factors. Development of multiplexed immunoassays requires validation of the assay structure and analytical performance to maximize precision and accuracy. Challenges associated with such multiplex assays include selection and immobilization of capture ligands, calibration, reagent–antibody compatibility, interference problems, dynamic range, and limits of detection (Ellington *et al.*, 2010). The MAP assays have already been applied in clinical investigations, including studies of sickle cell anemia (Lee *et al.*, 2008), ovarian cancer (Bertenshaw *et al.*, 2008; Clendenen *et al.*, 2011), scleroderma

(Duan *et al.*, 2008), coronary arterial disease (Gurbel *et al.*, 2008), inflammatory response after cardiopulmonary bypass (Aĝirbaşli *et al.*, 2010), schizophrenia (Schwarz *et al.*, 2010), and a phase II study of urothelial cancer therapy (Bellmunt *et al.*, 2011).

C. Strategies

It is now generally accepted that single biomarkers are not likely to be effective given the fact that the complexity of most diseases, particularly psychiatric disorders, cannot be represented by a single marker (Boja *et al.*, 2011). Also, most psychiatric conditions appear to be the result of a complex interaction between environmental and genetic factors (Dick, 2011). Therefore, a panel of biomarkers must be employed to reflect this complexity and to add specificity to the measurements. However, it is now becoming clear that such biomarker panels must consist of rigorously validated molecules in multiple centers and across different time points in order to provide a reproducible and accurate test. However, in most cases, biomarker candidates have failed at this hurdle (Alymani *et al.*, 2010).

Biomarker panels must also be disease specific, at least relative to other diseases which have similar symptoms. Again, this is particularly difficult for psychiatric disorders as these have many areas of overlap of subjectively assessed behavioral symptoms. Examples for this are the overlap of negative symptoms between schizophrenia and major depressive disorder (Fleischhacker, 2000), the similarity in psychotic symptoms between manic bipolar disorder and schizophrenia (Dunayevich and Keck, 2000), and the shared cognitive deficits across all of these conditions (Ferrier *et al.*, 1999). In addition, conditions such as bipolar disorder are characterized by having multiple stages such as the cycling of manic and depressed moods (Dunayevich and Keck, 2000). Identification of valid biomarkers which could predict the switch from one stage to another would be invaluable.

Psychiatric conditions also come with the special challenge that they have traditionally been considered to be disorders of the mind. Thus, convincing clinicians and the regulatory agencies that blood-based assays are sensible and can be predictive is most likely the biggest challenge of all. However, the evidence is accumulating that such conditions are essentially systemic disorders and not restricted to the brain. For example, several hormones secreted from the diffuse neuroendocrine system are now known to be altered even in first-onset schizophrenia patients. These include hormones from pancreatic islets, the pituitary, gonads, adrenal glands, the gastrointestinal system, and adipose tissue (Yang *et al.*, 2008; Schanze *et al.*, 2008; Venkatasubramanian *et al.*, 2010; Guest *et al.*, 2010, 2011).

Finally, the tests introduced must be in a format that is high throughput, accurate, and user friendly such that they can be used by clinicians, hospital staff, and scientists alike. Currently, no consensus exists on this should be done. Mass spectrometry and two-dimensional gel electrophoresis techniques would be too cumbersome and require too much expertise to be considered as realistic possibilities, although these are traditionally used in the discovery phase of biomarker research (see Chapter "Proteomic technologies for biomarker studies in psychiatry: Advances and needs" by Martins-de-Souza *et al.*). Instead, automated platforms such as the multiplexed immunoassay system (see Chapter "The application of multiplexed assay systems for molecular diagnostics" by Schwarz *et al.*) and selected reaction monitoring approach (Gallien *et al.*, 2011) are more likely candidates for clinically friendly platforms which have already shown some promise. Also holographic sensors have already been employed for detection of biological materials (Bhatta *et al.*, 2007) and molecules (Kabilan *et al.*, 2005; Sartain *et al.*, 2006; Tan and Lowe, 2009) and could therefore be adapted as a robust and comprehensive readout of a biomarker signature in clinical applications (see Chapter "The future: Biosensors and e-neuroscience" by Lowe). Further, the biomaterial of choice should be readily accessible with minimal discomfort to test subjects such as blood serum/plasma or urine. Brain material is not possible in living subjects for obvious reasons.

D. The Problem of Size

Although HUPO developed after the completion of the Human Genome Project, scientists attempting to decode the proteome understand that key differences exist considering the enormous task ahead (Mueller *et al.*, 2007). The Human Genome Project had a clear finishing line of sequencing the 3.2 billion nucleotides comprising the genome and also pinning down the location of the genes (Lander *et al.*, 2001; Ventner *et al.*, 2001). However, proteome scientists wonder if their task is even possible. "Genes were easy," stated Samir Hanash who headed the HUPO from its inception in 2001 (Pandey, 2001). However, a complete inventory of all the proteins in the human body is a quest that could take hundreds of years, if it is even possible. The human genome is now known to contain approximately 35,000 genes (Lander *et al.*, 2001; Ventner *et al.*, 2001). But the number of proteins is expected to be at least one order of magnitude greater. Although DNA is essentially the same in all cells, distinct proteins can be expressed depending on the nature of the tissue or cell types. Also proteins can come in many shapes and sizes, with different physical properties which are important for understanding their function.

Other reasons for the large size and complexity of the human proteome include the occurrence of alternative splicing of genes (Nakao *et al.*, 2005) along with posttranslational modifications such as limited proteolysis (Lopez-Otin and Overall, 2002), glycosylation (Pan *et al.*, 2011), and phosphorylation (Ozlu *et al.*, 2010). All of these modifications can give rise to multiple forms of the same gene product. Another factor which renders the process of deciphering the human proteome difficult is the fact that proteins do not work alone; they work together in complexes and pathways. Many proteins form complexes with other proteins, RNA, and various cofactors, all of which can affect function. This has been termed the interactome and adds another layer of complexity to the unraveling gene function (Stein *et al.*, 2011). Most scientists think that progress on deciphering the proteome will depend on new breakthroughs in technology (Fig. 3).

E. The Problem of Acceptance

Protein and gene expression approaches (see Chapters "Proteomic technologies for biomarker studies in psychiatry: Advances and needs" by Martins-de-Souza *et al.* and "Converging evidence of blood-based biomarkers for schizophrenia: An update" by Chan *et al.*) have the advantage that, if linked to disease presentation, they should capture both hereditary and environmental abnormalities of a given disease. However, they may not reveal the true etiology of the disease since abnormalities correlated with disease state may be adaptive rather than causative. This is a frequent argument against the use of biomarkers in psychiatric disorders, as described in the next section. This appears to be due to the apparent global desire of the psychiatric community to "separate the chicken from the egg." The fact, that both chicken and egg can be useful tools in the diagnosis, sub-stratification, and treatment of mental illness is a difficult concept to get across to practicing clinicians. This is in stark contrast to other fields such as the cancer and endocrinology, in which these concepts appear to cause little or no resistance.

Most psychiatrists have no problem accepting that schizophrenia has a genetic component. Indeed, the search for genetic variants has been a key focus of schizophrenia research over the past 20 years. The fact that genetic disorders often have consequences in multiple systems has been more difficult to accept in

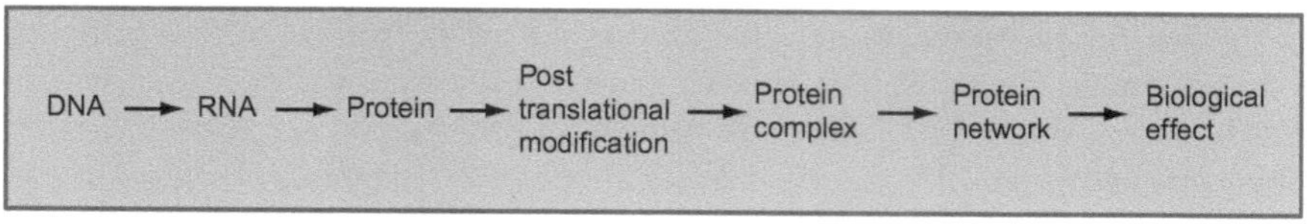

Fig. 3. The complexity of the human proteome in the production of biological effects.

the field of psychiatric disorders than in other medical disciplines. This is despite the fact that schizophrenia patients are more likely to develop other conditions such type II diabetes mellitus (15.8%) compared to the general population (2–3%), and this has been known for the past 40 years (Dynes, 1969; Tsuang *et al.*, 1983). Alterations in the prevalence of various infectious and immunological disorders, such as celiac disease have also been reported for approximately the same duration (Dohan, 1970).

At the time of the launch of VeriPsych™ in the United States, leading psychiatrists were asked whether they believe a blood test can really detect a mental illness. The following general attitudes and answers were provided: (http://www.msnbc.msn.com/id/39686973/ns/health-mental_health/). Some physicians were skeptical of the test. While recognizing the need and potential for relying on methods other than patient reports or observations, "the science is not there yet," said Dr. Gregory Light, an associate professor of psychiatry at the University of California, San Diego. "Blood-based tests, at the moment, are far away from being useful for an individual patient," Light said. "For example, genetic tests raise more questions than they answer, account for only a small number of cases, and the advances in technology are rapidly outpacing our ability to interpret the results." *There appears to have been a misunderstanding that the test under discussion is based on genes, when it is actually based on proteins and small molecules as a multiplexed biomarker blood test.* "The biomarkers that some research has suggested are associated with schizophrenia might be a consequence of the illness, or long-term use of antipsychotic medication," said Dr. Irving Gottesman, a professor emeritus of psychology at the University of Minnesota. *As mentioned above, biomarkers that are a response to underlying illness can lead to development of valuable biomarker tests, such as prostate-specific antigen for prostate cancer and troponin for cardiac injury.* "It takes, often, a period of observation, both with and without medication of 6 months or more, to get a better "feel" for what may be the proper diagnosis," Gottesman said. *However, by the time the psychiatrist has a better feel for the disorder, the patient's life and life quality may have been irreversibly devastated, as described above.*

IX. Development of a Molecular Blood Test for Schizophrenia

Rules Based Medicine in collaboration with Psynova Neurotech and our laboratory has launched the first biomarker test for psychiatric disorders in 2010, called VeriPsych™, as described above (Schwarz *et al.*, 2010). The test was launched as a CLIA-approved test as an aid to the diagnosis of schizophrenia and measures 51 protein and small molecule biomarkers using the multiplexed immunoassay platform (MAP; see Chapter "The application of multiplexed assay systems for molecular diagnostics" by Schwarz *et al.*). This assay employs a proprietary algorithm to achieve a sensitivity and specificity of approximately 83%, respectively (see Chapter "Algorithm development for diagnostic biomarker assays" by Izmailov *et al.*). It is anticipated that this biomarker panel may be employed at multiple stages of the schizophrenia disease process to improve patient lives (Fig. 4).

Decision-modeling analyses were used to construct a socioeconomic case for a biomarker-based test such as VeriPsych^{TM} for early diagnosis of schizophrenia and to determine the prospective market and economic values (Millan, 2007). This showed that the cost of each patient in the United Kingdom who has been diagnosed after the first psychotic episode, would be approximately £182,100 over a 5-year period. Importantly, the cost for a patient diagnosed early would be approximately £27,252 suggesting that this could potentially save 6.7-fold in costs to society and the healthcare services. This suggests that there is a strong and positive socioeconomic case for introducing better diagnostic tools for detection of schizophrenia during the prodromal phase.

Psychiatrists and health professionals have, so far, met this new test with a mixed reception. Most agreed that in principle a sensitive and specific blood test for psychiatric disorders would be a welcome and major advance in the field. However, many are resistant to actually using such a blood test in clinical practice. A specific and justified criticism of the current blood test is that it was developed to

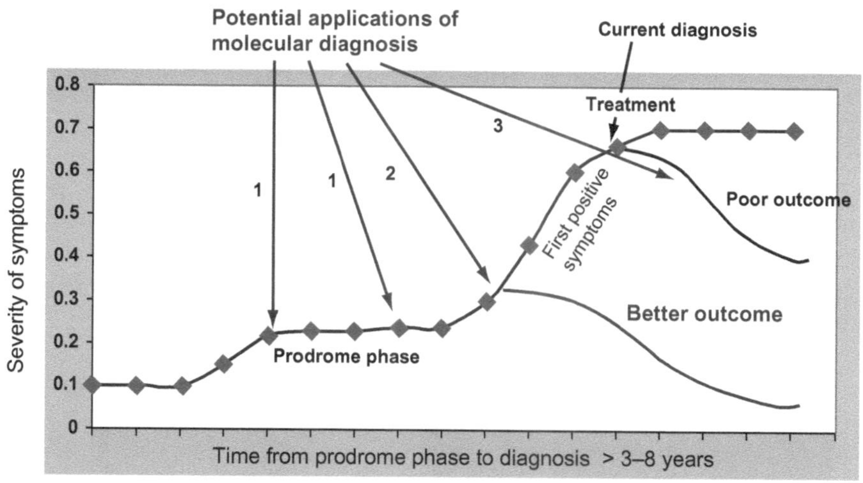

FIG. 4. Schematic representation of disease progression in schizophrenia and potential targets of novel molecular-based tests. (A) Disease risk—estimating risk of developing schizophrenia. (B) Early treatment—accurately identifying subjects who will benefit from antipsychotic treatment. (C) Patient monitoring—monitoring treatment response (efficacy and side effects).

distinguish schizophrenia patients from healthy controls and not as a differential diagnostic of schizophrenia from other psychiatric disorders. However, the next version of VeriPsych^{TM} is targeted at addressing this shortcoming by building in the differential diagnostic capability for schizophrenia against major depressive disorder and bipolar disorder. Market research found that most psychiatrists believe that they are very good at diagnosing schizophrenia patients using a basic clinical interview. However, they routinely felt that colleagues were less expert at achieving the correct diagnosis. Another resistance was the high price of $2500 at which the current test is marketed.

X. Conclusions

This chapter has described the many challenges associated with the discovery and development of blood-based biomarkers for psychiatric disorders. The current diagnostic process and strategies for developing novel pharmaceutical compounds are in need of an overhaul. As stated by the Nobel Laureate Lee Hartwell at the 2004 HUPO meeting in Beijing, "Biomarkers for early diagnosis will revolutionize the pharmaceutical industry allowing diseases to be treated at an earlier stage—increasing survival rate." Despite this, there is a reluctance to accept the idea that biomarkers will be of any help at all (Poste, 2011). It is true that only a handful of the thousands of promising biomarkers identified have lived up to the initial hype. However, the regulatory health authorities now consider biomarkers as important for the future of drug development and have called for efforts to modernize methods, tools, and techniques to achieve this. As U.S. President John Fitzgerald Kennedy said in 1962, "we will go to the moon because it is hard." It is interesting that in the early 1960s, many of the technologies required had not even been invented. Not only that, a massive integration of technologies was required to produce what many have claimed is the finest achievement of the twentieth century. There is now reason for optimism that further technological advancements, and interdisciplinary approaches will overcome current limitations in the field biomarkers to help usher medicine fully into the twenty-first century.

Acknowledgments

This research was supported by the Stanley Medical Research Institute (SMRI), the European Union FP7 SchizDX research programme (grant reference 223427), and the NEWMEDS Innovative Medicines Initiative. We also thank Enrique

Millan for research gathered through his MBA individual project entitled *The Value of the Schizophrenia Diagnostic Market*, carried out with the Judge Business School at Cambridge University and Psynova Neurotech in Cambridge, UK.

References

Ağirbaşli, M., Nguyen, M.L., Win, K., Kunselman, A.R., Clark, J.B., Myers, J.L., and Undar, A. (2010). Inflammatory and hemostatic response to cardiopulmonary bypass in pediatric population: feasibility of seriological testing of multiple biomarkers. *Artif. Organs* **34**, 987–995.

Agius, M., Shah, S., Ramkisson, R., Murphy, S., and Zaman, R. (2007). Three year outcomes of an early intervention for psychosis service as compared with treatment as usual for first psychotic episodes in a standard community mental health team. Preliminary results. *Psychiatr. Danub.* **19**, 10–19.

Aisen, P.S., Andrieu, S., Sampaio, C., Carrillo, M., Khachaturian, Z.S., Dubois, B., Feldman, H.H., Petersen, R.C., Siemers, E., Doody, R.S., Hendrix, S.B., Grundman, M., *et al.* (2011). Report of the task force on designing clinical trials in early (predementia) AD. *Neurology* **76**, 280–286.

Alymani, N.A., Smith, M.D., Williams, D.J., and Petty, R.D. (2010). Predictive biomarkers for personalised anti-cancer drug use: discovery to clinical implementation. *Eur. J. Cancer* **46**, 869–879.

Bagby, R.M., Rogers, R., Buis, T., Nicholson, R.A., Cameron, S.L., Rector, N.A., Schuller, D.R., and Seeman, M.V. (1997). Detecting feigned depression and schizophrenia on the MMPI-2. *J. Pers. Assess.* **68**, 650–664.

Ball, R.H., Caughey, A.B., Malone, F.D., Nyberg, D.A., Comstock, C.H., Saade, G.R., Berkowitz, R. L., Gross, S.J., Dugoff, L., Craigo, S.D., Timor-Tritsch, I.E., Carr, S.R., *et al.* (2007). First- and second-trimester evaluation of risk for Down syndrome. *Obstet. Gynecol.* **110**, 10–17.

Bellmunt, J., González-Larriba, J.L., Prior, C., Maroto, P., Carles, J., Castellano, D., Mellado, B., Gallardo, E., Perez-Gracia, J.L., Aguilar, G., Villanueva, X., Albanell, J., *et al.* (2011). Phase II study of sunitinib as first-line treatment of urothelial cancer patients ineligible to receive cisplatin-based chemotherapy: baseline interleukin-8 and tumor contrast enhancement as potential predictive factors of activity. *Ann. Oncol.* Mar 21 (Epub ahead of print).

Bertenshaw, G.P., Yip, P., Seshaiah, P., Zhao, J., Chen, T.H., Wiggins, W.S., *et al.* (2008). Multianalyte profiling of serum antigens and autoimmune and infectious disease molecules to identify biomarkers dysregulated in epithelial ovarian cancer. *Cancer Epidemiol. Biomarkers Prev.* **17**, 2872–2881.

Bhatta, D., Christie, G., Madrigal-González, B., Blyth, J., and Lowe, C.R. (2007). Holographic sensors for the detection of bacterial spores. *Biosens. Bioelectron.* **23**, 520–527.

Boja, E.S., Jortani, S.A., Ritchie, J., Hoofnagle, A.N., Tezak, Z., Mansfield, E., Keller, P., Rivers, R.C., Rahbar, A., Anderson, N.L., Srinivas, P., and Rodriguez, H. (2011). The journey to regulation of protein-based multiplex quantitative assays. *Clin. Chem.* **57**, 560–567.

Bondy, B. (2011). Genetics in psychiatry: are the promises met? *World J. Biol. Psychiatry* **12**, 81–88.

Cheniaux, E., Landeira-Fernandez, J., and Versiani, M. (2009). The diagnoses of schizophrenia, schizoaffective disorder, bipolar disorder and unipolar depression: interrater reliability and congruence between DSM-IV and ICD-10. *Psychopathology* **42**, 293–298.

Clendenen, T.V., Lundin, E., Zeleniuch-Jacquotte, A., Koenig, K.L., Berrino, F., Lukanova, A., Lokshin, A.E., Idahl, A., Ohlson, N., Hallmans, G., Krogh, V., Sieri, S.A., *et al.* (2011). Circulating inflammation markers and risk of epithelial ovarian cancer. *Cancer Epidemiol. Biomarkers Prev.* **20**, 799–810.

Cronin, M., Sangli, C., Liu, M.L., Pho, M., Dutta, D., Nguyen, A., Jeong, J., Clark-Langone, K.M., and Watson, D. (2007). Analytical validation of the OncotypeDX genomic test for recurence prognosis and therapeutic response prediction in node-negative, estrogen receptor-positive breast cancer. *Clin. Chem.* **53**, 1084–1091.

Dauncey, M.J., and Bicknell, R.J. (1999). Nutrition and neurodevelopment: mechanisms of developmental dysfunction and disease in later life. *Nutr. Res. Rev.* **12**, 231–253.

Davies, L.M., and Drummond, M.F. (1990). The economic burden of schizophrenia. *Psychiatric Bull.* **14**, 522–525.

Desmedt, C., Sperinde, J., Piette, F., Huang, W., Jin, X., Tan, Y., Durbecq, V., Larsimont, D., Giuliani, R., Chappey, C., Buyse, M., Winslow, J., *et al.* (2009). Quantitation of HER2 expression or HER2:HER2 dimers and differential survival in a cohort of metastatic breast cancer patients carefully selected for trastuzumab treatment primarily by FISH. *Diagn. Mol. Pathol.* **18**, 22–29.

Dick, D.M. (2011). Gene-environment interaction in psychological traits and disorders. *Annu. Rev. Clin. Psychol.* **7**, 383–409.

Dieterle, F., Sistare, F., Goodsaid, F., Papaluca, M., *et al.* (2010). Renal biomarker qualification submission: a dialog between the FDA-EMEA and Predictive Safety Testing Consortium. *Nat. Biotechnol.* **28**, 455–462.

DiMasi, J.A., and Faden, L.B. (2011). Competitiveness in follow-on drug R&D: a race or imitation? *Nat. Rev. Drug Discov.* **10**, 23–27.

Dohan, F.C. (1970). Coeliac disease and schizophrenia. *Lancet* **1**(7652), 897–898.

Duan, H., Fleming, J., Pritchard, D.K., Amon, L.M., Xue, J., Arnett, H.A., *et al.* (2008). Combined analysis of monocyte and lymphocyte messenger RNA expression with serum protein profiles in patients with scleroderma. *Arthritis Rheum.* **58**, 1465–1474.

Dunayevich, E., and Keck, P.E. Jr. (2000). Prevalence and description of psychotic features in bipolar mania. *Curr. Psychiatry Rep.* **2**, 286–290.

Dynes, J.B. (1969). Diabetes in schizophrenia and diabetes in nonpsychotic medical patients. *Dis. Nerv. Syst.* **30**, 341–344.

Ellington, A.A., Kullo, I.J., Bailey, K.R., and Klee, G.G. (2010). Antibody-based protein multiplex platforms: technical and operational challenges. *Clin. Chem.* **56**, 186–193.

Ferrier, I.N., Stanton, B.R., Kelly, T.P., and Scott, J. (1999). Neuropsychological function in euthymic patients with bipolar disorder. *Br. J. Psychiatry* **175**, 246–251.

Fleeman, N., Dundar, Y., Dickson, R., Jorgensen, A., Pushpakom, S., McLeod, C., Pirmohamed, M., and Walley, T. (2010). Cytochrome P450 testing for prescribing antipsychotics in adults with schizophrenia: systematic review and meta-analyses. *Pharmacogenomics J.* **11**, 1–14.

Fleischhacker, W. (2000). Negative symptoms in patients with schizophrenia with special reference to the primary versus secondary distinction. *Encéphale* **26**(Spec No 1), 12–14.

Flood, D.G., Marek, G.J., and Williams, M. (2011). Developing predictive CSF biomarkers-a challenge critical to success in Alzheimer's disease and neuropsychiatric translational medicine. *Biochem. Pharmacol.* **81**, 1422–1434.

Follette, W.C., and Houts, A.C. (1996). Models of scientific progress and the role of theory in taxonomy development: a case study of the DSM. *J. Consult. Clin. Psychol.* **64**, 1120–1132.

Gallien, S., Duriez, E., and Domon, B. (2011). Selected reaction monitoring applied to proteomics. *J. Mass Spectrom.* **46**, 298–312.

Gjertson, C.K., and Albertsen, P.C. (2011). Use and assessment of PSA in prostate cancer. *Med. Clin. North Am.* **95**, 191–200.

Gonzalez-Pinto, A., Gutierrez, M., Mosquera, F., Ballesteros, J., Lopez, P., Ezcurra, J., Figuerido, J.L., and de Leon, J. (1998). First episode in bipolar disorder: misdiagnosis and psychotic symptoms. *J. Affect. Disord.* **50**, 41–44.

Goodsaid, F., and Frueh, F. (2006). Process map proposal for the validation of genomic biomarkers. *Pharmacogenomics* **7**, 773–782.

Goodsaid, F., and Frueh, F.W. (2007). Implementing the U.S. FDA guidance on pharmacogenomic data submissions. *Environ. Mol. Mutagen.* **48**, 354–358.

Guest, P.C., Wang, L., Harris, L.W., Burling, K., Levin, Y., Ernst, A., Wayland, M.T., Umrania, Y., Herberth, M., Koethe, D., van Beveren, J.M., Rothermundt, M., *et al.* (2010). Increased levels of circulating insulin-related peptides in first-onset, antipsychotic naive schizophrenia patients. *Mol. Psychiatry* **15**, 118–119.

Guest, P.C., Schwarz, E., Krishnamurthy, D., Harris, L.W., Leweke, F.M., Rothermundt, M., van Beveren, N.J., Spain, M., Barnes, A., Steiner, J., Rahmoune, H., and Bahn, S. (2011). Altered levels of circulating insulin and other neuroendocrine hormones associated with the onset of schizophrenia. *Psychoneuroendocrinology* **36**, 1092–1096.

Gurbel, P.A., Kreutz, R.P., Bliden, K.P., DiChiara, J., and Tantry, U.S. (2008). Biomarker analysis by fluorokine multianalyte profiling distinguishes patients requiring intervention from patients with long-term quiescent coronary artery disease: a potential approach to identify atherosclerotic disease progression. *Am. Heart J.* **155**, 56–61.

Hafner, H., and Maurer, K. (2006). Early detection of schizophrenia: current evidence and future perspectives. *World Psychiatry* **5**, 130–138.

Hirschfeld, R.M. (2001). Bipolar spectrum disorder: improving its recognition and diagnosis. *J. Clin. Psychiatry* **62**(Suppl 14), 5–9.

Howey, J.E., Bennet, W.M., Browning, M.C., Jung, R.T., and Fraser, C.G. (1989). Clinical utility of assays of glycosylated haemoglobin and serum fructosamine compared: use of data on biological variation. *Diabet. Med.* **6**, 793–796.

Hunter, A.J. (2008). The Innovative Medicines Initiative: a pre-competitive initiative to enhance the biomedical science base of Europe to expedite the development of new medicines for patients. *Drug Discov. Today* **13**, 371–373.

Kabilan, S., Marshall, A.J., Sartain, F.K., Lee, M.C., Hussain, A., Yang, X., Blyth, J., Karangu, N., James, K., Zeng, J., Smith, D., Domschke, A., *et al.* (2005). Holographic glucose sensors. *Biosens. Bioelectron.* **20**, 1602–1610.

Kamel, N., Compton, C., Middelveld, R., Higenbottam, T., and Dahlén, S.E. (2008). The Innovative Medicines Initiative (IMI): a new opportunity for scientific collaboration between academia and industry at the European level. *Eur. Respir. J.* **31**, 924–926.

Kashshai, D., and Mate, B. (1961). The effect of nicotinic acid on the temperature of the skin of patients with schizophrenia in different states. *Zh. Nevropatol. Psikhiatr. Im. S S Korsakova* **61**, 1688–1698.

Kato, M., and Serretti, A. (2010). Review and meta-analysis of antidepressant pharmacogenetic findings in major depressive disorder. *Mol. Psychiatry* **15**, 473–500.

Keller, W.R., Fischer, B.A., and Carpenter, W.T. (2011). Revising the diagnosis of schizophrenia: where have we been and where are we going? *CNS Neurosci. Ther.* **17**, 83–88.

Khristov, V. (1986). Glucose clamping—a modern method for research on insulin secretion and resistance. *Vutr. Boles.* **25**, 32–39.

Klug, T.L., Bast, R.C. Jr., Niloff, J.M., Knapp, R.C., and Zurawski, V.R. Jr. (1984). Monoclonal antibody immunoradiometric assay for an antigenic determinant (CA 125) associated with human epithelial ovarian carcinomas. *Cancer Res.* **44**, 1048–1053.

Knapp, M., Mangalore, R., and Simon, J. (2004). The global costs of schizophrenia. *Schizophr. Bull.* **30**, 279–293.

Koenig, J.I., Kirkpatrick, B., and Lee, P. (2002). Glucocorticoid hormones and early brain development in schizophrenia. *Neuropsychopharmacology* **27**, 309–318.

Kopec, K.K., Bozyczko-Coyne, D., and Williams, M. (2005). Target identification and validation in drug discovery: the role of proteomics. *Biochem. Pharmacol.* **69**, 1133–1139.

Kuriyama, M., Wang, M.C., Papsidero, L.D., Killian, C.S., Shimano, T., Valenzuela, L., Nishiura, T., Murphy, G.P., and Chu, T.M. (1980). Quantitation of prostate-specific antigen in serum by a sensitive enzyme immunoassay. *Cancer Res.* **40**, 4658–4662.

Lander, E.S., Linton, L.M., Birren, B., Nusbaum, C., *et al.* (2001). Initial sequencing and analysis of the human genome. *Nature* **409**, 860–921.

Laughren, T.P. (2010). What's next after 50 years of psychiatric drug development: an FDA perspective. *J. Clin. Psychiatry* **71**, 1196–1204.

Lee, S.P., Ataga, K.I., Zayed, M., Manganello, J.M., Orringer, E.P., Phillips, D.R., *et al.* (2008). Phase I study of eptifibatide in patients with sickle cell anaemia. *Br. J. Haematol.* **139**, 612–620.

Lieberman, J.A., Stroup, T.S., McEvoy, J.P., Swartz, M.S., Rosenheck, R.A., Perkins, D.O., Keefe, R.S., Davis, S.M., Davis, C.E., Lebowitz, B.D., Severe, J., Hsiao, J.K., *et al.* (2005). Effectiveness of antipsychotic drugs in patients with chronic schizophrenia. *N. Engl. J. Med.* **353**, 1209–1223.

Lopez-Otin, C., and Overall, C.M. (2002). Protease degradomics: a new challenge for proteomics. *Nat. Rev. Mol. Cell Biol.* **3**, 509–519.

Lovatt, A., Mason, O., Brett, C., and Peters, E. (2010). Psychotic-like experiences, appraisals, and trauma. *J. Nerv. Ment. Dis.* **198**, 813–819.

Macphail, S.R. (1884). Clinical observations on the blood of the insane. *Br. J. Psychiatry* **30**, 378–389.

Marson, B. (2007). "Critical Path" is on the road forward; FDA reports industry activity is high. *Pink Sheet* **69**, 29.

McGue, M. (1992). When assessing twin concordance, use the probandwise not the pairwise rate. *Schizophr. Bull.* **18**, 171–176.

Millan, E. (2007). The Value of the Schizophrenia Diagnostic Market. Masters of Business Administration Individual Project. Judge Business School, Cambridge University, United Kingdom.

Mueller, M., Martens, L., and Apweiler, R. (2007). Annotating the human proteome: beyond establishing a parts list. *Biochim. Biophys. Acta* **1774**, 175–191.

Nakao, M., Barrero, R.A., Mukai, Y., Motono, C., Suwa, M., and Nakai, K. (2005). Large-scale analysis of human alternative protein isoforms: pattern classification and correlation with subcellular localization signals. *Nucleic Acids Res.* **33**, 2355–2363.

Noll, R. (2006). The blood of the insane. *Hist. Psychiatry* **17**, 395–418.

Noll, R. (2011). American madness: the rise and fall of dementia praecox. Harvard University Press, Cambridge.

Ovens, J. (2006). Funding for accelerating drug development initiative critical. *Nat. Rev. Drug Discov.* **5**, 271.

Ozer, J.S., Dieterle, F., Troth, S., Perentes, E., Cordier, A., Verdes, P., Staedtler, F., Mahl, A., Grenet, O., Roth, D.R., Wahl, D., Legay, F., *et al.* (2010). A panel of urinary biomarkers to monitor reversibility of renal injury and a serum marker with improved potential to assess renal function. *Nat. Biotechnol.* **28**, 486–494.

Ozlu, N., Akten, B., Timm, W., Haseley, N., Steen, H., and Steen, J.A. (2010). Phosphoproteomics. *Wiley Interdiscip. Rev. Syst. Biol. Med.* **2**, 255–276.

Pan, S., Chen, R., Aebersold, R., and Brentnall, T.A. (2011). Mass spectrometry based glycoproteomics—from a proteomics perspective. *Mol Cell Proteomics* **10**, R110.003251.

Pandey, A. (2001). 'Genes were easy'—the launch of the Human Proteome Organization. *Trends Genet.* **17**, 250.

Pegram, M.D., Pauletti, G., and Slamon, D.J. (1998). HER-2/neu as a predictive marker of response to breast cancer therapy. *Breast Cancer Res. Treat.* **52**, 65–77.

Pierre, J.M. (2008). Deconstructing schizophrenia for DSM-V: challenges for clinical and research agendas. *Clin. Schizophr. Relat. Psychoses* **2**, 166–174.

Plymoth, A., and Hainaut, P. (2011). Proteomics beyond proteomics: toward clinical applications. *Curr. Opin. Oncol.* **23**, 77–82.

Post, R.M. (2005). The impact of bipolar depression. *J. Clin. Psychiatry* **66**(Suppl. 5), 5–10.

Poste, G. (2011). Bring on the biomarkers. *Nature* **469**, 156–157.

Purcell, S.M., Wray, N.R., Stone, J.L., Visscher, P.M., O'Donovan, M.C., Sullivan, P.F., and Sklar, P. International Schizophrenia Consortium (2009). Common polygenic variation contributes to risk of schizophrenia and bipolar disorder. *Nature* **460**, 748–752.

Riedel, M., Strassnig, M., Schwarz, M.J., and Müller, N. (2005). COX-2 inhibitors as adjunctive therapy in schizophrenia: rationale for use and evidence to date. *CNS Drugs* **19**, 805–819.

Rush, A.J., Trivedi, M.H., Wisniewski, S.R., Nierenberg, A.A., Stewart, J.W., Warden, D., Niederehe, G., Thase, M.E., Lavori, P.W., Lebowitz, B.D., McGrath, P.J., Rosenbaum, J.F., *et al.* (2006). Acute and longer-term outcomes in depressed outpatients requiring one or several treatment steps: a STAR*D report. *Am. J. Psychiatry* **163**, 1905–1917.

Ryan, M.C., Collins, P., and Thakore, J.H. (2003). Impaired fasting glucose tolerance in first-episode, drug-naive patients with schizophrenia. *Am. J. Psychiatry* **160**, 284–289.

Salokangas, R.K., and McGlashan, T.H. (2008). Early detection and intervention of psychosis. A review. *Nord. J. Psychiatry* **62**, 92–105.

Sartain, F.K., Yang, X., and Lowe, C.R. (2006). Holographic lactate sensor. *Anal. Chem.* **78**, 5664–5670.

Satake, K., Kanazawa, G., Kho, I., Chung, Y.S., and Umeyama, K. (1985). A clinical evaluation of carbohydrate antigen 19-9 and carcinoembryonic antigen in patients with pancreatic carcinoma. *J. Surg. Oncol.* **29**, 15–21.

Scaglia, F. (2010). The role of mitochondrial dysfunction in psychiatric disease. *Dev. Disabil. Res. Rev.* **16**, 136–143.

Schaffer, D. (1996). A participant's observations: preparing DSM-IV. *Can. J. Psychiatry* **41**, 325–329.

Schanze, A., Reulbach, U., Scheuchenzuber, M., Groschl, M., Kornhuber, J., and Kraus, T. (2008). Ghrelin and eating disturbances in psychiatric disorders. *Neuropsychobiology* **57**, 126–130.

Schlotz, W., and Phillips, D.I. (2009). Fetal origins of mental health: evidence and mechanisms. *Brain Behav. Immun.* **23**, 905–916.

Schwarz, E., Izmailov, R., Spain, M., Barnes, A., Mapes, J.P., Guest, P.C., Rahmoune, H., Pietsch, S., Leweke, F.M., Rothermundt, M., Steiner, J., Koethe, D., *et al.* (2010). Validation of a blood-based laboratory test to aid in the confirmation of a diagnosis of schizophrenia. *Biomark. Insights* **5**, 39–47.

Schwarz, E., Guest, P.C., Rahmoune, H., Harris, L.W., Wang, L., Leweke, F.M., Rothermundt, M., Bogerts, B., Koethe, D., Kranaster, L., Ohrmann, P., Suslow, T., *et al.* (2011a). Identification of a biological signature for schizophrenia in serum. *Mol. Psychiatry* Apr 12 (Epub ahead of print).

Schwarz, E., Guest, P.C., Rahmoune, H., Martins-de-Souza, D., Niebuhr, D.W., Weber, N.S., Cowan, D.N., Torrey, E.F., Yolken, R.H., Spain, M., Barnes, A., and Bahn, S. (2011b). Identification of a blood-based biological signature in subjects with psychiatric disorders prior to clinical manifestation. *World J Biol. Psychiatry* (in press).

Shi, J., Levinson, D.F., Duan, J., Sanders, A.R., Zheng, Y., Pe'er, I., Dudbridge, F., Holmans, P.A., Whittemore, A.S., Mowry, B.J., Olincy, A., Amin, F., *et al.* (2009). Common variants on chromosome 6p22.1 are associated with schizophrenia. *Nature* **460**, 753–757.

Spadoni, L.R., Mclean, R.B., and Herrmann, W.L. (1964). A rapid immunological test for the detection of early pregnancy. *West. J. Surg. Obstet. Gynecol.* **72**, 92–97.

Spelman, L.M., Walsh, P.I., Sharifi, N., Collins, P., and Thakore, J.H. (2007). Impaired glucose tolerance in first-episode drug-naive patients with schizophrenia. *Diabet. Med.* **24**, 481–485.

Spitzer, R.L., Williams, J.B., and Skodol, A.E. (1980). DSM-III: the major achievements and an overview. *Am. J. Psychiatry* **137**, 151–164.

Stefansson, H., Ophoff, R.A., Steinberg, S., Andreassen, O.A., *et al.* (2009). Common variants conferring risk of schizophrenia. *Nature* **460**, 744–747.

Stein, A., Mosca, R., and Aloy, P. (2011). Three-dimensional modeling of protein interactions and complexes is going 'omics. *Curr. Opin. Struct. Biol.* **21**, 200–208.

Steiner, J., Jacobs, R., Panteli, B., Brauner, M., Schiltz, K., Bahn, S., Herberth, M., Westphal, S., Gos, T., Walter, M., Bernstein, H.G., Myint, A.M., *et al.* (2010). Acute schizophrenia is accompanied by reduced T cell and increased B cell immunity. *Eur. Arch. Psychiatry Clin. Neurosci.* **260**, 509–518.

Strakowski, S.M., Keck, P.E. Jr., Arnold, L.M., Collins, J., Wilson, R.M., Fleck, D.E., Corey, K.B., Amicone, J., and Adebimpe, V.R. (2003). Ethnicity and diagnosis in patients with affective disorders. *J. Clin. Psychiatry* **64**, 747–754.

Swain, A., Turton, J., Scudamore, C.L., Pereira, I., Viswanathan, N., Smyth, R., Munday, M., McClure, F., Gandhi, M., Sondh, S., and York, M. (2011). Urinary biomarkers in hexachloro-1:3-butadiene-induced acute kidney injury in the female Hanover Wistar rat; correlation of α-glutathione *S*-transferase, albumin and kidney injury molecule-1 with histopathology and gene expression. *J. Appl. Toxicol.* **31**, 366–377.

Swanson, B.N. (2002). Delivery of high-quality biomarker assays. *Dis. Markers* **18**, 47–56.

Szulc, A., Galińska, B., Konarzewska, B., Gudel-Trochimowicz, I., and Popławska, R. (2001). Immunological marker activity in first episode schizophrenic patients. *Pol. Merkur. Lekarski* **10**, 450–452.

Tan, E.V., and Lowe, C.R. (2009). Holographic enzyme inhibition assays for drug discovery. *Anal. Chem.* **81**, 7579–7589.

Tan, D.S., Thomas, G.V., Garrett, M.D., Banerji, U., de Bono, J.S., Kaye, S.B., and Workman, P. (2009). Biomarker-driven early clinical trials in oncology: a paradigm shift in drug development. *Cancer J.* **15**, 406–420.

Taube, S.E., Clark, G.M., Dancey, J.E., McShane, L.M., Sigman, C.C., and Gutman, S.I. (2009). A perspective on challenges and issues in biomarker development and drug and biomarker codevelopment. *J. Natl. Cancer Inst.* **101**, 1453–1463.

Thomas, P. (2004). The many forms of bipolar disorder: a modern look at an old illness. *J. Affect. Disord.* **79**(Suppl 1), S3–S8.

Torrey, E.F. (1992). Are we overestimating the genetic contribution to schizophrenia? *Schizophr. Bull.* **18**, 159–170.

Truglia, J.A., Livingston, J.N., and Lockwood, D.H. (1985). Insulin resistance: receptor and post-binding defects in human obesity and non-insulin-dependent diabetes mellitus. *Am. J. Med.* **79**, 13–22.

Tsuang, M.T. (1975). Heterogeneity of schizophrenia. *Biol. Psychiatry* **10**, 465–474.

Tsuang, M. (2000). Schizophrenia: genes and environment. *Biol. Psychiatry* **47**, 210–220.

Tsuang, M.T., Perkins, K., and Simpson, J.C. (1983). Physical diseases in schizophrenia and affective disorder. *J. Clin. Psychiatry* **44**, 42–46.

Vaddadi, K.S. (1981). Niacin flushing and schizophrenia. *Med. Hypotheses* **7**, 599–600.

van Nimwegen, L.J., Storosum, J.G., Blumer, R.M., Allick, G., Venema, H.W., de Haan, L., Becker, H., van Amelsvoort, T., Ackermans, M.T., Fliers, E., Serlie, M.J., and Sauerwein, H.P. (2008). Hepatic insulin resistance in antipsychotic naive schizophrenic patients: stable isotope studies of glucose metabolism. *J. Clin. Endocrinol. Metab.* **93**, 572–577.

van Venrooij, J.A., Fluitman, S.B., Lijmer, J.G., Kavelaars, A., Heijnen, C.J., Westenberg, H.G., Kahn, R.S., Gispen-de Wied, C.C., van Venrooij, J.A., Fluitman, S.B., Lijmer, J.G., Kavelaars, A., *et al.* (2010). Impaired neuroendocrine and immune response to acute stress in medication-naive patients with a first episode of psychosis. *Schizophr. Bull.* Jun 17 (Epub ahead of print).

Venkatasubramanian, G., Chittiprol, S., Neelakantachar, N., Shetty, T.K., and Gangadhar, B.N. (2010). A longitudinal study on the impact of antipsychotic treatment on serum leptin in schizophrenia. *Clin. Neuropharmacol.* **33**, 288–292.

Ventner, J.C., Adams, M.D., Myers, E.W., Li, P.W., *et al.* (2001). The sequence of the human genome. *Science* **291**, 1304–1351.

Wakefield, J.C. (2010). Misdiagnosing normality: Psychiatry's failure to address the problem of false positive diagnoses of mental disorder in a changing professional environment. *J. Ment. Health* **19**, 337–351.

Wilson, D.W., and Douglass, A.B. (1986). Niacin skin flush is not diagnostic of schizophrenia. *Biol. Psychiatry* **21**, 974–977.

Wu, E.Q., Birnbaum, H.G., Shi, L., Ball, D.E., Kessler, R.C., Moulis, M., and Aggarwal, J. (2005). The economic burden of schizophrenia in the United States in 2002. *J. Clin. Psychiatry* **66**, 1122–1129.

Yang, J., Wang, B.J., Ding, M., Pang, H., Sun, X.F., and Li, Z.J. (2008). The relationship between SNP of cholecystokinin gene and certain mental status and its forensic significance. *Fa Yi Xue Za Zhi* **24**, 284–287.

Yap, H.L. (2010). Early psychosis intervention. *Singapore Med. J.* **51**, 689–693.

Yolken, R.H., Dickerson, F.B., and Fuller Torrey, E. (2009). Toxoplasma and schizophrenia. *Parasite Immunol.* **31**, 706–715.

Zhang, Z., and Chan, D.W. (2010). The road from discovery to clinical diagnostics: lessons learned from the first FDA-cleared in vitro diagnostic multivariate index assay of proteomic biomarkers. *Cancer Epidemiol. Biomarkers Prev.* **19**, 2995–2999.

TOWARD PERSONALIZED MEDICINE IN THE NEUROPSYCHIATRIC FIELD

Erik H.F. Wong[1], Jayne C. Fox[2], Mandy Y.M. Ng[2] and Chi-Ming Lee[3]

[1]AstraZeneca Pharmaceuticals, External Science, CNS-Pain Innovative Medicine Unit, Wilmington, Delaware, USA

[2]AstraZeneca Pharmaceuticals, Personalized Health Care and Biomarkers, Alderley Park, Macclesfield, Cheshire, United Kingdom

[3]AstraZeneca Pharmaceuticals, Discovery Enabling Capabilities & Sciences, Alderley Park, Macclesfield, Cheshire, United Kingdom

Abstract

There are great expectations for the personalized medicine approach to address the therapeutic needs of patients in the twenty-first century. Advances in human genome science and molecular innovations in neuroscience have encouraged the pharmaceutical industry to focus beyond broad spectrum population therapeutics—the driving force behind the "blockbuster" product concept—to personalized medicine. For central nervous system (CNS) therapeutics, repeated failures in converting scientific discoveries to clinical trial successes and

DOI: 10.1016/B978-0-12-387718-5.00013-4

0074-7742/11 $35.00

regulatory approvals have precipitated a drug pipeline crisis and eroded confidence in the industry. This chapter describes how innovations in genomics and translational medicine can impact the future of neuropsychiatry and deconvolute the complexity of psychiatric diseases from symptoms biology. A targeted and consistent investment is needed to restore confidence in translating science into clinical success.

I. Introduction

Personalized medicine holds the promise to tailor appropriate therapy to the right types of patients. However, personalized medicine means different things to different people. According to the National Human Genome Research Institute, personalized medicine "is an emerging practice of medicine that uses an individual's genetic profile to guide decisions made in regard to the prevention, diagnosis, and treatment of disease. Knowledge of a patient's genetic profile can help doctors select the proper medication or therapy and administer it using the proper dose or regimen" (www.genome.gov/Glossary/). Today when people refer to personalized medicine, it is usually in the context of using genomics—the science of looking at all of the information in the human genome—to tailor medical care to individuals based on their genetic makeup. However, recognizing that factors in addition to the patient's genetic profile (such as diet, age, life styles, epigenetics, etc.) are at play in determining a patient's risk for certain diseases and response to certain treatments, the definition of personalized medicine has been broadened to "a form of medicine that uses information about a person's genes, proteins and environment to prevent, diagnose and treat disease" (www.cancer.gov/dictionary/).

We would like to define personalized medicine as a medical model emphasizing the systematic use of genetic, biologic, and clinical information about an individual patient to select or optimize that patient's preventative and therapeutic care. This approach is comprehensive and logical and should always be the aim of all medical practice. In this light, polypharmacy can be viewed as a form of personalized medicine. Inevitably limited in their understanding of disease biology, doctors have always prescribed what they deem to be the right combination of medicines to provide therapeutic needs for the individual patient's range of symptoms. In neuropsychiatry, this means prescribing a number of agents for the different symptoms (e.g., mood, sleep) of a complex disease like major depressive disorder (MDD), and since comorbidity of psychiatric diseases is common, the ensemble of agents are accordingly used (Wong *et al.*, 2008, 2010a).

So what does personalized medicine mean in the twenty-first century? The problem with this existing approach lies in the shortcomings of currently available agents. Their inadequate efficacy and issues with tolerability and compliance mean that combining agents does not remove but rather compounds these challenges (Wong *et al.*, 2008). Our industry has not met with high success in our efforts to develop better agents with novel mechanisms to drive improvement in efficacy and tolerability (Conn and Roth, 2008). The lack of success in the discovery and approval of new medication in the neuropsychiatric field has been discussed in many reviews (Agid *et al.*, 2007; Insel, 2009; Wong *et al.*, 2010b).

First, it has been argued that animal models lack appropriate face, construct, and predictive validity (Wong *et al.*, 2010a,b). Most of the preclinical models in the neuropsychiatric field were developed and optimized for the previous generation of agents (Cryan and Slattery, 2007). The use of these same models to characterize drugs targeting novel molecular mechanisms for the same diseases has met with low predictability of success in human clinical studies (Wong *et al.*, 2010a,b). Second, some suggest that psychiatric diseases are so complex and polygenetic that it is unrealistic to have a model that can capture all the complexity, pathology, and clinical presentation of diseases such as depression and schizophrenia (Klein, 2011). Third, patients in clinical trials are more heterogeneous than rodents by virtue of the lack of a clear biological classification, or pathological understanding. This might explain regardless of how well-intentioned scientists from industry are to pursue new targets and mechanisms to achieve differentiated medicine, the outcome has been disappointing. This includes the substance P antagonist (Herpfer and Lieb, 2005) for depression, the CRF1 antagonist (Binneman *et al.*, 2008) for anxiety or depression, and the histamine H3 inverse agonist (Egan *et al.*, 2009) for cognitive enhancement.

In contrast, oncology and neurology have enjoyed a much closer relationship between disease mechanism/biology and treatment approach. The successful identification of genetic form of Alzheimer's disease with specific forms of mutation has allowed an appropriate interrogation of the amyloid approach (Hardy and Selkoe, 2002). While this approach is controversial, there are clear paths to pursue. The neuropsychiatric field has not benefited from such "genetic rules of engagement." This chapter seeks to show innovation in biomarker technology and how this together with the progress made in "translational psychiatry" impact on the future of the neuropsychiatric field (Licinio, 2011). We discuss how genomic and translational medicine research, carried out under a framework of precompetitive consortia, offers opportunity to deconvolute the complex of psychiatry diseases from symptom-based to a biological foundation to support the personalized medicine vision.

II. Is Personalized Medicine All About Genetics? How Many Measures Are We Talking About?

The success of the Human Genome Project (HGP; International Human Genome Sequencing Consortium, 2004) and the completion of the subsequent human haplotype mapping (HapMap) project (International HapMap Consortium, 2005) have opened many opportunities to improve the biological understanding of health and disease states. An early prediction for applications of post-HGP-derived knowledge was the use of individual genetic data to optimize drug treatments (Roses, 2002). There is still belief that individual genetic profiles will play an important role in future medical practice as exemplified by direct-to-consumer genetic testing companies that offer DNA tests to assess disease risk, for example, 23 and me (https://www.23andme.com/) and the Navigenics Health Compass (http://www.navigenics.com/). However, predictive values and the complexity of genetic variation-based testing may limit anticipated treatment options, at least in the short term (Deverka, 2009; Ng *et al.*, 2009; Bloss *et al.*, 2011).

A. Heritability, Structural Genomics, and Personalized Medicine

Disease states are multifactorial in etiology and cannot be assessed solely using genetics. Uher (2009) reported that the heritability of different mental illnesses is variable, from 0.3 (anxiety disorders) to 0.9 (autism; see Table. I). However, the understanding of the interplay between genetic and nongenetic factors is still unclear for most mental illnesses, regardless of disease prevalence.

The assumption that heritability of a disease state will necessarily open doors for direct, genetics-based, drug treatment decisions is reductive because the space between clinical phenotype and genotype is exceedingly complex. However, the potential for genetics-based approaches to understand disease processes will provide important insights is indisputable (Green and Guyer, 2011). Such discovery will provide a route for the personalized approach of drug treatments. Genetic approaches may not provide direct routes for personalization, that is, via a "genetic test," as was originally proposed, but rather—particularly in the neuropsychiatric field—genetic approaches will help to unravel the complexity of the underlying diseases, thereby providing a scaffold for future biomarker approaches. This is exemplified by the evolving field of structural genomic variation. Structural variation was considered to be less important in the early days of the HGP than the potential role played by common, simple variations (single nucleotide polymorphisms, SNPs) either as direct markers or "tags" for coinherited sequence blocks (International HapMap Consortium, 2005). Technological advances have allowed genomic studies to move beyond common, simple genetic variations to identify

Table I
GENETIC EPIDEMIOLOGY OF SELECTED TYPES OF MENTAL ILLNESS.

	Prevalence (%)	Age onset	Mortality	Fertility	Heritability	Paternal age effect
Autism	0.30	1	2.0	0.05	0.90	1.4
Anorexia nervosa	0.60	15	6.2	0.33	0.56	–
Schizophrenia	0.70	22	2.6	0.40	0.81	1.4
Bipolar affective disorder	1.25	25	2.0	0.65	0.85	1.2
Unipolar depression	10.22	32	1.8	0.90	0.37	1
Anxiety disorders	28.80	11	1.2	0.90	0.32	–

Lifetime prevalence (percentage), median age of onset (years), mortality ratio (values of more than one indicate increased mortality compared with the general population), fertility ratios (values less than one indicate decreased fertility compared with the general population), heritability (estimated contribution of addictive genetic effects from twin studies), and an index of paternal age effect (risk ratio for 10-year increase in fathers aged above 30 years; no data are available for anorexia nervosa and anxiety disorders). All data are based on published report referenced in the "Building Blocks" section of Uher (2009).

genomic commonalities between, for instance, autism and schizophrenia (Guilmatre *et al.*, 2009). "Simple genetic" is being replaced by "complex genomic" modes of genome interrogation to better understand disease (Wain and Tobin, 2011).

B. GENETIC APPROACHES FOR POPULATION STRATIFICATION TO AVOID FALSE ASSOCIATIONS

The question of nature versus nurture, genetic versus environmental contribution to neuropsychiatric disease risk and disease course is a challenging problem for genetic studies. Care must be taken in study design to avoid the potential pitfalls inherent in such a complex area. For example, population stratification can be a source of misleading data in genetic association studies, where the reported association signal was driven by allele frequency differences between cases and controls due to systematic difference in ancestry, not disease-associated genes. Population stratification can be managed by statistical methods to control for inflation of test statistics (Devlin and Roeder, 1999) or by a principal component analysis to correct for differences between populations (Price *et al.*, 2006). Otherwise self-declared ethnicity can be employed (as a method "surrogate" for ancestry) to subgroup study participants prior to statistical analysis.

C. Gene and Environmental Approach for Personalized Medicine

Plasticity of brain structures and signaling systems plays an important role in personalized medicine. The brain's obvious sensitivity to a huge range of environmental factors suggests a genetic approach alone will not provide sufficient insight of likely drug response in all situations (Nestler and Hyman, 2010). Indeed, recent findings in smoking behaviors—a phenotype with close links to some psychiatric conditions—reinforce the likelihood of weak predictive power where genetic factors may modulate an intermediate phenotype (brain circuitry) but a behavioral or clinical phenotype is the end point of interest (Hong *et al.*, 2010). Nicotinic acetylcholine receptor α5-α3-β4 subunit gene variants correlated well with variation in brain connectivity, as measured by imaging, and explained approximately 10–12% of the nicotine addiction behavioral variance. This is three to four times "better" than the percentage of variance contributed by the gene variations if the smoking behavior was simply examined in the absence of the brain connectivity endophenotype. This is an excellent example of a genetic marker for behavioral variation leading to a better understanding of the biology underlying a phenotype, and supports the endophenotype approach which we discuss later.

D. Advanced Technical Development in Genomics

Fortunately, the technical developments which have supported the rapid expansion of genetic data are paralleled by developments in many other fields. Areas of advancement include those applicable to brain imaging as well as the other "omics" assay systems to investigate the rare genetic variations in individuals using DNA sequencing technologies (Durbin *et al.*, 2010), RNA species, and metabolite profiles. We will realize opportunities by considering integrated data derived from a variety of platforms. Indeed, although genetic analysis has many potential advantages (e.g., being a stable, easily accessible analyte), some of the most important genetic-related developments to date have been derived from an improved understanding of disease, principally via the study of rare, early-onset conditions and molecular insights yielded via genetic manipulation of model systems (Malenka and Malinow, 2011). Good examples of this are the early onset of Alzheimer's and Parkinson's diseases (Goate *et al.*, 1991; Rogaev *et al.*, 1995; Sherrington *et al.*, 1995; Paisan-Ruiz *et al.*, 2004; Zimprich *et al.*, 2004).

Major issues for treatment personalization in all common diseases, neuropsychiatric or otherwise, are presented by the current diagnostic/disease classification systems in use. The Diagnostic and Statistical Manual of Mental Disorders (DSM) is a valuable common framework for functional disease classification but may not be a good source of stratification options relevant to new or existing therapies. This has been recently reviewed and discussed (Klein, 2011; Grinker, 2010). Undoubtedly, if empirical analyses of DSM data give rise to meaningful treatment subgroups, then such opportunities should certainly be grasped. However, we believe that additional molecular or imaging-based biomarkers will be needed and large international consortia are now attempting to identify such markers. This is illustrated by the Sequenced Treatment Alternatives to Relieve Depression (STAR*D; Rush *et al.*, 2004) study funded by the National Institute of Mental Health (NIMH) in the United States, and the Innovative Medicine Initiative (IMI)-Novel Methods Leading to New Medications in Depression and Schizophrenia (NEWMEDS) study funded by the European Commission (Hughes, 2009; Abbott, 2010). By adopting an open approach to the exact nature of potential stratification tools—using the best platforms and systems available, but not focusing exclusively on a given technical approach—it seems feasible that markers deriving from different discovery modes could be combined and ultimately assayed via a single diagnostic platform, much as many common biochemical markers are measured in routine clinical decision making (Morrow *et al.*, 2007). Obviously, this approach would be a complex solution, and it would be preferable to devise simpler ways to personalize or segment neuropsychiatric disease populations and thus derive improved treatment outcomes. Attempts have been made in this area, in particular, to subtype depression (Parker *et al.*, 1999), although the application of an agreed common approach to personalize treatment choices, beyond current clinical practice, is not apparent.

E. From Biology Advance to Technology Challenge

The solution to one problem often unveils another. By applying the squeezed pipeline/moving bottle neck analogy, it is clear that the bottle neck has moved from sequence generation to data storage, computational power, and statistical analysis. Next-generation sequencing or contemporary SNP arrays are an intensely data-driven technology. For example, the Illumina Human Omni 2.5 can assay 2.5 million variants with minor allele frequencies down to 2.5% and the 1000 genome project has identified 20 million unique variants in 629 individuals (http://www.1000genomes.org/). The integration of this immense volume of data with clinical phenotypes presents significant challenges in statistics, data storage, and computational power. SNPs are often analyzed one at a time in the current GWAS studies. However, many researchers have advocated a more holistic approach that incorporates gene–gene and gene–environment interactions. The billions of comparisons of SNP pairs present a significant challenge in computation and interpretation of results due to problems of multiple testing. Substantial investments are required to build and maintain a computational infrastructure, including computing cluster and the relevant utilities for cluster operation.

The pharmaceutical industry is regulated by government bodies, and regulations require that data submitted for new drug applications are kept for a significant period of time. To keep all potential relevant genomic data requires a huge investment in data storage. In addition, access to publicly available genomic data by pharmaceutical companies is restricted. For example, the Wellcome Trust Consortium Case-Control Dataset has access control regulations which preclude easy use by the pharmaceutical industry. A more user-friendly precompetitive consortium framework similar in principle to that of IMI is called for (Hughes, 2009). Indeed, the formation of another European Commission-driven project called "Virtual Physiological Human" (VPH) has been timely—The VHP will revolutionize the way health knowledge is produced, stored, and managed as well as the way in which healthcare is currently delivered. Results will include personalized disease predictions, earlier diagnoses, better surgery planning and training, and a better understanding of the links between genes, diseases, and treatments (http://ec.europa.eu/information_society/events/ict_bio/2008).

III. Opportunities for Personalized Medicine in Neuropsychiatry for Drug Discovery

A. Biomarker Technology Beyond Genetics

It may be possible to exploit other data sources to improve treatment targeting and hence therapeutic outcome (Licinio, 2011). The discovery and exploitation of imaging-based biomarkers is a promising avenue to pursue for neuropsychiatry (see Chapter "Imaging brain microglial activation using positron emission tomography and translocator protein-specific radioligands" by Owen and Matthews). The various noninvasive imaging approaches provide a potentially rich source of data for determining treatment decisions. If robust clinical utility can be demonstrated, then some of these imaging techniques may have direct application in personalized medicine. The direct imaging of structural brain variation, metabolic activity, or even receptor levels/occupancy (Farde, 1996) could yield personalized medicine approaches, economic and infrastructure considerations permitting. Even the application of relatively simple and noninvasive approaches, such as electroencephalography (EEG), may yield methods to select patients more likely to respond to treatment (Losifescu *et al.*, 2009).

The selection of treatments for schizophrenia, with the arguable exception of attempts to improve treatment via dosing adjustment, guided by genetic information, is not currently augmented by molecular-based approaches. Neither have molecular-based subtypes of schizophrenia been identified, nor robust therapy specific markers of treatment response. However, there is promising recent findings that relate to enrichment for structural genomic variants in both schizophrenia and other neuropsychiatric disorders (Grozeva *et al.*, 2010). Although these new findings suggest some overlap in the genetic basis of neuropsychiatric conditions, the frequency of the individual genetic changes in disease populations, together with the impact of other genetic and nongenetic factors (Girirajan and Eichler, 2010), likely indicates their best use as tools to unravel biological pathways rather than stand-alone biomarkers for treatment decisions.

B. The Challenges of PM in Neuropsychiatry—A Lesson from Other Fields

There are some success stories for personalized medicine in oncology including Herceptin in HER2+ breast cancer, Gleevec in chronic myeloid leukemia, CML, and Rituxan in chronic lymphocytic leukemia, CLL. Herceptin was approved for breast cancer patients with an overexpression of HER2, human epidermal growth factor receptor ErbB2 (Ross *et al.*, 2009), and response to Gleevec is indicated for CML patients expressing BCR-ABL, the fusion protein created by a translocation between the long arms of chromosomes 9 and 22 (Druker, 2008). However, this technological progress in solid and liquid tumors has not been mirrored in personalized medicine in neuropsychiatry.

The success of personalized medicine in oncology is dependent on matching drugs with patients whose cancer is caused by the mechanism targeted by the drug. However, in psychiatry, the etiology and pathophysiology of many disorders are largely unknown and the categorization of psychiatric diseases using DSM/ICD criteria is largely based on assessment of symptoms and generally lacks proven biological validity (Klein, 2011). Most pharmacotherapy in psychiatry is discovered empirically based on serendipitous clinical observation and reversed pharmacology (Wong *et al.*, 2010b). Thus, there is insufficient understanding of disease biology to inform diagnosis and guide treatment selection for most neuropsychiatric disorders. Further, traditional pharmacotherapy in psychiatry treats patients with a "one-size fits all" broad spectrum population approach. It is perhaps not surprising that there is a low-treatment response rate; for any given psychiatric drug on the market works on average for only about half of the patients who receive it. For instance, the remission rate for MDD and bipolar disorder after conventional pharmacotherapy is only about 30% (Warden *et al.*, 2007; Thase, 2008).

It is also increasingly recognized that not all psychiatric patients with the same DSM/ICD diagnosis respond equally to the same treatment. The difference in response of different patients to a particular drug may be due to different genetic factors and/or patients' age, diet, or other environmental issues (such as psychological stress), and models are being used to tease apart these potential contributing components (Oliver and Davies, 2009). It is not unusual to have a heterogeneous syndrome described as a single psychiatric disorder, and even the subphenotypes commonly used are probably heterogeneous. For example, schizophrenia is characterized by positive symptoms, negative symptoms, and cognitive deficits, all of which may have different etiology and pathophysiological mechanisms. This artificial grouping of a heterogeneous syndrome into one disorder may be another reason for the low response rate.

To date, personalized medicine in neuropsychiatry has mostly operated by "trial and error," that is, if the patient does not respond to one treatment or has too many side effects, the physician will give a different drug in the same class (e.g., selective serotonin reuptake inhibitors (SSRI)) or switch to a drug with a different mechanism (e.g., from SSRI to Serotonin and noradrenaline reuptake inhibitor—SNRI). Attempts to define subtypes of depression based on symptomology, such as melancholia (Perry, 1996) and anxiety (Fava *et al.*, 2008) have been inconsistent in predicting treatment outcomes (Russell *et al.*, 2001; Brown, 2007; McGrath *et al.*, 2008; Nelson, 2010). Thus, more reliable predictors with a sound biological rationale to guide treatment selection are needed (Samuels *et al.*, 2011). We envision that stratifying patients into more clinically distinct and homogeneous subgroups, based on an understanding of their unique disease pathophysiology rather than nonspecific symptoms, would result in a more uniform and higher response rate to pharmacotherapy.

C. Endophenotypes—Leveraging Biology for Psychiatric Drug Discovery and Development

One way of deconstructing the biology of complex neuropsychiatric diseases is to adopt an endophenotype approach and use specific endophenotypes (rather than clinical syndromes) as end points in genetic and other "omics" biomarker discovery work (Dick *et al.*, 2006; Kendler and Neale, 2010). An endophenotype (also known as intermediate phenotype) is a quantitative biological trait that is reliable in reflecting the function of a discrete biological system and is reasonably heritable, and as such is more closely related to the root cause of the disease than the broad clinical phenotype (Gottesman and Gould, 2003; Cannon and Keller, 2006; Meyer-Lindenberg and Weinberger, 2006; Tan *et al.*, 2008). An example of an endophenotype in neuropsychiatry is that of mismatch negativity (MMN), which is an auditory event-related potential (ERP) component elicited in the context of an auditory oddball paradigm (Naatanen *et al.*, 2011). Deficits in MMN have been consistently reported in schizophrenia (Javitt *et al.*, 2000; Umbricht and Krljes, 2005). Schizophrenia-like deficits in MMN generation have been observed in human healthy volunteers following treatment with N-methyl-D-aspartate receptor (NMDAR) antagonists (Umbricht *et al.*, 2000; Heekeren *et al.*, 2008), suggesting that glutamate/NMDAR dysfunction may underlie the MMN deficits in schizophrenia. Indeed, the role of glutamate/NMDAR dysfunction in schizophrenia (Coyle, 1996; Javitt, 2007; Kantrowitz and Javitt, 2010) is increasingly supported by converging pharmacological (Javitt and Zukin, 1991; Krystal *et al.*, 1994), neurochemical (Coyle, 1996; Bergeron *et al.*, 2007), and genetic (Norton *et al.*, 2006; Li and He, 2007) data. It is therefore conceivable that, using objective measurement of deficits in MMN, one could potentially identify more homogeneous subgroups of schizophrenic patients with glutamate/NMDAR dysfunction. This could enable the more efficient testing of drugs that potentiate NMDAR function (i.e., right treatment for the right patient) as well as augment the discriminating power of genetic association and linkage studies in this subset of schizophrenic patients.

Another integrative mechanism-based endophenotype approach to facilitate personalized medicine in depression can be illustrated with the serotonin system. Fisher *et al.* (2006) observed an inverse relationship of 5HT1A autoreceptor density with amygdala reactivity. They suggested that an increase in 5HT1A autoreceptor availability may contribute to the risk of depression and the down-regulation of 5HT1A receptors may contribute to the therapeutic efficacy of antidepressant drugs (Richardson-Jones *et al.*, 2010; Samuels *et al.*, 2011). A relatively common polymorphism (−1018C/G) in the promoter region of 5-HTR1A is known to alter 5HT1A receptor gene expression (Lemonde *et al.*, 2003). The G/G allele, which is associated with higher 5HT1A autoreceptor expression, is overrepresented in MDD patients, especially those with delayed response to SSRIs and nonresponders. Thus by combining imaging studies of amygdala with genetic data, it may be possible to predict a person's risk for depression and response to treatment. It is conceivable that 5HT1A autoreceptor expression (which can be monitored by genetics, PET, or functional response) could be used to stratify patients to include those with lower receptor expression who are more likely to be responders to SSRIs (Richardson-Jones *et al.*, 2010; Samuels *et al.*, 2011).

An advantage of studying endophenotypes is that early clinical research for proof of mechanism and proof of principle can be performed on the nonaffected relatives of patients, as opposed to the healthy population not bearing the disease phenotype or on the patient population which may be confound by medication (Flordellis, 2005). The NIMH also advocated the use of endophenotypes as one way to bridge the genetic complexity and heterogeneity underlying neuropsychiatric disorders (Insel and Cuthbert, 2009). The NIMH has initiated a Research Domain Criteria (RDoC) project with the aim of classifying clinical subjects based on "the status of a gene polymorphism, a particular response in a neuroimaging task, or scores on a cognitive task" (Insel and Cuthbert, 2009; Report of the National Advisory Mental Health Councils Workgroup, 2010). It is envisaged that a better nosology of neuropsychiatric disorders could be obtained by using endophenotypes as well as objective, biologically meaningful, and predictive biomarkers to define patient subgroups.

Although endophenotypes have been useful in defining better end points for heritability and association studies, another challenge for genetics-based biomarker identification is the nature of genetic variation itself. As mentioned earlier, until recently, the primary focus of most genetic studies has been common SNPs as heritable markers of genetic diversity. Here, the common (>5–10% frequency in a defined population) SNP has been used first to map common diversity and then applied in attempts to locate regions where disease causative genetic variants reside within the genomes of unrelated individuals. The common disease-common variant (CDCV) approach has had success with the identification of compelling candidate genes for schizophrenia and autism with findings replicated in independently genome-wide association studies (O'Donovan *et al.*, 2008; Ma *et al.*, 2009; Wang *et al.*, 2009; Riley *et al.*, 2010). However, most variants individually or collectively explain only a small proportion of the heritability for schizophrenia (International Schizophrenia Consortium *et al.*, 2009).

Clearly, the common SNP-based approach has had limited success in unraveling the biology of neuropsychiatric disease, although it has fared much better in other high-heritability diseases. This conundrum (high heritability, but inability to identify robust associations/causative variants) is slowly being unraveled, but it is becoming obvious that a whole range of factors play a part in explaining the perceived slow progress to date. First, many studies have been underpowered. Recent work in schizophrenia has required huge consortia to pool resources to identify a robust but small overall contribution to inherited disease risk (Abbott, 2010). Second, the nature and frequency of heritable variation are only beginning to be clarified. The fundamental mutation rates of common/rare SNPs and small/large structural genomic variations are still not fully mapped. Third and perhaps most important, gene regulation and nonprimary sequence-based modes of inheritance (epigenetics) are likely to play an important role. Needless to say, the task of teasing apart the human genome is reliant on appropriate tools and technologies which themselves evolve to meet new discovery needs. Perhaps with the arrival of next-generation sequencing and the associated staggering fall in sequencing costs, we may at last see significant steps forward (Cirulli and Goldstein, 2010).

D. Pharmacokinetics and Personalized Medicine

Many CNS drugs are substrates of P-glycoprotein (P-gp) and cytochrome P450 enzymes. It is plausible that the genetic variations (polymorphism) of drug-metabolizing enzymes and transporters could impact on drug levels in plasma and in target organs (e.g., brain; Bertilsson *et al.*, 2002; Uhr *et al.*, 2008;

Franke *et al.*, 2010). A review of Medline databases (1970–2003) on the impact of genetic polymorphisms on positive and adverse reactions to antidepressants and antipsychotics found evidence of genetic variation of CYP2D6 or CYP2C19 that would require at least doubling of the dose in extensive metabolizers in comparison to poor metabolizers to achieve the same response (Llerena *et al.*, 1993; Kirchheiner *et al.*, 2004; Dorado *et al.*, 2007). However, recognizing that the functional impact of many genetic alleles of cytochrome P450 enzymes remain to be determined and genetic variabilities of drug transporters and drug targets might confound the impact of CYP variants on drug levels, the actual benefit of PM based on genetic variations of cytochrome P450 enzymes will have to be supported by prospective studies in order to validate dose recommendation (Gonzalez *et al.*, 2008; Kirchheiner *et al.*, 2010). Such studies will be facilitated by the availability of tests (such as Amplichip) to monitor the genetic variants of CYP2D6 and CYP2C19 (Jain, 2005).

Interestingly, higher brain concentration of several antidepressants (doxepin, venlafaxine, and paroxetine) have been reported in P-gp KO mice, which support a role of P-pg in regulating the levels of some neuropsychiatric drugs. A polymorphism has been found in the P-gp (also known as MDR-1 or ABCB1) gene that is associated with differences in gene expression and drug disposition (Franke *et al.*, 2010). There is also some early promise of polymorphism of P-gp being related to antidepressant response (Uhr *et al.*, 2008). Thus, genetic alleles of P-gp could be monitored in future neuropsychiatric drug development for drug candidates which are P-gp substrates. Genetic biomarkers that allow the selection of a more homogeneous population in relation to drug disposition (e.g., cytochrome P450) might enable patient enrichment in clinical trials and facilitate the selection of the right doses for the right patients.

The adoption of new, genetics-based dosing algorithms is not a common practice in neuropsychiatry. However, it is possible that, with the increased drive for improved cost/benefits as well as an integrated personalized medicine approach for new drugs and the Food and Drug Administration's (FDA's) hopes for PGx (Goodsaid and Frueh, 2007; Orr *et al.*, 2007), the situation could change. Genetic testing specifically targeted at those genes encoding absorption, distribution, metabolism, and excretion (ADME) proteins could still be embraced if robust cost–benefit data is generated (Williams *et al.*, 2008).

E. Biomarker Discovery/Qualification/Validation—The Partnership Model

To identify objective biomarkers which differentiate between psychiatric patients and normal individuals has been a major endeavor in biological psychiatry. However, as discussed earlier, such efforts have been hampered by the lack of understanding of the etiology and pathophysiology of most psychiatric diseases. Rapid advances in proteomics, metabonomics, and mass spectrometry methodologies which allow the specific detection of very low levels of multiple analytes in biological fluids (such as blood, plasma, urine, and CSF samples) have generated some optimism for differential detection of complex diseases based on such pattern analysis (Schwarz *et al.*, 2011; see Chapters "The application of multiplexed assay systems for molecular diagnostics" by Schwarz *et al.* and "Algorithm development for diagnostic biomarker assays" by Izmailov *et al.*). For instance, recent multianalyte proteomic approaches using plasma samples from large case–control collections of patients diagnosed according to DSM criteria have identified a set of analytes [such as insulin and matrix metalloproteinase (MMP)-9 for depression and brain-derived neurotrophic factor (BDNF), epidermal growth factor (EGF)] and a number of chemokines and biomarkers associated with inflammation, disturbances in amino acid and lipid metabolism, and signaling pathways for schizophrenia) as candidate biomarker signatures for depression and schizophrenia, respectively (McNally *et al.*, 2008; Domenici *et al.*, 2010).

In addition to innovative discovery efforts, promising biomarkers need to be "qualified" and "validated" before they can be approved for diagnostic and/or prognostic uses. According to Janet Woodcock at the FDA (2005), "qualification" refers to whether the biomarker being assayed is shown to be reliable in a biological sense (e.g., serum glucose as a marker of diabetes) and "validation" refers to assurance that the assay being used is able to monitor the analyte(s) in a reproducible manner (Woodcock, 2009). For example, in the genomics area, the Microarray Quality Control (MAQC) consortium has reported good reproducibility of gene expression measurements using multiple platforms and test sites (Shi *et al.*, 2006). Such consistency and reproducibility of assay methodology is a prerequisite for the characterization of biomarkers prior to their eventual qualification and application in personalized medicine (Casciano and Woodcock, 2006; Goodsaid and Frueh, 2006). Similar efforts have been undertaken by other public–private partnerships such as the validation and qualification of proteomic and imaging (volumetric MRI and PET) biomarkers for Alzheimer's disease by the Alzheimer's Disease Neuroimaging Initiative (ADNI) and Coalition Against Major Diseases (CAMD; Kauwe *et al.*, 2011). It is envisaged that biomarkers that predict the likelihood of a disease event or more rapid disease progression could increase the likelihood of detecting therapeutic benefits with fewer subjects and shorter time.

Notwithstanding the considerable challenge, the importance of personalized medicine for neuropsychiatry cannot be overemphasized for the following reasons: (i) The high morbidity, mortality, and societal cost of poorly served patients, some of whom receive little or no benefit from available treatments (Fava *et al.*, 2008), and (ii) individual and societal burdens related to nonbeneficial treatment exposures. When this situation is coupled with a third element—the current failure of new pharmaceutical developments, employing a nonindividualized treatment approach and existing clinical trial designs—the need for personalized medicine is obvious and urgent. This latter point is an important driver for the real cross-pharma (often referred to as "precompetitive") and globally coordinated efforts for better, therapy-linked patient segmentation, as exemplified by EU Innovative Medicine, Novel Methods leading to New Medications in Depression and Schizophrenia, that is, IMI-NEWMEDS (Abbott, 2010). In addition, efforts by individual companies suggest an opportunity for fresh thinking is now being embraced.

IV. Conclusion

It is clear that the "moment in the sun" for personalized medicine remains elusive for neuropsychiatry. The complex genetic and nongenetic factors that influence psychiatric disorders point to a need for a number of technologic breakthroughs in biology and organizational approaches. It has taken a crisis of confidence in neuropsychiatric research for psychiatric drug discovery and development to drive the realization that these challenges cannot be overcome at the individual lab or company level. The message here is clear; the scale of the challenge, and consequent investment requirement, has driven the need for precompetitive consortia that leverage open innovation to address the issue mentioned above. We remain optimistic that the formation of IMI, and similar, efforts signal a turnaround in approach to enable this form of "BIG" science and that step by step, the rules of engagement that connect neuropsychiatric phenotype to molecular etiology will be revealed to allow our research community to innovate back to the future!

Acknowledgments

We sincerely express our appreciation to Ed Pierson, Dennis McCarthy, Frank Yocca, Tim Piser, and Michael Quirk for their constructive comments during the preparation of this chapter. We also like to acknowledge that the motivation of this chapter arose from fruitful discussion with members from WP7, 8, 9 of IMI-NEWMEDS.

References

Abbott, A. (2010). Schizophrenia: the drug deadlock. *Nature* **468**, 158–159.

Agid, Y., Buzsaki, G., Diamond, D.M., Frackowiak, R., Giedd, J., Girault, J.A., Grace, A., Lambert, J.J., Manji, H., Mayberg, H., Popoli, M., Prochiantz, A., *et al.* (2007). How can drug discovery for psychiatric disorders be improved? *Nat. Rev. Drug Discov.* **6**, 189–201.

Bergeron, R., Imamura, Y., Frangioni, J.V., Greene, R.W., and Coyle, J.T. (2007). Endogenous N-acetylaspartylglutamate reduced NMDA receptor-dependent current neurotransmission in the CA1 area of the hippocampus. *J. Neurochem.* **100**, 346–357.

Bertilsson, L., Dahl, M.L., Dalen, P., and Al-Shurbaji, A. (2002). Molecular genetics of CYP2D6: clinical relevance with focus on psychotropic drugs. *Br. J. Clin. Pharmacol.* **53**, 111–122.

Binneman, B., Feltner, D., Kolluri, S., Shi, Y., Qiu, R., and Stiger, T. (2008). A 6-week randomized, placebo-controlled trial of CP-316,311 (a selective CRH1 antagonist) in the treatment of major depression. *Am. J. Psychiatry* **165**, 617–620.

Bloss, C.S., Schork, N.J., and Topol, E.J. (2011). Effect of direct-to-consumer genome wide profiling to assess disease risk. *N. Engl. J. Med.* **364**, 524–534.

Brown, W.A. (2007). Treatment response in melancholia. *Acta Psychiatr. Scand.* **115**, 125–129.

Cannon, T.D., and Keller, M.C. (2006). Endophenotypes in the genetic analyses of mental disorders. *Annu. Rev. Clin. Psychol.* **2**, 267–290.

Casciano, D.A., and Woodcock, J. (2006). Empowering microarrays in the regulatory setting. *Nat. Biotechnol.* **24**, 1103.

Cirulli, E.T., and Goldstein, D.B. (2010). Uncovering the roles of rare variants in common disease through whole-genome sequencing. *Nat. Rev. Genet.* **11**, 415–425.

Conn, P.J., and Roth, B.L. (2008). Opportunities and challenges of psychiatric drug discovery: roles for scientists in academic, industry, and government settings. *Neuropsychopharmacology* **33**, 2048–2060.

Coyle, J.T. (1996). The glutamatergic dysfunction hypothesis for schizophrenia. *Harv. Rev. Psychiatry* **3**, 241–253.

Cryan, J.F., and Slattery, D.A. (2007). Animal models of mood disorders: recent developments. *Curr. Opin. Psychiatry* **20**, 1–7.

Deverka, P.A. (2009). Pharmacogenomics, evidence, and the role of payers. *Public Health Genomics* **12**, 149–157.

Devlin, B., and Roeder, K. (1999). Genomic control for association studies. *Biometrics* **55**, 997–1004.

Dick, D.M., Jones, K., Saccone, N., Hinrichs, A., Wang, J.C., Goate, A., Bierut, L., Almasy, L., Schuckit, M., Hesselbrock, V., Tischfield, J., Foroud, T., *et al.* (2006). Endophenotypes successfully lead to gene identification: results from the collaborative study on the genetics of alcoholism. *Behav. Genet.* **36**, 112–126.

Domenici, E., Wille, D.R., Tozzi, F., Prokopenko, I., Miller, S., McKeown, A., Brittain, C., Rujescu, D., Giegling, I., Turck, C.W., Holsboer, F., Bullmore, E.T., *et al.* (2010). Plasma protein biomarkers for depression and schizophrenia by multi analyte profiling of case-control collections. *PLoS One* **5**, e9166.

Dorado, P., Penas-Lledo, E.M., and Llerena, A. (2007). CYP2D6 polymorphism: implications for antipsychotic drug response, schizophrenia and personality traits. *Pharmacogenomics* **8**, 1597–1608.

Druker, B.J. (2008). Translation of the Philadelphia chromosome into therapy for CML. *Blood* **112**, 4808.

Durbin, R.M., Abecasis, G.R., Altshuler, D.L., Auton, A., Brooks, L.D., Durbin, R.M., Gibbs, R.A., Hurles, M.E., and McVean, G.A. (2010). A map of human genome variation from population-scale sequencing. *Nature* **467**, 1061–1073.

Egan, M.F., Harper-Mozely, L., Gottswald, R., and Snavely, D. (2009). Effect of H3 inverse agonist MK-0249 on cognitive performance in patients with schizophrenia. American College of Neuropsychopharmacology (ACNP) 48th Annual Meeting. December 6 - 10, 2009; Hollywood, Florida. Abstract 206.

Farde, L. (1996). The advantage of using positron emission tomography in drug research. *Trends Neurosci.* **19**, 211–214.

Fava, M., Rush, A.J., Alpert, J.E., Balasubramani, G.K., Wisniewski, S.R., Carmin, C.N., Biggs, M.M., Zisook, S., Leuchter, A., Howland, R., Warden, D., and Trivedi, M.H. (2008). Difference in treatment outcome in outpatients with anxious versus nonanxious depression: a STAR*D report. *Am. J. Psychiatry* **165**, 342–351.

Fisher, P.M., Meltzer, C.C., Ziolko, S.K., Price, J.C., Moses-Kolko, E.L., Berga, S.L., and Hariri, A.R. (2006). Capacity for 5-HT1A-mediated autoregulation predicts amygdala reactivity. *Nat. Neurosci.* **9**, 1362–1363.

Flordellis, C.S. (2005). The emergence of a new paradigm of pharmacogenomics. *Pharmacogenomics* **6**, 515–526.

Franke, R.M., Gardner, E.R., and Sparreboom, A. (2010). Pharmacogenetics of drug transporters. *Curr. Pharm. Des.* **16**, 220–230.

Girirajan, S., and Eichler, E.E. (2010). Phenotypic variability and genetic susceptibility to genomic disorders. *Hum. Mol. Genet.* **19**, R176–R187.

Goate, A., Chartier-Harlin, M.C., Mullan, M., Brown, J., Crawford, F., Fidani, L., Giuffra, L., Haynes, A., Irving, N., and James, L. (1991). Segregation of a missense mutation in the amyloid precursor protein gene with familial Alzheimer's disease. *Nature* **349**, 704–706.

Gonzalez, I., Penas-Lledo, E.M., Perez, B., Dorado, P., Alvarez, M., and LLerena, A. (2008). Relation between CYP2D6 phenotype and genotype and personality in healthy volunteers. *Pharmacogenomics* **9**, 833–840.

Goodsaid, F., and Frueh, F. (2006). Process map proposal for the validation of genomic biomarkers. *Pharmacogenomics* **7**, 773–782.

Goodsaid, F., and Frueh, F.W. (2007). Implementing the U.S. FDA guidance on pharmacogenomic data submissions. *Environ. Mol. Mutagen.* **48**, 354–358.

Gottesman, I.I., and Gould, T.D. (2003). The endophenotype concept in psychiatry: etymology and strategic intentions. *Am. J. Psychiatry* **160**, 636–645.

Green, E.D., and Guyer, M.S. (2011). Charting a course for genomic medicine from base pairs to bedside. *Nature* **470**, 204–213.

Grinker, R.R. (2010). In retrospect: the five lives of the psychiatry manual. *Nature* **468**, 168–170.

Grozeva, D., Kirov, G., Ivanov, D., Jones, I.R., Jones, L., Green, E.K., St Clair, D.M., Young, A.H., Ferrier, N., Farmer, A.E., McGuffin, P., Holmans, P.A., *et al.* (2010). Rare copy number variants: a point of rarity in genetic risk for bipolar disorder and schizophrenia. *Arch. Gen. Psychiatry* **67**, 318–327.

Guilmatre, A., Dubourg, C., Mosca, A.L., Legallic, S., Goldenberg, A., Drouin-Garraud, V., Layet, V., Rosier, A., Briault, S., Bonnet-Brilhault, F., Laumonnier, F., Odent, S., *et al.* (2009). Recurrent rearrangements in synaptic and neurodevelopmental genes and shared biologic pathways in schizophrenia, autism, and mental retardation. *Arch. Gen. Psychiatry* **66**, 947–956.

Hardy, J., and Selkoe, D.J. (2002). The amyloid hypothesis of Alzheimer's disease: progress and problems on the road to therapeutics. *Science* **297**, 353–356.

Heekeren, K., Daumann, J., Neukirch, A., Stock, C., Kawohl, W., Norra, C., Waberski, T.D., and Gouzoulis-Mayfrank, E. (2008). Mismatch negativity generation in the human 5HT2A agonist and NMDA antagonist model of psychosis. *Psychopharmacology (Berl.)* **199**, 77–88.

Herpfer, I., and Lieb, K. (2005). Substance P receptor antagonists in psychiatry: rationale for development and therapeutic potential. *CNS Drugs* **19**, 275–293.

Hong, L.E., Hodgkinson, C.A., Yang, Y., Sampath, H., Ross, T.J., Buchholz, B., Salmeron, B.J., Srivastava, V., Thaker, G.K., Goldman, D., and Stein, E.A. (2010). A genetically modulated, intrinsic cingulate circuit supports human nicotine addiction. *Proc. Natl. Acad. Sci. USA* **107**, 13509–13514.

Hughes, B. (2009). Novel consortium to address shortfall in innovative medicines for psychiatric disorders. *Nat. Rev.* **8**, 523–524.

Insel, T.R. (2009). Translating scientific opportunity into public health impact: a strategic plan for research on mental illness. *Arch. Gen. Psychiatry* **66**, 128–133.

Insel, T.R., and Cuthbert, B.N. (2009). Endophenotypes: bridging genomic complexity and disorder heterogeneity. *Biol. Psychiatry* **66**, 988–989.

International HapMap Consortium (2005). A haplotype map of the human genome. *Nature* **437**, 1299–1320.

International Human Genome Sequencing Consortium (2004). Finishing the euchromatic sequence of the human genome. *Nature* **431**, 931–945.

International Schizophrenia Consortium. Purcell, S.M., Wray, N.R., Stone, J.L., Visscher, P.M., O'Donovan, M.C., Sullivan, P.F., and Sklar, P. (2009). Common polygenic variation contributes to risk of schizophrenia and bipolar disorder. *Nature* **460**, 748–752.

Jain, K.K. (2005). Applications of AmpliChip CYP450. *Mol. Diagn.* **9**, 119–127.

Javitt, D.C. (2007). Glutamate and schizophrenia: phencyclidine, N-methyl-D-aspartate receptors, and dopamine-glutamate interactions. *Int. Rev. Neurobiol.* **78**, 69–108.

Javitt, D.C., and Zukin, S.R. (1991). Recent advances in the phencyclidine model of schizophrenia. *Am. J. Psychiatry* **148**, 1301–1308.

Javitt, D.C., Shelley, A.M., Silipo, G., and Lieberman, J.A. (2000). Deficits in auditory and visual context-dependent processing in schizophrenia: defining the pattern. *Arch. Gen. Psychiatry* **57**, 1131–1137.

Kantrowitz, J.T., and Javitt, D.C. (2010). N-methyl-d-aspartate (NMDA) receptor dysfunction or dysregulation: the final common pathway on the road to schizophrenia? *Brain Res. Bull.* **83**, 108–121.

Kauwe, J.S., Cruchaga, C., Karch, C.M., Sadler, B., Lee, M., Mayo, K., Latu, W., Su'a, M., Fagan, A.M., Holtzman, D.M., and Morris, J.C., Alzheimer's Disease Neuroimaging Initiative. (2011). Fine mapping of genetic variants in BIN1, CLU, CR1 and PICALM for association with cerebrospinal fluid biomarkers for Alzheimer's disease. *PLoS One* **6**(2), e15918.

Kendler, K.S., and Neale, M.C. (2010). Endophenotype: a conceptual analysis. *Mol. Psychiatry* **15**, 789–797.

Kirchheiner, J., Nickchen, K., Bauer, M., Wong, M.L., Licinio, J., Roots, I., and Brockmoller, J. (2004). Pharmacogenetics of antidepressants and antipsychotics: the contribution of allelic variations to the phenotype of drug response. *Mol. Psychiatry* **9**, 442–473.

Kirchheiner, J., Seeringer, A., and Viviani, R. (2010). Pharmacogenetics in psychiatry—a useful clinical tool or wishful thinking for the future? *Curr. Pharm. Des.* **16**, 136–144.

Klein, D.F. (2011). Causal thinking for objective psychiatric diagnostic criteria—a programmatic approach in therapeutic context. In: *Causality and Psychopathology.* Oxford University Press, London, p. 321.

Krystal, J.H., Karper, L.P., Seibyl, J.P., Freeman, G.K., Delaney, R., Bremner, J.D., Heninger, G.R., Bowers, M.B.J., and Charney, D.S. (1994). Subanesthetic effects of the noncompetitive NMDA antagonist, ketamine, in humans. Psychotomimetic, perceptual, cognitive, and neuroendocrine responses. *Arch. Gen. Psychiatry* **51**, 199–214.

Lemonde, S., Turecki, G., Bakish, D., Du, L., Hrdina, P.D., Bown, C.D., Sequeira, A., Kushwaha, N., Morris, S.J., Basak, A., Ou, X.M., and Albert, P.R. (2003). Impaired repression at a 5-hydroxytryptamine 1A receptor gene polymorphism associated with major depression and suicide. *J. Neurosci.* **23**, 8788–8799.

Li, D., and He, L. (2007). G72/G30 genes and schizophrenia: a systematic meta-analysis of association studies. *Genetics* **175**, 917–922.

Licinio, J. (2011). Translational Psychiatry: leading the transition from the cesspool of devastation to a place where the grass is really greener. *Transl. Psychiatry* **1**.

Llerena, A., Edman, G., Cobaleda, J., Benitez, J., Schalling, D., and Bertilsson, L. (1993). Relationship between personality and debrisoquine hydroxylation capacity. Suggestion of an endogenous neuroactive substrate or product of the cytochrome P4502D6. *Acta Psychiatr. Scand.* **87**, 23–28.

Losifescu, D.V., Greenwald, S., Devlin, P., Mischoulon, D., Denninger, J.W., Alpert, J.E., and Fava, M. (2009). Frontal EEG predictors of treatment outcome in major depressive disorder. *Eur. Neuropsychopharmacol.* **19**, 772–777.

Ma, D., Salyakina, D., Jaworski, J.M., Konidari, I., Whitehead, P.L., Andersen, A.N., Hoffman, J.D., Slifer, S.H., Hedges, D.J., Cukier, H.N., Griswold, A.J., McCauley, J.L., *et al.* (2009). A genome-wide association study of autism reveals a common novel risk locus at 5p14.1. *Ann. Hum. Genet.* **73**, 263–273.

Malenka, R.C., and Malinow, R. (2011). Alzheimer's disease: recollection of lost memories. *Nature* **469**, 44–45.

McGrath, P.J., Khan, A.Y., Trivedi, M.H., Stewart, J.W., Morris, D.W., Wisniewski, S.R., Miyahara, S., Nierenberg, A.A., Fava, M., and Rush, J.A. (2008). Response to a selective serotonin reuptake inhibitor (citalopram) in major depressive disorder with melancholic features: a STAR*D report. *J. Clin. Psychiatry* **69**, 1847–1855.

McNally, L., Bhagwagar, Z., and Hannestad, J. (2008). Inflammation, glutamate, and glia in depression: a literature review. *CNS Spectr.* **13**, 501–510.

Meyer-Lindenberg, A., and Weinberger, D.R. (2006). Intermediate phenotypes and genetic mechanisms of psychiatric disorders. *Nat. Rev. Neurosci.* **7**, 818–827.

Morrow, D.A., Cannon, C.P., Jesse, R.L., Newby, L.K., Ravkilde, J., Storrow, A.B., Wu, A.H.B., Christenson, R.H., Christenson, R.H., Apple, F.S., Cannon, C.P., Francis, G., *et al.* (2007). National academy of clinical biochemistry laboratory medicine practice guidelines: clinical characteristics and utilization of biochemical markers in acute coronary syndromes. *Clin. Chem.* **53**, 552–574.

Naatanen, R., Kujala, T., and Winkler, I. (2011). Auditory processing that leads to conscious perception: a unique window to central auditory processing opened by the mismatch negativity and related responses. *Psychophysiology* **48**, 4–22.

Nelson, J.C. (2010). Anxiety does not predict response to duloxetine in major depression: results of a pooled analysis of individual patient data from 11 placebo-controlled trials. *Depress. Anxiety* **27**, 12–18.

Nestler, E.J., and Hyman, S.E. (2010). Animal models of neuropsychiatric disorders. *Nat. Neurosci.* **13** (10), 1161–1169.

Ng, P.C., Murray, S.S., Levy, S., and Venter, J.C. (2009). An agenda for personalized medicine. *Nature* **461**, 724–726.

Norton, N., Williams, H.J., and Owen, M.J. (2006). An update on the genetics of schizophrenia. *Curr. Opin. Psychiatry* **19**, 158–164.

O'Donovan, M.C., Craddock, N., Norton, N., Williams, H., Peirce, T., Moskvina, V., Nikolov, I., Hamshere, M., Carroll, L., Georgieva, L., Dwyer, S., Holmans, P., *et al.* (2008). Identification of loci associated with schizophrenia by genome-wide association and follow-up. *Nat. Genet.* **40**, 1053–1055.

Oliver, P.L., and Davies, K.E. (2009). Interaction between environmental and genetic factors modulates schizophrenic endophenotypes in the Snap-25 mouse mutant blind-drunk. *Hum. Mol. Genet.* **18**, 4576–4589.

Orr, M.S., Goodsaid, F., Amur, S., Rudman, A., and Frueh, F.W. (2007). The experience with voluntary genomic data submissions at the FDA and a vision for the future of the voluntary data submission program. *Clin. Pharmacol. Ther.* **81**, 294–297.

Paisan-Ruiz, C., Jain, S., Evans, E.W., Gilks, W.P., Simon, J., van der Brug, M., Lopez de Munain, A., Aparicio, S., Gil, A.M., Khan, N., Johnson, J., Martinez, J.R., *et al.* (2004). Cloning of the gene containing mutations that cause PARK8-linked Parkinson's disease. *Neuron* **44**, 595–600.

Parker, G., Wilhelm, K., Mitchell, P., Roy, K., and Hadzi-Pavlovic, D. (1999). Subtyping depression: testing algorithms and identification of a tiered model. *J. Nerv. Ment. Dis.* **187**, 610–617.

Perry, P.J. (1996). Pharmacotherapy for major depression with melancholic features: relative efficacy of tricyclic versus selective serotonin reuptake inhibitor antidepressants. *J. Affect. Disord.* **39**, 1–6.

Price, A.L., Patterson, N.J., Plenge, R.M., Weinblatt, M.E., Shadick, N.A., and Reich, D. (2006). Principal components analysis corrects for stratification in genome-wide association studies. *Nat. Genet.* **38**, 904–909.

Report of the National Advisory Mental Health Councils Workgroup (2010). From discovery to cure: accelerating the development of new and personalized interventions for mental illnesses. http://www.nimh.nih.gov/about/advisory-boards-and-groups/namhc/reports/fromdiscoverytocure.pdf.

Richardson-Jones, J.W., Craige, C.P., Guiard, B.P., Stephen, A., Metzger, K.L., Kung, H.F., Gardier, A.M., Dranovsky, A., David, D.J., Beck, S.G., Hen, R., and Leonardo, E.D. (2010). 5-HT1A autoreceptor levels determine vulnerability to stress and response to antidepressants. *Neuron* **65**, 40–52.

Riley, B., Thiselton, D., Maher, B.S., Bigdeli, T., Wormley, B., McMichael, G.O., Fanous, A.H., Vladimirov, V., O'Neill, F.A., Walsh, D., and Kendler, K.S. (2010). Replication of association between schizophrenia and ZNF804A in the Irish Case-Control Study of Schizophrenia sample. *Mol. Psychiatry* **15**, 29–37.

Rogaev, E.I., Sherrington, R., Rogaeva, E.A., Levesque, G., Ikeda, M., Liang, Y., Chi, H., Lin, C., Holman, K., and Tsuda, T. (1995). Familial Alzheimer's disease in kindreds with missense mutations in a gene on chromosome 1 related to the Alzheimer's disease type 3 gene. *Nature* **376**, 775–778.

Roses, A.D. (2002). Genome-based pharmacogenetics and the pharmaceutical industry. *Nat. Rev. Drug Discov.* **1**, 541–549.

Ross, J.S., Slodkowska, E.A., Symmans, W.F., Pusztai, L., Ravdin, P.M., and Hortobagyi, G.N. (2009). The HER-2 receptor and breast cancer: ten years of targeted anti-HER-2 therapy and personalized medicine. *Oncologist* **14**, 320–368.

Rush, A.J., Fava, M., Wisniewski, S.R., Lavori, P.W., Trivedi, M.H., Sackeim, H.A., Thase, M.E., Nierenberg, A.A., Quitkin, F.M., Kashner, T.M., Kupfer, D.J., Rosenbaum, J.F., *et al.* (2004). Sequenced treatment alternatives to relieve depression (STAR*D): rationale and design. *Control. Clin. Trials* **25**, 119–142.

Russell, J.M., Koran, L.M., Rush, J., Hirschfeld, R.M., Harrison, W., Friedman, E.S., Davis, S., and Keller, M. (2001). Effect of concurrent anxiety on response to sertraline and imipramine in patients with chronic depression. *Depress. Anxiety* **13**, 18–27.

Samuels, B.A., Leonardo, E.D., Gadient, R., Williams, A., Zhou, J., David, D.J., Gardier, A.M., Wong, E.H.F., and Hen, R. (2011). Modelling treatment resistant depression. *Neuropharmacology* **61**, 408–413.

Schwarz, E., Guest, P.C., Rahmoune, H., Harris, L.W., Wang, L., Leweke, F.M., Rothermundt, M., Bogerts, B., Koethe, D., Kranaster, L., Ohrmann, P., Suslow, T., *et al.* (2011). Identification of a biological signature for schizophrenia in serum. *Mol. Psychiatry* 2011 Apr 12 (Epub ahead of print).

Sherrington, R., Rogaev, E.I., Liang, Y., Rogaeva, E.A., Levesque, G., Ikeda, M., Chi, H., Lin, C., Li, G., Holman, K., Tsuda, T., Mar, L., *et al.* (1995). Cloning of a gene bearing missense mutations in early-onset familial Alzheimer's disease. *Nature* **375**, 754–760.

Shi, L., Reid, L.H., Jones, W.D., Shippy, R., Warrington, J.A., Baker, S.C., Collins, P.J., de Longueville, F., Kawasaki, E.S., Lee, K.Y., Luo, Y., Sun, Y.A., *et al.* (2006). The MicroArray Quality Control (MAQC) project shows inter- and intraplatform reproducibility of gene expression measurements. *Nat. Biotechnol.* **24**, 1151–1161.

Tan, H.Y., Callicott, J.H., and Weinberger, D.R. (2008). Intermediate phenotypes in schizophrenia genetics redux: is it a no brainer? *Mol. Psychiatry* **13**, 233–238.

Thase, M.E. (2008). Maintenance therapy for bipolar disorder. *J. Clin. Psychiatry* **69**, e32.

Uher, R. (2009). The role of genetic variation in the causation of mental illness: an evolution-informed framework. *Mol. Psychiatry* **14**, 1072–1082.

Uhr, M., Tontsch, A., Namendorf, C., Ripke, S., Lucae, S., Ising, M., Dose, T., Ebinger, M., Rosenhagen, M., Kohli, M., Kloiber, S., Salyakina, D., *et al.* (2008). Polymorphisms in the drug transporter gene ABCB1 predict antidepressant treatment response in depression. *Neuron* **57**, 203–209.

Umbricht, D., and Krljes, S. (2005). Mismatch negativity in schizophrenia: a meta-analysis. *Schizophr. Res.* **76**, 1–23.

Umbricht, D., Schmid, L., Koller, R., Vollenweider, F.X., Hell, D., and Javitt, D.C. (2000). Ketamine-induced deficits in auditory and visual context-dependent processing in healthy volunteers: implications for models of cognitive deficits in schizophrenia. *Arch. Gen. Psychiatry* **57**, 1139–1147.

Wain, L.V., and Tobin, M.D. (2011). Copy number variation. *Methods Mol. Biol.* **713**, 167–183.

Wang, K., Zhang, H., Ma, D., Bucan, M., Glessner, J.T., Abrahams, B.S., Salyakina, D., Imielinski, M., Bradfield, J.P., Sleiman, P.M., Kim, C.E., Hou, C., *et al.* (2009). Common genetic variants on 5p14.1 associate with autism spectrum disorders. *Nature* **459**, 528–533.

Warden, D., Rush, A.J., Trivedi, M.H., Fava, M., and Wisniewski, S.R. (2007). The STAR*D Project results: a comprehensive review of findings. *Curr. Psychiatry Rep.* **9**, 449–459.

Williams, J.A., Andersson, T., Andersson, T.B., Blanchard, R., Behm, M.O., Cohen, N., Edeki, T., Franc, M., Hillgren, K.M., Johnson, K.J., Katz, D.A., Milton, M.N., *et al.* (2008). PhRMA white paper on ADME pharmacogenomics. *J. Clin. Pharmacol.* **48**, 849–889.

Wong, E.H.F., Nikam, S.S., and Shahid, M. (2008). Multi- and single-target agents for major psychiatric diseases: therapeutic opportunities and challenges. *Curr. Opin. Investig. Drugs* **9**, 28–36.

Wong, E.H.F., Tarazi, F.I., and Shahid, M. (2010a). The effectiveness of multi-target agents in schizophrenia and mood disorders: relevance of receptor signature to clinical action. *Pharmacol. Ther.* **126**, 173–185.

Wong, E.H.F., Yocca, F., Smith, M.A., and Lee, C.M. (2010b). Challenges and opportunities for drug discovery in psychiatric disorders: the drug hunters' perspective. *Int. J. Neuropsychopharmacol.* **13**, 1269–1284.

Woodcock, J. (2009). Chutes and Ladders on the Critical Path: Comparative Effectiveness, Product Value, and the Use of Biomarkers in Drug Development. *Clin. Pharm. & Ther.* **86**, 12–14.

Zimprich, A., Biskup, S., Leitner, P., Lichtner, P., Farrer, M., Lincoln, S., Kachergus, J., Hulihan, M., Uitti, R.J., Calne, D.B., Stoessl, A.J., Pfeiffer, R.F., *et al.* (2004). Mutations in LRRK2 cause autosomal-dominant parkinsonism with pleomorphic pathology. *Neuron* **44**, 601–607.

CLINICAL UTILITY OF SERUM BIOMARKERS FOR MAJOR PSYCHIATRIC DISORDERS

Nico J.M. van Beveren and Witte J.G. Hoogendijk

Department of Psychiatry, Erasmus University Medical Center, Rotterdam, The Netherlands

Abstract

There is a major unmet clinical need for molecular blood-based biomarkers in studies of major psychiatric disorders. Thus far, identification of such biomarkers has been sparse, most likely due to the fact that this is reliant on long-standing diagnostic concepts used in psychiatry, which are notoriously heterogeneous. Also, identification of biomarkers for a syndrome that has already been categorized based on clinical phenomenology is not useful in the clinic. This chapter describes the need for innovative approaches for identification of biomarkers which can been used to classify at-risk patients such as youngsters with prodromal symptoms

DOI: 10.1016/B978-0-12-387718-5.00014-6

0074-7742/11 $35.00

for psychosis and existing patients who are likely to progress to more severe states. The authors argued for the use of broader categories of related patients and to deconstruct the traditional diagnoses in favor of molecular biomarker profiles.

I. Introduction and Scope of the Subject

A. What is a Biomarker?

The term biomarker refers to a characteristic that can be objectively measured and evaluated as an indicator of normal biological processes, pathogenic processes, a response to a therapeutic intervention, or which reflects a specific natural history of a (medical) condition (Biomarkers Definitions Working Group *et al.*, 2001). The Food and Drug Administration defines *biomarkers* as "measurable characteristics that reflect physiological, pharmacological, or disease processes in animals or humans". So, in medical practice, a valid biomarker reliably identifies a (sub)syndrome, a treatment response, or a clinical course (the latter includes the concept of a prodromal state preceding a full-blown clinical syndrome). For practical purposes, a biomarker should be easily measurable with a high degree of reproducibility, against acceptable costs, and within an acceptable time frame. Ideally, a biomarker is somehow related to the etiology of the disorder.

Obviously, the biomarker concept in its strictest sense is not new at all, as measuring biological parameters such as blood pressure or body temperature is part of everyday medical practice. Therefore, the current use of the term biomarker is usually reserved for a more or less complex parameter, being either composite (i.e., a profile consisting of several variables) or highly specific (i.e., a protein specifically produced by a tumor).

B. Focus on Serum Markers for Major Psychiatric Disorders

In psychiatry, the term biomarker has been used for virtually every nonsubjective measurement, ranging from imaging data [functional or structural magnetic resonance imaging (MRI)], neuropsychological tests (i.e., the test for sustained attention), electrophysiological responses (i.e., the P300 event related potential wave), or even skin flush after chemical provocation (Horrobin, 1980; Smesny *et al.*, 2007). In this chapter, we focus on the use of molecules identified in peripheral blood (i.e., proteins or gene-expression data). Further, we focus on the major psychiatric disorders: schizophrenia, bipolar disorder, and major depressive disorder.

C. Clinical Examples

1. *An Adolescent with an Adjustment Disorder*

A 17-year-old high school student was referred to a psychologist because of decreased performance at school. The student was frequently absent during lessons, tests scores were lower than before, and he seemed to have problems with attention. The psychologist saw an anxious and depressed young man. A diagnosis of adjustment disorder with alteration in mood was made with the possible presence of a major depressive episode. Uncertainty about future studies and conflicts with his father and younger brother were identified as predisposing factors. Treatment with an antidepressant and with cognitive behavior therapy was started. Initially, some improvement was observed; the patient attended lessons more regularly, although he performed under his previous level.

Six months later, after the summer holidays and upon entering the final year of his high school study, the patient stayed at home and stated that he did not want to go to school because "it is useless." Also, he did not want to go the psychologist. His parents were confused about what to do. On one occasion, his father forced the youth into the car to take him to school, but the youth escaped from the car and returned home several hours later. After this incident, he stayed in his room. The family doctor wanted an admission in a psychiatric hospital, but the patient refused this adamantly.

Finally, some weeks later, the patient stopped eating because "the food is poisoned with radio-active material." A forced admission in a psychiatric hospital followed. Over a couple of days the patient related several delusional beliefs concerning the scenario that his school was a center of nuclear scientists involved in "covering up worldwide nuclear accidents." He was afraid that they wanted to kill him because he had discovered this fact. A diagnosis of schizophrenia was made.

After unsuccessful treatment with the antipsychotic drugs risperidone and aripiprazole, this patient finally improved on olanzapine, although some limitations in attention and working memory performance remained. In collaboration with the school, a reentry program was made. The goal was to resume classes after the next summer, on a slightly lower level.

The patient, however, showed a considerable weight gain of 25 kg over a period of several months, resulting in a body mass index (BMI) increase from 23 to 30 kg/m^2. This is a well-known side effect of olanzapine. After careful discussions and consideration between the patient, parents, and psychiatrist, the treatment with olanzapine was continued because of the improvement of symptoms.

2. *A Young Female with Schizophrenia*

A 26-year-old female bank manager was referred to a psychiatrist because she developed "strange ideas and convictions." This patient thought she was a "female recreation of 'the prophet'." The patient stated that the "future of mankind is threatened" and that "everybody should listen to her warnings and improve their lives." She also said that she received messages from the president of the United States through the television and heard his and other voices giving her advice about the future of the world. She was diagnosed with schizophrenia and treated with the antipsychotic quetiapine. This resulted in disappearance of the delusions and hallucinations, but the patient became slower and psychomotorically retarded, and talked less, apart from that "there is no future anymore." A second diagnosis was then made: "negative symptoms of schizophrenia, possibly depression." Consequently, an antidepressant was added to the medication. After 1 week, the patient became agitated and restarted her statements about being a prophet. She also showed sexual disinhibition and picked up men in bars. After admission in a psychiatric hospital, she wore excessive makeup and could not stop talking. The diagnosis was then changed to a manic episode in the course of either a bipolar or a schizoaffective disorder. The antidepressant was stopped and a mood stabilizer was given. After 3 weeks, there was almost complete recovery.

3. *Another Young Female with Schizophrenia*

A 19-year-old college student with no psychiatric history and in general good health developed auditory and visual hallucinations over the course of 2 weeks. She was referred to a general mental health-care facility by her general practitioner. A provisional diagnosis of "psychotic disorder not-otherwise-specified (NOS)," possibly with concomitant "personality disorder," was made and treatment with olanzapine was started. After 1 week, hallucinations were still present and delusions developed; the patient thought that her parents were poisoning her food. A couple of days later, the patient developed acute anxiety and severe agitation, prompting an acute forced admission into the psychiatric hospital.

At admission, several clinical diagnoses were considered: "psychotic disorder NOS," schizophreniform disorder/early phase schizophrenia, major depressive disorder, and psychotic/mood phenomena accompanying a personality disorder. Routine neurological and laboratory examination was performed, but no aberrations were apparent. Over the next weeks, a gradual amelioration occurred, although the patient showed impaired concentration and (according to her parents) "altered personality" and "odd mood swings." A psychological test after partial remission showed deficits in working memory and sustained attention. A final diagnosis of "paranoid schizophrenia" was made. Seven months after the first symptoms, the patient again developed anxiety, and odd movement disorders described as "catatonia." She was readmitted into the psychiatric hospital with a "second psychotic episode." After a few days, she displayed "disorientation" and "disorganized speech." An electroencephalogram testing showed signs of "encephalopathy" but no clear epileptic activity. Brain MRI results appeared normal. The diagnosis was reformulated in "psychosis/schizophrenia" with status epilepticus, possibly a "somatic syndrome."

After extensive laboratory testing, the presence of antibodies against the N-Methyl-D-aspartate (NMDA) receptor was found in both serum and cerebrospinal fluid of the patient. Based on the presence of these antibodies, a final diagnosis of anti-NMDA receptor encephalitis was made. A teratoma of the ovary was identified and surgically removed. After therapy with prednisone and plasmaferese, the patient slowly improved, although speech impediments, loss of concentration, and altered personality persisted for more than 8 months.

One year after the final diagnosis was made all symptoms had disappeared, and the patient was able to restart her college education.

D. The Potential Use of Biomarkers in These Cases

The first case is a classical example of the development of schizophrenia and will be familiar to clinicians. With respect to biomarker research, two prominent features of the case merit attention. The first feature is that the patient initially showed atypical mood and motivational alterations, which can be seen in a large number of adolescents. It is not unusual that in the prodromal phases of the schizophrenia syndrome these symptoms are initially attributed to adolescence-related existential problems. Recently, there has been increasing interest in sorting out young patients truly at risk for serious mental disorders (like schizophrenia) from milder disorders, which usually have a self-limiting character. These former states are referred to as "at-risk mental states" in the clinical literature (Yung *et al.*, 2008; Woods *et al.*, 2009). At present, "at-risk mental states" can only be identified using key symptoms which have limited reliability, combined with familial (genetic) risk factors (Cannon *et al.*, 2008). However, a biomarker indicating a likely illness trajectory (i.e., "development toward schizophrenia," "development toward major depressive disorder," or "limited chance of development toward any major psychiatric disorder") would clearly be useful. The second feature of this first case is that after a diagnosis of schizophrenia was established, serious metabolic side effects occurred, probably in relationship with the use of an atypical antipsychotic. Therefore, a biomarker predicting treatment response, or the future occurrence of drug side effects, would also be clinically useful.

The second case will also have been familiar to clinicians: a patient with symptoms suggestive for one disorder (in this case, schizophrenia), which "switches sides" and finally shows the symptoms of and improves with treatment for another condition, namely, bipolar disorder. The general issue at hand here is that current psychiatric diagnoses are "fuzzy concepts," showing considerable clinical, and probably also biological, overlap. This issue is extensively discussed in the section below.

The third and last case was the description of a recently identified, probably rare, syndrome: anti-NMDA receptor encephalitis (Dalmau and Rosenfeld, 2008;

Dalmau *et al.*, 2007). Although considered to be a "neurological" syndrome, the clinical presentation of this condition often starts with "psychiatric" symptoms, which may include panic attacks, anxiety, hallucinations, and delusions, and often an initial psychiatric diagnosis is made as in the case described. The anti-NMDA receptor syndrome is a key example of a syndrome which can be diagnosed by the presence of a specific biomarker, namely, anti-NMDA receptor antibodies. From a biomarker perspective, it is straightforward syndrome, with the biomarker being a single, specific compound, which is probably also the causative agent for the syndrome itself in the same way as antiglutamatergic compounds (such as the drug of abuse phencyclidine or the anesthetic ketamine) are known to induce psychosis-like phenomena.

II. About the Concept of Diagnosis in (Current) Psychiatry

To fully grasp the possibilities as well as the issues related to the possible use of biomarkers in psychiatric practice, it is important to have an understanding of the concept of a diagnosis in medicine and, in particular, psychiatry. It will also be important to have an understanding of recent developments related to diagnostic concepts in psychiatry.

Any form of rational medicine depends on the existence of a valid method to group similar patients under a diagnosis. At present, the major psychiatric diagnostic system is the widely used Diagnostic and Statistical Manuel (DSM) system (Sadler, 2005), with the 1994 DSM-IV as the most recent edition. The revised DSM-V is expected in 2012.

Although the influential DSM-III was introduced in 1980, its fundamental structure tracks back to the late nineteenth and early twentieth centuries, when Kraepelin (1971) made his influential distinction between dementia praecox and mania. Dementia praecox developed into the schizophrenia concept, whereas the manias formed the basis for manic-depression and depressive disorder. Finally, the distinction was incorporated into the DSM-III system, with the group of psychotic disorders (incorporating schizophrenia), on the one hand, and broad group of mood disorders (incorporating bipolar disorder and depression), on the other.

Apart from mood and psychotic disorders, three other major diagnostic categories can be found in the DSM system: autism spectrum disorders (comprising patients with self-oriented behavior and abnormal social contacts), anxiety disorders, and the broad group of personality disorders. It is paramount for any researcher to be aware of the fact that the DSM-III and DSM-IV systems do not take into account any (hypothesized) cause or process which underlies the diagnostic categories. The DSM system defines "diagnostic categories" by defining cutoff criteria based on the presence or absence of symptoms, without reference to supposed psychological or biological processes associated with the diagnostic categories. As such, the DSM categories do not resemble mature medical diagnostic concepts reflecting an underlying biological alteration. This does not imply that DSM concepts are completely arbitrary. They have been chosen because of their clinical validity, established over one century of psychiatric thinking and practice (see chapter "Challenges of introducing new biomarker products into the market" by Bahn *et al.*).

Since their introduction, the DSM-III and DSM-IV systems have been praised for introducing a rigorous, methodologically sound approach to psychiatric diagnostics, which improved greatly on the quality and quantity of psychiatric research. Indeed, the burgeoning field of psychiatric research as it is today can hardly be imagined without the DSM system. However, over the years, the DSM approach has also met with severe criticism (Van Praag, 1997, 2000, 2001; Sadler, 2005; Galatzer-Levy and Galatzer-Levy, 2007). It has been argued that research into the biological determinants of abnormal behavior exacts particular standards upon psychiatric diagnosing and that the DSM system falls short in several respects. Clearly, diagnosis is the principal rate-limiting step in biological psychiatric research, and when researchers use invalid diagnostic categories, it is obvious that the results of research will be absent, clouded by inclusion noise, spurious, or, in the worst case, misleading.

The main disadvantage of the DSM diagnostic categories is that they do not reflect a true medical diagnosis but are to a certain extent arbitrary categories (Van Praag, 1997). Although with training, especially in the use of standardized interviews, DSM produces acceptable interrater validity, this does not necessarily imply validity of the identified constructs. It has been argued repeatedly that the validity of DSM constructs is limited with respect to biologically reproducible underpinnings, delimitation from other disorders, and follow-up. Specifically, DSM categories are notoriously heterogeneous, as the system allows for several combinations of symptoms to be arranged into one category. For example, the schizophrenia concept can be generated out of 23 different combinations of symptoms and phenomena.

A specific development of recent years has also contributed to the heterogeneity found in DSM-IV categories. Some researchers argue that there exists a continuum between core psychiatric symptoms and syndromes, and normal functioning (i.e., see Van Os *et al.*, 2000). This has lead to a situation in which the border between mental distress and mental illness is vaguely marked. Van Praag (1997) likened this situation to searching for the pathogenesis of tuberculosis but not making a diagnostic distinction with a common cold. He found this issue specifically prominent in depression research, where the distinction between sorrow and depression is inadequate.

Finally, another issue is that DSM categories show considerable overlap between one another. For example, recent findings show evidence for overlap between schizophrenia and bipolar disorder on the clinical, epidemiological, and genetic levels (Craddock *et al.*, 2006; Williams *et al.*, 2011). As stated above, the clear distinction between schizophrenia, on the one hand, and bipolar disorder, on the other hand, stems from Kraepelin's observations on the phenomenology and temporal course of symptoms. It is of note, however, that Kraepelin himself pointed in later work to the similarities in the course of both disorders. There is now also some evidence that autism spectrum disorders and schizophrenia show similar connectivity deficits genetic variations (King and Lord, 2011). Indeed, some patients with (mild) developmental symptoms (i.e., satisfying DSM criteria for "pervasive developmental disorder, NOS") develop schizophrenia at adolescence.

In conclusion of this section, biological psychiatry and the search for relevant biomarkers aim at elucidation of the biological underpinnings of discrete disorders. However, one may question their construct validity, as many of the disorders presently distinguished appear to represent a variety of more or less comparable, but in many ways, dissimilar, conditions. As a result, it is hard to believe that the search for particular brain dysfunctions underlying such heterogeneous diagnostic constructs stands much chance of success.

When we realize that in clinical practice, the majority of psychiatric patients do not meet the criteria of one particular disorder as currently defined, but instead shows signs and symptoms of a multitude of disorders or display a patchwork of parts of different disorders, what does this mean for biomarker research? Obviously, this suggests that we should not (fully) rely on traditional psychiatric diagnostics, because there is some fundamental uncertainty with respect to their validity.

III. What Should/Might a Useful Biomarker Predict/Distinguish in Psychiatry

A. A Diagnostic Category

A straightforward approach in biomarker discovery would be to investigate an association between a possible marker and a given diagnostic category. As outlined above, the main diagnostic system at present is the DSM system, and thus this approach would involve identifying a marker which separates a DSM diagnostic category from either a control group or another relevant diagnostic group. Indeed, this is the approach that the majority of all present biological research uses. However, from a clinical utility perspective there are two main and severe limitations to this approach.

The first issue is that, as outlined in the section above, the DSM categories are to a certain extent arbitrary categories, which probably do not fully reflect a specific underlying biological reality. As a result, it is not likely that any biomarker will reliably define a single DSM diagnostic category. This aspect is underscored by the fact that most psychiatric research efforts result in identification of associations showing considerable overlap with the control group used. In other words, although an association may be significant, the effect size is usually small which leads to limits in the clinical utility. A recent example of this phenomenon was seen in a study by Stefansson *et al.* (2009), which was published in *Nature.* This study presents the results of a whole-genome association study of a very large group of patients ($N = 12{,}945$) with DSM-IV-defined schizophrenia compared to a control population ($N = 34{,}591$). Associations in the patient population with seven gene variants were found [mainly related to the human leukocyte antigen (HLA) system], with high levels of significance. However, the odds ratios of these associations were small, ranging from 1.15 to 1.24. From a scientific point of view, this is an important study, but it has little or no use for clinical diagnostics.

The second limitation related to the identification of biomarkers in DSM-defined diagnostic categories is of a practical nature. If a biomarker was shown to be specifically related to a DSM diagnostic category, then this would be a major scientific breakthrough. However, such a finding would be trivial from a clinical diagnostic point of view, because then the diagnostic category would already be defined by the DSM criteria itself. This does not imply that DSM categories are completely useless for identifying biomarkers. For example, they may serve as a starting point for identifying subgroups within the DSM categories, as outlined in the next subsection.

B. A Subgroup Within a Diagnostic Category

A possible means of identifying subgroups within DSM categories is what might be called "the narrow-to-broad" approach. This method seeks to overcome the heterogeneity present in DSM categories by going back to the roots of the system and assumes that there is only evidence for the existence of a limited set of fundamental psychiatric syndromes which can be identified by disabling core symptoms. Such fundamental syndromes might be schizophrenia with severe negative symptoms and "first-rank symptoms" as defined by Schneider (Koehler *et al.*, 1977), bipolar mania, severe depression with autonomic dysregulation (formerly called endogeneous depression), and the classical autism syndrome as described by Kanner and Eisenberg (1957). These syndromes are relatively rare but can be recognized easily when encountered because of their severity and "classical" presentation of symptoms. As a first step, one would then need to identify biomarkers reliably associated with each of these few syndromes. The following step would be to take a large group of patients presenting with less specific psychiatric symptoms, irrespective of their severity of DSM classification, measure in these patients the biomarkers associated with the fundamental syndromes, and classify patients according to their resemblance, if any, with either of the fundamental biomarkers. The next step would be to investigate whether patients in whom a biomarker profile sufficiently resembles one of the fundamental profiles have a prognosis or treatment response fitting one of the fundamental syndromes.

This is, in part, the approach our group has taken when identifying a biomarker profile for schizophrenia and by showing that a similar profile was present in subjects who developed schizophrenia later on (Schwarz *et al.*, 2010, 2011).

C. A Specific Treatment Response

The previous two subsections dealt with the subject of diagnosis of psychiatric illnesses. Biomarkers may also be used for prediction of treatment response or the risk associated with psychiatric drug treatment or, conversely, to indicate a low likelihood of spontaneous improvement (Laughren, 2010). The main goal of biomarker application in predicting efficacy and risk is to subgroup the population into responders and nonresponders and into subgroups containing those who are at risk or not at risk for some adverse event of interest. Examples of possible biomarkers include imaging measures, serum assays, genetic assays (genomic markers), physiological measures, histopathological findings, psychological tests, and demographic variables (age, gender, race).

There are two principal ways a biomarker could be used to subdivide the population, such as on the basis of differences in exposure or differences in pharmacodynamic response. In either case, the differences could divide patients on the basis of either efficacy or risk (Laughren, 2010). For example, if biomarker-positive patients differ from biomarker-negative patients by having higher exposures to a drug, this difference could translate into a difference in efficacy (e.g., better efficacy in biomarker-positive patients) or a difference in risk (e.g., a greater risk in biomarker-positive patients). Similarly, a pharmacodynamic difference between biomarker-positive and -negative patients, unrelated to exposure, could be reflected by differences in efficacy or risk.

There are already some examples of genomic biomarkers that predict exposure such as pharmacokinetic differences based on the different activities in metabolizing enzymes. These enzymes include CYP2C9, CYP2B6, CYP2C19, and CYP2D6 (Laughren, 2010). Atomoxetine, a selective norepinephrine reuptake inhibitor approved for the treatment of attention deficit hyperactivity disorder (ADHD), is cleared predominantly by CYP2D6, and subjects who are poor at metabolizing 2D6 have 10-fold higher plasma levels of atomoxetine compared to those who are good 2D6 metabolizers (Sauer *et al.*, 2005). It is also known that 2D6 poor metabolizers have approximately eightfold increases in plasma levels of desipramine after exposure to this drug, compared to 2D6 high metabolizers (Brosen *et al.*, 1986).

There are fewer examples of biomarkers that predict differences in pharmacodynamic responses and most of these are in the oncology area in which the molecular nature of the disease is often well understood. There are several drugs used in oncology for which there are biomarkers that predict better efficacy for those subjects who are biomarker positive. For example, the *HER2* gene expresses a cell surface receptor that is needed for growth of breast cancer cells, and this gene is overexpressed in about 20% of subjects with breast cancers (Burstein, 2005). Trastuzumab (Herceptin) is an antibody that blocks this cell surface receptor. Data from clinical trials and other studies suggest that it is primarily this HER2 receptor subgroup that benefits from Herceptin treatment (Desmedt *et al.*, 2009).

In the case of psychiatric drugs, there are some early findings suggesting that biomarkers may help in predicting responsiveness to drugs. One such example is for selective serotonin reuptake inhibitors (SSRIs) and serotonin genes. Several studies suggest that an allele of the polymorphic serotonin transporter gene (5-*HTTLPR*) is associated with an SSRI response in Caucasians (Serretti *et al.*, 2007). Data from the STAR*D trial suggest that a polymorphism in the *HTR2A* receptor gene is associated with a positive response to the SSRI citalopram (McMahon *et al.*, 2006). Although these findings are not as robust as those for several oncology drugs, they nevertheless give some encouragement that searching for biomarkers of psychiatric drug response may be fruitful.

D. Novel Diagnostic Categories Overlapping the Boundaries of Traditional Diagnostic Categories

To circumvent the DSM-related issues, a more radical approach would be to abandon the DSM categories altogether and start instead with the broad array of symptoms and problems associated with psychiatric disorders. This approach assumes that current diagnostic categories, whether these are based on DSM criteria, historically defined archetypical syndromes, or dimensionally defined psychological functions, have limited value with respect to their underlying biological validity and should therefore not be used to steer biomarker research. Instead, the starting point should be the clinical reality that patients come into broad "problem basins" such as mental and behavioral disorders in young children, adolescents, or adults. Such basins would hold syndromes that share some symptoms but differ in others and that are heterogeneous as far as course, outcome, and, presumably, the etiology and pathogenesis are concerned.

Following this approach, a first step would be to collect a large group of patients from one of the problem basins, such as those from the adolescent-onset mental and behavioral disorder group. This group would probably comprise current DSM diagnoses including psychotic disorders such as schizophrenia, bipolar disorder, depressive disorder, conduct disorder, and various developing personality disorders. The next step would be to measure a multiplex biomarker profile (i.e., a gene-expression profile, serum proteome profile, etc.), assuming that alterations in serum molecules will be present in these patients as compared to healthy youngsters, even if the exact nature of the alterations is unknown. Then, clusters of patients can be identified who have similar biomarker profiles, if these are present. Patients belonging to a single cluster will, by definition, have the same biological profile (at least with respect to the biomarkers and tissue investigated). Such a cluster may represent patients exhibiting the specific signs and symptoms of a traditional diagnostic category. It will, however, be more likely that these clusters will consist of mixture of patients exhibiting the signs and symptoms of a variety of DSM-defined disorders. Finally, it should be investigated whether the patients of a single cluster show a common, meaningful clinical characteristic such as a similar prognosis, developmental trajectory, response to treatment, or alteration in biological or psychological function.

An initial form of this approach has been advocated by Van Praag (2000). Table I shows the problem basins as suggested by Van Praag.

Obviously, a number of assumptions underlie this "problem-basin" approach. The most prominent of these is the assumption that peripheral molecular alterations are present and that such alterations reflect underlying biological alterations common to a subgroup of patients. This second approach works in the opposite way as the narrow-to-broad approach described above and might therefore be called "the broad-to-narrow" approach. In essence, this approach seeks to

Table I

Problem Basins According to Van Praag (2000).

Syndromes characterized by disturbed reality testing and clear consciousness (a basin containing, among others, the group of schizophrenia psychoses)
Syndromes characterized by disturbed reality testing and lowered consciousness (including the group of so-called organic psychoses)
Syndromes characterized by disturbances in affect regulation among which the emphasis might be on mood, anxiety, or aggression dysregulation
Syndromes characterized by disturbed cognition, among which information storage and/or information retrieval and/or information appraisal may be the main seat of impairment
Conditions in which social adaptation and affiliative abilities are disturbed; a basin containing the various personality disorders
Conditions with disturbed impulse regulation, comprising, among others, the eating disorders and certain disorders in aggression regulation
Syndromes characterized by "termination pathology"—that is, inability to terminate behaviors at an appropriate point in time—a group holding, among others, obsessive-compulsive and the addiction disorders
Somatic syndromes without manifest somatic pathology—a basin that would house, among others, the somatiform and sexual disorders

deconstruct problem basins into clinically meaningful subclusters based on molecular biomarker profiles. Below, we show how such an approach has been used in oncology to deconstruct acute leukemia (Section IV).

E. Specific Developmental Trajectories (i.e., Clinical Staging Based on Biomarker Profiles)

Recent years have seen an increase in interest in the notion that psychiatric disorders do not come in stable "cross-sectional" entities but that they are dynamic in nature and not unlike the stages that can be identified in the development of malignancies. Conceptualizing schizophrenia as a neurodevelopmental disorder implies the notion of trajectory of illness. Figure 1 (Insel, 2010) shows this trajectory and the supposed underlying biological alterations.

Based on the neurodevelopmental model of schizophrenia, four stages of schizophrenia can be hypothesized from risk to prodrome, psychosis, and chronic disability (McGorry *et al.*, 2008). At present, diagnosis is based on the symptoms and signs of psychosis. The advent of biomarkers and new cognitive tools as well as the identification of subtle clinical features may enable the detection of earlier stages of risk and prodrome (Nestler and Hyman, 2010) and the identification of specific illness trajectories.

The earliest stage is risk, before detectable deficits occur. McGorry and his colleagues (Henry *et al.*, 2010) have established that the prodrome of schizophrenia is a valid second stage of the illness before psychosis. The prodrome is identified based on changes in thoughts (e.g., bizarre ideas falling short of psychotic ideation), social isolation, and impaired functioning (e.g., reduced performance at school). Some of these features seem endemic to adolescence and the problem of distinguishing a high risk for psychosis from more common adolescent angst remains a challenge. Although a structured interview was developed to aid in the identification of high risk for psychosis (Woods *et al.*, 2009), the addition of biomarkers may enhance detection and increase the predictive power. Given the high rate of behavioral distress in adolescence and the likelihood that many with prodromal symptoms will either grow out of these or develop other disorders, the challenge is to increase sensitivity for detecting ultrahigh risk while not sacrificing specificity. Specificity is a challenge as many of those who seek help for prodromal symptoms will actually develop other forms of psychopathology and not schizophrenia.

IV. Intermezzo: Lessons from Oncology

In the oncology field, a key study was published by Valk *et al.* (2004), entitled "Prognostically useful gene-expression profiles in acute myeloid leukemia." The issues present in oncology, as well as the methodology used, might be relevant for

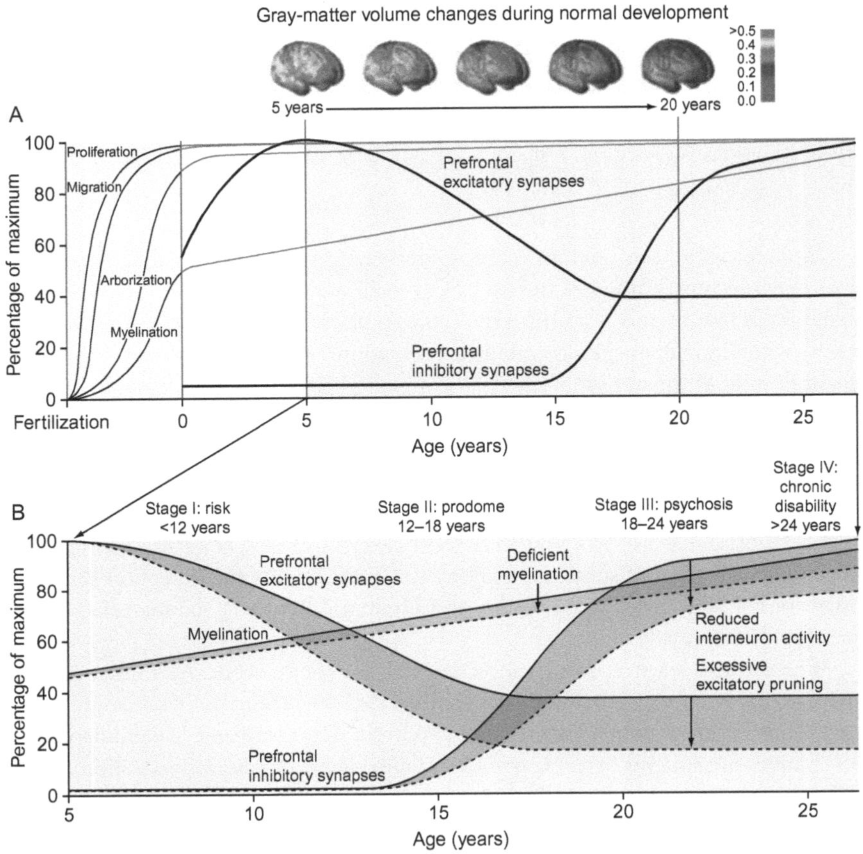

FIG. 1. Neurodevelopmental model of schizophrenia (Insel, 2010; reprinted with permission). (A) Normal cortical development involves proliferation, migration, arborization (circuit formation), and myelination. The first two processes occur mostly during prenatal life, and the latter two continue through the first two decades of life. The combined effects of pruning of the neuronal arbor and myelin deposition are thought to account for the progressive reduction of gray matter observed with longitudinal neuroimaging. (B) The trajectory in children developing schizophrenia could include reduced elaboration of inhibitory pathways and excessive pruning of excitatory pathways leading to an altered excitatory–inhibitory balance in the prefrontal cortex. Reduced myelination would alter connectivity. Although some data support each of these possible neurodevelopmental mechanisms for schizophrenia, none has been proven to cause the syndrome. Detection of prodromal neurodevelopmental changes could permit early intervention with potential prevention or preemption of psychosis. (For interpretation of the references to color in this figure legend, the reader is referred to the Web version of this chapter.)

overcoming some of those encountered in contemporary psychiatry as outlined above with regard to diagnostics and the search for biomarkers.

Like many psychiatric disorders, acute myeloid leukemia (AML) refers to a heterogeneous disorder, as specified in the opening sentence of their paper:

"Acute myeloid leukemia is not a single disease but a group of neoplasms with diverse genetic abnormalities and variable responses to treatment." The authors addressed an important issue in the field, namely, that the methods used at the time for (sub)classifying AML had limited value for prognosis and guiding treatment selection. It was known that the outcome of AML could in part be predicted by assessing the cytological characteristics of the tumor cells. However, these traditionally used characteristics correlated only partially with outcome. Although it was recognized that these different outcome trajectories were probably dependent on differential underlying molecular characteristics, the mechanisms were poorly recognized. Thus, this situation shows some relationship with psychiatric disorders if we substitute "major psychiatric syndrome" for AML, and "phenotypical criteria as used in DSM-IV" for cytological characteristics.

The Valk *et al.* (2004) paper addressed this issue by taking the gene-expression patterns of the tumor cells as a starting point, without an *a priori* hypothesis about the specific alterations. Using a group of 285 patients, the gene-expression patterns were clustered, based on their overall similarity. This allowed the identification of subclusters of patients with similar gene-expression patterns. A total of 16 subgroups were thus identified. These subgroups redefined the previous subclassifications and showed better predictive value with respect to prognosis.

This approach resembles the "broad-to-narrow" method that we described earlier and exemplifies the utility of unsupervised "bottom-up" clustering methods in identifying novel subcategories in heterogeneous syndromes, based on molecular profiles. Although there are clear limitations to this approach when using serum proteins or gene-expression patterns for psychiatric disorders, some conceptual elements may prove useful in psychiatry.

V. Overview of Current Results on Serum Biomarkers

A. Single Molecular Measurements

Over the past decade, converging results from *postmortem* research, neuroimaging, genetic association studies, and measurements of peripheral blood status have pointed to the presence of several "biological themes" within the broad context of the schizophrenia syndrome and, to a lesser extent, within the bipolar disorder syndrome. These themes can be summarized as "altered glucose metabolism" (Ryan *et al.*, 2003; Spelman *et al.*, 2007; Venkatasubramanian *et al.*, 2007; Van Nimwegen *et al.*, 2008; Fernandez-Egea *et al.*, 2008, 2009; Guest *et al.*, 2010), "immunological alterations" (for a review, see Drexhage *et al.*, 2010), "altered presence of growth factors" (for a review, see Van Beveren *et al.*, 2006), "alterations in myelination" (Davis *et al.*, 2003; Karoutzou *et al.*, 2008; Takahashi *et al.*, 2011), and "alterations in (cortical) cytoarchitecture" (Lewis and Mirnics, 2006; Garey, 2010; for an integrating overview, see Insel, 2010). In schizophrenia, there is also abundant evidence for alterations in dopaminergic and glutamatergic signaling (Coyle, 2006). In the case of depression, these alterations are less clear, but there is long-standing evidence for aberrant hypothalamic–pituitary–adrenal (HPA)-axis signaling (Holsboer and Barden, 1996; Nemeroff and Vale, 2005).

Increased serum levels of the neurotrophic proteins S100B and decreased levels of brain-derived neurotrophic factor (BDNF) have been identified repeatedly in schizophrenia (for a review, see Van Beveren *et al.*, 2006). Interestingly, these alterations seem to be partly dependent on the clinical phase of the disorder, with BDNF levels being decreased during the acute psychotic phase, and being restored to normal levels after remission. Increased levels of S100B may be related to the presence of negative symptoms (Van Beveren *et al.*, 2006).

Altered levels of cytokines have been studied extensively in schizophrenia. Generally, there are robust findings of increased levels of interleukin (IL)-1 (Theodoropoulou *et al.*, 2001; Maes *et al.*, 2000; Kim *et al.*, 2009; Naudin *et al.*, 1997), IL-6 (Frommberger *et al.*, 1997; Lin *et al.*, 1998; Kim *et al.*, 2009; Monteleone *et al.*, 1997), and tumor necrosis factor (TNF) (Kim *et al.*, 2009; Monteleone *et al.*, 1997). These findings, and their relationship with schizophrenia etiology, were described in a recent review by Drexhage *et al.* (2010).

An early report on the incidence of impaired energy metabolism in schizophrenia was published approximately 90 years ago (Kooy, 1919). In an extensive analysis of different psychiatric disorders, Kooy observed hyperglycemia in schizophrenia patients. Because many studies using schizophrenia patient samples have been obtained from treated subjects, it was hypothesized first that antipsychotic drugs exert an effect on glucose metabolism and the insulin response (Holt *et al.*, 2004; Liebzeit *et al.*, 2001). Indeed, antipsychotic drugs such as clozapine and olanzapine can lead to body-weight gain, type 2 diabetes mellitus, and hyperlipidemia (Meyer, 2002). However, it was shown recently that schizophrenia patients can show signs of insulin resistance, such as increased levels of insulin-related peptides, independent of antipsychotic treatment (Steiner *et al.*, 2010; Guest *et al.*, 2010; see chapter "Abnormalities in metabolism and hypothalamic–pituitary–adrenal axis function in schizophrenia" by Guest *et al.*). Indeed, over the past 10-year period, numerous studies have demonstrated the occurrence of hyperglycemia, impaired glucose tolerance, hyperinsulinemia, and/or insulin resistance in first-onset, antipsychotic naive schizophrenia patients (Ryan *et al.*, 2003; Spelman *et al.*, 2007; Venkatasubramanian *et al.*, 2007; Van Nimwegen *et al.*, 2008; Fernandez-Egea *et al.*, 2008, 2009; Guest *et al.*, 2010). An extensive overview on glucose metabolism abnormalities can be found in Martins-de-Souza *et al.* (2010). In bipolar disorder, BDNF has also been shown to be decreased in acute mania and depressive episodes (Cunha *et al.*, 2006; de Oliveira *et al.*, 2009). As seen in schizophrenia, BDNF levels appear to return to normal in successfully treated individuals (Tramontina *et al.*, 2009).

In major depression, there are long-standing observations of alterations in HPA axis function (Holsboer and Barden, 1996; Nemeroff and Vale, 2005). Evidence for hyperactivity of the HPA-axis was shown by higher daytime cortisol levels, nonsuppression after dexamethasone ingestion and higher corticotropin-releasing hormone, and adrenocorticotropin hormone levels among persons with depression (Carroll *et al.*, 1976; Gold *et al.*, 1995; Pfohl *et al.*, 1985; Stokes *et al.*, 1984), and recently confirmed by Vreeburg *et al.* (2009). However, inconsistencies are present in these studies, which most likely results from the heterogeneity of the depression syndrome. Higher cortisol levels have been reported frequently in studies of medicated inpatients with severe melancholic or psychotic depression. Also in major depression, convincing evidence for a (pro)inflammatory status has also been observed which includes increased levels of IL-1 and IL-6 (Catena-Dell'Osso *et al.*, 2011). The relationship between inflammation and depression has been underscored by the prominence of depressive symptoms following the acute and chronic administration of cytokines such as interferon-alpha. Altered levels of cytokines have also been described in bipolar disorder, although to a lesser extent as seen in schizophrenia. Elevated levels of TNF have been described, as well as some evidence for increased levels of IL-1 and IL-6 (O'Brien *et al.*, 2006; Brietzke *et al.*, 2009; Kunz *et al.*, 2011).

Therefore, there are robust findings showing alterations in peripheral molecules in major mental disorders. However, the measurement of single molecules has limited value as clinically useful biomarkers due to the substantial degree of overlap between patients and control, which is typically seen, resulting in limited sensitivity and specificity. Nevertheless, there are important proof-of-principle aspects related to these findings. The first of these is that alterations in several peripheral molecules are indeed present and with sufficient statistical effect sizes. The second of these is that alterations of some molecules covary with the severity of symptom dimensions (i.e., BDNF with the presence of positive psychotic symptoms), suggesting a causative relationship and emphasizing the possible utility of peripheral markers to identify symptoms dimensions. The final aspect is that the altered molecules reflect several of the biological themes that have been previously identified in genetic, *postmortem*, and neuroimaging research on the etiology of major psychiatric disorders, suggesting that subclustering of major psychiatric disorders based on these biological themes may be feasible.

To achieve this, multiple molecules should be investigated simultaneously, in a high-throughput manner, enabling functional subclustering within the patient groups. Thus far, a limited number of publications have described such a multi-analyte approach.

B. Multiplex Molecular Measurements

A more comprehensive, and possibly more fruitful, approach has been facilitated by a relatively recent development, namely, the investigation of multiple serum/plasma molecules at the same time. This is the approach of using multiprotein arrays or mass-spectroscopy profiling platforms (see chapters, "Proteomic technologies for biomarker studies in psychiatry: advances and needs" by Martins-de-Souza *et al.* and "The application of multiplexed assay systems for molecular diagnostics" by Schwarz *et al.*). It is expected that the results of these approaches will allow for a more reliable separation of patients with a specific diagnosis from controls and from other diagnostic groups compared to the use of single molecule biomarkers. Additionally, it opens up the possibility for the nonhypothesis driven subclustering of patients originating from broad diagnostic categories, such as the approach described in the section above (Section IV).

Multi-analyte and array profiling techniques enable the simultaneous detection of hundreds of proteins with high sensitivity and accuracy and can be successfully applied to identify biomarkers (or clusters of markers) that correlate with disease (see Ray *et al.*, 2007 for an application in Alzheimer's disease). At present, a limited number of studies have taken this approach. Domenici *et al.* (2010) have shown the results of a study investigating 245 patients with major depression, 229 patients with schizophrenia, and 254 controls, using a platform that allows for the simultaneous detection of 79 molecules including several cytokines, neurotrophins, and metabolic proteins. They reported identification of a cluster of molecules which gave a good degree of specificity for major depression versus controls, and "superior discriminative power" for schizophrenia versus controls. However, a major limitation of this study was that patients were medicated and subgroup analysis was not performed.

Schwarz *et al.* (2010, 2011) used a similar but extended platform of molecular assays, allowing for the simultaneous measurement of 181 molecules. In a more elaborate approach, involving a majority of never-medicated schizophrenia patients, they measured the levels of these molecules in serum from 250 first and recent onset schizophrenia patients, as well as patients with major depressive disorder, euthymic bipolar disorder, Asperger syndrome, and 280 control subjects. These analyses resulted in identification of a signature comprising 34 analytes allowing for the separation of over 80% of the schizophrenia subjects from controls across five independent cohorts. The authors concluded that a biological signature for schizophrenia can be identified in serum (Schwarz *et al.*, 2011). This study laid the groundwork for development of the first commercially available test (VeriPsych^{TM}, see chapter "Algorithm development for diagnostic biomarker assays" by Izmailov *et al.*) aiding in the diagnosis of schizophrenia, and for distinguishing schizophrenia patients from healthy controls and from those affected by related psychiatric illnesses with overlapping symptoms. However, this study also did not investigate the presence of subgroups.

VI. Future Prospects, Research Agenda

In this overview, we described the potential clinical utility of biomarkers for use in studies of major psychiatric disorders with a focus on peripheral blood-based molecules. At present, there are a limited number of clinically valid biomarkers available. This may be due to the fact that these are linked to the diagnostic concepts used in psychiatric biomarker research. We therefore focused on issues related to the current diagnostic system used in psychiatry. At present, most research efforts are aimed at identifying biomarkers for specific DSM-defined diagnostic categories. However, it is not likely that these efforts will be completely successful due to the heterogeneity inherent in these categories. Moreover, identifying a biomarker for a syndrome that has already been identified based on clinical phenomenology is not useful from a clinical perspective. Therefore, we argued for innovative approaches. First, we suggested the need for identification of biomarkers that can be measured in at-risk patients (i.e., youngsters with prodromal symptoms for psychosis) and which can be used as chance markers for a possible development toward more severe states, thus indicating the optimal intervention for that stage. This approach is closely linked with efforts aimed at disease profiling and clinical staging (Yung *et al.*, 2008; Woods *et al.*, 2009) based on more readily observable clinical characteristics. Second, we argued for the use of broader categories of related patients, and to deconstruct the traditional diagnoses of these patients using molecular biomarker profiles. This approach has already proven successful in oncology (i.e., see Valk *et al.*, 2004). Third, a more traditional approach is to identify markers predicting an optimal treatment response or risk for developing side effects.

Acknowledgments

This chapter was written while the authors were employed by the Erasmus MC Medical Center Rotterdam, The Netherlands. There was no other funding source. Many thanks to the members of the Bahn group (especially Sabine Bahn and Emanuel Schwarz) and to Lieuwe de Haan, Wim van den Brink, and Aart-jan Beekman for their interesting insights on issues of clinical diagnostics. Thanks to Peter Sillevis Smitt for his data on the relationship between anti-NMDA receptor encephalitis and psychiatric symptoms.

References

American Psychiatric Association (1994). Diagnostic and Statistical Manual of Mental Disorders. 4th ed. American Psychiatric Press, Washington, DC.

Biomarkers Definitions Working GroupAtkinson, A.J., Colburn, W.A., DeGruttola, V.G., DeMets, D.L., Downing, G.J., Hoth, D.F., Oates, J.A., Peck, C.C., Schooley, R.T., Spilker, B.A., Woodcock, J., *et al.* (2001). Biomarkers and surrogate endpoints: preferred definitions and conceptual framework. *Clin. Pharmacol. Ther.* **69**, 89–95.

Brietzke, E., Stertz, L., Fernandes, B.S., Kauer-Sant'anna, M., Mascarenhas, M., Escosteguy Vargas, A., Chies, J.A., and Kapczinski, F. (2009). Comparison of cytokine levels in depressed, manic and euthymic patients with bipolar disorder. *J. Affect. Disord.* **116**, 214–217.

Brosen, K., Otton, S.V., and Gram, L.F. (1986). Imipramine demethylation and hydroxylation: impact of the sparteine oxidation phenotype. *Clin. Pharmacol. Ther.* **40**, 543–549.

Burstein, H.J. (2005). The distinctive nature of HER2-positive breast cancers. *N. Engl. J. Med.* **20**, 1652–1654.

Cannon, T.D., Cadenhead, K., Cornblatt, B., Woods, S.W., Addington, J., Walker, E., Seidman, L.J., Perkins, D., Tsuang, M., McGlashan, T., and Heinssen, R. (2008). Prediction of psychosis in youth at high clinical risk: a multisite longitudinal study in North America. *Arch. Gen. Psychiatry* **65**, 28–37.

Carroll, B.J., Curtis, G.C., and Mendels, J. (1976). Neuroendocrine regulation in depression. I. Limbic system-adrenocortical dysfunction. *Arch. Gen. Psychiatry* **33**, 1039–1044.

Catena-Dell'Osso, M., Bellantuono, C., Consoli, G., Baroni, S., Rotella, F., and Marazziti, D. (2011). Inflammatory and neurodegenerative pathways in depression: a new avenue for antidepressant development? *Curr. Med. Chem.* **18**, 245–255.

Coyle, J.T. (2006). Glutamate and schizophrenia: beyond the dopamine hypothesis. *Cell. Mol. Neurobiol.* **26**, 363–382.

Craddock, N., O'Donovan, M.C., and Owen, M.J. (2006). Genes for schizophrenia and bipolar disorder? Implications for psychiatric nosology. *Schizophr. Bull.* **32**, 9–16.

Cunha, A.B., Frey, B.N., Andreazza, A.C., Goi, J.D., Rosa, A.R., Gonclaves, C.A., Santin, A., and Kapczinski, F. (2006). Serum brain-derived neurotrophic factor is decreased in bipolar disorder during depressive and manic episodes. *Neurosci. Lett.* **398**, 215–219.

Dalmau, J., and Rosenfeld, M.R. (2008). Paraneoplastic syndromes of the CNS. *Lancet Neurol.* **7**, 327–340.

Dalmau, J., Tüzün, E., Wu, H.Y., Masjuan, J., Rossi, J.E., Voloschin, A., Baehring, J.M., Shimazaki, H., Koide, R., King, D., Mason, W., Sansing, L.H., *et al.* (2007). Paraneoplastic anti-N-methyl-D-aspartate receptor encephalitis associated with ovarian teratoma. *Ann. Neurol.* **61**, 25–36.

Davis, K.L., Stewart, D.G., Friedman, J.I., Buchsbaum, M., Harvey, P.D., Hof, P.R., Buxbaum, J., and Haroutunian, V. (2003). White matter changes in schizophrenia: evidence for myelin-related dysfunction. *Arch. Gen. Psychiatry* **60**, 443–456.

de Oliveira, G.S., Ceresér, K.M., Fernandes, B.S., Kauer-Sant'Anna, M., Fries, G.R., Stertz, L., Aguiar, B., Pfaffenseller, B., and Kapczinski, F. (2009). Decreased brain-derived neurotrophic factor in medicated and drug-free bipolar patients. *J. Psychiatr. Res.* **43**, 1171–1174.

Desmedt, C., Sperinde, J., Piette, F., Huang, W., Jin, X., Tan, Y., Durbecq, V., Larsimont, D., Giuliani, R., Chappey, C., Buyse, M., Winslow, J., *et al.* (2009). Quantitation of HER2 expression or HER2:HER2 dimers and differential survival in a cohort of metastatic breast cancer patients carefully selected for trastuzumab treatment primarily by FISH. *Diagn. Mol. Pathol.* **18**, 22–29.

Domenici, E., Willé, D.R., Tozzi, F., Prokopenko, I., Miller, S., McKeown, A., Brittain, C., Rujescu, D., Giegling, I., Turck, C.W., Holsboer, F., Bullmore, E.T., *et al.* (2010). Plasma protein biomarkers for depression and schizophrenia by multi analyte profiling of case-control collections. *PLoS One* **5**(2), e9166.

Drexhage, R.C., Knijff, E.M., Padmos, R.C., Heul-Nieuwenhuijzen, L., Beumer, W., Versnel, M.A., and Drexhage, H.A. (2010). The mononuclear phagocyte system and its cytokine inflammatory networks in schizophrenia and bipolar disorder. *Expert Rev. Neurother.* **10**, 59–76.

Fernandez-Egea, E., Bernardo, M., Parellada, E., Justicia, A., Garcia-Rizo, C., Esmatjes, E., Conget, I., and Kirkpatrick, B. (2008). Glucose abnormalities in the siblings of people with schizophrenia. *Schizophr. Res.* **103**, 110–113.

Fernandez-Egea, E., Bernardo, M., Donner, T., Conget, I., Parellada, E., Justicia, A., Esmatjes, E., Garcia-Rizo, C., and Kirkpatrick, B. (2009). Metabolic profile of antipsychotic-naive individuals with non-affective psychosis. *Br. J. Psychiatry* **194**, 434–438.

Frommberger, U.H., Bauer, J., Haselbauer, P., Fräulin, A., Riemann, D., and Berger, M. (1997). Interleukin-6-(IL-6) plasma levels in depression and schizophrenia: comparison between the acute state and after remission. *Eur. Arch. Psychiatry Clin. Neurosci.* **247**, 228–233.

Galatzer-Levy, I.R., and Galatzer-Levy, R.M. (2007). The revolution in psychiatric diagnosis: problems at the foundations. *Perspect. Biol. Med.* **50**, 161–180.

Garey, L. (2010). When cortical development goes wrong: schizophrenia as a neurodevelopmental disease of microcircuits. *J. Anat.* **217**, 324–333.

Gold, P.W., Licinio, J., Wong, M.L., and Chrousos, G.P. (1995). Corticotropin releasing hormone in the pathophysiology of melancholic and atypical depression and in the mechanism of action of antidepressant drugs. *Ann. N. Y. Acad. Sci.* **771**, 716–729.

Guest, P.C., Wang, L., Harris, L.W., Burling, K., Levin, Y., Ernst, A., Wayland, M.T., Umrania, Y., Herberth, M., Koethe, D., van Beveren, J.M., Rothermundt, M., *et al.* (2010). Increased levels of circulating insulin-related peptides in first-onset, antipsychotic naive schizophrenia patients. *Mol. Psychiatry* **15**, 118–119.

Henry, L.P., Amminger, G.P., Harris, M.G., Yuen, H.P., Harrigan, S.M., Prosser, A.L., Schwartz, O.S., Farrelly, S.E., Herrman, H., Jackson, H.J., and McGorry, P.D. (2010). The EPPIC follow-up study of first-episode psychosis: longer-term clinical and functional outcome 7 years after index admission. *J. Clin. Psychiatry* **71**, 716–728.

Holsboer, F., and Barden, N. (1996). Antidepressants and hypothalamic-pituitary-adrenocortical regulation. *Endocr. Rev.* **17**, 187–205.

Holt, R.I., Peveler, R.C., and Byrne, C.D. (2004). Schizophrenia, the metabolic syndrome and diabetes. *Diabet. Med.* **21**, 515–523.

Horrobin, D.F. (1980). Niacin flushing, prostaglandin E3 and evening primrose oil. A possible objective test for monitoring therapy in schizophrenia. *J. Orthomolec. Psychiatr.* **9**, 33–34.

Insel, T.R. (2010). Rethinking schizophrenia. *Nature* **468**, 187–193.

Kanner, L., and Eisenberg, L. (1957). Early infantile autism, 1943–1955. *Psychiatr. Res. Rep. Am. Psychiatr. Assoc.* **7**, 55–65.

Karoutzou, G., Emrich, H.M., and Dietrich, D.E. (2008). The myelin-pathogenesis puzzle in schizophrenia: a literature review. *Mol. Psychiatry* **13**, 245–260.

Kim, Y.K., Myint, A.M., Verkerk, R., Scharpe, S., Steinbusch, H., and Leonard, B. (2009). Cytokine changes and tryptophan metabolites in medication-naive and medication-free schizophrenic patients. *Neuropsychobiology* **59**, 123–129.

King, B.H., and Lord, C. (2011). Is schizophrenia on the autism spectrum? *Brain Res.* **1380**, 34–41.

Koehler, K., Guth, W., and Grimm, G. (1977). First-rank symptoms of schizophrenia in Schneider-oriented German centers. *Arch. Gen. Psychiatry* **34**, 810–813.

Kooy, F.H. (1919). Hyperglycemia inmental disorders. *Brain* **42**, 214–289.

Kraepelin, E. (1971). Dementia Praecox and Paraphrenia. Robert E Krieger Publishing Co., Huntington, NY.

Kunz, M., Ceresér, K.M., Goi, P.D., Fries, G.R., Teixeira, A.L., Fernandes, B.S., Belmonte-de-Abreu, P.S., Kauer-Sant'anna, M., Kapczinski, F., and Gama, C.S. (2011). Serum levels of

IL-6, IL-10 and TNF-α in patients with bipolar disorder and schizophrenia: differences in pro- and anti-inflammatory balance. *Rev. Bras. Psiquiatr.* Mar 18 [Epub ahead of print].

Laughren, T.P. (2010). What's next after 50 years of psychiatric drug development: an FDA perspective. *J. Clin. Psychiatry* **71**, 1196–1204.

Lewis, D.A., and Mirnics, K. (2006). Transcriptome alterations in schizophrenia: disturbing the functional architecture of the dorsolateral prefrontal cortex. *Prog. Brain Res.* **158**, 141–152.

Liebzeit, K.A., Markowitz, J.S., and Caley, C.F. (2001). New onset diabetes and atypical antipsychotics. *Eur. Neuropsychopharmacol.* **11**, 25–32.

Lin, A., Kenis, G., Bignotti, S., Tura, G.J., De Jong, R., Bosmans, E., Pioli, R., Altamura, C., Scharp, S., and Maes, M. (1998). The inflammatory response system in treatment-resistant schizophrenia: increased serum interleukin-6. *Schizophr. Res.* **32**, 9–15.

Maes, M., Bocchio Chiavetto, L., Bignotti, S., Battisa Tura, G., Pioli, R., Boin, F., Kenis, G., Bosmans, E., de Jongh, R., Lin, A., Racagni, G., and Altamura, C.A. (2000). Effects of atypical antipsychotics on the inflammatory response system in schizofrenia patients resistant to treatment with typical neuroleptics. *Eur. Neuropsychopharmacol.* **10**, 119–124.

Martins-de-Souza, D., Harris, L.W., Guest, P.C., and Bahn, S. (2010). The role of energy metabolism dysfunction and oxidative stress in schizophrenia revealed by proteomics. *Antioxid. Redox Signal.* Dec 2 [Epub ahead of print].

McGorry, P.D., Yung, A.R., Bechdolf, A., and Amminger, P. (2008). Back to the future: prediking and reshaping the course of psychotic disorder. *Arch. Gen. Psychiatry* **65**, 25–27.

McMahon, F.J., Buervenich, S., Charney, D., Lipsky, R., Rush, A.J., Wilson, A.F., Sorant, A.J., Papanicolaou, G.J., Laje, G., Fava, M., Trivedi, M.H., Wisniewski, S.R., *et al.* (2006). Variation in the gene encoding the serotonin 2A receptor is associated with outcome of antidepressant treatment. *Am. J. Hum. Genet.* **78**, 804–814.

Meyer, J.M. (2002). A retrospective comparison of weight, lipid, and glucose changes between risperidone- and olanzapine-treated inpatients: metabolic outcomes after 1 year. *J. Clin. Psychiatry* **63**, 425–433.

Monteleone, P., Fabrazzo, M., Tortorella, A., and Maj, M. (1997). Plasma levels of interleukin-6 and tumor necrosis factor alpha in chronic schizophrenia: effects of clozapine treatment. *Psychiatry Res.* **7**, 11–17.

Naudin, J., Capo, C., Giusano, B., Moge, J.L., and Azorin, J.M. (1997). A differential role for interleukin-6 and tumor necrosis factor-alpha in schizophrenia? *Schizophr. Res.* **26**, 227–233.

Nemeroff, C.B., and Vale, W.W. (2005). The neurobiology of depression: inroads to treatment and new drug discovery. *J. Clin. Psychiatry* **66**(Suppl. 7), 5–13.

Nestler, E.J., and Hyman, S.E. (2010). Animal models of neuropsychiatric disorders. *Nat. Neurosci.* **13**, 1161–1169.

O'Brien, S.M., Scully, P., Scott, L.V., and Dinan, T.G. (2006). Cytokine profiles in bipolar affective disorder: focus on acutely ill patients. *J. Affect. Disord.* **90**, 263–267.

Pfohl, B., Sherman, B., Schlechte, J., and Winokur, G. (1985). Differences in plasma ACTH and cortisol between depressed patients and normal controls. *Biol. Psychiatry* **20**, 1055–1072.

Ray, S., Britschgi, M., Herbert, C., Takeda-Uchimura, Y., Boxer, A., Blennow, K., Friedman, L.F., Galasko, D.R., Jutel, M., Karydas, A., Kaye, J.A., Leszek, J., *et al.* (2007). Classification and prediction of clinical Alzheimer's diagnosis based on plasma signaling proteins. *Nat. Med.* **13**, 1359–1362.

Ryan, M.C., Collins, P., and Thakore, J.H. (2003). Impaired fasting glucose tolerance in first-episode, drug-naive patients with schizophrenia. *Am. J. Psychiatry* **160**, 284–289.

Sadler, J.Z. (2005). Values and Psychiatric Diagnosis Oxford University Press, Oxford, UK.

Sauer, J.M., Ring, B.J., and Witcher, J.W. (2005). Clinical pharmacokinetics of atomoxetine. *Clin. Pharmacokinet.* **44**, 571–590.

Schwarz, E., Izmailov, R., Spain, M., Barnes, A., Mapes, J.P., Guest, P.C., Rahmoune, H., Pietsch, S., Leweke, F.M., Rothermundt, M., Steiner, J., Koethe, D., *et al.* (2010). Validation of a blood-based laboratory test to aid in the confirmation of a diagnosis of schizophrenia. *Biomark. Insights* **5**, 39–47.

Schwarz, E., Guest, P.C., Rahmoune, H., Harris, L.W., Wang, L., Leweke, F.M., Rothermundt, M., Bogerts, B., Koethe, D., Kranaster, L., Ohrmann, P., Suslow, T., *et al.* (2011). Identification of a biological signature for schizophrenia in serum. *Mol. Psychiatry* Apr 12 [Epub ahead of print].

Serretti, A., Kato, M., De Ronchi, D., and Kinoshita, T. (2007). Meta-analysis of serotonine transporter gene promoter polymorphism (5-HTTLPR) association with selective serotonin reuptake inhibitor efficacy in depressed patients. *Mol. Psychiatry* **12**, 247–257.

Smesny, S., Klemm, S., Stockebrand, M., Grunwald, S., Gerhard, U.J., Rosburg, T., Sauer, H., and Blanz, B. (2007). Endophenotype properties of niacin sensitivity as marker of impaired prostaglandin signalling in schizophrenia. *Prostaglandins Leukot. Essent. Fatty Acids* **77**, 79–85.

Spelman, L.M., Walsh, P.I., Sharifi, N., Collins, P., and Thakore, J.H. (2007). Impaired glucose tolerance in first-episode drug-naïve patients with schizophrenia. *Diabet. Med.* **24**, 481–485.

Stefansson, H., Ophoff, R.A., Steinberg, S., Andreassen, O.A., Cichon, S., Rujescu, D., Werge, T., Pietiläinen, O.P., Mors, O., Mortensen, P.B., Sigurdsson, E., Gustafsson, O., *et al.* (2009). Common variants conferring risk of schizophrenia. *Nature* **460**, 744–747.

Steiner, J., Walter, M., Guest, P., Myint, A.M., Schiltz, K., Panteli, B., Brauner, M., Bernstein, H.G., Gos, T., Herberth, M., Schroeter, M.L., Schwarz, M.J., *et al.* (2010). Elevated S100B levels in schizophrenia are associated with insulin resistance. *Mol. Psychiatry* **15**, 3–4.

Stokes, P.E., Stoll, P.M., Koslow, S.H., Maas, J.W., Davis, J.M., Swann, A.C., and Robins, E. (1984). Pretreatment DST and hypothalamic-pituitary-adrenocortical function in depressed patients and comparison groups. A multicenter study. *Arch. Gen. Psychiatry* **41**, 257–267.

Takahashi, N., Sakurai, T., Davis, K.L., and Buxbaum, J.D. (2011). Linking oligodendrocyte and myelin dysfunction to neurocircuitry abnormalities in schizophrenia. *Prog. Neurobiol.* **93**, 13–24.

Theodoropoulou, S., Spanakos, G., Baxevanis, C.N., Economou, M., Gritzapis, A.D., Papamichail, M. P., and Stefanis, C.N. (2001). Cytokine serum levels, autologous mixed lymphocyte reaction and surface marker analysis in never medicated and chronically medicated schizophrenic patients. *Schizophr. Res.* **47**, 13–25.

Tramontina, J.F., Andreazza, A.C., Kauer-Sant'anna, M., Stertz, L., Goi, J., Chiarani, F., and Kapczinski, F. (2009). Brain-derived neurotrophic factor serum levels before and after treatment for acute mania. *Neurosci. Lett.* **452**, 111–113.

Valk, P.J., Verhaak, R.G., Beijen, M.A., Erpelinck, C.A., van Waalwijk, Barjesteh, van Doorn-Khosrovani, S., Boer, J.M., Beverloo, H.B., Moorhouse, M.J., van der Spek, P.J., Lowenberg, B., and Delwel, R. (2004). Prognostically useful gene-expression profiles in acute myeloid leukemia. *N. Engl. J. Med.* **350**, 1617–1628.

van Beveren, N.J., van der Spelt, J.J., de Haan, L., and Fekkes, D. (2006). Schizophrenia-associated neural growth factors in peripheral blood. A review. *Eur. Neuropsychopharmacol.* **16**, 469–480.

Van Nimwegen, L.J., Storosum, J.G., Blumer, R.M., Allick, G., Venema, H.W., de Haan, L., Becker, H., van Amelsvoort, T., Ackermans, M.T., Fliers, E., Serlie, M.J., and Sauerwein, H.P. (2008). Hepatic insulin resistance in antipsychotic naive schizophrenic patients: stable isotope studies of glucose metabolism. *J. Clin. Endocrinol. Metab.* **93**, 572–577.

Van Os, J., Hanssenm, M., Bijl, R.V., and Ravelli, A. (2000). Straus (1969) revisited; a psychosis continuum in the general population? *Schizophr. Res.* **29**, 11–20.

Van Praag, H.M. (1997). Over the mainstream: diagnostic requirements for biological psychiatric research. *Psychiatry Res.* **72**, 201–212.

Van Praag, H.M. (2000). Nosologomania: a disorder of psychiatry. *World J. Biol. Psychiatry* **1**, 151–158.

Van Praag, H.M. (2001). Past expectations, present disappointments, future hopes or psychopathology as the rate-limiting step of progress in psychopharmacology. *Hum. Psychopharmacol.* **16**, 3–7.

Venkatasubramanian, G., Chittiprol, S., Neelakantachar, N., Naveen, M.N., Thirthall, J., Gangadhar, B.N., and Shetty, K.T. (2007). Insulin and insulin-like growth factor-1 abnormalities in antipsychotic-naive schizophrenia. *Am. J. Psychiatry* **164**, 1557–1560.

Vreeburg, S.A., Hoogendijk, W.J., van Pelt, J., Derijk, R.H., Verhagen, J.C., van Dyck, R., Smit, J.H., Zitman, F.G., and Penninx, B.W. (2009). Major depressive disorder and hypothalamic-pituitary-adrenal axis activity: results from a large cohort study. *Arch. Gen. Psychiatry* **66**, 617–626.

Williams, H.J., Craddock, N., Russo, G., Hamshere, M.L., Moskvina, V., Dwyer, S., Smith, R.L., Green, E., Grozeva, D., Holmans, P., Owenm, M.J., and O'Donovan, M.C. (2011). Most genome-wide significant susceptibility loci for schizophrenia and bipolar disorder reported to date cross-traditional diagnostic boundaries. *Hum. Mol. Genet.* **20**, 387–391.

Woods, S.W., Addington, J., Cadenhead, K.S., Cannon, T.D., Cornblatt, B.A., Heinssen, R., Perkins, D.O., Seidman, L.J., Tsuang, M.T., Walker, E.F., and McGlashan, T.H. (2009). Validity of the prodromal risk syndrome for first psychosis: findings from the north American prodrome longitudinal study. *Schizophr. Bull.* **35**, 894–908.

Yung, A.R., Nelson, B., Stanford, C., Simmons, M.B., Cosgrave, E.M., Killackey, E., Phillips, L.J., Bechdolf, A., Buckby, J., and McGorry, P.D. (2008). Validationof "prodromal" criteria to detect individuals at ultra high risk of psychosis: 2 year follow-up. *Schizophr. Res.* **105**, 10–17.

THE FUTURE: BIOMARKERS, BIOSENSORS, NEUROINFORMATICS, AND E-NEUROPSYCHIATRY

Christopher R. Lowe

Department of Chemical Engineering and Biotechnology, Institute of Biotechnology, University of Cambridge, Cambridge, United Kingdom

Abstract

The emergence of molecular biomarkers for psychological, psychiatric, and neurodegenerative disorders is beginning to change current diagnostic paradigms for this debilitating family of mental illnesses. The development of new genomic, proteomic, and metabolomic tools has created the prospect of sensitive and specific biochemical tests to replace traditional pen-and-paper questionnaires. In the future, the realization of biosensor technologies, point-of-care testing, and the fusion of clinical biomarker data, electroencephalogram, and MRI data with the patient's past medical history, biopatterns, and prognosis may create personalized bioprofiles or fingerprints for brain disorders. Further, the application of mobile communications technology and grid computing to support data-, computation- and knowledge-based tasks will assist disease prediction, diagnosis, prognosis, and compliance monitoring. It is anticipated that, ultimately, mobile devices could become the next generation of personalized pharmacies.

INTERNATIONAL REVIEW OF
NEUROBIOLOGY, VOL. 101
DOI: 10.1016/B978-0-12-387718-5.00015-8

I. Introduction

A mental illness is a psychological or behavioral pattern associated with subjective distress or disability which lies outside normal development or culture of the afflicted individuals. There are multiple types of mental illness where different facets of human behavior, emotion, and personality can become disordered (Gazzaniga and Heatherton, 2006). Psychological, psychiatric, and neurodegenerative disorders are three manifestations of this diverse group of debilitating behavioral and mental health diseases which can strike at various ages and gravely impair the sufferer's quality of life, social well-being, and productivity. Psychological disorders lie at the mild end of the spectrum and include atypical expressions of anxiety, panic, obsession, compulsion, personality, and behavior and are generally treated by psychopharmacy and psychotherapy. They can strike at any age, whereas the more severe neuropsychiatric disorders, such as schizophrenia, bipolar disorder, major depression, and autism, afflict the young and can be ameliorated to some extent with prescription medicines. For example, major depression is a severe neuropsychiatric disorder which affects approximately 10% of the global population and is characterized by low mood and self-esteem, misguided guilt, harboring thoughts of death and suicide, reduced concentration, anhedonia, and disturbance of sleep and appetite. Only 60% of patients respond to current lengthy antidepressant regimes with a high rate of relapse and marked resistance to treatment (Martins-de-Souza *et al.*, 2010). Similarly, schizophrenia and bipolar affective disorder affect at least 2% of the population worldwide and cost hundreds of billions of dollars in health-care provision, treatments, and lost earnings. The World Health Organization (WHO) found schizophrenia to be the world's fourth leading cause of disability accounting for 1.1% of the total DALYs (disability adjusted life years) and 2.8% of YLDs (years of life lived with disability). These severe psychiatric disorders often manifest themselves with psychotic states characterized by disruption of basic perceptual, cognitive, affective, and judgmental processes. Typically, schizophrenia has its onset in late adolescence or early adulthood and presents as a constellation of positive (hallucination, delusions, disorganization of thought, and bizarre behavior), negative (loss of motivation, restricted range of emotional experience and expression, and reduced hedonic capacity), and cognitive impairments with extensive variation between individuals. Like many neuropsychiatric disorders, no single symptom is unique to schizophrenia and/or is present in every case. Psychotic episodes, for example, are also not uncommon in cases of brain injury, learning disability, substance abuse, and a range of metabolic disorders, and may occur after chronic psychological stress and vary in duration between individuals. In addition, many patients with schizophrenia experience comorbid difficulties with depression and substance abuse contributing to the 10–15% lifetime incidence of suicide. Not surprisingly, up to 90% of schizophrenia patients are unemployed.

Neurodegenerative disorders, on the other hand, are characterized by senile dementia and include Alzheimer's (AD), Huntington's (HD), and Parkinson's diseases (PD) plus a plethora of other medical or neurological disorders characterized by chronic inflammation of the brain. It is estimated that nearly 25 million people currently suffer from dementia with the number of sufferers projected to increase to more than 80 million by 2040 (Ferri *et al.*, 2005). The most common dementia, AD, is an age-related and insidious onset neurodegenerative disease which afflicts approximately 12.5% of people over 65 years (Song *et al.*, 2009). Disease pathogenesis may affect the brains of patients possibly years or decades prior to expression of the clinical symptoms (Snowdon *et al.*, 1996).

II. Current Diagnostic Tools for Mental Illness

In conventional medical practice, the physician renders a diagnosis based on observation of particular overt combinations of symptoms that patients present, identifies an underlying disorder, and prescribes a specific treatment. However, with mental illness, psychologists, psychiatrists, and other mental health workers are confronted with a plethora of ethereal and even covert cognitive, behavioral, and emotional symptoms which collectively may form a particular syndrome. Diagnosis of mental disorders is generally based on judgment of the patient's appearance and behavior, self-reported symptoms, mental health history, and current lifestyle. Psychiatric diagnosis is usually performed via paper-and-pen or computerized questionnaires based on a standard diagnostic proforma guided by ICD-10, Chapter V: "Mental and behavioral disorders, part of the International Classification of Diseases" produced by the WHO, and the Diagnostic and Statistical Manual of Mental Health (DSM-IV) produced by the American Psychiatric Association (APA). However, in reality, these diagnostic evaluations are often incomplete, unstructured, open-ended, and prone to inaccuracy or misdiagnosis of the patient (Shear *et al.*, 2000).

These current diagnostic protocols for mental disorders rely on phenomenological and behavioral paradigms which systematically exclude observations from biological or pathophysiological sources. However, recent advances in clinical neuroscience are beginning to unravel the neurobiochemical underpinnings of such disorders. Advances in molecular imaging techniques, such as positron emission tomography (PET), magnetic resonance imaging (MRI), functional magnetic resonance imaging (fMRI), ultrasound, and single photon emission computerized tomography (SPECT), have made enlightening contributions to the understanding of the pathophysiology of neuropsychiatric disorders (Bressan *et al.*, 2007; see chapter "Imaging brain microglial activation using positron emission tomography and translocator protein specific radioligands" by Owen and Matthews). These minimally invasive technologies exploit radiolabeled tracers based most commonly on ^{11}C ($t_{1/2} = 20$ min), ^{13}N ($t_{1/2} = 10$ min), ^{15}O ($t_{1/2} = 2$ min), and ^{18}F ($t_{1/2} = 110$ min) which when injected *in vivo* undergo positron emission decay to produce a pair of photons moving in approximately opposite directions and detected with a photomultiplier or silicon avalanche photodiode in a dedicated PET scanner. The use of 2-[^{18}F]-fluorodeoxy-D-glucose (2FDG), a nonmetabolizable analogue of glucose, permits regional glucose uptake, and hence tissue metabolic activity to be assessed, in 3D or 4D space (Young *et al.*, 1999). PET neuroimaging is based on the assumption that areas of high radioactivity are associated with brain activity and this has been widely used in aiding early diagnosis of AD and for differentiating AD from other dementias (Klunk *et al.*, 2004; see Chapter "Imaging brain microglial activation using positron emission tomography and translocator protein-specific radioligands" by Owen and Matthews). Similarly, numerous compounds that bind selectively to neuroreceptors of interest, including dopamine (D1, D2), serotonin (5HT1A, 5HT2A), and opioid (μ) receptors, have been labeled with ^{11}C or ^{18}F and used in human subjects for investigating the differences between patients and healthy controls in dementia, schizophrenia, substance abuse, mood disorders, and other psychiatric disorders (Bressan *et al.*, 2001; Ravina *et al.*, 2005; Small *et al.*, 2006). Molecular imaging techniques have furnished compelling pathophysiological evidence of dopaminergic dysfunction *in vivo* in psychiatric disorders such as schizophrenia (Bressan *et al.*, 2001). Several studies showed increased DOPA decarboxylase activity using [^{18}F]-fluoro-DOPA or [^{11}C]-DOPA and higher dopamine release after amphetamine α-methyl-paratyrosine challenges (Bressan *et al.*, 2001). However, despite being of interest scientifically, such techniques still have limited clinical utility due to the complexity of psychiatric disorders and the low sensitivity and specificity of the current data. Nevertheless, a more powerful strategy of combining several diagnostic techniques at the same time is beginning to emerge. For example, coregistration of PET scans alongside CT or MRI scans allows both anatomic and metabolic information to be spatiotemporally colocated, while inclusion of genetic algorithms, structural neuroimaging, neuropsychology, neuroendocrinology, and molecular imaging allows a more confident interpretation of clinical data (Bressan *et al.*, 2007).

III. The Emergence of Molecular Biomarkers

Current dogma on the diagnosis and follow up of most psychological, neuropsychiatric, and neurodegenerative disorders is based on the aggregation of a cluster of symptoms and scales, which makes identification of individuals at risk, the severity of the disorder, and even an accurate diagnosis quite difficult. This is primarily due to the fact that the fundamental lesions underlying diseases of the central nervous system (CNS) are ill established and likely to be accentuated by complex perturbations and dysregulation of regulatory mechanisms, protein expression profiles, and metabolic pathways. Consequently, any understanding of these disorders should be accompanied with a systems level view of the dysregulated factors contributing to the pathogenesis. It is imperative to construct disease- or symptom-specific molecular fingerprints or panels of molecular biomarkers by integrating multiple "omics" databases established from genomics, proteomics, and metabolomics studies on disease versus control samples (see chapters "The utility of gene expression in blood cells for diagnosing neuropsychiatric disorders" by Woelk *et al.* and "Proteomic technologies for biomarker studies in psychiatry: advances and needs" by Martins-de-Souza *et al.*).

A molecular biomarker is a chemical, physical, or biological parameter that can be objectively measured and evaluated quantitatively as an indicator of normal or pathogenic states, to measure the onset or progress of disease, of compliance or response to a therapeutic intervention (Quinones and Kaddurah-Daouk, 2009). Biomarkers can be specific cells, genes, gene products, such as mRNA transcripts, proteins, and their posttranslational modifications, enzymes, hormones, peptides, or metabolites (Craig-Shapiro *et al.*, 2008).

The discovery of authenticated biomarkers for psychological, neuropsychiatric, and neurodegenerative disorders could dramatically change the future delivery of mental health care if they are incorporated into standard operating systems and clinical decision making. Appropriate genomic, proteomic, and metabolomic biomarkers should allow precise prediction of disease susceptibility and risk, diagnosis and prognosis, patient and therapeutic stratification, patient response, and adverse drug reactions, and ensure compliance (Fig. 1).

Genes abnormally regulated during neuropsychiatric disorders or affected by drug treatments may help to define aberrant cellular processes and serve as diagnostics and novel druggable targets. Several recent studies using cDNA microarray and differential display techniques have investigated changes in gene expression in AD (Hata *et al.*, 2001; Pasinetti, 2001), seizures and epilepsy (Aronica *et al.*, 2001; French *et al.*, 2001), PD (Chun *et al.*, 2001), schizophrenia (Mirnics *et al.*, 2000), bipolar disorder (Sun *et al.*, 2001; Le-Niculescu *et al.*, 2007), and other neurological disorders (Kontkanen and Castrén, 2002). However, despite progress on functional genomics for neuropsychiatric disorders, significant challenges remain before expression profiling can form a reliable basis for specific and sensitive biomarkers. Neuropsychiatric diseases are complex polygenic disorders with variable penetrance, whose phenotypic heterogeneity, overlap, and interdependence are not well understood and which are greatly influenced by epigenetic modifications, stress, lifestyle, infections, and medications (Danziger *et al.*, 2005; Crow, 2007 (Fig. 2).

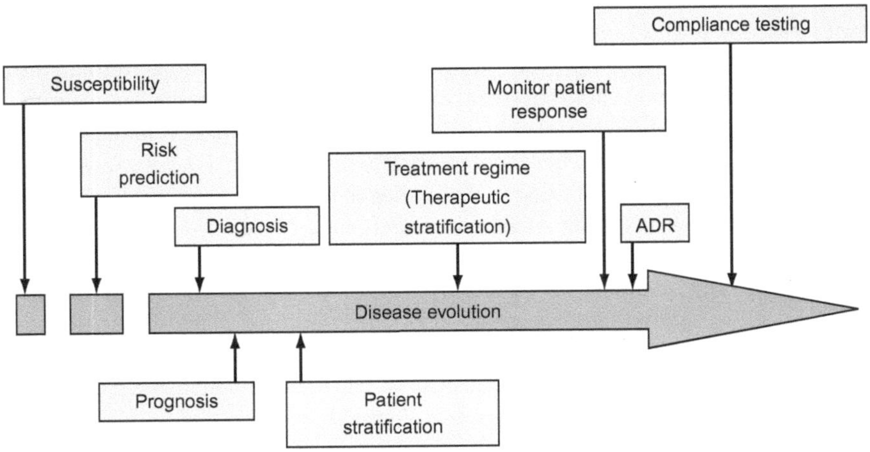

FIG. 1. Implication of biomarkers and biosensors in the disease evolution of psychological, neuropsychiatric, and neurodegenerative disorders.

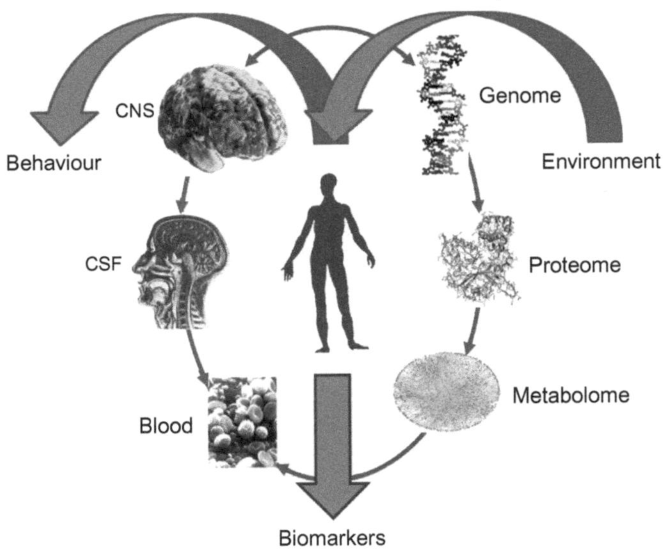

FIG. 2. Schematic of the interactions between genetic and epigenetic factors and behavior and their effect on the discovery of genomic, proteomic, and metabolomic biomarkers from brain biopsies, cerebrospinal fluid (CSF), and blood.

Genomic studies are able to identify genes conferring susceptibility to a particular disease, although the functional abnormality is ultimately reflected in the posttranslational proteome. In recent years, proteomics has been used as a tool for the discovery of biomarkers for diagnosis, monitoring disease progression, treatment response, and for the identification of novel therapeutic targets (Martins-de-Souza *et al.*, 2010). However, for proteome analysis of CNS disorders, the brain is not readily accessible for invasive diagnostic purposes and sources such as cerebrospinal fluid (CSF), serum, plasma, saliva, urine, and peripheral blood cells are more appropriate (Schwarz and Bahn, 2008; Taurines *et al.*, 2011). CSF is of particular relevance to biomarker discovery in psychiatric disorders since it is in direct contact with the brain and most likely reflects aberrations in brain function more closely than any other fluid (Rohlff, 2001). The protein concentration (0.18–0.58 mg ml^{-1}) in CSF is two to three orders of magnitude lower than in serum ($\sim$75 mg ml^{-1}; Schwarz and Bahn, 2008) and has presented problems with analytical techniques.

Serum presents the most accessible fluid for proteome analysis, although it is inherently complex, has a dynamic range exceeding 10 orders of magnitude, and the 22 most abundant proteins account for 99% of the proteome. The concentration of proteins in plasma ranges from 5 pg ml^{-1} (interleukin-6) to 50 mg ml^{-1} (albumin; Anderson and Anderson, 2002). Proteomics platforms must cope with multiple samples, high throughput, substantial complexity, a wide dynamic range, and a plethora of posttranslational modifications. Key techniques with promise include traditional 2-DE MS, 2D-DIGE, SELDI-TOF MS, iTRAQ-based LC–MALDI–MS/MS, and label-free LC–MS/MS which balance accuracy and high throughput with varying degrees of success (Anderson and Anderson, 2002; see Chapter "Proteomic technologies for biomarker studies in psychiatry: Advances and needs" by Martins-de-Souza et al.). Application of these platforms in CSF and serum has identified multiple up- and downregulated protein biomarkers in AD (Song *et al.*, 2009; Flood *et al.*, 2011), schizophrenia (Schwarz and Bahn, 2008; Madaan *et al.*, 2010; Stanta *et al.*, 2010), depression (Gudmundsson *et al.*, 2010; Martins-de-Souza *et al.*, 2010), and Asperger's syndrome (Schwarz *et al.*, 2010a,b).

The study of metabolism at the global level also promises to impact on the identification of biomarkers for neuropsychiatric diseases (Quinones and Kaddurah-Daouk, 2009). Several metabolic aberrations have been observed in CNS disorders; for example, impairments in neuronal survival, neurotransmitters, oxidative stress, free radical ratios, membrane composition, mitochondrial function, and immune function. Thousands of low molecular weight metabolites can be resolved and quantified in CSF, plasma, or urine samples using a variety of platforms such as a nuclear magnetic resonance (NMR) spectroscopy, gas chromatography– and liquid chromatography–mass spectroscopy (GC–MS, LC–MS), and liquid chromatography with electrochemical array detection (LCECA). However, there is no universal platform which can capture the entire metabolome, and generally speaking, combinations of such techniques will give a more complete picture of metabolite changes in the patient versus the healthy controls. Metabolic signatures have been identified in PD (glutathione, uric acid; Bogdanov *et al.*, 2008) and HD (glycerol, malonate; Underwood *et al.*, 2006), depression (fatty acids, 3-hydroxybutanoic acid, glycerol, γ-aminobutyric acid; Paige *et al.*, 2007), and schizophrenia (structural and energetic lipids, glucose; Holmes *et al.*, 2006; Wolf and Quinn, 2008).

IV. Clinical Impact of Biomarkers

In principle, simultaneously capturing biomarker data from genomic, proteomic, and metabolomic platforms should assist in determining how gene-expression patterns result in specific protein translation, pathway up -or downregulation, and changes in the metabolome. Such an approach has demonstrated compromised brain metabolism in bipolar disorder and schizophrenia (Prabakaran *et al.*, 2004; Khaitovich *et al.*, 2008). However, currently, clinicians capture only a small fraction of the information contained in the global bionome, usually via a standard set of blood analytes, including gases, ions, metabolites, enzymes, and antigens/antibodies, to define healthy and diseased states. In the near future, it is conceivable that such a restricted blood analysis will be replaced with a much more extensive bionomic signature that captures global biochemical changes in health, disease, and upon medication. A carefully selected multiparameter diagnostic should enhance the sensitivity (the ability of a diagnostic to identify all patients afflicted with the illness, i.e., few false negatives) and specificity (the ability to identify all patients without the illness, i.e., few false positives) of the test. Ideally, to be clinically relevant to diagnose a particular disorder, a multiparameter biomarker panel should have a sensitivity and specificity of $>85\%$ (Quinones and Kaddurah-Daouk, 2009). Multiplexed biomarker assays have proven useful in diagnosing breast cancer using gene-expression signatures (Sotiriou and Pusztai, 2009).

The avalanche of candidate biomarkers discovered by MS-based proteomics has created a market for high-throughput multiplexed immunoassays that allow simultaneous quantification of the analytes. Two basic assay formats have been developed to facilitate quantification of multiple antigens: planar array assays and microbead assays (Fu *et al.*, 2010). In the first format, different capture antibodies are spotted at defined positions on a 2D array. In the second, the capture antibodies are conjugated to different populations of microbeads, which can be distinguished by their fluorescence intensity in a flow cytometer. Commercial assay platforms such as MULTI-ARRAY (Meso Scale Discovery), Bio-Plex (Bio-Rad Laboratories), A^2 (Beckman Coulter), FAST Quant (Whatman Schleicher & Schuell BioScience), and FlowCytomix (Bender MedSystems) represent examples of these technologies currently used for high-throughput immunoanalysis (Fu *et al.*, 2010).

Recently, a serum-based diagnostic test has been developed to aid in the confirmation of a diagnosis of schizophrenia (Schwarz *et al.*, 2010a,b; see chapters "The application of multiplexed assay systems for molecular diagnostics" by Schwarz *et al.* and "Algorithm development for diagnostic biomarker assays" by Izmailov *et al*). In preliminary studies using a multiplexed immunoassay platform, the luminex xMAP technology, a disease signature comprising 51 metabolites, peptides, and proteins which could differentiate schizophrenia ($n = 250$) from control ($n = 230$) samples. These analytes were then used to construct a diagnostic decision rule by developing a refined 51-plex immunoassay panel, validated using a large independent cohort of samples from schizophrenia ($n = 577$) and matched control ($n = 229$) subjects, and shown to display an overall sensitivity of 83% and specificity of 83% while implemented in a Clinical Laboratory Improvement Amendments (CLIA)-certified laboratory in the United States (Schwarz *et al.*, 2010a,b). It is anticipated that a multiplexed immunoassay platform will eventually be able to distinguish between various neuropsychiatric diseases such as schizophrenia, bipolar disorder, and major depressive disorder.

The luminex technology comprises 100 color-coded microsphere sets which can be coated with a reagent specific to a particular bioassay, allowing the capture and detection of specific analytes from a sample. Lasers within the compact analyzer excite the internal dyes that identify both each microsphere particle and any reporter dye captured during the assay. In this way, the xMAP technology allows rapid and precise multiplexing of up to 100 unique assays within a single sample. Focused, flexible multiplexing in the range of 1–100 analytes meets the needs of a wide variety of applications, including protein and gene-expression profiling; autoimmune, genetic, and molecular infectious disease monitoring; and human leukocyte antigen (HLA) testing.

V. Point-of-Care Testing

A typical centralized hospital laboratory has evolved over the past 100 years into a fully automated system of bar-coded patient identification, sample collection, sample pretreatment, and passage through high-throughput multiplexed clinical chemistry and immunoassay platforms, with results becoming available minutes, hours, or days later. Such delays can hamper timely diagnosis, impede decision making, and affect outcomes (see chapter "Challenges of introducing new biomarker products for neuropsychiatric disorders into the market" by Bahn *et al.*). However, in recent years, there has been a quiet revolution in diagnostics practice, with a perceptible trend in taking tests to the patient, rather than the patient to the tests. Substantial technological advances in assay chemistry, sensor and transducer configurations, electronic processing and data manipulation, instrumentation and miniaturization have seen an upsurge of interest in "alternate site" diagnostic testing in ward, outpatients, physician's office, workplace, and home. This type of patient-proximal, or point-of-care (POC) testing as it is often referred to, can reduce the cost per test by as much as 35% with additional savings in skilled manpower. POC testing simplifies the steps involved in sample handling and pretreatment and can reduce the turnaround time from hours to minutes with proportionately rapid clinical decisions (Price *et al.*, 2010). These trends are illustrated schematically in Fig. 3 which highlights the movement toward making health care more patient centric as part of global health-care reforms which are being driven by the steady growth in health-care expenditure as a proportion of GDP and concerns about the quality of health care, particularly the increasing burden of chronic conditions such as psychiatric and neurodegenerative disorders, and their effective management (Price *et al.*, 2010).

The question is what technology can accommodate multiparameter diagnostic tests to generate rapid, precise, reliable, and foolproof data suitable for clinical decision rules for psychiatric and neurodegenerative patients. POC testing devices include a spectrum of systems covering *in vivo* implanted, wearable, and handheld

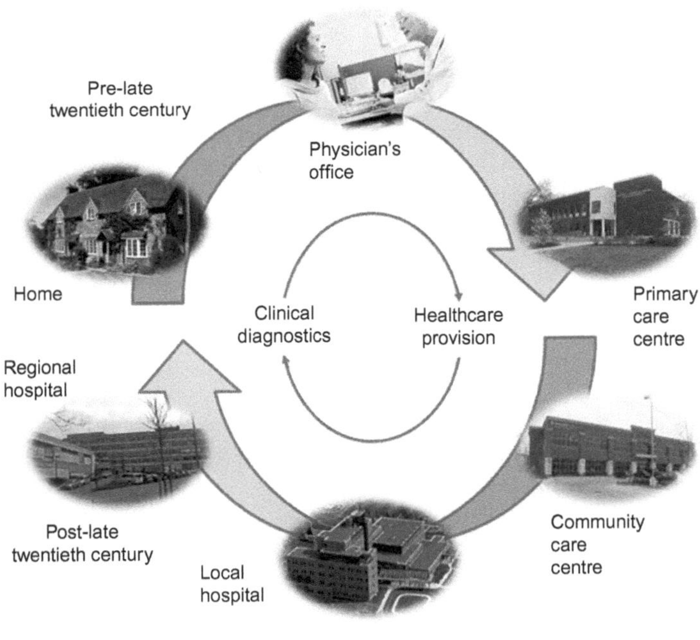

FIG. 3. Evolution of health-care provision from the home in the late twentieth century to the regional hospital via various community and local services in the late twentieth century and anticipated full circle back to the home in the early twenty-first century. Clinical diagnostics are following a similar pattern of development.

devices and the larger table- or bench-top instruments. The majority of handheld systems comprise a disposable reagent prefilled strip incorporated into a cartridge or cassette to facilitate addition of the sample, whence following development of the test, the resulting signal is interpreted visually or measured using an inexpensive reader or meter. Current POC tests cover critical care (blood gases, ions), emergency medicine (troponin), risk assessment (cholesterol), diabetes management (glucose, hemoglobin A1c), personal well-being (pregnancy, fertility), lifestyle (alcohol, drugs of abuse), and infectious diseases (human immunodeficiency virus, sexually transmitted infections) in a market estimated to be worth $18.7B by 2013 (Kricka and Thorpe, 2010). Such tests usually comprise a reagent-impregnated paper strip or pad, lateral flow devices, dipsticks, cards, slides, flow-through tubes, cartridges, and cassettes with inbuilt meters and displays. A good example of state-of-the-art consumer electronics technology for POC testing is the introduction of the Clearblue Digital Pregnancy test® by SPD, which provides both an unambiguous readout of the result (pregnant/not pregnant) and an indication of the number of weeks since conception.

VI. Biosensors

A key element for the development of POC systems is the biosensor. Biosensor technology is an analytical platform which can empower physicians or psychiatrists with the ability to confirm a suspected diagnosis on the spot when the patient presents him or herself and thereby obviate the requirement to attend hospital. A biosensor is an analytical device which combines a biorecognition system with a suitable physicochemical transducer to convert the recognition of the target analyte directly into an electrical signal (Lowe *et al.*, 1992; Gizeli and Lowe, 1996; Lowe, 2007a,b). Figure 4 shows a schematic biosensor comprising (i) a biorecognition system, such as a natural or an engineered enzyme, antibody, receptor, nucleic acid, aptamer, peptide, or other synthetic biomimetic, microorganism, organelle, or tissue, in close conjunction with (ii) a transducer, which transforms the signal resulting from the interaction of the analyte with the biorecognition system into an electrical system, and (iii) the associated electronics and signal processing to display the results in a user-friendly fashion. The transducer can exploit any physical principle based on electrochemical, optical, acoustic, magnetic, thermal, or microegineered devices (Lowe, 1999, 2007a,b).

Classic examples of biosensors include amperometric glucose sensors for diabetes management (Foulds and Lowe, 1988; Wolowacz *et al.*, 1992), conductimetric devices (Cullen *et al.*, 1990), surface plasmon resonance (SPR; Cullen *et al.*, 1987/1988), the resonant mirror (Buckle *et al.*, 1993; Davies *et al.*, 1994;

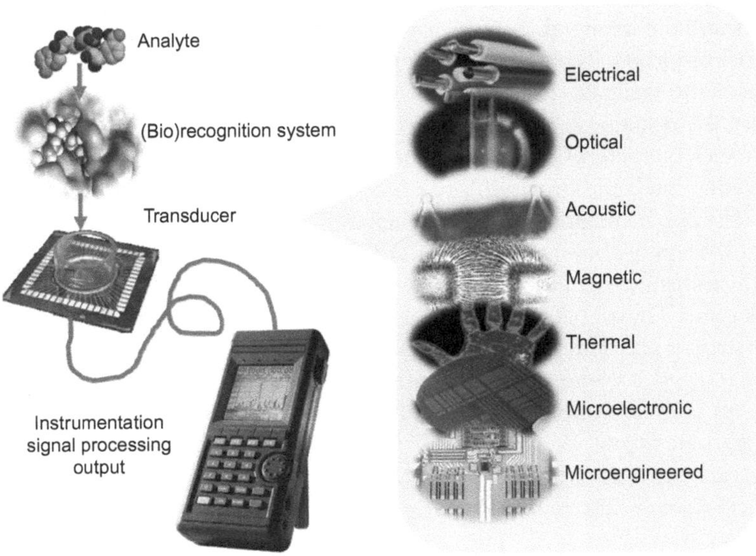

FIG. 4. Schematic of a generalized biosensor showing the (bio)recognition system identifying the analyte in a complex biological sample, the physic-chemical transducer for conversion of the recognition into an electrical signal, and the instrumentation for outputting the signal in the desired user-friendly fashion.

Watts *et al.*, 1994), fiber optic sensors (Carlyon *et al.*, 1992; Tubb *et al.*, 1995, 1997; Hale *et al.*, 1996; Schipper *et al.*, 1997), Mach-Zehnder interferometry (Schipper *et al.*, 1997), planar waveguides (Mayr *et al.*, 2009), and various optical grating (Erdélyi *et al.*, 2007) and acoustic and microcantilever devices (Gizeli *et al.*, 1992; Stevenson and Lowe, 1999; Sindi *et al.*, 2001; Stevenson *et al.*, 2001, 2003, 2006; Haefliger and Boisen, 2007). More recent trends in biosensor technology include the use of aptamers (Brody and Gold, 2000; Strehlitz *et al.*, 2008; Abe *et al.*, 2011), peptides (Huang and Koide, 2010), molecularly imprinted polymers (Haupt and Belmont, 2007), and genetically engineered binding proteins (Ge *et al.*, 2003) and enzymes (Campas *et al.*, 2009) as more durable, selective, and higher affinity recognition elements, the use of electropolymerization to immobilize biomolecules in thin films on sensor surfaces (Cosnier, 2007), metal (Elghanian *et al.*, 1997) and magnetic nanoparticles (Yellen and Erb, 2007), quantum dots (Abramowitz, 2007) to amplify signals, and conducting polymer nanowire- (Wanekaya *et al.*, 2007) and carbon nanotube-based sensors (Barone *et al.*, 2007) to aid in miniaturization of sensor formats.

Perhaps the largest single advance in biosensor technology for clinical and POC analysis involves microfluidics and lab-on-a-chip platforms (Li, 2010). This technology was first demonstrated with a fully integrated GC and thermal conductivity detector on a 4-in. silicon wafer in 1979 (Terry *et al.*, 1979), while the first liquid separation was exemplified in 1989 (Sethi *et al.*, 1989) and a high-performance liquid chromatography (HPLC) and conductimetric detector was fabricated on Si-Pyrex in 1990 (Manz *et al.*, 1990). A miniaturized chemical analysis platform based on capillary electrophoresis (CE) linked to laser-induced fluorescence (LIF) detection was fabricated on a glass chip with channels 10 μm deep and 30 μm wide and used to resolve biomolecules in 6 min (Manz *et al.*, 1992). Since then, a plethora of different clinical, cellular, protein, and nucleic acid analytes have been resolved and quantified using these techniques in devices fabricated in silicon, glass, plastics, and polydimethylsiloxane using on-chip CE, dielectrophoresis, optical trapping, filters, pumps, and valves (Li, 2010; Napoli *et al.*, 2010). For example, DNA extraction; amplification by thermal, isothermal, and rolling circle polymerase chain reaction (PCR); and subsequent hybridization, sequencing, and genotyping have been facilitated by the use of a microfluidic system implemented on a plastic monolithic platform (Liu *et al.*, 2002). Similarly, microfluidic devices have been actively developed for clinical diagnostics, protein separation and functional assay, proteomics and immunoassay (Li, 2010). Thus, homogeneous (Chen *et al.*, 2002), competitive (Schmalzing *et al.*, 1997), and heterogeneous (Ko *et al.*, 2003) immunoassays and multiplexed label-free assays (Carlborg *et al.*, 2010) have been developed.

VII. Biosensors in Neuroscience

The long cherished ambition to witness biosensors displacing all other analytical technologies within the health-care and biomedical sectors has yet to be realized after nearly three decades of intensive research and the introduction of a vast armamentarium of new devices and techniques. The most attractive features of the biosensor, that is, its small size, ruggedness, inexpensiveness, low power burden, rapid and real-time response, use by lay personnel, biocompatibility, and ready interface with computer and mobile technology has yet to be translated into reality. However, biosensors are beginning to emerge as a solution to resolving challenging issues in fundamental neuroscience and as a means to reduce the complexity of multiplexed genomic and proteomic biomarker assays and bring the diagnosis of multigenic neurological diseases closer to the patient (Bell and Kornguth, 2007).

Chemical signaling plays a key role in neural function although few investigators directly measure the concentrations of neurotransmitters and modulators in the extracellular space. Biosensors offer the prospect of measuring neurotransmitter production with adequate temporal (milliseconds) and spatial (100 nm–10 μm) resolution (Dale *et al.*, 2005). Key targets such as pH, pO_2, glutamate, dopamine, GABA, ATP, adenosine, acetylcholine, glucose, pyruvate, lactate, hydrogen peroxide (H_2O_2), and nitric oxide (NO) can be measured with single or cascade oxidase-based microamperometric electrodes, with direct electrochemical assay with carbon fiber microelectrodes (Llaudet *et al.*, 2003, 2005; Dale *et al.*, 2005), or by using additional enzymes to ameliorate interferences or amplify the sensor signal by introducing substrate or coenzyme recycling (Georganopoulou *et al.*, 2000; Llaudet *et al.*, 2003). Amperometric biosensors were developed originally in the early 1960s, although it is only the miniaturization of these electrodes that has made the study of chemical signaling in the brain possible, since they are less invasive than microdialysis sensors (Larsson *et al.*, 1998; Mitchell, 2004; Bell and Kornguth, 2007; Broderick *et al.*, 2008; Broderick and Kolodny, 2009). More recently, an *in vivo* biosensor for neurotransmitter release and *in situ* receptor activity has been devised which uses cell-based neurotransmitter fluorescent engineered reporters (CNiFERs) to address this challenge and monitor *in situ* neurotransmitter receptor activation (Quoc-Thang Nguyen *et al.*, 2010).

However, despite these advances in fundamental understanding of brain metabolism, metabolites alone offer poor definition for the diagnosis of psychological, neuropsychiatric, and neurodegenerative disorders. Thus a key challenge is to circumvent the fact that for most neurological disorders there is currently no single test that can provide an unequivocal diagnosis; instead, a combined multiparametric psychoanalytic, genomic, proteomic, and metabolomic approach is suggested with a panel of 1–100 biomarkers to obviate the well-established conundrum that biomarker signatures overlap various diseases, syndromes, and symptoms (Dale *et al.*, 2005). Protein- and peptide-based chips from Ciphergen Biosystems have been used to identify biomarker panels for AD from CSF samples using SELDI-TOF MS (Choe *et al.*, 2002; Carrette *et al.*, 2003). A panel of four biomarkers was able to distinguish mild AD from unaffected controls in a blinded test set with high sensitivity and specificity (Carrette *et al.*, 2003).

A second challenge facing the early detection of mental disorders is the low concentration of key biomarkers in accessible biological fluids. Biomarker panels can be interrogated via bead assays, as in the case of the schizophrenia diagnostic (Schwarz *et al.*, 2010a,b), or by various particulate systems designed to enhance sensitivity. For example, the biobarcode system is a nonenzymatic amplification protocol capable of detecting low concentrations of nucleic acid, protein, small molecules, and metal ions. The biobarcode system was able to detect the amyloid-β-derived AD marker in CSF at a concentration of 100 fM–100 aM (Georganopoulou *et al.*, 2005) and, when used in conjunction with a PCR amplification step, was able to detect prostate specific antigen (PSA) at a concentration six orders of magnitude higher sensitivity than conventional ELISA, that is, at 3 aM versus 3 pM, respectively (Nam *et al.*, 2004). The biobarcode assay has also proven useful for the detection of the tau protein AD biomarker (Bell and Kornguth, 2007).

The most popular platform for the detection and quantitation of neurodegenerative biomarkers uses the phenomenon of SPR (Cullen *et al.*, 1987/1988). For example, a nanoscale optical sensor based on SPR has been exploited to quantitate the interaction between the amyloid-β-derived diffusible ligands (ADDLs) and specific anti-ADDL antibodies (Haes *et al.*, 2005), while the Biacore® SPR system has been used to characterize the polyglutamine tracts as threshold biomarkers for the onset of AD (Bennett *et al.*, 2002). Similarly, fiber optic and SPR array technologies have been suggested as ways to multiplex multiple assays on the same biological sample (Gašparac and Walt, 2007; Suzuki *et al.*, 2007). More recently, a flexible new grating-coupled SPR sensor system that combines the joint advantages of discretely functionalized, encoded microparticles and label-free detection has been reported (Kastl *et al.*, 2008). This system offers the prospect of simultaneously investigating the real-time binding kinetics of a variety of molecular interactions in a single multiplexed platform; thus, one multiplexed assay could employ a wide range of immobilization chemistries, surface preparation methods, and formats. The new system offers a very high level of assay conformability to the end user, particularly when compared to fixed microarrays and permits the contemporaneous assay of both genomic and proteomic biomarkers (Kastl *et al.*, 2010).

The third, and most challenging option, for the diagnosis of psychological, neuropsychiatric, and neurodegenerative disorders, is to reduce the complexity and cost of these multiparametric assay panels to permit POC testing in the hands of frontline physician's and psychiatrists. There are a variety of simplified tests being developed for field use, that is, outside the controlled laboratory environment. One example of such a test is the development of simple reflection holograms that combine an analyte-selective "smart" polymer with optical interrogation by eye or instrumentation and a reporting diffraction grating transducer (Blyth *et al.*, 1996). They are fabricated by passing a single collimated laser beam through a silver halide emulsion coated onto a glass or plastic substrate backed with a silvered mirror. Interference between the ingoing and outgoing beams creates a standing wave interference pattern, which after development and fixing, forms a series of fringes of silver grains ~20 nm in diameter separated by a distance equal to $1/2\lambda$ of the laser light used in their construction and distributed within the thickness of the emulsion (~5–10 μm). Under white light illumination, the fringes act like a Bragg diffraction grating and reflect a narrow band of wavelength governed by Bragg's law:

$$m\lambda = 2n\partial\cos\theta$$

where m is the diffraction order, λ is the wavelength of light *in vacuo*, n is the average refractive index of the system, ∂ is the spacing between the fringes, and θ is the angle between the incident light and the diffracting planes. Any physical, chemical, or biological interaction which alters the fringe spacing (∂) by swelling or contraction, the average refractive index (*n*) or the total number or distribution of the fringes within the thickness of the film will result in observable changes in the wavelength (color) or intensity (brightness) of the hologram. It has been demonstrated that through the rational design of the hydrogel, which incorporates suitable receptors, that volumetric changes can be induced on binding the complementary target analyte. Thus, responsive holograms for gases, ions, metabolites, inhibitors, drugs, enzymes, antigens, antibodies, and whole cells have been created based on this principle (Mayes *et al.*, 1999, 2002; Marshall *et al.*, 2003, 2004; Lee *et al.*, 2004; Sartain *et al.*, 2006; Lowe, 2007a,b; Tan and Lowe, 2009; Martinez-Hurtado *et al.*, 2010). Responsive holograms of this type are ideally suited to POC applications since they are designed react to specific stimuli; are inexpensive to manufacture by mass producible techniques; are miniaturizable; generate a readout in real time; can be configured as planar, fiber optic, or particulate arrays; and display low or no power burden (Martinez-Hurtado *et al.*, 2010). It is anticipated that these devices will be configured as a handheld card for use by the physician or psychiatrist to aid in the diagnosis of psychological, neuropsychiatric, and neurodegenerative disorders.

Microfluidic devices also offer the prospect of inexpensive POC tests suitable for execution by lay personnel with small sample sizes (Whitesides, 2006). Such devices have been fabricated on thin and flexible films (Focke *et al.*, 2010) and have used manual torque-operated valves for sandwich immunoassays (Weibel *et al.*, 2005) and thread as a versatile material for low-cost microfluidic diagnostics (Li *et al.*, 2010; Reches *et al.*, 2010).

VIII. Neuroinformatics

Advances in experimental protocols have given neuroscientists and psychiatrists an increasingly powerful arsenal for acquiring data across multiple spatiotemporal scales, from the level of single biomarker molecules, cellular architectures, neural connectivity to complex, and interdependent metabolic pathway, physiological, and behavioral data (Kotter, 2001; Martone *et al.*, 2004). It is also evident that combining multiple "omics" data with matching detailed imaging, microscopic, physiological, behavioral, and psychiatric codata for complex multigeneic neuropsychiatric and neurodegenerative disorders is a task beyond even the best funded research groups. Thus, government agencies have promoted the concept of data sharing via schemes such as the Human Brain Project (Shepherd *et al.*, 1998) in order to support the development of computational tools and algorithms for visualizing, analyzing, and modeling of neuroscience and neuropsychiatric data. In addition, the Neuroscience Trust of the

National Science Foundation (NSF) supported an initiative, the National Partnerships for Advanced Computational Infrastructure (NPACI) which encouraged neuroinformatics and computer scientists to collaborate on new database architectures, grids, and networking algorithms (www.npaci.edu). Such initiatives are not without challenges since, aside from the difficulties in representing, storing, and accessing substantial quantities of nonstandardized complex data, there is a range of sociological, ethical, and commercial issues associated with sharing hard won data from different sponsoring organizations. However, if these issues could be resolved, a universal database would go some distance toward realizing the promise of a "global neuroscientific and neuropsychiatric forum" (Martone *et al.*, 2004). Nevertheless, even if this lofty goal could be achieved, it is important to realize that relatively unstructured data does not necessarily enhance knowledge or understanding of these complex disorders and thus data will have to be transmitted to competent interpreters in order to gain maximum value out of its content.

IX. e-Neuroscience and e-Neuropsychiatry

The final part of the jigsaw relates to mobile communication and the internet which will allow unstructured data to be organized, interpreted, evaluated, and reformatted for facile presentation to the end user, be they the physician, psychiatrist, or patient. Frontline analysis, that is, near or on-patient collection of data via real-time or multiplexed sensor arrays, coupled with artificial intelligence and mobile communication systems could bring diagnosis, application, and internet-based services back to the patient (Fig. 5; Ryhänen *et al.*, 2010). The prospect of ambient intelligence, in which sensing, computation, and communication are universally available and ready to serve the end user in an intelligent way, is predicated on intelligent mobile devices embedded in all aspects of the human environment, home, office, and public places. The concept of remote health care is particularly relevant to psychological, neurodegenerative, and neuropsychiatric disorders, since these are long-term debilitating conditions which require constant monitoring and treatment and afflict many of the more developed regions of the world. Telepsychiatry is already an established approach where the patient is in his own home or office, particularly in rural or underserved regions, and can access the psychiatrist via Webcam or high-speed internet. Similarly, there have been developments in emergency e-psychiatry to provide consultation for suicidal, homicidal, violent, psychotic, depressed, manic, and acutely anxious psychiatric patients (Shore *et al.*, 2007). Such emergency e-psychiatry services can be provided to hospital A&E departments, jails, community mental health centers, substance abuse clinics, and schools.

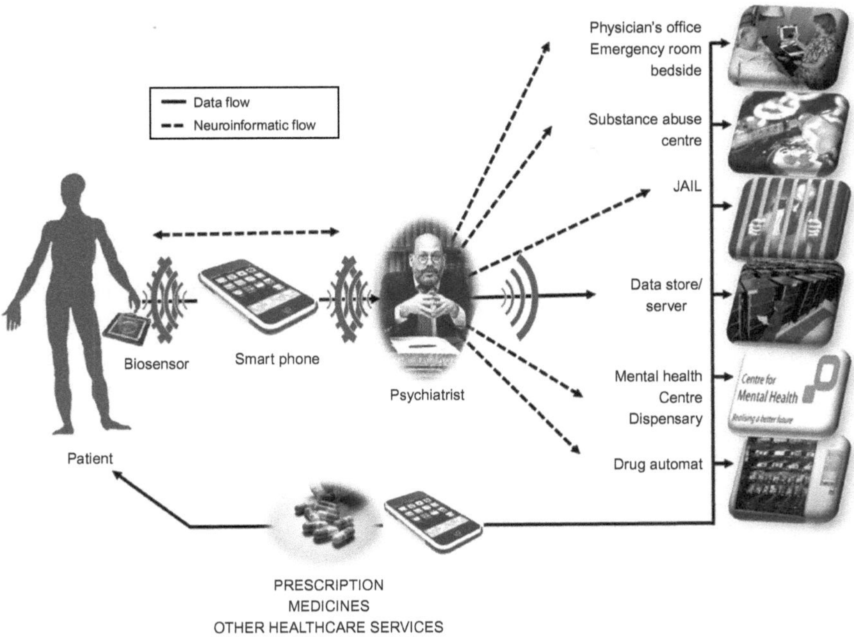

FIG. 5. e-Neuroscience: A future scenario in which psychiatric and neurodegenerative patients are monitored with selective wireless biosensors and the data is transmitted to the psychiatrist for interpretation and onward passage to key service providers and drug automats prior to release of prescription medicines and other health-care services to the patient.

There is an upsurge in interest in individualization of health care and the application of grid computing to support data-, computation-, and knowledge-based tasks to assist disease prediction, diagnosis, prognosis, and compliance. There is effort aimed at creating bioprofiles or personal fingerprints for brain disorders derived from fusing genomics, proteomics, electroencephalogram, and MRI data with the patient's past medical history, biopatterns, and prognosis (Sun *et al.*, 2006). It has been shown that the grid could be used for individual biopattern analysis and bioprofiling for the early detection of dementia and brain injury assessment.

Mobile technologies are likely to offer the most pervasive approach to e-psychiatry and e-neuromedicine, since such mobile integrated cognitive systems could include sensing, perception, cognition, learning, and in some cases where medicines need to be prescribed and delivered, actuation. Mobile devices could thus become the next generation of personalized pharmacies, which sample an accessible fluid from the patient, measure key analytes, communicate the unstructured data to a central interpretation point, return the diagnosis with an encrypted key to allow access to an automated drug dispenser, and then dose the appropriate drug treatment on an individualized basis. However, for this concept to become a reality, a key step is to ensure a correct diagnosis with reliable data coming from sensitive and selective sensors for analytes of interest to psychological, neurodegenerative, and neuropsychiatric disorders.

Acknowledgments

This work was partially supported by a grant awarded to C. R. Lowe by the BBSRC and the Home Office (CBRN).

References

Abe, K., Ogasawara, D., Yoshida, W., Sode, K., and Ikebukuro, K. (2011). Aptameric sensors based on structural change for diagnosis. *Faraday Discuss.* **149**, 93–106.

Abramowitz, S. (2007). Quantum dots: their use in biomedical research and clinical diagnostics. In: R. S. Marks, D.C. Cullen, I. Karube, C.R. Lowe, and H. Weetall (Eds.), *Handbook of Biosensors and Biochips*. Wiley Interscience, Sussex, UK, pp. 793–798.

Anderson, N.L., and Anderson, N.G. (2002). The human plasma proteome: history, character, and diagnostic prospects. *Mol. Cell. Proteomics* **1**, 845–867.

Aronica, E., Van Vliet, E.A., Hendricksen, E., *et al.* (2001). Cystatin C, a cysteine protease inhibitor, is persistently upregulated in neurons and glia in a rat model for mesial temporal lobe epilepsy. *Eur. J. Neurosci.* **14**, 1485–1491.

Barone, P.W., Jeng, E.S., Heller, D.A., and Strano, M.S. (2007). Biosensors based on single-walled carbon nanotube near-infrared fluorescence. In: R.S. Marks, D.C. Cullen, I. Karube, C.R. Lowe, and H. Weetall (Eds.), *Handbook of Biosensors and Biochips*. Wiley Interscience, Sussex, UK, pp. 843–854.

Bell, K.M., and Kornguth, S.E. (2007). Biosensors for neurological disease. In: R.S. Marks, D.C. Cullen, I. Karube, C.R. Lowe, and H. Weetall (Eds.), *Handbook of Biosensors and Biochips*. Wiley Interscience, Sussex, UK, pp. 1084–1098.

Bennett, M.J., Huey-Tubman, K.E., Herr, A.B., West, A.P., Ross, S.A., and Bjorkman, P.J. (2002). Inaugural article: a linear lattice model for polyglutamine in CAG-expansion diseases. *Proc. Natl. Acad. Sci. USA* **99**, 11634–11639.

Blyth, J., Millington, R.B., Mayes, A.G., Frears, E.R., and Lowe, C.R. (1996). A holographic sensor for water in solvents. *Anal. Chem.* **68**, 1089–1094.

Bogdanov, M., Matson, W.R., Wang, L., Matson, T., Saunders-Pullman, R., Bressman, S.S., and Flint Beal, M. (2008). Metabolomic profiling to develop blood biomarkers for Parkinson's disease. *Brain* **131**, 389–396.

Bressan, R.A., Bigliani, V., and Pilowsky, L.S. (2001). Neuroimaging of D2 dopamine receptors in schizophrenia. *Rev. Bras. Psiquiatr.* **23**, 46–49.

Bressan, R.A., Shih, M.C., Hoexter, M.Q., and Lacerda, A.L.T. (2007). Can molecular imaging techniques identify biomarkers for neuropsychiatric disorders? *Rev. Bras. Psiquiatr.* **29**, 102–104.

Broderick, P.A., and Kolodny, E.H. (2009). Real time imaging of biomarkers in the Parkinson's brain using mini-implantable biosensors. II. Pharmaceutical therapy with bromocriptine. *Pharmaceuticals* **2**, 236–249.

Broderick, P.A., Ho, H., Karyn Wat, K., and Vivek Murthy, V. (2008). Laurate biosensors image brain neurotransmitters *in vivo*: can an antihypertensive medication alter psychostimulant behaviour? *Sensors* **8**, 4033–4061.

Brody, E.N., and Gold, L. (2000). Aptamers as therapeutic and diagnostic agents. *Rev. Mol. Biotechnol.* **74**, 5–13.

Buckle, P.E., Davies, R.J., Kinning, T., Yeung, D., Edwards, P.R., Pollard-Knight, D.V., and Lowe, C.R. (1993). The resonant mirror: a novel optical sensor for direct sensing of biomolecular interactions part II: applications. *Biosens. Bioelectron.* **8**, 355–363.

Campas, M., Prieto-Simon, B., and Marty, J.-L. (2009). A review of the use of genetically engineered enzymes in electrochemical biosensors. *Semin. Cell Dev. Biol.* **20**, 3–9.

Carlborg, C.F., Gylfason, K.B., Kaźmierszak, A., Dortu, F., Banuls Polo, M.J., Maquieira Catala, A., Kresbach, G.M., Sohlström, H., Moh, T., Vivien, L., Popplewell, J., Ronan, G., *et al.* (2010). A packaged optical slot waveguide ring resonator sensor array for multiplex label-free assays in lab-on-chips. *Lab Chip* **10**, 281–290.

Carlyon, E.E., Lowe, C.R., Reid, D., and Bennion, I. (1992). A single-mode fibre optic evanescent wave biosensor. *Biosens. Bioelectron.* **7**, 141–146.

Carrette, O., Demalte, I., Scherl, A., Yalkinoglu, O., Corthals, G., Burkhard, P., Hochstrasser, D.F., and Sanchez, J.C. (2003). A panel of cerebral spinal fluid potential biomarkers for the diagnosis of Alzheimer's disease. *Proteomics* **3**, 1486–1494.

Chen, S., Lin, Y., Wang, L., Lin, C., and Lee, G. (2002). Flow-through sampling for electrophoresis-based microfluidic chips and their applications for protein analysis. *Anal. Chem.* **74**, 5146–5153.

Choe, L.H., Dutt, M.J., Relkin, N., and Lee, K.H. (2002). Studies on potential cerebral spinal fluid molecular markers for Alzheimer's disease. *Electrophoresis* **23**, 2247–2251.

Chun, H.S., Gibson, G.E., Degiorgio, L.A., Zhang, H., Kidd, V.J., and Son, J.H. (2001). Dopaminergic cell death induced by MPP+, oxidant and specific neurotoxicants shares the common mechanism. *J. Neurochem.* **76**, 1010–1021.

Cosnier, S. (2007). Immobilization of biomolecules by electropolymerised films. In: R.S. Marks, D. C. Cullen, I. Karube, C.R. Lowe, and H. Weetall (Eds.), *Handbook of Biosensors and Biochips.* Wiley Interscience, Sussex, UK, pp. 237–249.

Craig-Shapiro, R., Fagan, A.M., and Holtzman, D.M. (2008). Biomarkers of Alzheimer's disease. *Neurobiol. Dis.* .

Crow, T.J. (2007). How and why genetic linkage has not solved the problem of psychosis: review and hypothesis. *Am. J. Psychiatry* **164**, 13–21.

Cullen, D.C., Brown, R.G.W., and Lowe, C.R. (1987). Detection of immuno-complex formation on goldcoated diffraction gratings. *Biosensors* **3**, 211–225.

Cullen, D.C., Sethi, R.S., and Lowe, C.R. (1990). A multi-analyte miniature conductance biosensor. *Anal. Chim. Acta* **231**, 33–40.

Dale, N., Hatz, S., Tian, F., and Llaudet, E. (2005). Listening to the brain: microelectrode biosensors for neurochemicals. *Trends Biotechnol.* **23**, 420–428.

Danziger, R.S., You, M., and Akil, H. (2005). Discovering the genetics of complex disorders through integration of genomic mapping and transcriptional profiling. *Curr. Hypertens. Rev.* **1**, 21–34.

Davies, R.J., Edwards, P.R., Watts, H.J., Lowe, C.R., Buckle, P.E., Yeung, D., Kinning, T.M., and Pollard-Knight, D.V. (1994). The resonant mirror: a versatile tool for the study of biomolecular interactions. In: J.W. Crabb (Ed.), *Techniques in Protein Chemistry V.* Academic Press, New York, pp. 285–292.

Elghanian, R., Storhoff, R.C., Mucic, R., Letsinger, R.L., and Mirkin, C.A. (1997). Selective colorimetric detection of polynucleotides based on the distance-dependent optical properties of gold nanoparticles. *Science* **277**, 1078–1081.

Erdélyi, K., Frutos, A.G., Ramsden, J.J., and Szendrö, I. (2007). Grating-based optical biosensors. In: R.S. Marks, D.C. Cullen, I. Karube, C.R. Lowe, and H. Weetall (Eds.), *Handbook of Biosensors and Biochips*. Wiley Interscience, Sussex, UK, pp. 569–586.

Ferri, C.P., Prince, C.P., Brayne, C., Brodaty, H., Fratiglione, L., Ganguli, M., Hall, K., Hasegawa, K., Hendrie, H., Huang, Y., Jorm, A., Mathers, C., *et al.* (2005). Global prevalence of dementia: a Delphi consensus study. *Lancet* **366**, 2112–2117.

Flood, D.G., Marek, G.J., and Williams, M. (2011). Developing predictive CSF biomarkers: a challenge critical to success in Alzheimer's disease and neuropsychiatric translational medicine. *Biochem. Pharmacol.* **81**, 1422–1434. Available on-line 03 February 2011.

Focke, M., Kosse, D., Müller, C., Reinecke, H., Zengerle, R., and von Stetten, F. (2010). Lab-on-a-Foil: microfluidics on thin and flexible films. *Lab Chip* **10**, 1365–1386.

Foulds, N.C., and Lowe, C.R. (1988). Immobilisation of glucose oxidase in ferrocene-modified pyrrole polymers. *Anal. Chem.* **60**, 2473–2478.

French, P.J., O'Connor, V., Voss, K., Stean, T., Hunt, S.P., and Bliss, T.V. (2001). Seizure-induced gene expression in the area CA1 of the mouse hippocampus. *Eur. J. Neurosci.* **14**, 2037–2041.

Fu, Q., Jie Zhu, J., and Van Eyk, J.E. (2010). Comparison of multiplex immunoassay platforms. *Clin. Chem.* **56**, 314–318.

Gašparac, R., and Walt, D.R. (2007). Fibre-optic array biosensors. In: R.S. Marks, D.C. Cullen, I. Karube, C.R. Lowe, and H. Weetall (Eds.), *Handbook of Biosensors and Biochips*. Wiley Interscience, Sussex, UK, pp. 895–916.

Gazzaniga, M.S., and Heatherton, T.F. (2006). Psychological Science. W.W. Norton & Co., Inc., New York.

Ge, X., Tolosa, L., Simpson, J., and Rao, G. (2003). Genetically engineered binding proteins as biosensors for fermentation and cell culture. *Biotechnol. Bioeng.* **84**, 723–731.

Georganopoulou, D.G., Carley, R., Jones, D.A., and Boutelle, M.G. (2000). Development and comparison of biosensors for *in-vivo* applications. *Faraday Discuss.* **116**, 291–303.

Georganopoulou, D.G., Chang, L., Nam, J.M., Thaxton, C.S., Mufson, E.J., Klein, W.L., and Mirkin, C.A. (2005). Nanoparticle-based detection in cerebral spinal fluid of a soluble pathogenic biomarker for Alzheimer's disease. *Proc. Natl. Acad. Sci. USA* **102**, 2273–2276.

Gizeli, E., and Lowe, C.R. (1996). Immunosensors. *Curr. Opin. Biotechnol.* **7**, 66–79.

Gizeli, E., Stevenson, A.C., Goddard, N.J., and Lowe, C.R. (1992). A love-plate biosensor utilising a polymer layer. *Sens. Actuators* **6**, 131–137.

Gudmundsson, P., Skoog, I., Waern, M., Blennow, K., Zetterberg, H., Rosengren, L., and Gustafson, D. (2010). Is there a CSF biomarker profile related to depression in elderly women? *Psychiatry Res.* **176**, 99–102.

Haefliger, D., and Boisen, A. (2007). Microcantilever array devices. In: R.S. Marks, D.C. Cullen, I. Karube, C.R. Lowe, and H. Weetall (Eds.), *Handbook of Biosensors and Biochips*. Wiley Interscience, Sussex, UK, pp. 949–959.

Haes, A.J., Chang, L., Klein, W.L., and Van Duyne, R.P. (2005). Detection of a biomarker for Alzheimer's disease from synthetic and clinical samples using a nanoscale optical biosensor. *J. Am. Chem. Soc.* **127**, 2264–2271.

Hale, Z.M., Payne, F.P., Marks, R.S., Lowe, C.R., and Levine, M.M. (1996). The single mode tapered optical fibre immunosensor. *Biosens. Bioelectron.* **11**, 137–148.

Hata, R., Masumara, M., Akatsu, H., Li, F., Fujita, H., Nagai, Y., Yamamoto, T., Okada, H., Kosaka, K., Sakanaka, M., and Sawada, T. (2001). Upregulation of calcineurin Aβ mRNA in the Alzheimer's disease brain: assessment by cDNA microarray. *Biochem. Biophys. Res. Commun.* **284**, 310–316.

Haupt, K., and Belmont, A.-S. (2007). Molecularly imprinted polymers as recognition elements in sensors. In: R.S. Marks, D.C. Cullen, I. Karube, C.R. Lowe, and H. Weetall (Eds.), *Handbook of Biosensors and Biochips*. Wiley Interscience, Sussex, UK, pp. 199–215.

Holmes, E., Tzang, T.M., Huang, J.T.-J., Leweke, F.M., Koethe, D., Gerth, C.W., Nolden, B.M., Gross, S., Schreiber, D., Nicholson, J.K., and Bahn, S. (2006). Metabolic profiling of CSF: evidence that early intervention may impact on disease progression and outcome in schizophrenia. *PLoS Med.* **3**, e327.

Huang, J., and Koide, S. (2010). Rational conversion of affinity reagents into label-free sensors for peptide motifs by designed allostery. *ACS Chem. Biol.* **5**, 273–277.

Kastl, K.F., Lowe, C.R., and Norman, C.E. (2008). A novel encoded and multiplexed SPR sensor platform. *Anal. Chem.* **80**, 7862–7869.

Kastl, K.F., Lowe, C.R., and Norman, C.E. (2010). Label-free genetic and proteomic marker detection within a single flow-cell assay. *Biosens. Bioelectron.* **26**, 1719–1722.

Khaitovich, P., Lockstone, H.E., Wayland, M.T., Tsang, T.M., Jayatilaka, S.D., Guo, A.J., Somel, M., Harris, L.W., Holmes, E., Pääbo, S., and Bahn, S. (2008). Metabolic changes in schizophrenia and human brain evolution. *Genome Biol.* **9**, R124.

Klunk, W.E., Engler, H., Nordberg, A., Wang, Y., Blomqvist, G., Holt, D.P., Bergstrom, M., Savitcheva, I., Huang, G.F., Estrada, S., Ausen, B., Debnath, M.L., *et al.* (2004). Imaging brain amyloid in Alzheimer's disease with Pittsburgh compound-B. *Ann. Neurol.* **55**, 306–319.

Ko, J.S., Yoon, H.C., Yang, H., Pyo, H.B., Chung, K.H., Kim, S.J., and Kim, Y.T. (2003). A polymer-based microfluidic device for immunosensing biochips. *Lab Chip* **3**, 106–113.

Kontkanen, O., and Castrén, E. (2002). Functional genomics in neuropsychiatric disorders and in neuropharmacology. *Expert Opin. Ther. Targets* **6**, 363–374.

Kotter, R. (2001). Neuroscience databases: tools for exploring brain structure–function relationships. *Philos. Trans. R. Soc. Lond. B Biol. Sci.* **356**, 1111–1120.

Kricka, L.J., and Thorpe, G.H.G. (2010). Technology for handheld devices for point-of-care testing. In: C.P. Price, A. St John, and L.J. Kricka (Eds.), *Point-of-Care Testing: Needs, Opportunity and Innovation*. AACC Press, Washington DC, USA, pp. 27–41.

Larsson, N., Ruzgas, T., Gorton, L., Kokaia, M., Kissinger, P., and Csoregi, E. (1998). Design and development of an amperometric biosensor for acetylcholine determination in brain microdialysates. *Electrochim. Acta* **43**, 3541–3554.

Lee, M.-C., Kabilian, S., Hussain, A., Yang, X., Blyth, J., and Lowe, C.R. (2004). Glucose-sensitive holographic sensors for monitoring bacterial growth. *Anal. Chem.* **76**, 5748–5755.

Le-Niculescu, H., McFarland, M.J., Mamidipalli, S., Ogden, C.A., Kuczensli, R., Kurian, S.M., Salomon, D.R., Tsuang, M.T., Nurnberger, J.I., and Niculescu, A.B. (2007). Convergent functional genomics of bipolar disorder: from animal model pharmacogenomics to human genetics and biomarkers. *Neurosci. Biobehav. Rev.* **31**, 897–903.

Li, P.C.H. (2010). Fundamentals of Microfluidics and Lab on a Chip for Biological Analysis and Discovery. CRC Press, Taylor & Francis Group, Boca Raton, FL.

Li, X., Tian, J., and Shen, W. (2010). Thread as a versatile material for low cost microfluidic diagnostics. *Appl. Mater. Interfaces* **2**, 1–6.

Liu, Y., Rauch, C.B., Stevens, R.L., Lenigk, R., Yang, J., Rhine, D.B., and Grodzinski, P. (2002). DNA amplification and hybridization assays in integrated monolithic plastic devices. *Anal. Chem.* **74**, 3063–3070.

Llaudet, E., Botting, N.P., Crayston, J.A., and Dale, N. (2003). A three-enzyme microelectrode sensor for detecting purine release from central nervous system. *Biosens. Bioelectron.* **18**, 43–52.

Llaudet, E., Hatz, S., Droniou, M., and Dale, N. (2005). Microelectrode biosensor for real-time measurement of ATP in biological tissue. *Anal. Chem.* **77**, 3267–3273.

Lowe, C.R. (1999). Chemoselective biosensors. *Curr. Opin. Chem. Biol.* **3**, 106–111.

Lowe, C.R. (2007a). Overview of biosensor and bioarray technologies. In: R.S. Marks, D.C. Cullen, I. Karube, C.R. Lowe, and H. Weetall (Eds.), *Handbook of Biosensors and Biochips*. Wiley Interscience, Sussex, UK, pp. 7–22.

Lowe, C.R. (2007b). Holographic sensors. In: R.S. Marks, D.C. Cullen, I. Karube, C.R. Lowe, and H. Weetall (Eds.), *Handbook of Biosensors and Biochips*. Wiley Interscience, Sussex, UK, pp. 587–596.

Lowe, C.R., Cullen, D.C., Gizeli, E., Goddard, N.J., Gray Stephens, L.D., Maynard, P., and Yon Hin, B.F.Y. (1992). Biosensors. *Clin. Biochem. Rev.* **13**, 22–26.

Madaan, V., Bestha, D.P., and Kolli, V.B. (2010). Biological markers in schizophrenia: an update. *Drugs Today* **46**, 661–669.

Manz, A., Miyahara, Y., Miura, J., Watanabe, Y., Miyagi, H., and Sato, K. (1990). Design of an open-tubular column liquid chromatograph using silicon chip technology. *Sens. Actuators B* **1**, 249–255.

Manz, A., Harrison, J.D., Verpoorte, E.M.J., Fettinger, J.C., Paulus, A.L.H., and Widmer, H.M. (1992). Planar chip technology for miniaturization and integration of separation techniques into monitoring system. *J. Chromatogr.* **593**, 253–258.

Marshall, A.J., Blyth, J., Davidson, C.A.B., and Lowe, C.R. (2003). pH-sensitive holographic sensors. *Anal. Chem.* **75**, 4423–4431.

Marshall, A.J., Young, D.S., Blyth, J., Kabilan, S., and Lowe, C.R. (2004). Metabolite-sensitive holographic sensors. *Anal. Chem.* **76**, 1518–1523.

Martinez-Hurtado, J.L., Davidson, C.A.B., Blyth, J., and Lowe, C.R. (2010). Holographic sensors for hydrocarbon gases and other volatile organic compounds. *Langmuir* **26**, 15694–15699.

Martins-de-Souza, D., Harris, L.W., Guest, P.C., Turck, C.W., and Bahn, S. (2010). The role of proteomics in depression research. *Eur. Arch. Psychiatry Clin. Neurosci.* **260**, 499–506.

Martone, M.E., Gupta, A., and Ellisman, M.H. (2004). e-Neuroscience: challenges and triumphs in integrating data from molecules to brains. *Nat. Neurosci.* **7**, 467–472.

Mayes, A.G., Blyth, J., Kyröläinen-Reay, M., Millington, R.B., and Lowe, C.R. (1999). A holographic alcohol sensor. *Anal. Chem.* **71**, 3390–3396.

Mayes, A.G., Blyth, J., Millington, R.B., and Lowe, C.R. (2002). Metal ion-sensitive holographic sensors. *Anal. Chem.* **74**, 3649–3657.

Mayr, T., Abel, T., Enko, B., Borisov, S., Konrad, C., Köstler, S., Lamprecht, B., Sax, S., List, E.J.W., and Klimant, I. (2009). A planar waveguide optical sensor employing simple light coupling. *Analyst* **134**, 1544–1547.

Mirnics, K., Middleton, F.A., Marquez, A., Lewis, D.A., and Levitt, P. (2000). Molecular characterization of schizophrenia viewed by microarray analysis of gene expression in prefrontal cortex. *Neuron* **28**, 40–48.

Mitchell, K.M. (2004). Acetylcholine and choline amperometric enzyme sensors characterised in vitro and in vivo. *Anal. Chem.* **76**, 1098–1106.

Nam, J.M., Stoeva, S.I., and Mirkin, C.A. (2004). Biobarcode-based DNA detection with PCR-like sensitivity. *J. Am. Chem. Soc.* **126**, 5932–5933.

Napoli, M., Eijkel, J.C.T., and Pennathur, S. (2010). Nanofluidic technology for biomolecule applications: a critical review. *Lab Chip* **10**, 957–985.

Paige, L.A., Mitchell, M.W., Krishman, R.R., Kaddurah-Daouk, R., and Steffens, D.C. (2007). A preliminary metabolomic analysis of older adults with and without depression. *Int. J. Geriatr. Psychiatry* **22**, 418–423.

Pasinetti, G.M. (2001). Use of cDNA microarray in the search for molecular markers involved in the onset of Alzheimer's disease dementia. *J. Neurosci. Res.* **65**, 471–476.

Prabakaran, S., Swatton, E., Ryan, M.M., Huffaker, S.J., Huang, J.T., Griffin, J.L., Freeman, T., Dudbridge, F., Lilley, K.S., Karp, N.A., Hester, S., Tkachev, D., *et al.* (2004). Mitochondrial dysfunction in schizophrenia: evidence for compromised brain metabolism and oxidative stress. *Mol. Psychiatry* **9**, 684–697.

Price, C.P., St John, A., and Kricka, L.J. (2010). Putting point-of-care testing into context: moving beyond innovation to adoption. In: C.P. Price, A. St John, and L.J. Kricka (Eds.), *Point-of-Care Testing: Needs, Opportunity and Innovation*. AACC Press, Washington DC, USA, pp. 1–20.

Quinones, M.P., and Kaddurah-Daouk, R. (2009). Metabolomics tools for identifying biomarkers for neuropsychiatric diseases. *Neurobiol. Dis.* **35**, 165–176.

Quoc-Thang Nguyen, Q.-T., Schroeder, L.F., Mank, M., Muller, A., Taylor, P., Griesbeck, O., and Kleinfeld, D. (2010). An *in vivo* biosensor for neurotransmitter release and *in situ* receptor activity. *Nat. Neurosci.* **13**, 127–132.

Ravina, B., Eidelberg, D., Ahlskog, J.E., Albin, R.L., Brooks, D.J., Carbon, M., Dhawan, V., Feigin, A., Fahn, S., Guttman, M., Gwinn-Hardy, K., McFarland, H., *et al.* (2005). The use of radiotracer imaging in Parkinson disease. *Neurology* **64**, 208–215.

Reches, M., Mirica, K.A., Dasgupta, R., Dickey, M.D., Butte, M.J., and Whitesides, G.M. (2010). Thread as a matrix for biomedical assays. *Appl. Mater. Interfaces* **2**, 1722–1728.

Rohlff, C. (2001). Proteomics in neuropsychiatric disorders. *Int. J. Neuropsychopharm.* **4**, 93–102.

Ryhänen, T., Uusitalo, M.A., and Kärkkäinen, A. (2010). When everything is connected. In: T. Ryhänen, M.A. Uusitalo, O. Ikkala, and A. Kärkkäinen (Eds.), *Nanotechnologies for Future Mobile Devices*. Cambridge University Press, Cambridge, UK.

Sartain, F., Yang, X., and Lowe, C.R. (2006). A holographic lactate sensor. *Anal. Chem.* **78**, 5664–5670.

Schipper, E.F., Brugman, A.M., Dominguez, C., Lechuga, L.M., Kooyman, R.P.H., and Greve, J. (1997). The realisation of an integrated Mach-Zehnder waveguide immunosensor in silicon technology. *Sens. Actuators B* **40**, 147–153.

Schmalzing, D., Koutny, L.B., Taylor, T.A., Nashabeh, W., and Fuchs, M. (1997). Immunoassay for thyroxine (T4) in serum using capillary electrophoresis and micromachined devices. *J. Chromatogr. B* **697**, 175–180.

Schwarz, E., and Bahn, S. (2008). Biomarker discovery in psychiatric disorders. *Electrophoresis* **29**, 2884–2890.

Schwarz, E., Guest, P.C., Rahmoune, H., Wang, L., Levin, Y., Ingudomnukul, E., Ruta, L., Kent, L., Spain, M., Baron-Cohen, S., and Bahn, S. (2010a). Sex-specific serum biomarker patterns in adults with Asperger's syndrome. *Mol. Psychiatry* Original article.

Schwarz, E., Izmailov, R., Spain, M., Barnes, A., Mapes, J.P., Guest, P.C., Rahmoune, H., Pietsch, S., Leweke, F.M., Rothermindt, M., Steiner, J., Koethe, D., *et al.* (2010b). Validation of a blood-based laboratory test to aid in the confirmation of a diagnosis of schizophrenia. *Biomark. Insights* **5**, 39–47.

Sethi, R.S., Brettle, J., and Lowe, C.R. (1989). Chromatographic microseparation and detection on a silicon wafer. Proceedings of the Fifth International Conference on Solid State Sensors and Actuators and Eurosensors III, Montreux, Switzerland, p. 45.

Shear, M.K., Greeno, C., Kang, J., Ludewig, D., Frank, E., Swartz, H.A., and Hanekamp, M. (2000). Diagnosis of non-psychotic patients in community clinics. *Am. J. Psychiatry* **157**, 581–587.

Shepherd, G.M., Mirsky, J.S., Healy, M.D., Singer, M.S., Skoufos, E., Hines, M.S., Nadkarni, P.M., Miller, P.L., and Shepherd, G.M. (1998). The human brain project: neuroinformatics tools for integrating, searching and modeling multidisciplinary neuroscience data. *Trends Neurosci.* **21**, 460–468www.nimh.nih.gov/neuroinformatics.

Shore, J.H., Hilty, D.M., and Yellowlees, P. (2007). Emergency management guidelines for telepsychiatry. *Gen. Hosp. Psychiatry* **29**, 199–206.

Sindi, H.S., Stevenson, A.C., and Lowe, C.R. (2001). A strategy for chemical sensing based on frequency tunable acoustic devices. *Anal. Chem.* **73**, 1577–1586.

Small, G.W., Kepe, V., Ercoli, L.M., Siddarth, P., Bookheimer, S.Y., Miller, K.J., Lavretsky, H., Bruggren, A.C., Cole, G.M., Vinters, H.V., Thompson, P.M., Huang, S.C., *et al.* (2006). PET of brain amyloid and tau in mild cognitive impairment. *N. Engl. J. Med.* **355**, 2652–2663.

Snowdon, D.A., Kemper, S.J., Mortimer, J.A., Greiner, L.H., Wekstein, D.R., and Markesbery, W.R. (1996). Linguistic ability in early life and cognitive function and Alzheimer's disease in late life. Findings from the Nun study. *JAMA* **275**, 528–532.

Song, F., Poljak, A., Smythe, G.A., and Sachdev, P. (2009). Plasma biomarkers for mild cognitive impairment and Alzheimer's disease. *Brain Res. Rev.* **61**, 69–80.

Sotiriou, C., and Pusztai, L. (2009). Gene-expression signatures in breast cancer. *N. Engl. J. Med.* **19**, 790–800.

Stanta, J.L., Saldova, R., Struwe, W.B., Byme, J.C., Leweke, F.M., Rothemund, M., Rathmoune, H., Levin, Y., Guest, P.C., Bahn, S., and Rudd, P.M. (2010). Identification of N-glycosylation changes in the CSF and serum in patients with schizophrenia. *J. Proteome Res.* **9**, 4476–4489.

Stevenson, A.C., and Lowe, C.R. (1999). Magnetic-acoustic-resonator sensors (MARS): a new sensing methodology. *Sens. Actuators A* **72**, 32–37.

Stevenson, A.C., Mehta, H.M., Sethi, R.S., Cheran, L.-E., Thompson, M., Davies, I., and Lowe, C.R. (2001). Gigaherz surface acoustic wave probe for chemical analysis. *Analyst* **126**, 1619–1624.

Stevenson, A.C., Araya-Kleinsteuber, B., Sethi, R.S., Mehta, H.M., and Lowe, C.R. (2003). The acoustic spectrophonometer: a novel bioanalytical technique based on multifrequency acoustic devices. *Analyst* **128**, 1222–1227.

Stevenson, A.C., Araya-Kleinsteuber, B., Kioupritzi, E., Roque, A.C.A., and Lowe, C.R. (2006). Non-contact excitation of quartz crystal resonator chips. *Appl. Phys. Lett.* **89**, 083516.

Strehlitz, B., Nikolaus, N., and Stoltenburg, R. (2008). Protein detection with aptamer biosensors. *Sensors* **8**, 4296–4307.

Sun, Y., Zhang, L., Johnston, N.L., Torrey, E.F., and Yolken, R.H. (2001). Serial analysis of gene expression in the frontal cortex of patients with bipolar disorder. *Br. J. Psychiatry* **178**, S137–S141.

Sun, L., Hu, P., Goh, C., Hamadicharef, B., Ifeachor, E., Barbounalis, I., Zervakis, M., Nurminen, N., Varri, A., Fontanelli, R., Di Bona, S., Guerri, D., *et al.* (2006). Bioprofiling over grid for healthcare. InfoScale '06 Proceedings of the First International Conference on Scalable Information Systems, ACM, New York, NY, USA1-59593-428-6.

Suzuki, M., Iribe, Y., and Tobita, T. (2007). Surface plasmon resonance array devices. In: R.S. Marks, D.C. Cullen, I. Karube, C.R. Lowe, and H. Weetall (Eds.), *Handbook of Biosensors and Biochips*. Wiley Interscience, Sussex, UK, pp. 917–924.

Tan, E.V., and Lowe, C.R. (2009). Holographic enzyme inhibition assays for drug discovery. *Anal. Chem.* **81**, 7579–7589.

Taurines, R., Dudley, E., Grassl, J., Wamke, A., Gerlach, M., Googan, A.N., and Thome, J. (2011). Proteomic research in psychiatry. *J. Psychopharmacol.* **25**, 151–196.

Terry, S.C., Jerman, J.H., and Angell, J.B. (1979). A gas chromatographic air analyser fabricated on a silicon wafer. *IEEE Trans. Electron Devices* **26**, 1880–1886.

Tubb, A.J.C., Payne, F.P., Millington, R.B., and Lowe, C.R. (1995). Single mode optical fibre surface plasma wave chemical sensor. *Electronics Lett.* **31**, 1770–1771.

Tubb, A.J.C., Payne, F.P., Millington, R.B., and Lowe, C.R. (1997). Single-mode optical fibre surface plasma wave chemical sensor. *Sens. Actuators B Chem.* **41**, 71–79.

Underwood, B., Broadhurst, D., Dunn, W.B., Ellis, D.I., Michell, A.W., Vacher, C., Mosedale, D.E., Kell, D.B., Barker, R.A., Grainger, D.J., and Rubinsztein, D.C. (2006). Huntington disease patients and transgenic mice have similar pro-catabolic serum metabolite profiles. *Brain* **129**, 877–886.

Wanekaya, A.K., Chen, W., Myung, N.V., and Mulchandani, A. (2007). Conducting polymer nano-wire-based biosensors. In: R.S. Marks, D.C. Cullen, I. Karube, C.R. Lowe, and H. Weetall (Eds.), *Handbook of Biosensors and Biochips*. Wiley Interscience, Sussex, UK, pp. 831–842.

Watts, H.J., Lowe, C.R., and Pollard-Knight, D.V. (1994). An optical sensor for monitoring whole microbial cells. *Anal. Chem.* **66**, 2465–2470.

Weibel, D.B., Kruithof, M., Potenta, S., Sia, S.K., Lee, A., and Whitesides, G.M. (2005). Torque-operated valves for microfluidics. *Anal. Chem.* **77**, 4726–4733.

Whitesides, G.M. (2006). The origins and the future of microfluidics. *Nature* **442**, 368–373.

Wolf, C., and Quinn, P.J. (2008). Lipidomics: practical aspects and applications. *Prog. Lipid Res.* **47**, 15–36.

Wolowacz, S.E., Yon Hin, B.F.Y., and Lowe, C.R. (1992). Covalent electropolymerisation of glucose oxidase in polypyrrole. *Anal. Chem.* **64**, 1541–1545.

Yellen, B.B., and Erb, R.M. (2007). Manipulation and detection of magnetic nanoparticles for diagnostic applications. In: R.S. Marks, D.C. Cullen, I. Karube, C.R. Lowe, and H. Weetall (Eds.), *Handbook of Biosensors and Biochips*. Wiley Interscience, Sussex, UK, pp. 799–809.

Young, H., Baum, R., Cremerius, U., Herholtz, K., Hoekstra, O., Lammertma, A.A., Pruim, J., and Price, P. (1999). Measurement of clinical and subclinical tumour response using [18F]-fluorodeoxyglucose and positron emission tomography: review and 1999 EORTC recommendations. *Eur. J. Cancer* **35**, 1773–1782.

Subject Index

N

R

S

T

CONTENTS OF RECENT VOLUMES

Volume 39

Volume 40

Volume 41

Volume 42

Volume 43

Volume 44

Volume 45

Volume 46

Volume 47

Volume 48

Volume 49

Volume 50

Volume 51

Volume 52

Volume 53

Volume 54

Volume 55

Volume 56

Volume 57

Volume 58

Volume 59

Volume 60

Volume 61

Volume 62

Volume 63

Volume 67

Volume 68

Volume 69

Volume 70

Volume 71

Volume 72

Volume 73

Volume 74

Volume 75

Volume 76

Volume 77

Volume 78

Volume 79

Volume 79

Volume 81

Volume 82

Volume 83

Volume 84

Volume 85

Volume 86

Volume 87

Volume 88

Volume 89

Volume 90

Volume 91

Volume 92

Volume 93

Volume 94

Volume 95

Volume 96

Volume 97

Volume 98

Volume 99

Volume 100

Kolibri 360 GmbH
Ferdinand-Jühlke-Straße 7
99095 Erfurt, Deutschland
produktsicherheit@kolibri360.de